1402703516

CRANFIELD UNIVERSITY
LIBRARY

AF616189

BIOTECHNOLOGY FOCUS 3

Preface

This is the third volume in a series of biotechnology reviews published every 12 to 18 months. Filling need for responsible overviews, flooded as we are with information, FOCUS shows highlights and interprets key developments with appeal to both industrial and academic readers.

All different areas of biotechnology are considered before selection of the most important and timely topics which have reached a stage of maturity with which one can sum up the state of the art. In every volume FOCUS treats three main sections:

1) Basic fundamentals of biotechnology – covered in six to eight chapters for chemical engineers, microbiologists, applied biochemists and genetic engineers.
2) Protocols – from which after several years we expect to develop a sound laboratory techniques guide for chemical engineers. Some of these will be apparatus-oriented, some recipes for isolating enzymes, etc. All will be new procedures.
3) Information – dealing with regulations, editorial perspectives, research programs, company profiles, industry standards, and data base resources.

The editors' goal is to bridge the gap among each of the subdisciplinary groups comprising biotechnologists. The aim is to give FOCUS readers general information about other specialists' work. Coverage is not critical, but provides synthesis on topic themes which receive rotating treatment every several years. So a reader should not look here just at his or her own field, but also at the others.

We try to get the best people to write so we can provide the best reviews on current topics. Thus FOCUS is a kind of progress report or series of "advances" unlike a conference proceeding. As a resource it is intended to provide value so that readers do not have to go to ten different journals for perspective.

As the series grows, users can gain a "broad brush" view of what has happened in the field. On close view what might appear haphazard topic choice in any given volume actually follows a structure which becomes apparent. The cumulative effect of several volumes, each with a constant framework of section headings, gives a generally comprehensive yet evolving picture.

Comments and suggestions for forthcoming editions are welcomed and can be addressed to the editors c/o Hanser Publishers.

December 1991 The Editors

BIOTECHNOLOGY FOCUS 3

Fundamentals · Applications · Information

Edited by

R. K. Finn, P. Präve, M. Schlingmann
W. Crueger, K. Esser, R. Thauer
F. Wagner

With contributions from

S. Bertram, C. K. Biebricher, J. Bode, F. Bunge, G. Drews,
H. Erdmann, K. Esser, K. Fritsche, H. G. Gassen, S. Grabley,
H. Hauser, M. Kordel, S. Lang, J. R. Lemke, W. Lokotsch,
M. Markweg, G. Mohr, S. Mohr, R. Müller-Hurtig, P. Präve,
N. Rau, W. Scheirer, R. D. Schmid, W. Schmidt, H. Schneckenburger,
M. Schneider, J. Schwedes, R. Stiller, J. Stollwerk, C. Syldatk,
J. Thiem, W. Tischer, F. Wagner, E. Wasserbauer, F. Widdel,
J. Wink, M. Wrann, A. Zeeck

Hanser Publishers, Munich, Vienna, New York, Barcelona

Distributed in the United States of America
and in Canada by
Oxford University Press, New York

Editors:
Prof. Dr. Robert K. Finn, Cornell University, Ithaca, USA
Prof. Dr. Paul Präve, Hoechst AG, Frankfurt, FRG
Prof. Dr. Merten Schlingmann, Hoechst AG, Frankfurt, FRG
Dr. Wulf Crueger, Bayer AG, Wuppertal, FRG
Prof. Dr. Karl Esser, Ruhr-Universität Bochum, FRG
Prof. Dr. Rudolf Thauer, Philipps-Universität Marburg, FRG
Prof. Dr. Fritz Wagner, Technische Universität Braunschweig, FRG

Distributed in USA and in Canada by
Oxford University Press
200 Madison Avenue, New York, N.Y. 10016

Distributed in all other countries by
Carl Hanser Verlag
Kolbergerstr. 22
D-8000 München 80

Translated from German by
Dr. Elfriede Linsmaier-Bednar, Western Springs, IL 60558, USA

Copy editing by
Dr. Helen D. Haller, Ithaca, N.Y. 14850, USA

Front cover: The Hoechst Biohoch-Reaktor

CIP-Titelaufnahme der Deutschen Bibliothek

Biotechnology focus : fundamentals, applications, information. – Munich ; Vienna ; New York : Hanser ; New York ; Don Mills : Oxford Univ. Pr.
Einheitssacht.: Jahrbuch Biotechnologie <engl.>
Erscheint jährl. – Aufnahme nach 1 (1988)
ISSN 0935–1043

NE: EST

1 (1988) –

ISSN 0935-1043
ISBN 3-446-15957-6 Carl Hanser Verlag Munich Vienna New York Barcelona
ISBN 0-19-520931-1 Oxford University Press

Cover design: Kaselow Design, Munich
Printed in Germany by Druck- und Verlagsanstalt Konrad Triltsch, Würzburg

Contents

Industrial Biotechnology

Industrial and Technical Information

Codes and Regulations

Research Initiatives and Funding

Contents of Focus 1

Information

New Developments

Contents of Focus 2

Research and Development

Research and Development

Microbial Chemistry and Biochemistry

Mobile Genetic Elements in Eukaryotes: Principles and Applications in Biotechnology

by S. MOHR and K. ESSER

Contents

Dipl.-Biol. S. MOHR and Prof. Dr. Dr. h. c. K. ESSER,
Lehrstuhl für Allgemeine Botanik,
Ruhr-Universität Bochum,
Postfach 10 21 48,
D-4630 Bochum 1, Fed. Rep. Germany

1 Introduction

One of the most interesting discoveries of recent decades is the fact that prokaryotic and eukaryotic genomes are less stable than classical genetic investigations had led us to believe. Genetic elements are responsible for these instabilities; they are able to change their location within the genome. They are collectively spoken of as *mobile genetic elements*.

Barbara McClintock postulated the existence of mobile genetic elements in 1947, based on her studies with corn (*Zea mays*). (In 1983, she received the Nobel Prize in medicine/biology for her pioneering work.) During the past 20 years, similar elements have been discovered in many prokaryotic and eukaryotic organisms [6, 21, 45, 61, 80]. Many of these elements have since been isolated and their molecular biology has been studied in detail. They are natural constituents of their host organisms' genomes. Their role in evolution is still unclear, so we can only speculate as to their origins.

Mobile genetic elements can impart a variety of effects upon their hosts [61]. For example, they can cause sequence divergences, that is, changes in nucleotide sequence within a given DNA, if an element is imprecisely excised. They can also participate in establishing new regulatory units within a genome [77] or play a role in genomic rearrangements [67, 70].

In this review, we will deal exclusively with the known characteristics of eukaryotic mobile genetic elements.

The mobile genetic elements of eukaryotes can be divided into two classes, depending upon their mode of replication:

1. elements that use a DNA-dependent DNA polymerase to multiply: examples are transposons, plasmids, viruses, and introns; and
2. elements that use an RNA-dependent DNA polymerase to multiply: the retroposons.

In the following sections we discuss typical elements of both classes, and we give special emphasis to relationships between mobile genetic elements based upon structure and function. We conclude with a hypothesis, based on current research data, on the evolution of mobile genetic elements.

2 Principles

2.1 Mobile genetic elements with DNA-dependent DNA-polymerase replication

2.1.1 Transposons

Transposons are linear DNA molecules that can be dispersed throughout the genome by an enzymatically governed process. This process is called *transposition*, whence the elements are called *transposons* (Figure 1), a term in common use today.

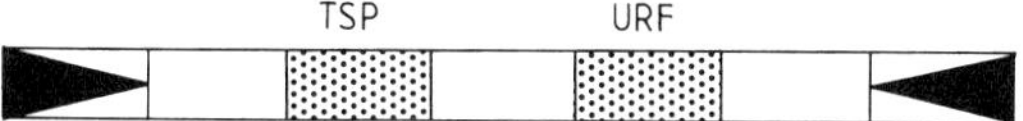

Fig. 1 Schematic representation of a transposon. Black triangles = repeat sequences, arrows indicating orientation; dotted areas = genes; TSP = transposase gene; URF = unidentified reading frame (non-obligatory)

Transposons have *two structural commonalities*.

1. They have *terminal direct repeats of base sequences* (LTDRs, long terminal direct repeats) or *inverse repeats of base sequences* (LTIRs, long terminal inverted repeats), and
2. they generate *base duplications upon insertion.*

Many transposons have been discovered to date. Those best characterized are listed in Table 1; additional transposons in higher plants are described by Nevers *et al.* [61].

Table 1 Transposons in eukaryotes*

Organism	Element designation	Size (kb)	Reference
Drosophila melanogaster	P		
Zea mays	AC	4.563	Mueller-Neumann *et al.* 1984 [60]
	Ds	2.04	Doering *et al.* 1984 [17]
	En1	8.287	Pereira *et al.* 1986 [66]
	Spm18	2.242	Gierl *et al.* 1985 [28]
Antirrhinum majus	Tam 1	17	Bonas *et al.* 1984 [7]
	Tam 2	5	Upadhyaya *et al.* 1985 [90]
	Tam 3	3.5	Coen *et al.* 1986 [13]
Bombyx mori	K-1.4	1.4	Ueda *et al.* 1986 [89]
Caenorhabditis elegans	Tc1	1.6	Ruan & Emmons 1984 [73]

* Because so many elements have been described, only those most studied are included in this table

We will describe two elements in detail:

1. the P element of *Drosophila melanogaster* (fruit fly), because of its use in gene technology, and
2. the *Ac*, *Ds* system of *Zea mays* (maize or corn), the plant system that has been most studied.

The P element is responsible for the hybrid dysgenesis syndrome in the fruit fly [35]. Its primary effect is sterility due to gonadal dystrophy; there are some additional incidental effects [5]. Two different types of *Drosophila* strains can be identified on the basis of the occurrence of hybrid dysgenesis:

1. P strains (= paternally contributed), and
2. M strains (= maternally contributed).

The syndrome manifests itself only upon interstrain mating of an M female with a P male. A reciprocal cross (M male × P female) leads to normal progeny; furthermore, this phenomenon cannot be observed in intrastrain matings. Therefore, the transposition of an element is activated extrachromosomally, as soon as P elements enter the oocyte of an M female via the sperm. A *transposase*, which is an enzyme encoded by the internal region of the P element, confers the capability for transposition. This transposase is inactivated in the cytoplasm of P strains, but it is active in the cytoplasm of M strains and induces the mobility of the P elements. This explains the non-Mendelian heredity.

The appearance of striking phenotypes was also responsible for the discovery of plant transposons. On studying the pigmentation of maize kernels, McClintock observed that some pigmentation genes behaved unpredictably, while others lost their influence. Data obtained with crossings led her to postulate the existence of mobile genetic elements that elicit color changes in maize kernels. Molecular genetic analyses ultimately led to the discovery of several transposons responsible for these observations. In plants, transposons often exist in pairs, an *autonomous* element and a *nonautonomous* element. The autonomous element is capable of integration, excision, and transposition, and is designated a *regulator* [52, 53]. The nonautonomous element lacks these traits [61]; however, it can be activated by the regulator element (*trans-activation*) and is subsequently also capable of integration, excision, and transposition. This element is called a *receptor*.

The first such system described was the *Ac* (= *activator*), *Ds* (= *dissociation*) system in maize. In this case, the *Ac* element is the regulator and the *Ds* element is the receptor.

The factor that governs the receptor elements in *Zea mays* has not yet been identified, but we know that the regulator element *Ac* is inactivated upon methylation [11]; perhaps the methylation affects a transposase gene that consequently becomes inactive [76].

The most striking structural characters of all transposons, the terminal repeats, are of decisive importance for the transposition mechanism [61]. If the complementary ends of an element are paired, *hairpin structures* that make transposition possible can be formed (Figure 2) [60]. If a transposon is imprecisely excised, changes in the base sequence of the DNA can result; the element may thus leave its "footprint". Transpo-

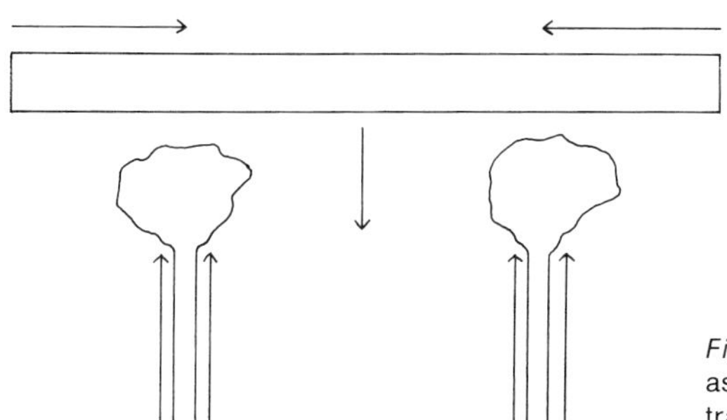

Fig. 2 Formation of hairpin structures as a consequence of LTIR pairing in a transposon

sons are thus responsible for generating sequencing deviations over the course of evolution [77]. We have only a rudimentary understanding of the total impact of transposons on their hosts, and can only guess at their evolutionary origin.

2.1.2 *Viruses*

Viruses are ultramicroscopic particles consisting of genetic information encapsulated in a protein coat. The genetic information can be single-stranded DNA, double-stranded DNA, or RNA [79, 92]. Viruses require the physiological machinery of their host cell in order to replicate. Adenoviruses (ADN = acide desoxyribonucleique, french for DNA) are representative of the wide array of viruses, because they show structural parallels to linear plasmids (see Section 2.1.3). Adenoviruses have been the subjects of considerable study because they cause acute human respiratory illnesses. Figure 3 is a schematic representation of an adenovirus.

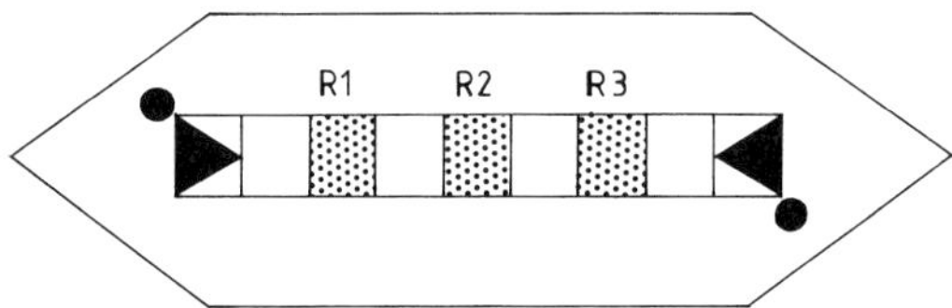

Fig. 3 Schematic representation of an adenovirus. Hexagon = virus coat; black triangles = repeat sequences, arrows indicating orientation; large black dots = 5′-bound proteins; dotted areas = genes for replication (R1–R3)

The genome of adenoviruses is flanked by two inverted repeat sequences (IRs) that carry 5′-bound proteins. The proteins and the IRs all participate in virus replication [32, 95]. The proteins and the 45 terminal base pairs are critical for replication and are therefore highly conserved in all adenoviruses [82].

2.1.3 *Plasmids*

Plasmids are linear or circular extrachromosomal DNA molecules. The first eukaryotic plasmid was discovered in 1967 in bakers' yeast, *Saccharomyces cerevisiae*; since then, numerous other plasmids have been found and a great deal of information on them has

Table 2 Linear plasmids in eukaryotes*

Organism	Plasmid designation	Size (kb)	Reference
Yeasts			
Kluyveromyces lactis	pGKL 2 pGKL 1	13.4 8.9	Gunge *et al.* 1981 [29]
Saccharomycopsis crataegensis	pScr 1 pScr 2 pScr 3	15.0 7.0 5.0	Sheperd *et al.* 1987 [81]
Saccharomyces kluyveri	pSKL	14.2	Kitada & Hishinuma 1987 [44]
Filamentous fungi			
Agaricus bitorquis	pEM pMPJ	7.4 3.7	Mohan *et al.* 1984 [58]
Ascobolus immersus	pA 1 pAI 1 pAI 2	6.4 7.9 5.6	Francou 1981 [24] Meinhardt *et al.* 1986 [55]
Ceratocystis fimbriata	pCF 637 pFQ 501	8.2 6.0	Giasson & Lalonde 1987 [27] Normand *et al.* 1986 [62]
Claviceps purpurea	pCLK 1 pCLK 2 pCLK 3	6.7 5.5 1.1	Tudzynski & Esser 1986 [86] Tudzynski *et al.* 1983 [85]
Fusarium oxysporum	pFOXC 1 pFOXC 2	1.9 1.9	Kistler & Leong 1986 [43]
Gaeomannomyces graminis	E 1 E 2	8.4 7.2	Honeyman & Currier 1983 [36]
Morchella conica	pMC31 pMC32	8.0 6.0	Meinhardt & Esser 1984 [54]
Morchella elata	pME 141-1 pME 141-2	6.7 6.0	Meinhardt & Esser 1987 [54]
Morchella hortensis	pMH1	7.8	Meinhardt & Esser 1987 [54]
Neurospora intermedia	kalilo	9.0	Bertrand et al. 1985 [4]
Rhizoctonia solani	pRS 64	2.6	Hashiba et al. 1984 [33]
Higher Plants			
Brassica campestris	–	11.3	Erickson *et al.* 1985 [19], Turpen *et al.* 1987 [87]
Brassica napus	–	11.3	Palmer *et al.* 1983 [65]
Sorghum bicolor cms	N 1 N 2	5.7 5.3	Pring *et al.* 1982 [68]

Table 2 (cont.) Linear plasmids in eukaryotes*

Organism	Plasmid designation	Size (kb)	Reference
Zea diploperennis	D 1	7.5	TIMOTHY *et al.* 1982 [83]
	D 2	5.4	
Zea mays N	—	2.3	KEMBLE *et al.* 1980 [40]
C	—	2.3	
S	S 1	6.4	PRING *et al.* 1977 [68]
	S 2	5.5	
	—	2.3	KEMBLE *et al.* 1980 [40]
T	—	2.3	
RU	R 1	7.5	WEISSINGER *et al.* 1982 [94]
	R 2	5.5	

* Because the number of linear plasmids reported is still manageable, this list is complete at the time of writing

been amassed. The *linear* plasmids will be used here to illustrate the possible relationship of plasmids to other mobile genetic elements.

All currently known linear plasmids (Table 2) consist of double-stranded DNA and are flanked by long terminal inverted repeat sequences (LTIRs). All plasmids that have been examined bear 5′-bound proteins, as do adenoviruses [20]. Furthermore, they appear able to replicate autonomously [64]. Figure 4 is a schematic representation of the structure of a linear plasmid.

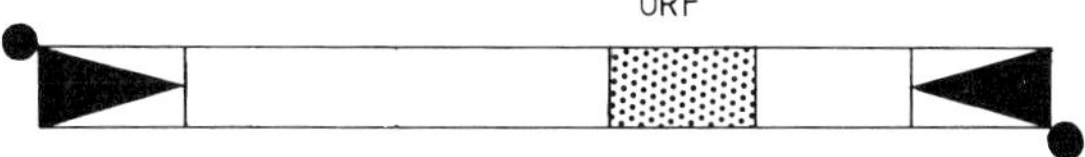

Fig. 4 Schematic representation of a linear plasmid. Black triangles = repeat sequences, arrows indicating orientation; large black dots = 5′-bound proteins; dotted area = gene, non-obligatory; URF = unidentified reading frame, non-obligatory

It seems likely that plasmids can replicate autonomously because copies of plasmids are generally present in higher copy numbers in the cell than are other individual gene sequences. For example, plasmids S-1 and S-2 of *Zea mays* are five times as frequent as other individual sequences [64]. Since linear plasmids, like the previously mentioned adenoviruses, bear 5′-bound proteins, we hypothesize that they may also replicate autonomously [41, 64, 78]. This hypothesis, however, cannot yet be documented experimentally, because replication mechanisms in linear plasmids have not been investigated.

In only a few cases is the function of linear plasmids fully or even partially understood; we will now discuss this in detail.

In strains of the fungus *Neurospora crassa* isolated in Hawaii (*Neurospora intermedia*), the linear plasmid kalDNA (*kalilo* is Hawaiian for destined to die) is the trigger for senescence [3]. The kalDNA integrates into the mitochondrial genome. The result of this insertion is that the cells cannot form functional small or large subunits of mitochondrial ribosomes, and consequently die. Nonsenescent strains lack kalDNA. The precursor of the mitochondrial kalDNA is a structurally homologous linear plasmid present in high copy numbers in the cell nucleus. Presumably this precursor penetrates the mitochondrion and acts, by insertion, as a sort of mutagen within the mitochondrial genome [4].

In the yeast *Kluyveromyces lactis*, the presence of the linear plasmids pGKL1 and pGKL2 correlates with the *killer* phenomenon [30]. Cells that contain both plasmids produce a *killer toxin* that kills other cells while they remain immune to it [29]. Plasmid pGKL1 codes for the toxin and the immunity, and plasmid pGKL2 is responsible for the replication of both plasmids and for the expression of the immunity trait [84].

Linear plasmids S-1 and S-2 in *Zea mays* seem to be associated with cytoplasmic male sterility (cms). The plants are unable to produce fertile pollen, a trait transferred maternally. The cms phenotype comes about because of conformational changes in mitochondrial DNA (mtDNA); plasmids are involved in these changes [74], but their precise function is not completely understood [88].

We have only the barest understanding of the significance of linear plasmids in their host organisms, with the exception of the killer plasmids. Plasmid structural characteristics are suggestive of a relationship with other mobile genetic elements.

2.1.4 Introns

Introns are portions of the primary transcripts of certain genes that are excised from primary RNA during transcription [69, 75]. They occur, as far as is known, in archaebacteria, in bacteriophage T4, and in eukaryotes, in quite different genes [96]. The process of removing the intron from the precursor RNA is called *splicing*, and has been studied intensively in recent years. A significant amount of data is available about introns in organelles; therefore, an intron from an organelle was selected for the schematic illustration in Figure 5.

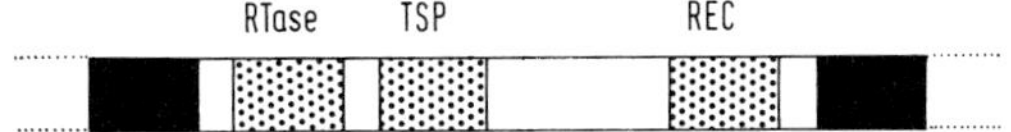

Fig. 5 Schematic representation of intron in an organelle. This diagram is a synthesis of what we know about all genes in which introns have been found at the time of writing; it is therefore unlikely that an intron of exactly this form exists. Solid line = intron region; dotted line = exon region; black areas = consensus sequences at the 5′ and 3′ ends of the intron; dotted areas = postulated genes; RTase = reverse transcriptase; TSP = transposase; REC = recombinase; see text for additional details

We are here emphasizing those characteristics of introns that suggest a potential relationship with other mobile genetic elements, but we must note that the most important commonality of almost all mobile genetic elements, the terminal repeat sequence, is absent in introns. However, introns of the nuclear mRNA gene and of organelle genes include conserved sequences at the 5′ and 3′ ends [10] that one could envision as analogs of terminal repeat sequences.

1. Introns, like the transposons mentioned earlier, can encode proteins that have *transposase* and *integrase* functions; the transposase is responsible for mobility (transposition), and the integrase for integration of the DNA sequence. These functions have been demonstrated in bakers' yeast, *Saccharomyces cerevisiae* [14, 18, 38, 51]. Therefore, it may be that introns operate at the DNA level and can arbitrarily insert at other genomic loci. Additional evidence in support of this hypothesis comes from the finding that related genes of different organisms can have completely different arrangements of exons and introns. A typical example of this is the gene for subunit I of cytochrome oxidase in *Saccharomyces cerevisiae* [34], *Schizosaccharomyces pombe* [49], *Aspergillus nidulans* [91], and race s of *Podospora anserina* [39, 47].

2. Introns may possibly code for *reverse transcriptase*, an enzyme that transcribes RNA into DNA. Since information flow is normally from DNA into RNA, the transcription in this case is designated as *reverse*. The enzyme is operative predominantly in retroposons (Section 2.2) and is essential for virus replication.
   A *mobile intron*, p1DNA, is associated with early senescence in the fungus *Podospora anserina*; in its original location it corresponds to the first intron of the cytochrome oxidase gene [48]. It can be excised and then exists in the cell as a free, circular, double-stranded DNA plasmid [46]. It may possibly trigger the senescence phenomenon upon its insertion into essential regions of mtDNA [48]. The plasmid contains a long open reading frame, the gene product of which shares considerable identity with reverse transcriptase [57].

3. Structurally, some introns appear to be related to viroids and virusoids [16, 59]. *Viroids* and *virusoids* are low molecular weight, single-stranded RNA phytopathogens, and can be either circular or linear [23]. We postulate the relationship because viroid and virusoid sequences show strong similarities to those sequences that are important to the secondary structure of introns; viroids, virusoids, and introns can therefore fold into similar secondary structures [16]. Because introns may possibly encode integrases and transposases, we can further postulate a relationship between them and transposons, plasmids, and viruses. However, since some introns code for a reverse transcriptase they can be grouped with the retroposons, discussed in Section 2.2.

2.2 Mobile genetic elements with RNA-dependent DNA-polymerase replication (retroposons)

Transposition, discussed earlier, is an enzymatically governed process by which sequences can be distributed or multiplied within genomes. The transfer of information from the site of origin to the target site can, in principle, be accomplished with either DNA or RNA intermediates. While DNA-mediated transposition has been described for prokaryotes as well as eukaryotes, transposition by RNA intermediates has been reported only for eukaryotes. These two modes of transcription are referred to by different names: the reverse flow of genetic information from RNA to DNA is called *retroposition*, and the informational sequence transported is called a retroposon [72, 93].

Table 3 Structural characteristics of retroposons

	Retroviruses and retrotransposons	Retropseudogenes
Elements are distributed throughout the genome	+	+
Long terminal direct repeat sequences (LTDRs) flank the elements	+	−
Active transposition occurs	+	−
Base duplication is generated upon insertion	yes, 4 to 6 bp	yes, 7 to 21 bp
Poly-A region is present	−	+
Introns are present	possibly	generally no

Adapted from WEINER *et al.* 1986 [93]

Retroposons include the retroviruses, retrotransposons, and retropseudogenes; what they have in common is reverse transcriptase activity. Table 3 is a summary of their structural characteristics; retroviruses and retrotransposons are listed together, and we will discuss them together, because they have the same essential characteristics. We will discuss the retropseudogenes briefly in Section 2.2.2.

2.2.1 Retroviruses and retrotransposons

The genome of *retroviruses* and *retrotransposons* contains an internal coding region that is framed by long terminal direct repeat sequences (LTDRs), generally flanked by short terminal inverse repeats (STIRs). Figure 6 is a schematic representation of this.

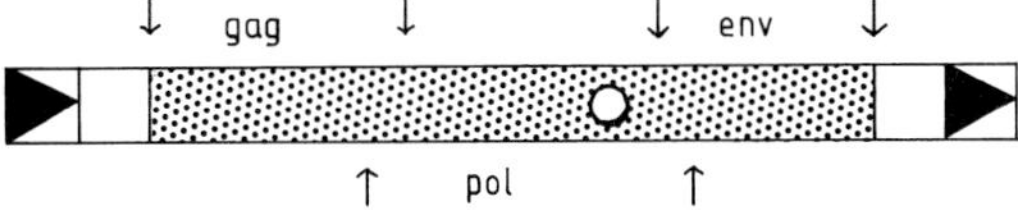

Fig. 6 Schematic representation of a retroposon. Black triangles = repeat sequences, arrows indicating orientation (the STIRs have been omitted for simplicity); dotted area = genes, with small arrows indicating start and termination of single genes; *gag* = gene for viral capsid protein; *pol* = gene for reverse transcriptase and (circle) for integrase; *env* = gene for viral envelope protein; see text for additional details

Besides the genes identified with *gag*, *pol*, and *env* that we discuss below, there are virus-specific genes; for example, we currently know of four additional genes in the HIV virus [26].

The *gag* (*group-specific antigen*) gene encodes a polyprotein that is processed into the four mature proteins that constitute the most important components of the inner virus coat (capsid). Capsid-forming proteins are apparently necessary for retroviruses to be infectious, but it is surprising that retrotransposons, which do not occur extracellularly, also encode these proteins, as has been demonstrated for the yeast transposon Ty *1* [1, 42, 56].

The *pol* (*polymerase*) gene of all mobile elements encodes a polypeptide that has several functions, the most important being that of a reverse transcriptase. The 3′ end codes for an additional integrase that governs the integration of the DNA into the host genome. The *env* (*envelope*) gene encodes two proteins that are fission products of a polyprotein; they become part of the lipoprotein membrane that envelopes the retrovirus. Because retrotransposons do not occur extracellularly, it is not surprising that they lack the *env* gene, either completely or partially [25].

Besides their structural similarities, retroviruses and retrotransposons have rather similar life cycles. The *gag* and *pol* genes are always translated simultaneously, thus generating a fusion protein that is processed subsequently. (*Processing* is the excising of noncoding nucleotide sequences of RNA and the adding of termination-specific nucleotide sequences to mRNA.) Because the two genes are located in different reading frames, the precursor protein can be produced only by a frame shift during translation; such frame shifts have been reported for both retroviruses and retrotransposons [12, 37, 42].

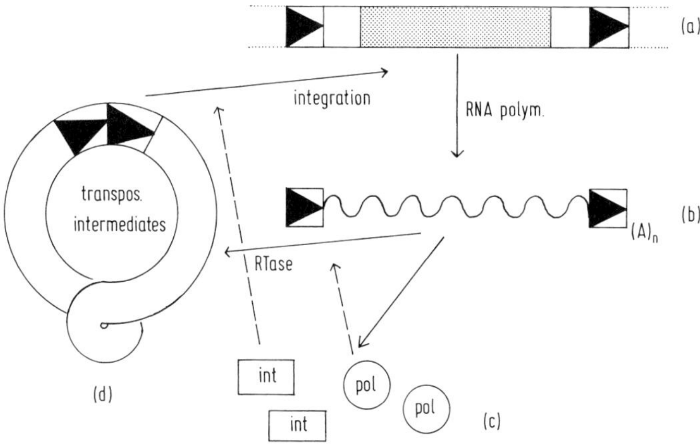

Fig. 7 Schematic representation of the replication cycle of retroviruses and retrotransposons. (a) Provirus or retrotransposon, integrated in a chromosome (dotted line); (b) RNA transcript of a viral retroposon; (c) protein translation products of a viral retroposon; (d) transposition intermediate.
Black triangles = repeat sequences, arrows indicating orientation; dotted area = gene; wavy line = RNA; circles = gene products participating in reverse transcription (*pol*); *Int* = integrase; RTase = reverse transcriptase; see text for additional details

Retroviruses and retrotransposons are also similar in their mechanisms of replication. If we can assume that retrotransposons are analogous to integrated proviruses in a retrovirus [31], the replication cycle shown in Figure 7 applies to both. The integrated DNA of a retroviral provirus or of a retrotransposon is transcribed into RNA. The RNA transcript is translated into the proteins referred to earlier, which participate either in reverse transcription or in integration. The RNA transcript further serves as a template for structural proteins and for the synthesis of circular cDNAs that may insert at new sites along the genome.

These examples are evidence of a close relationship between retroviruses and retrotransposons. Possible relationships with other mobile genetic elements will be discussed in Section 2.3.

2.2.2 Retropseudogenes

Retropseudogenes are a more or less complete DNA copy of a cellular mRNA species [31, 50, 71, 93]. The DNA copy is integrated into the genome and is usually immobile. The properties of retropseudogenes, which we will discuss shortly, are such that we hypothesize that a reverse transcriptase was involved in their evolution [22, 31]. For example, they contain no introns, but do have a poly-A region and the "footprints" of reverse transcriptase mRNA. Furthermore, retropseudogenes of the mammalian L1 sequences and the F elements and I factors in *Drosophila* contain long open reading frames whose potential gene products have similarities to reverse transcriptases [15]. Insertion of a retropseudogene into DNA causes a sequence duplication of this DNA at the insertion locus, generally involving 7 to 21 base pairs [93]; this observation is also typical of retroviruses and retrotransposons (Table 3).

We can only speculate as to the function of retropseudogenes. Because they are present in large numbers in the genome, it seems possible that they may participate in genomic rearrangements [8, 31, 71], but the question is still open.

2.3 Interrelationships among mobile genetic elements

Because mobile genetic elements have such similar structural features, we believe they are related entities. They fall into two groups, differentiated by their mode of replication. The elements replicate either by the action of a DNA-dependent DNA polymerase or of a reverse transcriptase (RNA-dependent DNA polymerase); similar elements developed in parallel in the two groups. Introns may be intermediate between the two groups, since both replication mechanisms have been postulated for them.

One hypothesis on the interrelationships is illustrated in Figure 8. We will discuss the relationships individually.

2.3.1 Relationships among mobile genetic elements with DNA-dependent DNA-polymerase replication

This group includes introns, transposons, plasmids, and viruses (Section 2.1). They are depicted on the right in Figure 8.

All these elements except introns have LTIRs. The 5′ and 3′ consensus sequences of many introns may be considered analogs of LTIRs, and are important to the gene splicing process. Likewise, the LTIRs of transposons participate in excision and integration processes. On this basis we assume that introns and transposons are related. The assumption receives further support from the fact that some introns can code for a transposase [14], as is true of the P transposons of *Drosophila* [35] and *Ac* in maize [76].

Furthermore, some introns and transposons share the ability to integrate arbitrarily into the genome. The integration process is precise: the integrated DNA always has the same nucleotide sequences at both ends of the insertion. This applies to the consensus

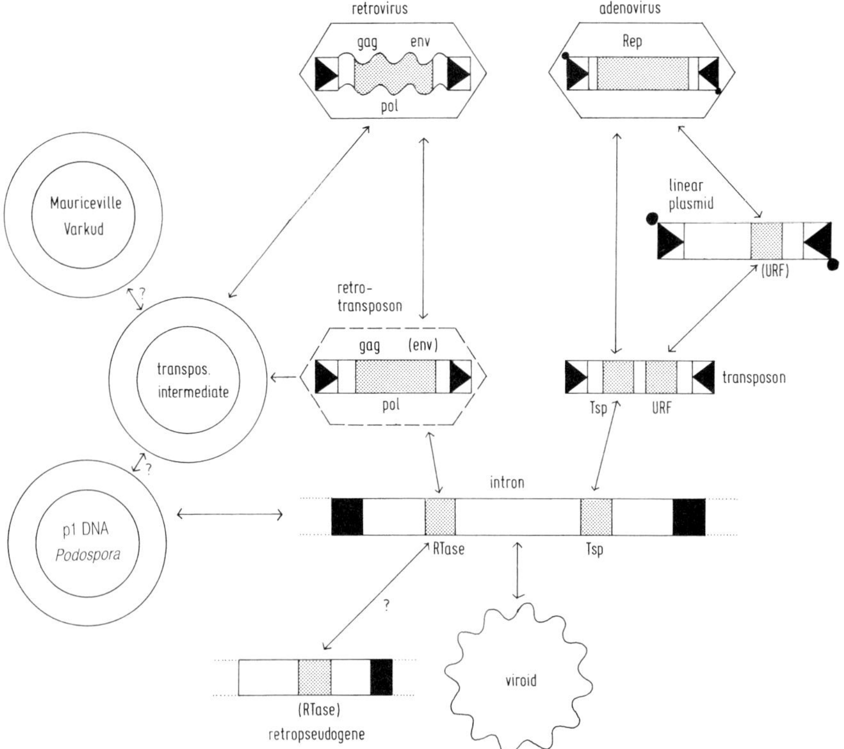

Fig. 8 Schematic representation of the hypothetical relationships among mobile genetic elements. Black triangles = repeat sequences, arrows indicating orientation; large black dots = 5′-bound proteins; dotted areas = genes; black squares = consensus sequences of introns, or poly-A regions in retropseudogenes; hexagon, solid line = viral capsid, genetic element existing intra- and extracellularly; hexagon, dashed line = viral envelope, existing only intracellularly; wavy lines = RNA; *gag* = gene for viral capsid protein; *pol* = gene for reverse transcriptase (RTase)

sequences of introns as well as to the TIRs of transposons.

Transposons share features with adenoviruses, as well as with introns. Common to both are the TIR sequences and their mobility. Adenoviruses differ from transposons in that they can replicate autonomously and are packaged into a viral capsid that permits their extracellular existence. Furthermore, adenoviruses have properties also found in all linear plasmids studied to date: covalently 5′-bound proteins, and the ability to replicate autonomously. Some linear plasmids, however, seem also to be related to transposons because, like transposons, they are mobile.

On the basis of the data summarized above, we can outline the following hypothetical interrelationships for this group of eukaryotic mobile genetic elements that replicate via a DNA-dependent DNA polymerase (Figure 8).

- Introns are the elements with the least autonomy, but they can also be classed with those elements that replicate via an RNA-dependent DNA polymerase, those with the fewest capabilities. For this reason, they can be considered intermediate between the two groups.
- Since they lack LTIRs, introns may perhaps be precursors or relics of transposons, although only few introns can code for transposase. At the end of Section 2.3.2 we will look at the question of whether introns arose early during evolution and other elements may have evolved from them, or whether they are the most primitive form of all elements.
- Adenoviruses appear to be the infectious versions of transposons. Linear plasmids, to the extent they have been studied, appear to be related to adenoviruses as well as to transposons and may possibly represent a lateral branch on our hypothetical tree.

2.3.2 Relationships among mobile genetic elements with RNA-dependent DNA-polymerase replication

Replication by a reverse transcriptase mechanism is common to all the elements in the central portion of Figure 8, but their properties are otherwise quite different.

An intron (p1DNA) of the fungus *Podospora anserina* (Figure 8, left) may have reverse transcriptase activity, in addition to being mobile. In its excised form it resembles a circular transposition intermediate of retroviruses and retrotransposons.

The mitochondrial plasmids Mauriceville and Varkud of *Neurospora crassa* (Figure 8, left) also resemble these transposition intermediates because they consist of circular dsDNA, encode a protein very similar to reverse transcriptase, and can integrate into DNA. They are structurally comparable to introns of class I [2] and may, therefore, be related to them.

Retroviruses and retrotransposons have additional properties, besides those they share with the introns and plasmids mentioned above: they include several genes and, usually, virus capsids, so that at least the retroviruses are able to exist extracellularly.

Retropseudogenes are more difficult to classify. On the one hand, they have base duplications at their terminal ends, a feature that also occurs when retroviruses or retrotransposons are inserted, but, on the other hand, nobody has yet found evidence that they are mobile. Some retropseudogenes may possibly encode a reverse transcriptase. These two properties of pseudogenes are shared with introns; however, we note that most retropseudogenes are found in mammals, while introns occur also in lower organisms; therefore, retropseudogenes may represent an evolutionary advance.

On the basis of these data, we can establish the following hypothetical relationships for the elements just discussed.

- Retroviruses have the greatest autonomy.
- Retrotransposons may be considered precursors or degenerate forms of retroviruses, since they do not have an extracellular state and are in essence genetic parasites.
- The introns are the elements with the fewest capabilities.
- The plasmids Mauriceville and Varkud could be a lateral branch since they are structurally related to introns; however, their ability to replicate via reverse transcriptase means that they are also related to the other elements. On the one hand, they

resemble transposition intermediates of retroviruses and may, therefore, be thought of as frozen intermediates; on the other hand, they could also represent escaped circular introns, such as p1DNA.

This raises the question of the direction of evolutionary development among the mobile genetic elements. In principle, we can imagine two scenarios.

1. The first element to develop was autonomous (a virus). This would mean that the overall process is degenerative, and the other elements are more or less functional derivatives of the virus; the final product, as far as we can now tell, is the intron.
2. The mobile genetic elements are derivatives of escaped introns. Highly specialized elements could have evolved from different intermediates; those in the present final state (the viruses) are at the threshold of life as independent beings.

In the first scenario, the evolution of introns is the result of an integration of degenerate mobile genetic elements. The second scenario begs the question of the origin of introns.

Recent research has brought forth the first attempts at elucidation; whether life originated at the level of RNA or DNA seems currently to be answered in favor of RNA [9, 63, 92]. A current hypothesis is that life began with RNA viruses and that present RNA viruses may be "'living fossils' of early stages of evolution" [63]. Since RNA viroids and RNA virusoids bear a structural relationship to many introns, it seems possible that viroids or virusoids were at one time transcribed reversely and subsequently integrated into DNA. They could have been the first introns. Parallel and similar mobile genetic elements, differing only in their method of replication, could have evolved from these introns. This discussion of the potential interrelationships of mobile genetic elements is at the moment hypothetical; only with further detailed characterization of mobile genetic elements can the question of their origin be resolved.

3 Relevance to biotechnology

Recent information about the structure and function of mobile genetic elements has not only deepened our basic knowledge but has also, especially in recent years, given it a relevance with regard to biotechnology. Targeted improvements in production systems and in strain stability are particularly likely results.

Mobile genetic elements can serve as a useful tool in these efforts, because they constitute "in-house" vectors for transferring the desired genetic information to the recipient. For example, the P element of *Drosophila* is a vector used for this purpose in basic research [35].

Furthermore, the hereditary infrastructure can be altered with mobile genetic elements, since "jumping elements" are used as mutagens. Transposons are used for this purpose, and the technique has become established as *transposon mutagenesis*.

Only time will tell whether and to what extent knowledge from basic research will be useful in the applied sector; perhaps a new technique for genetic engineering based on vectoring may be devised, in addition to what we already do with genetic transformation.

4 References

[1] ADAMS, S. E.; DAWSON, K. M.; GULL, K.; KINGSMAN, S. M.; KINGSMAN, A. J.: The Expression of Hybrid HIV:Ty-Virus Like Particles in Yeast; *Nature (London)* **329** (1987) 68–70.

[2] AKINS, R. A.; KELLY, R. L.; LAMBOWITZ, A. M.: Mitochondrial Plasmids of *Neurospora*: Integration into Mitochondrial DNA and Evidence for Reverse Transcription in Mitochondria; *Cell* **47** (1986) 505–516.

[3] BERTRAND, H.; CHAN, B. S. S.; GRIFFITHS, A. J. F.: Insertion of a Foreign Nucleotide Sequence into Mitochondrial DNA Causes Senescence in *Neurospora intermedia*; *Cell* **41** (1985) 877–884.

[4] BERTRAND, H.; GRIFFITHS, A. J. F.; COURT, D. A.; CHENG, C. K.: An Extrachromosomal Plasmid in the Etiological Precursor of kalDNA Insertion Sequences is the Mitochondrial Chromosome of Senescent *Neurospora*; *Cell* **47** (1986) 829–837.

[5] BINGHAM, M. D.; ROWAN, R. G.; DICKINSON, W. J.; KIDWELL, M. G.; RUBIN, G. M.: The Molecular Basis of P-M Hybrid Dysgenesis: The Role of the P Element, a P-Strain Specific Transposon Family; *Cell* **29** (1982) 995–1004.

[6] BIRNBOIM, H. C.; DOLY, J.: A Rapid Alkaline Extraction Procedure for Screening Recombinant Plasmid DNA; *Nucl. Acids Res.* **7** (1979) 1513–1522.

[7] BONAS, U.; SOMMER, H.; HARRISON, B. J.; SAEDLER, H.: The Transposable Element Tam 1 of *Antirrhinum majus* Is 17 kb Long; *Mol. Gen. Genet.* **194** (1984) 138–143.

[8] CALABRETTA, B.; ROBBERSON, D. L.; BARRERA-SALDANA, H. A.; LAMBROU, T. P.; SAUNDERS, G. F.: Genome Instability in a Region of Human DNA Enriched in Alu Repeat Sequences; *Nature (London)* **296** (1982) 219–225.

[9] CECH, T. R.; BASS, B. L.: Biological Catalysis by RNA; *Ann. Rev. Biochem.* **55** (1986) 599–629.

[10] CECH, T. R.: RNA als Enzym; *Spektrum der Wissenschaft* **1** (1987) 42–51.

[11] CHOMET, P. S.; WESSLER, S.; DELLAPORTA, S. L.: Inactivation of the Maize Transposable Element Activator (*Ac*) Is Associated with Its DNA Modification; *EMBO* **6** (1987) 295–302.

[12] CLARE, J.; FARABAUGH, P.: Nucleotide Sequence of a Yeast Ty Element: Evidence for Unusual Mechanism of Gene Expression; *PNAS USA* **82** (1985) 2829–2833.

[13] COEN, E. S.; CARPENTER, R.; MARTIN, C. R.: Transposable Genetic Elements and Genetic Instability in *Antirrhinum*; *John Innes Report* (1986) 52–64.

[14] COLLEAUX, L.; D'AURIOL, L.; BETERMIER, M.; COTTAREL, G.; JACQUIER, A.; GALIBERT, F.; DUJON, B.: A Universal Code Equivalent of a Yeast Mitochondrial Intron Reading Frame Is Expressed into *E. coli* As a Specific Double-Strand Endonuclease; *Cell* **44** (1986) 521–533.

[15] DINOCERA, P. P.; CASARI, G.: Related Polypeptides Are Encoded by *Drosophila* F Elements, I Factors and Mammalian L1 sequences; *PNAS USA* **84** (1987) 5843–5847.

[16] DINTER-GOTTLIEB, G.: Viroids and Virusoids Are Related to Group I Introns; *PNAS USA* **83** (1986) 6250–6254.

[17] DÖRING, H. P.; TILLMANN, E.; STARLINGER, P.: DNA Sequence of the Maize Transposable Element Dissociation; *Nature (London)* **307** (1984) 127–130.

[18] DUJON, B.; SLONIMSKI, P. P.; WEILL, L.: Mitochondrial Genetics IX. A Model for Recombination and Segregation of Mitochondrial Genomes in *Saccharomyces cerevisiae*; *Genetics* **78** (1974) 415–437.

[19] ERICKSON, L.; BEVERSDORF, W. D.; PAULS, K. P.: Linear Mitochondrial Plasmid in *Brassica* Has Terminal Protein; *Curr. Genet.* **9** (1985) 679–682.

[20] Esser, K.; Kempken, F.: Structure and Function of Linear Extrachromosomal DNA in Eukaryotes; *Process. Biochem.* **21** (1986) 69–76.
[21] Esser, K.; Kück, U.; Lang-Hinrichs, C.; Lemke, P.; Osiewacz, H. D.; Stahl, U.; Tudzynski, P.: *Plasmids of Eukaryotes*, Heidelberg Science Library, Springer Verlag (1986).
[22] Flores, S. C.; Moore, T. K.; Gaubatz, J. W.: Dispersed Repetitive Sequences of the Mouse Genome Are Differentially Represented in Extrachromosomal Circular DNAs *in Vivo*; *Plasmid* **17** (1987) 257–260.
[23] Forster, A. C.; Symons, R. H.: Self-Cleavage of Virusoid RNA Is Performed by the Proposed 55–Nucleotide Active Site; *Cell* **50** (1987) 9–16.
[24] Francou, F.: Isolation and Characterization of a Linear DNA Molecule in the Fungus *Ascobolus immersus*; *Mol. Gen. Genet.* **184** (1981) 440–444.
[25] Fuetterer, J.; Hohn, T.: Involvement of Nucleocapsids in Reverse Transcription: A General Phenomenon? *TIBS* **12** (1987) 92–95.
[26] Gallo, R. C.: Das AIDS-Virus; *Spektrum der Wissenschaft* **3** (1987) 82–93.
[27] Giasson, L.; Lalonde, M.: Analysis of a Linear Plasmid Isolated from the Pathogenic Fungus *Ceratocystis fimbriata*; *Ell. & Halst. Curr. Genet.* **11** (1987) 331–334.
[28] Gierl, A.; Schwarz-Sommer, Z.; Saedler, H.: Molecular Interactions between the Components of the En-I Transposable Element System of *Zea mays*; *EMBO* **4** (1984) 579–583.
[29] Gunge, N.; Tamaru, A.; Ozawa, F.; Sakaguchi, K.: Isolation and Characterization of Linear Deoxyribonucleic Acid Plasmids from *Kluyveromyces lactis* and the Plasmid-Associated Killer Character; *J. Bacteriol.* **145** (1981) 382–390.
[30] Gunge, N.: Linear DNA Killer Plasmids from the Yeast *Kluyveromyces*; *Yeast* **2** (1986) 153–162.
[31] Hardman, N.: Structure and Function of Repetitive DNA in Eukaryotes; *Biochem. J.* **234** (1986) 1–11.
[32] Harris, M. P. G.; Hay, R. T.: DNA Sequences Required for the Initiation of Adenovirus Type 4 Replication *in Vitro*; *J. Mol. Biol.* **201** (1988) 57–67.
[33] Hashiba, T.; Homma, Y.; Hyakumachi, M.; Matsuda, I.: Isolation of a Plasmid in the Fungus *Rhizoctonia solani*; *J. Gen. Microbiol.* **130** (1984) 2067–2070.
[34] Hensgens, L. A. M.; Arnberg, A. C.; Roosendaal, E.; van der Horst, G.; van der Veen, R.; van Ommen, G. B.: Variation, Transcription and Circular RNAs of the Mitochondrial Gene for Subunit I of Cytochrome c Oxidase; *J. Mol. Biol.* **164** (1983) 35–38.
[35] Hess, O.: Das P-Element – Ein potenter Vektor für den Gentransfer bei *Drosophila*; *BioEngineering* **2** (1986).
[36] Honeyman, A. L.; Currier, T. C.: The Isolation and Characterization of Two Linear DNA Elements from *Gaeumannomyces graminis* var. *tritici*, the Causative Agent of "Take-All Disease" of Wheat; Abstracts, 83rd Ann. Meeting (New Orleans), Am. Soc. Microbiol., Washington, DC, (1983); Abstract # H175, p 135.
[37] Jacks, T.; Varmus, H. E.: Expression of the Rous Sarcoma Virus *pol* Gene by Ribosomal Frameshifting; *Science* **230** (1985) 1237–1242.
[38] Jacquier, A.; Dujon, B.: An Intron-Encoded Protein Is Active in a Gene Conversion Process That Spreads an Intron into a Mitochondrial Gene; *Cell* **41** (1985) 389–394.
[39] Jamet-Vierny, C.; Begel, O.; Belcour, L.: A 20 × 10/3–Base Mosaic Gene Identified on the Mitochondrial Chromosome of *Podospora anserina*; *Eur. J. Biochem.* **143** (1984) 389–394.
[40] Kemble, R. J.; Gunn, R. E.; Flavell, R. B.: Classification of Normal and Male-Sterile Cytoplasms in Maize. II. Electrophoretic Analysis of DNA Species in Mitochondria; *Genetics* **95** (1980) 451–458.

[41] Kikuchi, Y.; Hirai, K.; Gunge, N.; Hishinuma, F.: Hairpin Plasmid – A Novel Linear DNA of Perfect Hairpin Structure; *EMBO* **4** (1985) 1881–1886.

[42] Kingsman, A. J.; Kingsman, S. M.: Ty: A Retroelement Moving Forward; *Cell* **53** (1988) 333–335.

[43] Kistler, H. C.; Leong, S. A.: Linear Plasmidlike DNA in the Plant Pathogenic Fungus *Fusarium oxysporum* f. sp. *conglutinans*; *J. Bacteriol.* **167** (1986) 587–593.

[44] Kitada, K.; Hishinuma, F.: A New Linear DNA Plasmid Isolated from the Yeast *Saccharomyces kluyveri*; *Mol. Gen. Genet.* **206** (1987) 377–381.

[45] Kleckner, N.: Transposable Elements in Prokaryotes; *Ann. Rev. Genet.* **15** (1981) 341–404.

[46] Kück, U.; Stahl, U.; Tudzynski, P.; Esser, K.: A Mitochondrial Plasmid in *Podospora anserina*; *Neurospora Newsl.* **29** (1982) 7.

[47] Kück, U.; Kappelhoff, B.; Esser, K.: Despite mtDNA Polymorphism the Mobile Intron (plDNA) of the COI Gene Is Present in Ten Different Races of *Podospora anserina*; *Curr. Genet.* **10** (1985) 59–67.

[48] Kück, U.; Osiewacz, H. D.; Schmidt, U.; Kappelhoff, B.; Schulte, E.; Stahl, U.; Esser, K.: The Onset of Senescence Is Affected by DNA Rearrangements of a Discontinuous Mitochondrial Gene in *Podospora anserina*; *Curr. Genet.* **9** (1985) 373–382.

[49] Lang, F. B.: The Mitochondrial Genome of the Fission Yeast *Schizosaccharomyces pombe*: Highly Homologous Introns Are Inserted at the Same Position of the Otherwise Less Conserved cox1 Genes in *Schizosaccharomyces pombe* and *Aspergillus nidulans*; *EMBO* **3** (1984) 2129–2136.

[50] Long, E. O.; Dawid, J. B.: Repeated Genes in Eukaryotes; *Ann. Rev. Biochem.* **49** (1980) 727–764.

[51] Macreadi, I. G.; Scott, R. M.; Linn, A. R.; Butow, R. A.: Transposition of an Intron in Yeast Mitochondria Requires a Protein Encoded by That Intron; *Cell* **41** (1985) 395–402.

[52] McClintock, B.: Some Parallels between Gene Control Systems in Maize and in Bacteria; *Am. Nat.* **95** (1961) 265–277.

[53] McClintock, B.: The Control of Gene Action in Maize; *Brookhaven Symp. Biol.* **18** (1965) 162–184.

[54] Meinhardt, F.; Esser, K.: The Plasmids of the Morels: Characterization and Prerequisites for Vector Development; *Appl. Microbiol. Biotechnol.* **27** (1987) 276–282.

[55] Meinhardt, F.; Kempken, F.; Esser, K.: Proteins Are Attached to the Ends of a Linear Plasmid in the Filamentous Fungus *Ascobolus immersus*; *Curr. Genet.* **11** (1986) 243–246.

[56] Mellor, J.; Fulton, S. M.; Dobson, M. J.; Wilson, W.; Kingsman, S. M.; Kingsman, A. J.: A Retrovirus-Like Strategy for Expression of a Fusion Protein Encoded by Yeast Transposon Ty *1*; *Nature (London)* **313** (1985) 243–246.

[57] Michel, F.; Lang, F.: Mitochondrial Class II Introns Encode Proteins Related to the Reverse Transcriptases of Retroviruses; *Nature (London)* **316** (1985) 641–643.

[58] Mohan, M.; Meyer, R. J.; Anderson, J. B.; Horgen, P. A.: Plasmid-Like DNAs in the Commercially Important Mushroom Genus *Agaricus*; *Curr. Genet.* **8** (1984) 615–619.

[59] Mühlbach, H. P.: Viroide: Freie infektiöse RNA-Moleküle als Erreger von Pflanzenkrankheiten; *BiuZ* **3** (1987) 65–78.

[60] Müller-Neumann, M.; Joder, J.; Starlinger, P.: The DNA Sequence of the Transposable Element *Ac* of *Zea mays* L.; *Mol. Gen. Genet.* **198** (1984) 19–24.

[61] Nevers, P.; Shepherd, N. S.; Saedler, H.: Plant Transposable Elements. In: Callow, J. A., Ed.; *Advances in Botanical Research* **12**, Academic Press, New York (1986); pp 103–203.

[62] NORMAND, P.; SIMONET, P.; GIASSON, L.; RAVEL-CHAPIUS, P.; FORTIN, J. A.; LALONDE, M.: Presence of a Linear Plasmid-Like Molecule in the Fungal Pathogen *Ceratocystis fimbriata; Ell. & Halst. Curr. Genet.* **11** (1987) 335–338.

[63] NORTH, G.: Back to the RNA World–and Beyond; *Nature (London)* **328** (1987) 18–19.

[64] PAILLARD, M.; SEDEROFF, R. R.; LEVINGS, III, C. S.: Nucleotide Sequence of the S-1 Mitochondrial DNA from the S Cytoplasm of Maize; *EMBO* **4** (1985) 1125–1128.

[65] PALMER, J. D.; SHIELDS, C. R.; COHEN, D. B.; ORTON, T. J.: An Unusual mtDNA Plasmid in the Genus *Brassica*; *Nature (London)* **301** (1983) 725–728.

[66] PEREIRA, A.; CUYPERS, H.; GIERL, A.; SCHWARZ-SOMMER, Z.; SAEDLER, H.: Molecular Analysis of the En/Spm Transposable Element System of *Zea mays*; *EMBO* **5** (1986) 835–841.

[67] PITTLER, S. J.; SALZ, J. K.; DAVIS, R. L.: An Interchromosomal Gene Conversion of the *Drosophila* Dunce Locus Identified with Restriction Site Polymorphisms: A Potential Involvement of Transposable Elements in Gene Conversion; *Mol. Gen. Genet.* **208** (1987) 315–324.

[68] PRING, D. R.; CONDE, M. F.; SCHERTZ, K. F.; LEVINGS, III, C. S.: Plasmid-Like DNAs Associated with Mitochondria of Cytoplasmic Male-Sterile *Sorghum*; *Mol. Gen. Genet.* **186** (1982) 180–184.

[69] RÖDEL, G.; WOLF, K.: Mosaikgene; *BiuZ* **15** (1985) 179–185.

[70] RODITI, I.; CARRINGTON, M.; TURNER, M.: Expression of a Polypeptide Containing a Dipeptide Repeat Is Continued to the Insect Stage of *Trypanosoma brucei*; *Nature (London)* **325** (1987) 272–274.

[71] ROGERS, J.: Exon Shuffling and Intron Insertion in Serine Protease Genes; *Nature (London)* **315** (1985) 458–459.

[72] ROGERS, J. H.: The Origin and Evolution of Retroposons; *Int. Rev. Cytol.* **93** (1985) 187–279.

[73] RUAN, K.; EMMONS, S.: Extrachromosomal Copies of Transposon Tc1 in the Nematode *Caenorhabditis elegans*; *PNAS USA* **81** (1984) 4018–4022.

[74] SCHARDL, C. L.; LONSDALE, D. M.; PRING, D. R.; ROSE, K. R.: Linearization of Maize Mitochondrial Chromosomes by Recombination with Linear Episomes; *Nature (London)* **310** (1984) 292–296.

[75] SCHMIDT, F. J.: RNA Splicing in Prokaryotes: Bacteriophage T4 Leads the Way; *Cell* **41** (1985) 339–340.

[76] SCHWARTZ, D.; DENNIS, E.: Transposase Activity of the *Ac* Controlling Element in Maize Is Regulated by Its Degree of Methylation; *Mol. Gen. Genet.* **205** (1986) 476–482.

[77] SCHWARZ-SOMMER, Z.; SAEDLER, H.: Can Plant Transposable Elements Generate Novel Regulatory Systems? *Mol. Gen. Genet.* **209** (1987) 207–209.

[78] SEDEROFF, R. R.; RONALD, P.; BEDINGER, P.; RIVIN, C.; WALBOT, V.; BLAND, M.; LEVINGS, III, C. S.: Maize Mitochondrial Plasmid S-1 Sequences Share Homology with Chloroplast Gene psBA; *Genetics* **113** (1986) 469–482.

[79] SHAPIRO, J. A.: Molecular Model for the Transposition and Replication of Bacteriophage Mu and Other Transposable Elements; *PNAS USA* **76** (1979) 1933–1973.

[80] SHAPIRO, J. A., Ed.: *Mobile Genetic Elements*, Academic Press, New York (1983).

[81] SHEPERD, H. S.; LIGON, J. M.; BOLEN, P. L.; KURTZMANN, C. P.: Cryptic DNA Plasmids of the Heterothallic Yeast *Saccharomyces crataegensis*; *Curr. Genet.* **12** (1987) 297–304.

[82] SHINAGAWA, M.; IIDA, Y.; MATSUDA, A.; TSUKIYAMA, T.; SATO, G.: Phylogenic Relationships between Adenoviruses As Inferred from Nucleotide Sequences of Inverted Terminal Repeats; *Gene* **55** (1987) 85–93.

[83] TIMOTHY, D. H.; LEVINGS, III, C. S.; HU, W. W. L.; GOODMAN, M. M.: *Zea diploperennis* May Have Plasmid-Like Mitochondrial DNAs; *Maize Genet. Crop News Lett.* **56** (1982) 133–140.

[84] TOKUNAGA, M.; WADA, N.; HISHINUMA, F.: Expression and Identification of Immunity Determinants on Linear Killer Plasmids pGKL1 and pGKL2 in *Kluyveromyces lactis*; *Nucleic Acids Research* **15** (1987) 1031–1046.

[85] TUDZYNSKI, P.; DÜVELL, A.; ESSER, K.: Extrachromosomal Genetics of *Claviceps purpurea*. I. Mitochondrial DNA and Mitochondrial Plasmids; *Curr. Genet.* **7** (1983) 145–150.

[86] TUDZYNSKI, P.; ESSER, K.: Extrachromosomal Genetics of *Claviceps purpurea*. II. Plasmids in Various Wild Strains and Integrated Plasmid Sequences in Mitochondrial Genomic DNA; *Curr. Genet.* **10** (1986) 463–467.

[87] TURPEN, P.; GARGER, S. J.; MARKS, D.; GRILL, L. K.: Molecular Cloning and Physical Characterization of a *Brassica* Linear Mitochondrial Plasmid; *Mol. Gen. Genet.* **209** (1987) 227–233.

[88] TRAYNOR, P. L.; LEVINGS, III, C. S.: Transcription of the S-2 Maize Mitochondrial Plasmid; *Plant Mol. Biol.* **7** (1986) 255–263.

[89] UEDA, H.; MIZUNO, S.; SHIMURA, K.: Transposable Genetic Element Found in the 5′-Flanking Region of the Fibroin H-Chain Gene in a Genomic Clone from the Silkworm *Bombyx mori*; *J. Mol. Biol.* **190** (1986) 319–327.

[90] UPADHYAYA, K. C.; SOMMER, H.; KREBBERS, E.; SAEDLER, H.: The Paramutagenic Line *niv*-44 Has a 5 kb Insert, Tam 2, in the Chalcone Synthase Gene of *Antirrhinum majus*; *Mol. Gen. Genet.* **199** (1985) 201–207.

[91] WARING, R. B.; BROWN, T. A.; RAY, J. A.; SCAZZOCCHIO, C.; DAVIES, R. W.: Three Variant Introns of the Same General Class in the Mitochondrial Gene for Cytochrome Oxidase Subunit 1 in *Aspergillus nidulans*; *EMBO* **3** (1984) 2121–2128.

[92] WATSON, J. D.; HOPKINS, N. H.; ROBERTS, J. W.; STEITZ, J. A.; WEINER, A. M.: *Molecular Biology of the Gene*, 4th ed., The Benjamin/Cummings Publishing Co., Inc., Menlo Park, CA (1987).

[93] WEINER, A. M.; DEINIGER, P. L.; EFSTRATIADIS, A.: Nonviral Retroposons: Genes, Pseudogenes, and Transposable Elements Generated by the Reverse Flow of Genetic Information; *Ann. Rev. Biochem.* **55** (1986) 631–661.

[94] WEISSINGER, A. K.; TIMOTHY, D. H.; LEVINGS, III, C. S.; HU, W. W. L.; GOODMAN, M. M.: Unique Plasmid-Like Mitochondrial DNAs from Indigenous Maize Races of Latin America; *PNAS USA* **79** (1982) 1–5.

[95] WIMMER, E.: Genome-Linked Proteins of Viruses; *Cell* **28** (1982) 199–201.

[96] WOLF, K.: Sind Viroide entkommene Intronen? *BiuZ* **1** (1987) 29–30.

Ribozymes: Biocatalysts Composed of Ribonucleic Acid

by C. K. Biebricher

Contents

Dr. Christof K. Biebricher,
Max-Planck-Institut für Biophysikalische Chemie,
Am Fassberg,
D-3400 Göttingen, Fed. Rep. Germany

1 Introduction

Ribonucleic acids with enzymatic activity, or *ribozymes*, were discovered a few years ago. All of them are involved in RNA metabolism: they catalyze intramolecular and intermolecular RNA cleavage as well as nucleotidyl phosphoryl transfer to RNA. Even though the sequences of many ribozymes have been determined, nothing is known about their three-dimensional structure and what groups participate at the active centers. Since RNA molecules of defined sequence are easy to synthesize and are stable, ribozyme design analogous to the efforts in protein design seems feasible. How practical is such a program with the current state of the art?

WÖHLER's synthesis of urea from inorganic salts was the first of many times that biochemical researchers have shown that the substances as well as the chemical reactions in living cells are in full accord with the laws of chemistry. However, the synthetic strategies of metabolic reactions differ fundamentally from the methods usually used in preparative organic chemistry:

a) exergonic reactions are used nearly exclusively, which is to say, the thermodynamic equilibrium is far on the side of the products, but
b) they proceed extremely slowly at physiological conditions.

Highly specific biocatalysts, the enzymes, accelerate the desired reactions by many orders of magnitude. The main component of nearly all enzymes is a specific protein, often aided by cofactors and coenzymes. Even though coenzymes are required for enzymatic activity, they are considered as mere cosubstrates that provide necessary chemical groups. Often chemical modification of the protein by glycosylation, phosphorylation, and nucleotidylation also contributes to enzymatic activity. The catalytic properties themselves, however, are attributed to the proteins, because the dynamic folding of the polypeptide chain is largely responsible for the enzymatic activity.

Metabolism of RNA includes all the many reactions involved in the expression of genetic information. Translation of the nucleotide sequence into an amino acid sequence in the biosynthesis of proteins takes place at the ribosomes, organelles composed of about equal amounts of RNA (rRNA, where *r* stands for *ribosomal*) and protein. Translation of the nucleotide sequence into a peptide sequence is achieved by tRNAs (*t* stands for *transfer*), molecules that carry a specific amino acid and recognize a specific base triplet at the mRNA (*m* stands for *messenger*). Francis C. H. CRICK [1] proposed in 1968 that rRNA and tRNA are by no means merely supporting players, but have direct and central roles in gene expression. He considered tRNA and rRNA to be living fossils: their ancient roles in catalysis could not be replaced by protein without losing the hitherto accumulated information. Furthermore, he proposed that RNA may have had a much larger role in catalysis in the early stages of the evolution of life.

CRICK's ideas have been shown to be largely correct: today we know that the genetic code is universal (with the exception of a few mitochondrial codons). We have learned from comparative sequence studies of tRNA and rRNA [2, 3] that their sequences are not randomized by mutation. Furthermore, rRNA is directly involved in all steps of translation (reviewed in [4]). Many other examples of the contribution of RNA to cellular processes have been found; I discuss some of them in this chapter.

2 Biochemical catalysis

Enzymes are linear peptide chains about 100–1000 amino acids long; they are stabilized by S–S cross-links after the optimal chain structure is established. The chain structure is by no means a random coil; the folding energy directs the formation of a defined three-dimensional structure, where every residue is assigned its position (if we ignore thermal motion) [5]. This is why most enzymes can be crystallized. The peptide residues in the backbone can readily form intramolecular hydrogen bonds with nearby residues, to produce the stabilizing structures called *α helices* and *β pleated sheets*; it is the side groups of the 20 coded amino acids, however, that contribute most to defining the structure formed by intramolecular interactions. Hydrophobic amino acids form a core in the interior of the proteins, while hydrophilic amino acids can either lie in hydrated form at the surface or be buried in the hydrophobic core, if their polar groups are saturated by intramolecular hydrogen bonds or salt bridges. The tertiary structures of several hundred proteins that have been determined have motifs common to many proteins, but, in spite of the wealth of information already available, direct calculation of the correct structures from determined amino acid sequences is only possible for short polypeptides; the determination of catalytic properties is even less tractable.

The specific binding and discrimination of substrates is often compared in the literature with a key fitting into a lock, but this static simile needs to be updated in light of recent investigations. Peptide chains are by no means as rigid as steel; rather, they are more or less viscous, in some sense of that word [6]. Access of some substrates to the specific pockets constituting the active sites requires motion of the peptide chains [7]. Binding of a substrate to an enzyme is accompanied by conformational changes in both enzyme and substrate. An important feature of catalysis is that the binding energy is used to distort substrate molecules in the direction of the structure of the active intermediate [5, 8, 9]. The side chains of several amino acids are particularly well-suited to efficient acid-base catalysis. Metal ions and other cofactors bind selectively to increase the repertoire of possible reactions.

Some of the properties essential to the catalytic activity of proteins are also shared by other macromolecules, especially RNA. Single strands of RNA also fold into defined tertiary structures. Base pairing and base stacking are the predominant forces for structure formation; in addition to the Watson-Crick base pairs in double-helical regions, other hydrogen bonding between bases, the 2′-hydroxyl of the ribose, and the phosphate also contribute to the stability of the tertiary structure [10]. Determination of tertiary structures by X-ray diffraction has so far been limited to a few tRNAs [11–14], but the secondary structures of RNA molecules can be calculated [15, 16]. Chemical modification experiments and other data allow us to guess at tentative tertiary structures [17]. However, all methods available so far are restricted to RNA with lengths of not more than a few hundred bases.

3 Hydrolytic cleavage of RNA

RNA hydrolyzes in aqueous solution, at neutral pH very slowly, but rapidly under alkaline conditions. The reaction starts with a nucleophilic attack of the 2′-OH on the phosphorus atom, with transient formation of a trigonal-bipyrimidal structure, which decomposes by releasing the ribose 5′-OH of the following nucleotide. The cleavage product has a 2′,3′-phosphodiester bond at its 3′ terminus; it may be hydrolyzed by acid catalysis into 2′- and 3′-terminal phosphates. The cleavage reaction itself is not a hydrolysis but rather a transesterification (Figure 1).

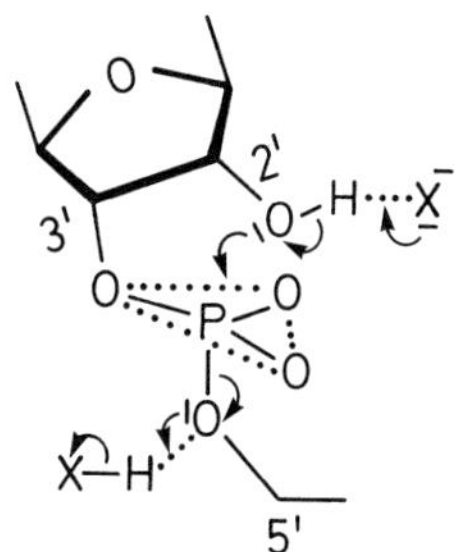

Fig. 1 Cleavage of the RNA backbone. Nucleophilic attack at the 3′-hydroxyl produces a trigonal-bipyramidal intermediate with the phosphorus atom at the center. The leaving group is the 3′-terminal piece of the polynucleotide chain carrying a 5′-hydroxyl; the 5′-terminal piece has a 3′-terminal 2′,3′-phosphodiester. The reaction is base-catalyzed

Phosphodiester bonds are broken at random only under conditions where the secondary structure is destroyed [18]. Cleavage in double-stranded regions occurs at greatly reduced rates [19]. Di- and trivalent metal ions catalyze RNA cleavage specifically at positions in the neighborhood of the complexed ion. This specific catalysis has been thoroughly investigated with $tRNA^{phe}$. In slightly alkaline Mg^{2+} solution, cleavage takes place between ^{16}**D** and ^{17}**D** [20], while Pb^{2+} ions catalyze cleavage of the phosphodiester bridge between ^{17}**D** and ^{18}**G** even in crystals and under neutral conditions [21–23]. Results of X-ray experiments are clear that the ions are indeed complexed to these nucleotides. The rate increase by and specificity of the Pb^{2+} cation are comparable to those of enzymic reactions. However, the active center of the complex is destroyed by the reaction and thus consumed, in contrast to the situation when a true enzyme catalyzes a reaction. The unusual bases in $tRNA^{phe}$ apparently do not contribute to the catalysis, because the reaction also takes place with a synthetic tRNA analog containing only the four main bases [24, 25]. Altering the nucleotide sequence at the active center of the tRNA, however, seriously affects the rate of the cleavage reaction. Cleavage of tRNA is unlikely to have biological significance since it requires unphysiological conditions to proceed at reasonable rates.

4 Spontaneous hydrolysis of virusoids

Spontaneous cleavage of virusoid precursor RNA [26] does have potential biological significance. These RNAs are what are called *satellites* of RNA viruses; they use the replication apparatus of the viruses for their own amplification, thus modifying the pathogenicity of the host viruses. During replication of virusoid RNA, genome *concatemers* are formed; these are cleaved spontaneously at specific positions to monomer genomes. Experimenters using recombinant DNA technology have synthesized tandem sequences of virusoid RNA that were suitable for *in vitro* studies of the cleavage mechanism [26]. The minimum sequence length that can still react is quite small. Its structure was tentatively proposed to be a "hammerhead"; the four strand segments constituting this structure may belong to different RNA strands (Figure 2) [24].

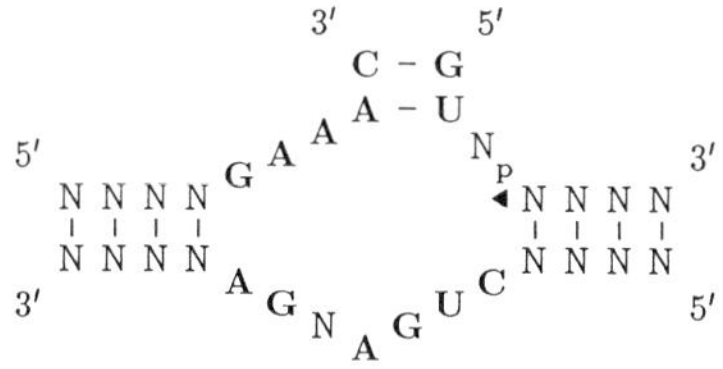

```
                 3'     5'
                  C - G
                A A - U
5'           A        N         3'
  N N N N  G           P
  | | | |              ◄N N N N
  N N N N                | | | |
           A          C N N N N
3'           G      U           5'
              N   G
                A
```

Fig. 2 Consensus sequence for a hammerhead structure. Bold symbols represent positions with invariant nucleotides. The three component chains may be parts of one and the same chain, as in the natural virusoid RNA concatemers (*cis*), or they may belong to different chains (*trans*). The structure shown is tentative. Cleavage occurs at the position designated with a triangle, between the phosphate and the 5′-OH of the following nucleotide

Transcripts of satellite DNA of the newt *Notophthalmus viridescens* [27] have homologous site-specific cleavage; the sequence features found were identical to those of virusoids. The homology may be coincidental, or it may be due to horizontal gene transfer [28]. However, other similar reactions do not seem to involve hammerhead structures [29]. Divalent cations, especially Mg^{2+}, are required for the cleavage reaction to proceed. Cleavage rates are low in comparison to those of ribonuclease-catalyzed reactions. The reaction products are RNA pieces with 5′-OH and 2′,3′-cyclic phosphodiester termini.

5 Ribonuclease P as an example of a ribozyme

Enzymes are true catalysts: they are not consumed by the reactions catalyzed, which is not true for the cleavage reactions described above. There are, however, RNA catalysts that meet all the criteria of enzymes. A typical example is *ribonuclease P*, an enzyme class found in prokaryotes, archaebacteria, and eukaryotes. It is responsible for cutting tRNA structures from precursor RNA transcripts to produce the correct 5′ termini. RNase P usually consists of a basic protein, C5, and an RNA subunit, M1 [30]. Sydney Altman and co-workers at Yale University have found that the enzymatic activity of

RNase P is associated with the RNA subunit; M1 alone can perform the RNase P function *in vitro* [31–34]. The reaction with M1 alone is rather slow and requires high concentrations of Mg^{2+} and monovalent cations. *In vivo*, the protein subunit C5 is also required if the host is to be viable. Ribonuclease P cleaves the phosphodiester bridge between the 5′-phosphate and the 3′-OH; the reaction is thus a true hydrolysis. It has not yet been possible to detect covalent intermediates between enzyme and substrate RNAs.

In catalyzing the reaction, ribonuclease P and the substrate precursor RNA play complementary catalytic roles, because both enzyme and substrate are polymers with defined folding patterns. Analogous complementary roles of enzyme and RNA substrate have also been found in RNA replication [35] and may be common to most reactions involved in maturing tRNAs and rRNAs. RNA subunits and protein subunits of RNase P isolated from different organisms may complement each other [36], even though different RNA subunits have been found upon comparison to have little sequence homology [36]. Great care was taken to prove that the residual enzymatic activity of the RNA moiety of ribonuclease is not due to contaminating traces of protein: not only was the RNA carefully isolated and the classical specific digestion experiments performed, but also the M1 RNA was cloned into plasmids and the RNA produced by transcription *in vitro* [36].

At the specific phosphodiester bond, the rate of hydrolysis is stimulated by M1 to 10 orders of magnitude greater than that of spontaneous hydrolysis; addition of the protein subunit increases the rate by another three orders of magnitude. The reaction with M1 follows Michaelis-Menten kinetics. Cleavage in the first reaction cycle is much more rapid; apparently the recycling of the RNA to the enzymatic active state, that is, the release of the cleaved RNA, is rate-determining [32]. This interpretation is further supported by the strong product inhibition observed with the M1 reaction.

Nothing we have so far learned about the primary and secondary structures of M1 RNA suggests a mechanism. Altering the RNA sequences removed by the action of RNase P revealed no specificity. The fact that little sequence homology is found among M1 subunits of P RNases isolated from different hosts is indicative of little selection pressure to maintain the sequence. It is thus the principle of having an RNA subunit, not the sequence of the RNA, that has been conserved in evolution. One is led to conclude that the RNA has a function that cannot be performed with equal efficiency by a protein. However, other RNases involved in the maturing of tRNA and rRNA are proteins, which suggests it is another, additional role that has prevented the RNA from being replaced by a protein. There is speculation that – in analogy to the roles of rRNA in ribosomes and snRNA in spliceosomes [37] – the role may be structural, a possible "maturosome" providing all the cutting and modifying reactions for maturing tRNA [32]. This suggestion must be subjected to experimental scrutiny, which will enable us to decide whether there is evidence for such a (probably unstable) complex.

6 Self-splicing RNA

Eukaryotic genes often have intervening sequences (or *introns*) between the sequences translated into protein (*exons*). Before the finished protein is exported from the nucleus into the cytoplasm, the intervening sequences have to be spliced out of the primary RNA transcripts; there are several mechanisms for this. In many cases (most "group II" introns), splicing is performed by a complicated complex between RNA and protein called a *spliceosome*, where the RNA of the complex has an active part similar to that in the ribosome [37]. *In vitro*, the splicing of "group I" RNA occurs without participation of enzymes: the RNA splices itself [38]. *In vivo* splicing may well be assisted by proteins. This reaction is not hydrolysis but transesterification; for each phosphodiester bond cleaved, another is formed [39]. Self-splicing was discovered and investigated by Thomas CECH and collaborators at the University of Colorado, in experiments with the precursor RNA transcript of rRNA in the ciliate *Tetrahymena thermophila*. *In vivo*, the reaction requires GTP as cofactor; *in vitro*, GMP or guanosine, but not dGTP, can substitute [38–40].

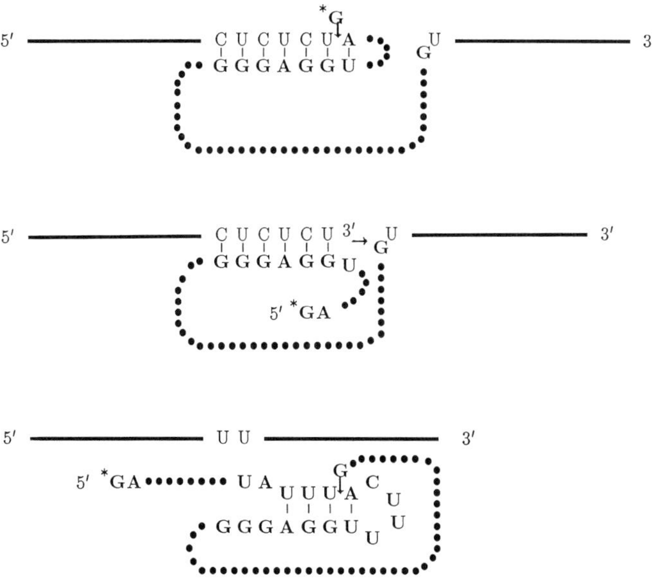

Fig. 3 Self-splicing of pre-rRNA in *Tetrahymena*. Exons are denoted by solid lines and plain letters, introns by dotted lines and bold letters. Lengths and structure of the schematic drawing are arbitrary. GTP (or guanosine) attacks the phosphodiester at the 5′ terminus of the intron. The newly formed terminus attacks the phosphodiester at the 3′ terminus of the intron. The spliced intron cyclizes and is released as an oligonucleotide. The ring reopens again at the newly formed link by spontaneous hydrolysis. The resulting core sequence can catalyze splicing reactions of other RNA chains (Fig. 4)

Fig. 4 Transfer of groups (R) containing 5′-terminal guanosine to acceptor 3′ termini of RNAs, catalyzed by the core *Tetrahymena* ribozyme. Constituents of the *internal guide sequence* are shown in bold letters. The R groups may be oligonucleotides, mononucleotides, or phosphate; acceptor groups may be 3′ termini of RNA, or water. The internal guide sequence may be altered to change the sequence specificity, or removed entirely to abolish sequence specificity of the acceptor RNA. Group transfer liberates the 5′-terminal guanosine residues

The first step is a nucleophilic attack of the cofactor upon the phosphodiester bond between leading exon and intron, followed by transesterification (Figure 3). Splicing is completed by an attack of the newly formed 3′ terminus of the leading exon upon the phosphodiester group between the intron and the trailing exon. The intron carries the active catalytic site [41], but because the splicing works only *cis*, which is to say, on the catalytic RNA itself, it cannot be considered a truly enzymatic reaction. The released intron is then converted by a cascade of hydrolysis and transesterification reactions to a core RNA which can no longer react with itself. However, the core can serve *trans* for the modification of other RNA and is thus a ribozyme [42–45]. The catalyzed reactions are transfers of groups at the 3′ end of one RNA to the 3′ end of another RNA (Figure 4) [46, 47]. The transferred groups are usually mono- or oligonucleotides, but phosphate transfer has also been reported. Instead of transferring to a 3′ terminus, the group may also be transferred, at greatly reduced rates, to water. The most interesting reaction is a sort of *polymerase* action [48], in which nucleotides are transferred from the 3′ ends of donor oligonucleotides to an acceptor that synthesizes a sequence complementary to an *internal guide sequence* [49] on the ribozyme, up to about 15 nucleotides. The internal guide sequence, and with it the sequence specificity, can be altered without interfering with the catalytic function [44, 49]. When the region with the internal guide sequence was removed, the resulting ribozyme was able to ligate, under special conditions, oligoribonucleotides hybridized to a template sequence, provided the oligonucleotides had at the junction an overhanging G, which was released in the ligation reaction [50]. The Michaelis constants and the turnover numbers are not impressive, but the reaction seems to be quite sequence-independent. Experimenters have been trying to find the active groups in the *Tetrahymena* ribozyme, but even though the tertiary structure is under investigation [51–53], the models presented are not yet supported by solid experimental evidence.

Self-splicing occurs in several RNAs in different organisms [55–58]. The evolutionary advantage of self-splicing is not yet clear. The intron carrying the ribozymic activity is missing in some *Tetrahymena* species; there is no strong selection pressure to conserve it. The splicing reaction shown in Figure 3 is irreversible, but the self-splicing intron of *Tetrahymena* can be inserted into a foreign RNA by reverse splicing [59]. This reaction has considerable interest as a possible origin of introns [28, 59].

7 The mechanism of RNA catalysis

Since no structure of a catalytic RNA is known, my discussion of mechanism focuses on the reaction at the substrate, in all cases a specific RNA. This specificity is a strong indication that the substrate itself is also intimately involved in the catalytic function of the ribozymic RNA. In all examples known so far, the mechanism at the substrate is a nucleophilic attack (S_N2) on the phosphorus atom at the modified phosphodiester bond, forming a trigonal-bipyramidal intermediate, with the attacking nucleophile and the leaving group in the axial positions (*in-line mechanism*). Alternatively, the leaving group can reach an axial position by a pseudorotation (*adjacent mechanism*) [5, 23]. In the alkali-catalyzed cleavage of RNA, the attacking nucleophile is the 2′-hydroxyl; in RNase P it is water; and in the self-splicing ribozymes it is the 3′-hydroxyl of an RNA. All these nucleophiles are too weak at physiological conditions to react at significant rates and require catalysis, most likely acid-base catalysis, as with RNases. However, except for the 5′-terminal phosphate (which has never been observed to participate in reactions known to date) none of the residues in RNA has a p*K* in the physiological pH range. Complexing with metal ions reduces the p*K* of complexed water considerably; indeed, all ribozymes require divalent cations at rather high concentrations to function [58]. Divalent cations may also increase the electrophilic activity of the phosphorus atom. Measurements of the atomic distances in Pb^{2+}-catalyzed cleavage of tRNA were evidence for the former mechanism: the distance of the phosphorus atoms of the cleaved phosphodiester is too great for an effective activation by the metal ion, while several nucleophilic side groups of bases are quite close [23]. Ribozyme activities are enhanced by raising the pH [32, 34]; this is also suggestive of a catalytic function for bases.

8 Prospects for synthetic RNA catalysts and their limitations

Ribozymes have an enormous potential for manipulating nucleic acids, in particular RNA. The device of (single strand) sequence-specific DNases, RNases, RNA ligases and RNA recombinases by modifying the internal guide sequences of ribozymes is particularly promising [60–63] and may revolutionize the RNA biochemistry; the new techniques may have even more impact for the RNA technology than the detection of the restriction endonucleases provided for the DNA technology. Furthermore, application of ribozyme action to design ribozymic agents to destroy (or activate) intracellular mRNA *in vivo*, for example serving as highly specific antiviral agents [64–66] offers great possibilities for the future.

The prospects for devising metabolic ribozymes seem less promising.

There can be no doubt that proteins are far more effective and versatile agents for catalysis. The ribozymes known to date do not reach the rates that protein enzymes do; *in vivo* their action is always reinforced by proteins. However, ribozymes have nucleotidyl transferase activity; this is most important for activating monomers, such as amino acids, sugars, or terpenes, to undergo condensation reactions. The activation of monomers by nucleotidylation is unlikely to be merely a "frozen accident" in the evolution of metabolism; rather, it is probably brought about by strong selection pressure. There appears to be no substitute for the precise stereospecificity in the reaction of phosphate esters [67, 68]. Furthermore, nucleosides or nucleotides are much better leaving groups than water or phosphate, respectively. In cellular metabolism, high-energy phosphate anhydrides such as ATP are used to activate monomers, to force the reaction to go to completion. However, the requirements of industrial catalysts with respect to thermodynamic equilibria or reaction rates are much less stringent. Thus the development of ribozymes for synthesis reactions, of terpenes for example, seems to be highly desirable, parallel to efforts in protein design. Efficient methods for synthesizing large amounts of specific RNA sequences are known. Furthermore, since nucleotides not contained in natural RNA are known to participate in biochemical reactions, examples being the flavin and nicotinamide nucleotides in redox reactions, we may take the opportunity to integrate them into possible synthetic catalytic RNA [69–72]. Indeed, we know from the occurrence of unusual nucleotides in tRNA that modifying the structural interactions among nucleotides is often a prerequisite for function.

However, with the current state of the art, there is not enough information available to suggest the design of ribozymes. The structures of several hundred proteins have been solved, but not a single tertiary structure of a ribozyme is known, even though substantial progress has been made in solving the secondary structures of ribozymes [72–74]. In all known cases, highly structured specific RNAs were the substrates of ribozymes, and it has not yet been shown that ribozymes are able to bind and discriminate low molecular weight substances. What should a specific pocket in the RNA look like? Double helical stems, which can be easily constructed, are quite stiff and are probably suitable only for providing a frame for the more subtle interactions that form a defined tertiary structure. The successful construction of a catalytic synthetic RNA certainly cannot be ruled out, but, given our limited knowledge, we would have to rely more on serendipity than on proper design. However, the study of the catalytic roles of RNA has revealed the importance of RNA structure in the expression of genetic information. In addition to "intrinsic" and "extrinsic" information, the former being the direct reading of a DNA sequence by recognition proteins and the latter the translation of nucleotide sequences into polypeptide sequences, a third important source for the coding of the cell's genetic program is the information in active RNA structures. The impact of this area of research on our ideas about the prebiotic and early biotic evolution of life has been described [68, 70]. There has been speculation that, in the beginning, life may have started as a pure RNA world [68, 70]. While this is not very plausible, we can now begin to investigate other functions of RNA structure also, such as a direct role of transesterification in RNA recombination. The accelerating pace of research in this field leads us to expect many exciting new revelations in the near future.

9 Acknowledgement

I thank Dr. McCaskill for correcting my English.

10 References

[1] CRICK, F. H. C.: The Origin of the Genetic Code; *J. Mol. Biol.* **38** (1968) 367–379.

[2] EIGEN, M.; LINDEMANN, B.; TIETZE, M.; WINKLER-OSWATITISCH, R.; DRESS, A.; VON HAESELER, A.: How Old is the Genetic Code? Statistical Geometry of tRNA Provides an Answer; *Science* **244** (1989) 673–679.

[3] PACE, N. R.; OLSEN, G. J.; WOESE, C. R.: Ribosomal RNA Phylogeny and the Primary Lines of Evolutionary Descent; *Cell* **45** (1986) 325–326.

[4] DAHLBERG, A. E.: The Functional Role of Ribosomal RNA in Protein Synthesis; *Cell* **57** (1989) 525–529.

[5] FERSHT, A.: *Enzyme Structure and Mechanism*, Freeman, New York (1985).

[6] KARPLUS, M.; MCCAMMON, J. A.: Dynamics of Proteins: Elements and Function; *Ann. Rev. Biochem.* **53** (1983) 263–300.

[7] FRAUENFELDER, H.; PETSKO, G. A.; TSERNOGLOU, D.: Temperature-Dependent X-ray Diffraction as a Probe of Protein Structural Dynamics; *Nature (London)* **280** (1979) 558–563.

[8] JENCKS, W. P.: Economics of Enzyme Catalysis; *Cold Spring Harbor Symp. Quant. Biol.* **52** (1987) 65–73.

[9] LERNER, R. A.; TRAMONTAUR, A.: Catalytic Antibodies; *Sci. Am.* **258** N3 (1988) 42–50.

[10] SAENGER, W.: *Principles of Nucleic Acid Structure*, Springer-Verlag, New York (1984).

[11] KIM, S. H.; SUDDATH, F. L.; QUIGLEY, F. L.; MCPHERSON, A.; SUSSMAN, J. L.; WANG, A. H. J.; SEEMAN, N. C.; RICH, A.: Three-Dimensional Tertiary Structure of Yeast Phenylalanine Transfer RNA; *Science* **185** (1974) 435–440.

[12] ROBERTUS, J. D.; LADNER, J. E.; FINCH, J. T.; RHODES, D.; BROWN, R. S.; CLARK, B. F. C.; KLUG, A.: Structure of Yeast Phenylalanine tRNA at 3 Å Resolution; *Nature (London)* **250** (1974) 546–551.

[13] WOO, W.; RICH, A.: The Three-Dimensional Structure of *E. coli* Initiator $tRNA^{met}$; *Nature (London)* **286** (1980) 346–351.

[14] MORAS, D.; COMARMOND, M. B.; FISHER, J.; WEISS, R.; THIERRY, J. C.; EBEL, J. P.; GEIGE, R.: Crystal Structure of Yeast $tRNA^{asp}$; *Nature (London)* **288** (1980) 669–674.

[15] ZUKER, M.; STIEGLER, P.: Optimal Computer Folding of Large RNA Sequences Using Thermodynamics and Auxiliary Information; *Nucleic Acid Res.* **9** (1981) 133–148.

[16] TURNER, D. H.; SUGIMOTO, N.; JAEGER, J. A.; LONGFELLOW, C. E.; FREIER, S. M.; KIERZEK, R.: Improved Parameters for Prediction of RNA Structure; *Cold Spring Harbor Symp. Quant. Biol.* **52** (1987) 123–133.

[17] WESTHOF, E.; ROMBY, P.; ROMANIUK, P. J.; EBEL, J.-P.; EHRESMANN, C.; EHRESMANN, B.: Computer Modeling from Solution Data of Spinach Chloroplast and of *Xenopus laevis* Somatic and Oocytic 5S rRNAs; *J. Mol. Biol.* **207** (1989) 417–431.

[18] STANLEY, J.; VASSILENKO, S.: A Different Approach to RNA Sequencing; *Nature (London)* **274** (1978), 87–89.

[19] DOCK-BREGEON, A. C.; MORAS, D.: Conformational Changes and Dynamics of tRNAs: Evidence from Hydrolysis Patterns; *Cold Spring Harbor Symp. Quant. Biol.* **52** (1987) 113–121.

[20] WINTERMEYER, W.; ZACHAU, H. G.: Mg^{2+}-katalysierte spezifische Spaltung von tRNA; *Biochim. Biophys. Acta* **299** (1973) 82–90.

[21] WERNER, C.; KREBS, B.; KEITH, G.; DIRHEIMER, G.: Specific Cleavages of Pure tRNAs by Plumbous Ions; *Biochim. Biophys. Acta* **432** (1976) 161–175.

[22] BROWN, R. S.; HINGERTY, B. E.; DEWAN, J. C.; KLUG, A.: Pb(II)-Catalyzed Cleavage of the Sugar-Phosphate Backbone of Yeast $tRNA^{phe}$ – Implications for Lead Toxicity and Self-Splicing RNA; *Nature (London)* **303** (1983) 543–546.

[23] Brown, R. S.; Dewan, J. C.; Klug, A.: Crystallographic and Biochemical Investigation of the Lead(II)-Catalyzed Hydrolysis of Yeast Phenylalanine tRNA; *Biochemistry* **24** (1985) 4785–4801.

[24] Sampson, J. E.; Sullman, F. X.; Behlen, L. S.; DiRenzo, A. B.; Uhlenbeck, O. C.: Characterization of Two RNA-Catalyzed RNA Cleavage Reactions; *Cold Spring Harbor Symp. Quant. Biol.* **52** (1987) 267–275.

[25] Hall, K. B.; Sampson, J. R.; Uhlenbeck, O. C.; Redfield, A. G.: Structure of an Unmodified tRNA Molecule; *Biochemistry* **28** (1989) 5794–5801.

[26] Forster, A. C.; Symons, R. H.: Self-Cleavage of Plus and Minus RNAs of a Virusoid and a Structural Model for the Active Sites; *Cell* **49** (1987) 211–220.

[27] Epstein, L. M.; Gall, J. G.: Self-Cleaving Transcripts of Satellite DNA from the Newt; *Cell* **48** (1987) 535–543.

[28] Weiner, A. M.; Deininger, P. L.; Efstradiadis, A.: Nonviral Retrotransposons: Genes, Pseudogenes, and Transposable Elements Generated by the Reverse Flow of Genetic Information; *Ann. Rev. Biochem.* **55** (1986) 631–661.

[29] Hampel, A.; Tritz, R.: RNA Catalytic Properties of the Minimum (−)sTRSV Sequence; *Biochemistry* **28** (1989) 4929–4933.

[30] Stark, B. C.; Kole, R.; Bowman, E. J.; Altman, S.: Ribonuclease P: an Enzyme with an Essential RNA Component; *PNAS USA* **75** (1978) 3717–3721.

[31] Reed, R. E.; Baer, M.; Guerrier-Takada, C.; Donis-Keller, H.; Altman, S.: Nucleotide Sequence of the Gene Encoding the RNA Subunit (M_1-RNA) of Ribonuclease P from *Escherichia coli*; *Cell* **30** (1982) 627–636.

[32] Pace, N. R.; Reich, C.; James, B. D.; Olsen, G. J.; Pace, B.; Augh, D. S.: Structure and Catalytic Function in Ribonuclease P; *Cold Spring Harbor Symp. Quant. Biol.* **52** (1987) 239–248.

[33] Guerrier-Takada, C.; Gardiner, K.; Marsh, T.; Pace, N.; Altman, S.: The RNA Moiety of Ribonuclease P Is the Catalytic Subunit of the Enzyme; *Cell* **35** (1983) 849–857.

[34] Guerrier-Takada, C.; Haydock, K.; Allen, L.; Altman, S.: Metal Ion Requirements and Other Aspects of the Reaction Catalyzed by M_1-RNA, the RNA subunit of Ribonuclease P from *Escherichia coli*; *Biochemistry* **25** (1986) 1509–1515.

[35] Biebricher, C. K.: Replication and Evolution of Short-Chained RNA Species Replicated by Qβ Replicase; *Cold Spring Harbor Symp. Quant. Biol.* **52** (1987) 299–306.

[36] Lawrence, N.; Weselowski, D.; Gold, H.; Bartkiewitz, M.; Guerrier-Takada, C.; McClain, W. H.; Altman, S.: Characteristics of Ribonuclease P from Various Organisms; *Cold Spring Harbor Symp. Quant. Biol.* **52** (1987) 233–238.

[37] Steitz, J. A.: "Snurps"; *Sci. Am.* **258** (6) (1988) 36–41.

[38] Kruger, K.; Grabowski, P. J.; Zaug, A. J.; Sands, J.; Gottschling, D. E.; Cech, T. R.: Self-Splicing RNA: Autoexcision and Autocyclization of the Ribosomal RNA Intervening Sequence of *Tetrahymena*; *Cell* **31** (1982) 147–157.

[39] Zaug, A. J.; Grabowski, P. J.; Cech, T. R.: Autocatalytic Cyclization of an Excised Intervening Sequence RNA Is a Cleavage-Ligation Reaction; *Nature (London)* **301** (1983) 578–583.

[40] Brehm, S. L.; Cech, T. R.: Fate of an Intervening Sequence Ribonucleic Acid: Excision and Cyclization of the *Tetrahymena* Ribonucleic Acid Intervening Sequence *in vivo*; *Biochemistry* **22** (1983) 2390–2397.

[41] Been, M. D.; Cech, T. R.: One Binding Site Determines Sequence Specificity of *Tetrahymena* pre-rRNA Self-Splicing, *Trans*-Splicing, and RNA Enzyme Activity; *Cell* **47** (1986) 207–216.

[42] Zaug, A. J.; Cech, T. R.: The Intervening Sequence RNA of *Tetrahymena* Is an Enzyme; *Science* **231** (1986) 470–475.

[43] CECH, T. R.: The Chemistry of Self-Splicing RNA and RNA Enzymes; *Science* **236** (1987) 1532–1539.
[44] SZOSTAK, J. W.: Enzymatic Activity of Conserved Core of a Group I Self-Splicing Intron; *Nature (London)* **322** (1986) 83–86.
[45] DOUDNA, J. A.; GERBER, A. S.; CHERRY, J. M.; SZOSTAK, J. W.: Genetic Dissection of an RNA Enzyme; *Cold Spring Harbor Symp. Quant. Biol.* **52** (1987) 173–180.
[46] BEEN, M. D.; BARFORD, E. T.; BURKE, J. M.; PRICE, J. V.; TANNER, N. K.; ZAUG, A. J.; CECH, T. R.: Structures Involved in *Tetrahymena* rRNA Self-Splicing and RNA Enzyme Activity; *Cold Spring Harbor Symp. Quant. Biol.* **52** (1987) 147–157.
[47] KAY, P. S.; INOUE, T.: Catalysis of Splicing-Related Reactions between Dinucleotides by a Ribozyme; *Nature (London)* **327** (1987) 343–346.
[48] CECH, T. R.: Conserved Sequences and Structures of Group I Introns: Building an Active Site for RNA Catalysis – A Review; *Gene* **73** (1988) 259–271.
[49] BEEN, M. D.; CECH, T. R.: RNA as an RNA Polymerase: Net Elongation of an RNA Primer Catalyzed by the *Tetrahymena* Ribozyme; *Science* **239** (1988) 1412–1415.
[50] DAVIES, R. W.; WARING, R. B.; TOWNER, P.: Internal Guide Sequence and Reaction Specificity of Group I Self-Splicing Introns; *Cold Spring Harbor Symp. Quant. Biol.* **52** (1987) 165–171.
[51] BURKE, J. M.; BELFORT, M., CECH, T. R.; DAVIES, R. W.; SCHWEYEN, R. J.; SHUB, D. A.; SZOSTAK, J. W.; TABAK, H. F.: Structural Conventions for Group I Introns; *Nucl. Acid Res.* **15** (1987) 7217–7221.
[50] DOUDNA, J. A.; SZOSTAK, J. W.: RNA-Catalysed Synthesis of Complementary Strand RNA; *Nature (London)* **339** (1989) 519–522.
[51] MCSWIGGEN, J. A.; CECH, T. R.: Stereochemistry of RNA Cleavage by the *Tetrahymena* Ribozyme and Evidence That the Chemical Step Is Not Rate-Limiting; *Science* **244** (1989) 679–683.
[52] RAJAGOPAL, J.; DOUDNA, J. A.; SZOSTAK, J. W.: Stereochemical Course of Catalysis by the *Tetrahymena* Ribozyme; *Science* **244** (1989) 692–694.
[53] LATHAM, J. A.; CECH, T. R.: Defining the Inside and Outside of a Catalytic RNA Molecule; *Science* **245** (1989) 276–282.
[54] SHUB, D. A.; XU, M.-Q.; GOTT, J. M.; ZEEH, A.; WILSON, L. D.: A Family of Autocatalytic Group I Introns in Bacteriophage T4; *Cold Spring Harbor Symp. Quant. Biol.* **52** (1987) 193–200.
[55] BURKE, J. M.: Molecular Genetics of Group I Introns: RNA Structures and Protein Factors Required for Splicing – a Review; *Gene* **73** (1988) 273–294.
[56] BELFORT, M.; CHAUDRY, P. S.; PEDERSON-LANE, S.: Genetic Delineation of Functional Components of the Group I Intron in the Phage T4td Gene; *Cold Spring Harbor Symp. Quant. Biol.* **52** (1987 181–192.
[57] VAN DER HORST, G.; TABAK, H. F.: Self-Splicing of Yeast Mitochondrial Precursor RNA; *Cell* **40** (1985) 759–766.
[58] CECH, T. R.; BASS, B. L.: Biological Catalysis by RNA; *Ann. Rev. Biochem.* **55** (1986) 599–629.
[59] WOODSON, S. A.; CECH, T. R.: Reverse Self-Splicing of the *Tetrahymena* Group I Intron: Implication for the Directionality of Splicing and for Intron Transposition; *Cell* **57** (1989) 335–345.
[60] MURPHY, F. L.; CECH, T. R. Alteration of Substrate Specificity for Endoribonucleolytic Cleavage of RNA by the *Tetrahymena* Ribozyme; *PNAS USA* **86** (1989) 9218–9222.
[61] HERSCHLAG, D.; CECH, T. R.: DNA Cleavage Catalyzed by the Ribozyme from Tetrahymena; *Nature (London)* **344** (1990) 405–409.
[62] ROBERTSON, D. L.; JOYCE, G. F.: Selection *in vitro* of an RNA Enzyme that Specifically Cleaves Single-Stranded DNA; *Nature (London)* **344** (1990) 467–468.

[63] Forster, A. C.; Altman, S.: External Guide Sequences for an RNA Enzyme; *Science* **249** (1990) 783–786.

[64] Cotten, M.; Birnstiel, M. L.: Ribozyme-Mediated Destruction of RNA *in vivo; EMBO J.* **8** (1989) 3861–3866.

[65] Cameron, F. H.; Jennings, P. A.: Specific Gene Suppression by Engineered Ribozymes in Monkey Cells; *PNAS USA* **86** (1989) 9139–9143.

[66] Sarver, N.; Cantin, E. M.; Chang, P. S.; Zaia, J. A.; Ladne, P. A.; Stephens, D. A.; Rossi, J. J.: Ribozymes as Potential Anti HIV-1 Therapeutic Agents; *Science* **247** (1990) 1222–1225.

[67] Westheimer, F. H.: Why Nature Chose Phosphates; *Science* **235** (1987) 1173–1178.

[68] Joyce, G. F.: RNA Evolution and the Origins of Life; *Nature (London)* **338** (1989) 217–224.

[69] Benner, S. A.; Allemann, R. K.; Ellington, A. D.; Ge, L.; Glasfeld, A.; Leanz, G. F.; Krauch, T.; MacPherson, L. J.; Moroney, S.; Piccirilli, J. A.; Weinhold, E.: Natural Selection, Protein Engineering and the Last Riboorganism: Rational Model Building in Biochemistry; *Cold Spring Harbor Symp. Quant. Biol.* **52** (1987 53–63.

[70] Orgel, L. E.: RNA Catalysis and the Origin of Life; *J. Theor. Biol.* **123** (1986) 127–149.

[71] Piccirilli J. A.; Krauch, T.; Moroney, S. E.; Benner, S. A.: Enzymatic incorporation of a new base pair into DNA and RNA extends the genetic alphabet; *Nature* **343** (1990) 33–37.

[72] Michel, F.; Hanna, M.; Green, R.; Bartel, D. P.; Szostak, J. W.: The guanosine binding site of the *Tetrahymena* ribozyme; *Nature* **342** (1989) 391–395.

[73] Beaudry, A. A.; Joyce, G. F.: Minimum secondary structure requirements for catalytic activity of a self-splicing group-I intron; *Biochemistry* **29** (1990) 6534–6539.

[74] Michel, F.; Westhof, E.: Modeling of the 3-dimensional architecture of group I catalytic itrons based on comparative sequence analysis; *J. Mol. Biol.* **216** (1990) 585–610.

Transduction of Light Energy into an Electrochemical Gradient of Protons at the Bacterial Photosynthetic Apparatus

by G. Drews

Contents

Prof. Dr. Gerhart Drews,
Institut für Biologie II, Mikrobiologie,
Schänzlestr. 1,
D-7800 Freiburg, Fed. Rep. Germany

1 Introduction

The transduction of light energy by photosynthesis into metabolically usable energy is the basis for life. Primary production by terrestrial plants and algae is ecologically the most important contribution, and that of phototrophic bacteria a minor component. However, the cyanobacteria, phototrophic purple and green bacteria, and halobacteria are important members of the ecological niches they populate. They also transduce light energy into chemical energy and are involved in the cycling of nutrients.

The photosynthetic apparatus of purple and green bacteria is simpler in composition and organization than that of cyanobacteria and higher plants, and is therefore an excellent model system to study the relationship of structure to function and the fundamental mechanisms of energy transduction. Close cooperation among biophysicists, biochemists, and geneticists in recent years has enabled us to make substantial progress in the analysis of the photosynthetic apparatus. I describe some exciting current findings in what follows. The literature has been summarized in many review articles [16, 17, 18, 27, 28, 49, 62]. In this chapter I restrict my remarks to the photosynthetic apparatus of green and purple bacteria.

2 Organization and function of the bacterial photosynthetic apparatus

2.1 Antenna systems

Antenna systems or, as they are more usually known in English, *light-harvesting systems* are well-organized pigment-protein complexes that absorb photons of light. The energy of the photons is converted into an increase in the orbital electronic energy of the excited singlet state of the pigment molecules. The bacterial light-harvesting pigments, which are *carotenoids* and *bacteriochlorophylls*, in their aggregated or protein-bound state absorb photons of wavelengths in the regions 380, 450–550, 715–750, 800–880 and 1020 nm (Figure 1). The energy is a function of the wavelength, with 350 nm corresponding to 342 KJ/einstein, and 1050 nm corresponding to 114 KJ/einstein. The bacterial light-harvesting molecules absorb photons in a portion of the visible spectrum that the antenna systems of higher plants rarely or never use. Particular taxa of bacteria have antenna pigments with characteristic effective absorption bands (Figure 1). The diversity of antenna systems developed over the course of evolution is of ecological importance. The light-harvesting molecules increase the absorption cross-section of the photosynthetic unit around the photochemical reaction center and allow photosynthetic growth even under low light intensities. The excited states of the pigment molecules can migrate over the antenna system and are finally trapped by the reaction center.

Carotenoids (Figure 2) and bacteriochlorophylls (Figure 3) have alternating single and double bonds as a feature in common. The pigments are noncovalently bound, by electrostatic and hydrophobic bonds, to polypeptides or to each other (as aggregates in chlorosomes, or as dimers in other antenna complexes). The binding of pigments to

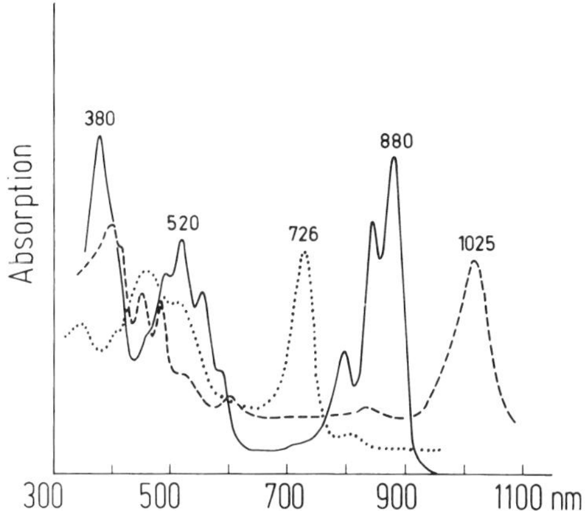

Fig. 1 Absorption spectra of membranes from photosynthetic bacteria.
—— *Rhodopseudomonas salexigens* (Bchl *a*);
---- *Rhodopseudomonas sulfoviridis* (Bchl *b*);
.... *Chlorobium phaeovibrioides* (Bchl *d*, *a*).
Carotenoids absorb in the region around 450–550 nm

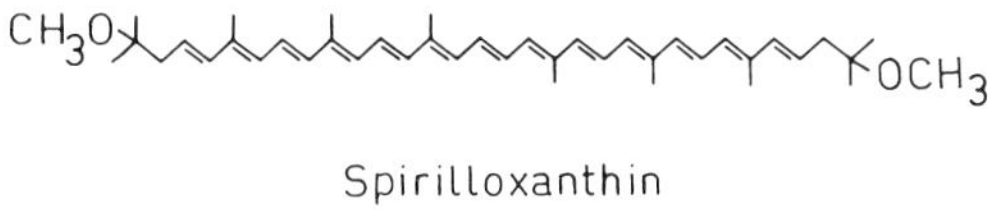

CH_3O ... OCH_3

Spirilloxanthin

Fig. 2 Structure of spirilloxanthin, a carotenoid of *Rhodospirillum rubrum*

polypeptides and the formation of exciton-coupled Bchl-Bchl dimers have three consequences. First, the electronic absorption band of protein-bound bacteriochlorophyll (Bchl) *a* shifts from 770 nm (measured in acetone-methanol) to 800–880 nm (protein-bound). Second, the orientation of the tetrapyrrole ring relative to the plane of the membrane is established. Third, the distance between pigment molecules is defined. Different complexes of antenna pigment and protein combine to make up a system of broad overlapping absorption bands; the importance of this system is that photons from different wavelength regions are absorbed, and excitation energy is transferred efficiently within the pigment-protein complexes and to the reaction center.

2.1.1 Integral membrane-bound antenna systems

All antenna complexes of purple sulfur and nonsulfur purple bacteria are integral membrane particles. The single unit of the antenna complex contains two different pigment-binding polypeptides (α, β) with a central hydrophobic α-helical domain, and shorter or longer *C*- and *N*-terminal domains. The central α-helical domain of the polypeptide traverses the membrane. The *N*-terminal regions are exposed on the cyto-

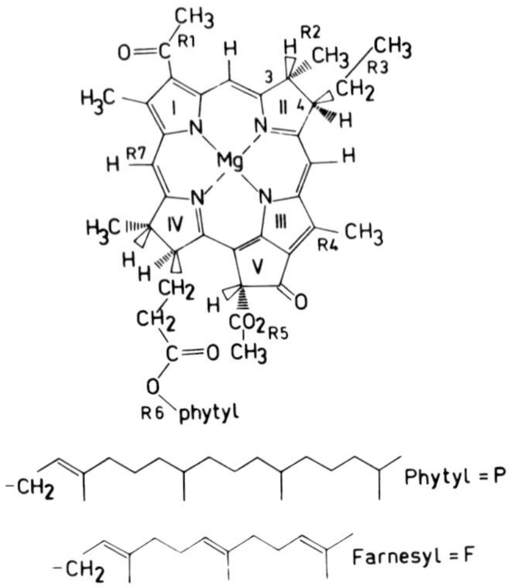

Fig. 3a Structure of chlorophylls

Pigment	R_1	R_2	R_3	R_4	R_5	R_6	R_7	*in vivo* absorption, nm, protein-bound
Chlorophyll *a*	$-CH=CH_2$	$-CH_3^*$	$-CH_2-CH_3$	$-CH_3$	$-CO-OCH_3$	phytyl	$-H$	618
Bacterio-chlorophyll *a*	$-CO-CH_3$	$-CH_3$	$-CH_2-CH_3$	$-CH_3$	$-CO-OCH_3$	phytyl or geranyl-geraniol	$-H$	800–880
Bchl *b*	$-CO-CH_3$	$-CH_3$	$=CH-CH_3$	$-CH_3$	$-CO-OCH_3$	phytyl or phytadienyl	$-H$	1020
Bchl *c*	$-CHOH-CH_3$	$-CH_3^*$	$-C_2H_5$ to $-C_4H_9$	$-C_2H_5$	$-H$	farnexal or stearyl	$-CH_3$	750
Bchl *d*	$-CHOH-CH_3$	$-CH_3^*$	$-C_2H_5$ to $-C_5H_{11}$	$-CH_3$	$-H$	farnesyl	$-H$	725–745
Bchl *e*	$-CHOH-CH_3$	$-CHO^*$	$-C_2H_5$ to $-C_4H_9$	$-C_2H_5$	$-H$	farnesyl	$-CH_3$	715–725
Bchl *g*	$-CH=CH_2$	$-CH_3^*$	$=CH-CH_3$	$-CH_3$	$-CO-OCH_3$	geranyl-geraniol	$-H$	788

* Bonds between C-3 and C-4 unsaturated. From *Biologie in unserer Zeit* **16** (1986) 113–123, (modified)

Fig. 3b Substituents of chlorophyll a and bacteriochlorophylls

plasmic surface of the membrane, while the *C*-termini point to the periplasmic space [53]. Histidine residues, always conserved, in the hydrophobic central domain are the ligands for the Mg atom in the tetrapyrrole ring. This and other bonds determine the localization and orientation of the Q_x and Q_y transition moments of the tetrapyrrole ring of Bchl (Figures 3, 4, and 5) relative to the plane of the membrane [5, 49]. Two (B870) or three (B800–850) molecules of Bchl and one molecule of carotenoid are bound to a pair of α and β polypeptides.

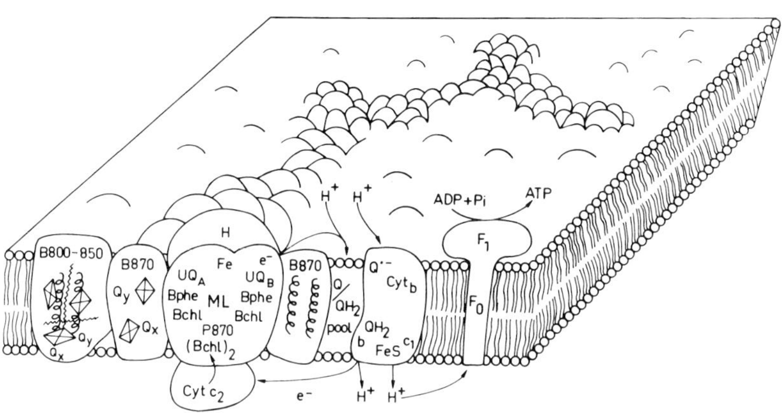

Fig. 4 Schematic view of the photosynthetic membrane of purple bacteria. Integral functional pigment-protein complexes are in the lipid double layer. The reaction center (labeled subunit H) includes the *M* and *L* subunits (labeled *ML*) with the primary acceptor (P870, $(Bchl)_2$), Bacteriopheophytin (Bphe), ubiquinone *A* and *B* (UQ_A, UQ_B), and iron (Fe). The RC is surrounded by the light-harvesting complex B870 (LH I); the α helices of pigment-binding polypeptides and the Q_x and Q_y transition moments of tetrapyrrole rings of Bchl are included in the sketch. The RC + LH I combination forms the core complex. The core complexes are interconnected by the LH II complex (B800–850) with 3 Bchls per subunit. The view is on to the cytoplasmic surface of the membrane.

In the ubiquinone-cytochrome b/c_1 complex, cytochrome *b* is labeled cyt *b* or *b* (high and low potential); cytochrome c_1 is labeled c_1; ubiquinone is *Q*; semiquinone is $Q^{\cdot -}$; ubiquinol is QH_2. The proton-ATPase is labeled F_1F_0

The pigment-protein subunits aggregate in the membrane, as well as in the isolated state, to form oligomeric structures with relative molecular mass, M_r, of about 80,000 to 100,000 [17]. Although the integral antenna complexes can move by diffusion within the lipid double layer of the membrane, they are organized as supercomplexes in the membrane (Figure 4). The B870 type of light-harvesting complexes surround the reaction centers in stoichiometric ratios of about 25–30 antenna Bchl molecules per reaction center, forming the *core complexes*. We know from kinetic spectroscopic studies that some of the B870 complexes have a specific function as direct donors of excitation energy (B896), and presumably have a specific state of organization [27]. The core complexes are interconnected directly or by a second type of antenna complex (B800–850). We also know that many core complexes are energetically coupled, that is, excitation energy that cannot be trapped by the reaction center of a particular core complex can be transferred to another core complex where the reaction center is open [49]. The pigment proteins are not only laterally organized but are also oriented relative to the perpendicular axis of the plane of the membrane [49, 59], as was shown upon chemical analysis of the *N*- and *C*-termini [53]. The orientation is established by the primary structure and by the process of inserting, translocating, and assembling polypeptides in the membrane.

2.1.2 Membrane-associated antenna systems: chlorosomes

All green phototrophic bacteria, such as *Chloroflexus* and *Chlorobium*, contain accessory Bchl pigments, for example, Bchl *c* (Figure 3). These pigments are localized in *chlorosomes* – ovoid structures, 30–50 nm in diameter and 100–200 nm long, that are attached on the cytoplasmic side of the cytoplasmic membrane [50, 51]. The Bchl *c* molecules seem to occur in aggregates with each other and to be oriented along polypeptides (A. R. HOLZWARTH and J. OLSON, in [16]); linear dichroism (LD) measurements have established that the pigments are oriented along the long axis [16, 50]. The Bchl *c* molecules are joined into oligomeric structures through the 2-hydroxyethyl and 9-keto groups of the tetrapyrroles [7, 60].

2.1.3 Transfer of excitation energy within the antenna and to the reaction center

When a carotenoid or a Bchl molecule absorbs a photon, the electrons in the molecule are redistributed and the orbital electronic energy of the excited state of the molecule increases. The probability that a photon will be absorbed depends on the physical cross section and the absorption band broadening of the Bchl molecule, and on the direction of incident light relative to the y axis (through rings I and III) and x axis (through rings II and IV) of the porphyrin (Figure 3). Incident light that is polarized in a plane parallel to the transient moments x or y is most efficiently absorbed, according to LD absorption spectra [58,59]. Electrical effects in the surroundings of the Bchl molecule, such as charged groups of the binding polypeptide, cause the absorption maxima to shift and to become broader [49]. The excitation energy can be exchanged between the pigment molecules of one subunit by delocalized excitons in 1–2 ps (1 ps = 10^{-12} s), or it can migrate within the population of the oligomers of one antenna system, by inductive resonance transfer (Förster mechanism), in 10–50 ps, or it can migrate to other pigment systems and finally be trapped by the reaction center (RC) in 50–100 ps [27]. Excitation energy usually migrates from the shorter to the longer wavelength absorption bands of pigments, for example:

Carot. ⟶ B800 ⟶ B850 ⟶ B870 ⟶ RC (Bchl *a*); or

↑ ↑ ↑ ↑ ↑

Bchl *c* (B740) ⟶ Bchl *a* (B795) ⟶ B808 ⟶ 866 ⟶ RC [54, 56, 57].

Förster transfer is effective up to distances of 8–10 nm. KNOX [30] has described the theory of exciton migration and trapping. Experiments with time-resolved absorption spectroscopy reveal that the different antenna systems and reaction centers are energetically coupled, that is, excitation energy can be communicated not only within one photosynthetic unit but between different photosynthetic units, so that energy that cannot be trapped by one reaction center can be delivered to the next one (Figure 4).

2.2 The photochemical reaction center

The reaction center receives excitation energy from the antenna or is directly excited by a photon. The primary donor (P, a special pair of Bchl molecules), after excitation, loses an electron by transferring it to the primary acceptor, a pheophytin molecule. As a result the electron and the hole are separated from one another and localized at different molecular species. The state of charge separation is stabilized by transfer of the electron to a quinone Q_A, so the excited state is transduced into an electrical potential difference over the membrane ($\Delta\Psi$) and a redox potential difference between reductant and oxidant (ΔE_h) [49].

In pioneering work, H. MICHEL, J. DEISENHOFER, and their co-workers have in recent years established the structure of the reaction center of *Rhodopseudomonas viridis*. They determined the primary structure of the polypeptides, and they crystallized the reaction center in a detergent micelle and determined its fine structure by high resolution X-ray analysis [13, 36, 39]. They located the exact position of the pigments relative to the amino acid residues of the M (medium) and L (light) subunits. G. FEHER and colleagues subsequently determined the structure of the reaction center of *Rhodobacter sphaeroides* [1, 63]. When the data of these groups are combined with results from experiments with femtosecond time-resolved absorption spectroscopy and with site-specific mutagenesis of amino acid residues in close proximity to pigments, and the theory of electronic states is also applied, the result is a much more advanced picture of the process of charge separation in the reaction center.

The pigments of the reaction center are bound to two amphipathic polypeptides in the membrane. The polypeptides are larger than those in antenna complexes (M_r of the L is 30,571 and of the M, 35,902). Hydrophobic α helices in the L and the M subunits traverse the membrane five times. The pigments bind to the polypeptides with hydrophobic and hydrogen bonds (Figure 5). The special pair of Bchl molecules, $(Bchl)_2$ (primary donor, P), is located on the periplasmic side of the membrane. The tetrapyrrole rings partially overlap, and the planes of the rings are separated by 0.3 nm. The pigments are located in a sequence on each side, parallel to the axis of symmetry through the reaction center, as two branches consisting of (in this order) single Bchl molecules ($Bchl_{A,B}$), bacteriopheophytin ($Bphe_{A,B}$), and ubiquinone ($Q_{A,B}$). An iron (Fe) atom is between the quinone molecules. The tetrapyrroles are about 10 Å apart and the other redox carriers are about 15 Å apart (Figure 5). Both branches of the chains of molecules have twofold rotational symmetry [37]. Workers using X-ray diffraction and resonance Raman spectroscopy have determined the ligands of the Mg atom of Bchl (His) and of the 2-acetyl and 9-keto carbonyl groups of the tetrapyrrole rings [44], and more recently others have investigated the amino acid ligands and their vibrational motion during excitation using time-resolved Fourier-transformed infrared spectroscopy [33]. When the histidine in position 200 was replaced by leucine, a heterodimer of the special pair (Bchl-Bphe) was formed [9], and the quantum yield of the light-driven reactions of this altered reaction center was reduced to 50% [29]. The bacteria were unable to grow photosynthetically. The Glu-212 was found by the same technique to have an anomalously high pK_a value of 9.5; this is evidence in support of the idea that Glu-212 is involved as a proton donor to reduced Q_B [16].

The pigments in the two branches are surrounded by different proteins, so the electron density distributions in the branches are different. This may be why the right branch, leading to Q_A, has most of the activity (Figure 5).

When the amino acid sequences of the M and L subunits of different reaction centers are lined up, it is obvious that the amino acids in certain positions are conserved; the

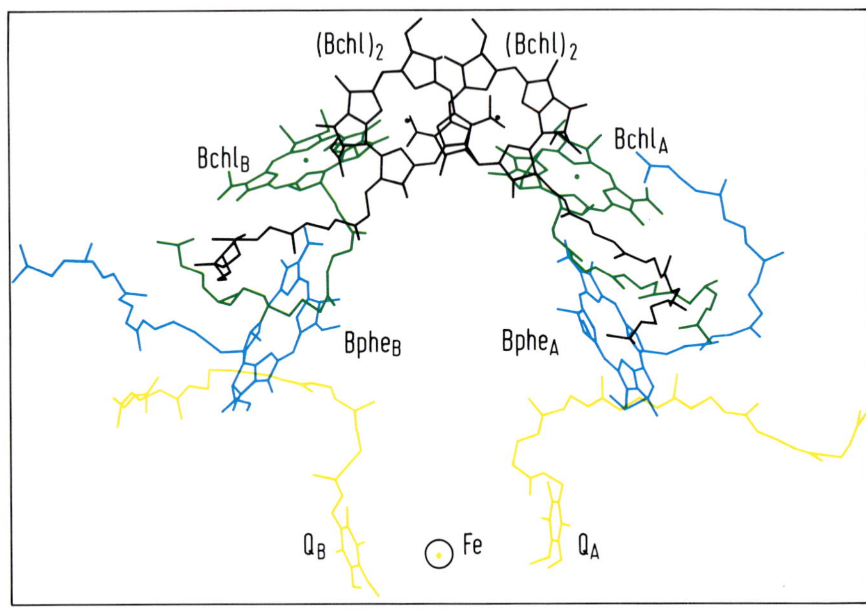

Fig. 5 Localization of pigment molecules in the reaction center of *Rhodobacter sphaeroides* (from ALLEN *et al.*, 1987, [1]; Fig. 1a; with kind permission of authors). Only the pigment molecules are drawn. At the top is the periplasmic side of the membrane, at the bottom the cytoplasmic side of the membrane. On either side of the axis of symmetry are the two branches of pigments; the L branch is the more active (see text). $(Bchl)_2$ is the special pair of Bchl, primary donor; $Bchl_{A,B}$ are the accessory bacteriochlorophylls; $Bphe_{A,B}$ are the bacteriopheophytins; $Q_{A,B}$ are the ubiquinones *A* and *B*; and Fe is an iron atom. The distances between the molecular centers are (in nm): $(Bchl)_2$: $Bchl_A = 1.1$; $Bchl_A$: $Bphe_A = 1.05$; $Bphe_A$: $Q_A = 1.3$; Q_A : $Q_B = 1.85$; Fe: $Q_A = 1.1$.

importance of these amino acids has been demonstrated in experiments using site-directed mutagenesis [8,9]. If these amino acids were replaced by others, function was often impaired. Strong similarities between the primary structures of *L* and *M*, on the one hand, and D_1 and D_2 of the plant photosystem II, on the other, also emerge when the sequences are compared; the conserved amino acids are those near the membrane surface [38]. The plant photosystem II and the bacterial RC are also similar in that they have a herbicide binding site and are associated with similar pigments [38].

The reaction centers of most bacteria have a third subunit, *H*, which contains only one hydrophobic domain by which it is anchored in the membrane. Most of the *H* subunit covers the *M* and *L* subunits on the surface of the cytoplasmic membrane and stabilizes the RC [36]. The H subunit may participate in the assembly process, as proposed in [28].

The RC of *Rhodopseudomonas viridis* is unique in having a subunit of cytochrome *c* containing four covalently bound hemes ($E_0 = -6$, $+20$, $+310$ and $+380$ mV

[15, 61]. The *N*-terminal amino acid of the cytochrome subunit is a cysteine, linked to a diglyceride by a thioether bond. Two fatty acids (18 : OH and 18 : 1) bind to glycerol with ester bonds, and seem to anchor the cytochrome subunit in the membrane [61].

2.3 Electron transfer and charge separation

We have learned from experiments in ESR spectroscopy that the excitation energy from antenna pigments is transferred to the special pair, and an excited state, P^*, is formed. P^* is a strong reductant. The electronically excited primary donor P^* decays within 3 ps and populates a very short-lived intermediate involving a reduced accessory Bchl molecule [26 a]. The electron is transferred to the neighboring Bpheo with a time constant of 0.9 ps, and from there to the quinone$_A$ within 200 ps [6, 26 a]. This step stabilizes the charge separation between P^* and Q_A^-. The subsequent electron transport from Q_A^- to Q_B is much slower (0.3 ms). Most of the electron transport flows over the L branch [40]. Energy is transduced with high efficiency because the electron acceptor is of high affinity, and because the back reaction is several orders of magnitude slower.

The primary donor is re-reduced directly by reduced cytochrome c_2, which can move freely in the periplasmic space (*Rhodobacter sphaeroides*), or via bound cytochrome (*Rhodopseudomonas viridis*). Additional excitation energy can be transferred to the special pair, and a second round of radical pair formation is initiated. In the second round Q_B accepts a second electron, and protons from the outside. The ubiquinol$_B$ exchanges with the quinone pool of the membrane.

The turnover rate of the RC, that is, the number of charge separations per unit time, depends on the energy flow through the antenna to the RC, the rate of reduction of P by cytochrome c_2, and the reoxidation of Q_A. Under medium and high light intensities the turnover rate of the RC is determined by the turnover of the cyclic electron flow.

2.4 Ubiquinone-cytochrome b/c_1 oxidoreductase and cyclic electron transport

In the RC the excitation energy is transduced into a redox potential difference across the membrane. The difference in E_h of about 0.35 V between QH_2/Q and ferro-/ferri- (reduced/oxidized) cytochrome is used to create both an additional charge separation and a proton gradient across the membrane at the ubiquinone-cytochrome b/c_1 oxidoreductase (b/c_1 complex). The RC and the b/c_1 complex are functionally connected by two mobile redox carriers, the quinone pool in the lipid phase of the membrane, and the freely movable cytochrome c_2 in the periplasmic space. The RC and the b/c_1 complex catalyze a cyclic electron transport process (right side of Figure 6).

The b/c_1 complex is a nearly universal constituent of energy-coupled membrane-bound electron transport chains for oxidative and light-driven phosphorylation, whether in mitochondria, in chloroplasts, or in bacterial systems [3]. The equivalent of the b/c_1 complex in plant chloroplasts is the b_6/f complex [26]. The b/c_1 complex of bacteria and cyanobacteria consists of three polypeptides: cytochrome f or c_1, cytochrome b, and the Rieske FeS protein (Figure 6).

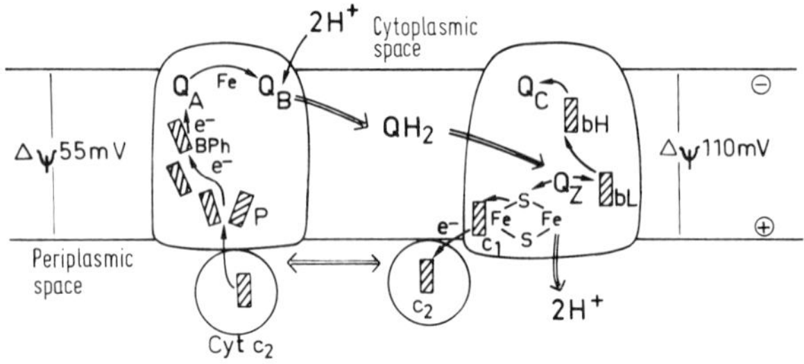

Fig. 6 Cyclic electron transport in the bacterial photosynthetic apparatus. The integral membrane complexes are shown with the reaction center RC on the left and the cytochrome b/c_1 complex on the right. The RC includes P (primary donor), Bphe (bacteriopheophytin), Q_A, Q_B (see Fig. 5). The cytochrome b/c_1 complex includes Q_C and Q_Z (quinone binding sites), b_H, b_L (high-potential and low-potential cytochrome *b*), c_1 (cytochrome c_1), FeS (Rieske iron center). Cyt c_2 is cytochrome c_2 and QH_2 is the quinol pool

The largest subunit of the b/c_1 complex of *Rhodobacter capsulatus* consists of 437 amino acids (M_r 49,357), has nine hydrophobic domains (presumably nine membrane-spanning α helices), and binds two hemes (cytochrome b_{566}, $E_0 = -90$ mV, also called b_L, and cytochrome b_{562}, $E_0 = +50$ mV, called b_H) at conserved amino acid residues such as four histidines in the α helices II and V [26]. Cytochrome b_L accepts one electron and one proton for reduction (pH 6–8).

The cytochrome c_1 protein of *Rhodobacter capsulatus* consists of 279 amino acid residues (M_r 30,326), with the heme bound covalently to the sequence Cys-X-Y-Cys-His [11, 12, 23, 24]. The E_0' of cytochrome c_1 is about + 300 mV [43].

The third polypeptide (191 amino acids, M_r 20,443) has three hydrophobic domains and binds the sulfur-iron cluster (Rieske FeS protein, $E_0 = +280$ mV).

When the respective subunits of the b/c_1 complexes isolated from bacteria, mitochondria, and chloroplasts are lined up and compared, their amino acid sequences, deduced from the DNA sequences for the three polypeptides, are very similar [22]. Active preparations were recently isolated from phylogenetically remote organisms [31]; the samples isolated from yeast and beef heart mitochondria contain up to seven polypeptides. The function of the proteins that do not bind heme or FeS is unknown.

The b/c_1 complex is believed to have two quinone-binding sites (Q_C and Q_Z). Antimycin inhibits electron transfer from cyt b_H to Q_C, and myxothiazole blocks electron flow from Q_Z to cyt b_L (Figure 6). Hydrophobic hydroxyquinones block electron transfer on the high-potential site of the Rieske iron center [43]. Kinetic measurements were used to deduce the redox reactions that operate in the b/c_1 complex, described as the *Q cycle* ([43], right side of Figure 6).

We have learned through experiments that 4 protons (H^+) for each 2 electrons (e^-) are translocated from the cytoplasmic side to the periplasmic side of the membrane through the b/c_1 complex. The complex therefore functions as a proton pump [3], and contributes to establishing the proton gradient across the membrane, coupled to the linear or cyclic electron flow through the photosynthetic or respiratory electron transport chain. Proton translocation is driven by two electron transport cycles. We believe one e^- is transferred from QH_2Q_Z to the Rieske FeS center and to cyt c_1, which reduces the cytochrome c_2 in the periplasmic space, reversibly bound to the b/c_1 complex (Figure 6). The semiquinone $Q^{\bullet -} Q_Z$ reduces cyt b_L and from there the electron is transferred to cyt b_H. The oxidized quinone $Q Q_Z$ is afterwards reduced, or exchanged with a QH_2 from the pool. $QH_2 Q_Z$ again reduces a Rieske FeS and cyt b_L. The b cytochromes reduce Q_C. The two electron transport cycles, which cause the oxidation of $Q_Z H_2$, act in concert to expel 4 H^+ on the periplasmic side and to reduce cyt c_2, which is the mobile e^- carrier between the RC and the b/c_1 complex (Figure 6). The membrane potential of about 55 mV created by the RC is increased to 110 mV after the second electron transfer round in the b/c_1 complex (Figure 6).

The localization of the various constituents is vital to generating the potential across the dielectric of the membrane: Q_Z, the FeS center, cyt c_1, and cyt b_L must be on the periplasmic side, and cyt b_H and Q_C in the hydrophobic layer, oriented towards the cytoplasmic side [45]. This proposed distribution of redox centers is supported by results of work with mutated complexes [46]. The membrane potential can be measured by the electrochromic bandshift of carotenoids bound to the antenna complex B800–850. The electrogenic step is presumably the electron transport from cyt b_L and b_H to Q_C [55]. The turnover rate of the b/c_1 complex depends on the size of the quinone pool and the concentration of cytochromes [25, 55]. If the quinone is partially extracted or if the quinone concentration in the membrane is diluted by fusing membrane vesicles to liposomes, the rate of electron transport is reduced [48, 55].

2.5 Proton-ATPase, photophosphorylation

The electrochemical proton gradient across the membrane can be used for various different energy-consuming processes, such as the transport of solutes against a concentration gradient (either H^+/Na^+ antiport or H^+/succinate symport) or the powering of the flagellar motor. During growth, most of the proton-motive force is used to synthesize ATP from ADP and inorganic phosphate. ATP is the universal energy currency that pays for endergonic reactions such as the biosynthesis of proteins and nucleic acids. Proton-ATPases have been isolated from the bacterial cytoplasmic membrane ($BF_1\,BF_0$) and from the membranes of chloroplasts ($CF_1\,CF_0$) and mitochondria. The primary structure of the catalytic polypeptides of ATPase is so similar in all these different species that we deduce that cell H^+-ATPases have a common origin. The proton-ATPases consist of a membrane-spanning *proton channel*, F_0, and the catalytic center, F_1, which sits on the F_0 and is exposed on the cytoplasmic side of the membrane (bacteria), the matrix side (P) of the inner mitochondrial membrane, or the stroma thylakoid membrane in chloroplasts. The $BF_1\,BF_0$ ATPase of *Escherichia coli* consists of eight protein subunits (Table 1). The coding genes are clustered in the *unc* operon. The sequence of genes for the F_1 subunit is the same in different bacteria (*E. coli*, *Rhodospirillum rubrum*, *Rhodopseudomonas blastica*), in chloroplasts, and cyanobacteria [19, 32, 64]. The amino acid sequences of the α and β polypeptides are very similar, but those of the δ, γ, and ε polypeptides are less so [21]. The genes for the F_0 subunit of the phototrophic bacteria are not in the *unc* operon.

According to the present model of the F_1, three α and three β polypeptides alternate, arranged in a quasi-hexagonal pattern and forming a knoblike structure 9.5 nm in

Table 1 Proteins of the F_0 F_1 ATPase

Escherichia coli subunits	Genes for the proteins (*unc* operon)	Stoichiometry of proteins in the ATPase	Mitochondria	M_r of proteins	
				Chloroplasts	*Rhodospirillum rubrum*, deduced from DNA of the *atp* operon (Falk *et al.*, 1985) [19]
1	1	ab_2c_{10-15}		I 18,000	
a	B			II 16,000	
F_0 *c*	E			III 8,000	
b	F				
δ	H	$\alpha_3\beta_3\gamma_1\delta_1\epsilon_1$	Oligomycin-sensitivity-conferring protein	δ 21,000	19,543
α	A		α	α 58,000	55,026
F_1 γ	G		γ	γ 37,000	32,437
β	D		β	β 57,000	50,852
ϵ	C		δ	ϵ 13,000	14,307

diameter around a central region of low density [32]. The nucleotide-binding domain of the β subunit consists of alternating α helices and β sheets. The γ subunit functions as a center of organization or anchoring point for the α and β subunits. The δ subunit connects F_1 to F_0. It corresponds to the oligomycin-sensitivity-conferring protein of mitochondria. The ε polypeptide is involved in coupling ATP synthesis to the proton channel F_0, but its exact function is not known.

The active BF_0 complex consists of one a, two b, and more than three c polypeptides. The a subunit probably spans the membrane with several helices, and may connect F_1 and F_0. The b polypeptide is anchored with its hydrophobic *N*-terminus in the membrane. The hydrophilic region presumably binds to BF_1. The c protein binds the uncoupler dicyclohexylcarbodiimide (DCCD), which blocks the proton flow through BF_0. DCCD binds to acid residues, such as Asp-61 and Glu. The replacement of these amino acids by other amino acids blocks proton transfer. These and other conserved amino acids contribute to the function of the proton channel [4].

The vectorial driving force of the electrochemical proton gradient causes a *proton pressure* on the connection between F_0 and F_1. We do not yet know how the electrochemical gradient is used to drive ATP synthesis. We suppose that the small subunits γ, δ, ε and the b polypeptide bind to the α and β subunits and cause a change in the conformation of the protein. The catalytic domain for ATP synthesis is on the β subunit or between α and β [52]. The three catalytic sites work in concert. Boyer [4] has proposed that a conformational change in the catalytic sites is responsible for energy-dependent variations in the substrate affinity of the active sites, that the coupling subunits between F_0 and F_1 transduce the energy of the proton gradient into a conformational change of the next $\alpha\,\beta$ unit, and that F_1 rotates with respect to F_0. The following processes may run sequentially [4, 10, 34, 52, 64]:

(i) Mg-ADP and inorganic phosphate bind, in the first subunit;
(ii) ATP is formed, in the second subunit; and
(iii) ATP is released, in the third subunit.

ATP is not released before new substrate (Mg·ADP + inorg. PO_4^{3-}) has been bound [4, 34]. The three processes cycle around the three subunits; each is correlated with a change in the binding affinity [4, 35, 52]. At least three protons have to be translocated for the synthesis of one mole of ATP. The release of ATP seems to be the energy-consuming step.

3 Regulation of the biosynthesis of the bacterial photosynthetic apparatus

Many of the purple, facultatively aerobic bacteria (*Rhodobacter capsulatus*, *Rhodospirillum rubrum*, *Rhodopseudomonas palustris*, and others) can grow chemotrophically in the dark using oxidative metabolism. Lowering the partial pressure of oxygen induces synthesis of the photosynthetic apparatus. The process is not light-dependent, but the intensity of the light can regulate the formation of the photosynthetic apparatus and the differentiation of the intracytoplasmic membrane system. The lowering of the oxygen tension has three consequences: it induces the formation of antenna and RC complexes; it brings about the synthesis and assembly of many redox carriers and functional complexes by various means; and it increases the rate of membrane synthesis. This very exciting model system for investigating membrane differentiation has been described quite recently [14, 16, 17, 18, 28, 47]. The assembly of the supramolecular pigment-protein complexes requires all the constituents. The genes for the polypeptides of the RC and the B870 antenna complex, and for the enzymes of Bchl and carotenoid synthesis, are clustered on a 50-kbp stretch of chromosomal DNA [2, 47]. The genes for the cytochromes and the B800–850 polypeptides are located at different sites on the genome [11, 28]. If the oxygen partial pressure is lowered to about 150 Pa, a signal chain (the nature of which is not yet known) induces the formation of mRNA for pigment-binding polypeptides. Some genes for pigment synthesis are regulated at the transcriptional level, as well as at the level of enzyme activity. The insertion and translocation of membrane proteins depends on the proton-motive force. As in other complex systems, there appears to be a cascade of regulatory processes [16, 17].

4 The bacteriorhodopsin of the halobacteria is a light-driven proton pump

Large distinct red areas of the cytoplasmic membrane of the extreme halophile *Halobacterium halobium* have been identified as the *purple membrane*. The purple membrane consists of, besides lipids, a single protein, *bacteriorhodopsin*. This hydrophobic protein spans the membrane seven times, and the pigment *retinal* is bound to the ε-amino group of lysine residue 216. In a light-induced cyclic process, all-*trans*-retinal is isomerized to

13-*cis*-retinal at the same time the protein undergoes a conformational change. One proton is pumped from the cytoplasm to the outside each time the cycle is completed. The proton gradient across the membrane is used for ATP synthesis or other endergonic processes [41].

5 Ecology and biotechnology of phototrophic bacteria

The contribution of the anoxigenic photosynthesis of bacteria to global energy production is minimal. However, photosynthetic bacteria are important members of some ecological niches. Purple sulfur bacteria, for example, inhabit a narrow zone between the oxygen-poor, H_2S-containing bottom waters of lakes and the illuminated oxygen-rich top layer. They use H_2S as the electron donor for the reductive assimilation of CO_2, with light as the energy source. Some green sulfur bacteria grow at the very low light intensities found at lake depths of more than 10 m. Other purple bacteria live in extreme biotopes, such as alkaline lakes with high salt concentrations.

Facultatively aerobic phototrophic bacteria live not only anaerobically in light, but also aerobically or anaerobically in the dark. Under aerobic conditions they oxidize low molecular weight substrates such as dicarboxylic acids, and under anaerobic dark conditions they use external electron acceptors such as dimethyl sulfoxide [20]. They live in eutrophic bodies of water. Phototrophic bacteria have not been used for biotechnological purposes on an industrial scale; the need to saturate dense cultures in large fermentors with light energy is particularly challenging. Furthermore, *E. coli*, species of *Bacillus*, and some other bacteria can easily be manipulated by gene technological methods to produce substances of interest.

All phototrophic bacteria have a nitrogenase and fix nitrogen. Nitrogenase reduces not only N_2 but also H^+, $HC{=}CH$, HCN, and N_2O; resting cultures can produce H_2 with nitrogenase if a suitable carbon source is present. Hydrogen gas is of interest as an alternative source of energy. Large-scale processes have been developed for the production of H_2 using cheap carbon sources and light energy, but to date no economically profitable process has been described. The same applies to the production of single-cell protein from purple bacteria: we are limited by the need to supply light energy to large fermentors.

Phototrophic bacteria are, however, excellent model systems for the study of the molecular mechanisms of light energy transduction and membrane differentiation. High-resolution X-ray diffraction analysis of crystallized photochemical reaction centers (J. Deisenhofer, R. Huber, and H. Michel won the Nobel prize in chemistry in 1988 for work in this area), the different time-resolved spectroscopic studies on primary photosynthetic processes, the theoretical assessment of these results in the light of current theories, and the genetic approach have given us a deep understanding of charge separation and the formation of electrochemical gradients in biological systems.

Donor-acceptor systems to study charge separation and water splitting have been synthesized by preparative chemical methods. Metal porphyrinates are good photosensitizers, and zinc porphyrinates have some of the features of semiconductors. Experiments have been reported in which electron transport from the pigment to metal electrodes was established [42].

6 Acknowledgement

My work has been supported by Deutsche Forschungsgemeinschaft and the Chemical Industry Fund (Fonds der Chemischen Industrie).

7 References

[1] ALLEN, J. P.; FEHER, G.; YEATES, T. O.; KOMIYO, H.; REES, D. C.: Structure of the Reaction Center from *Rhodobacter sphaeroides* R-26: The Protein Subunits; *PNAS USA* **84** (1987) 6162–6166.

[2] BAUER, C. E.; YOUNG, D. A.; MARRS, B. L.: Analysis of the *Rhodobacter capsulatus puf* Operon; *J. Biol. Chem.* **263** (1988) 4820–4827.

[3] BEATTIE, D. S.: Is Cytochrome b/c_1 Complex a Proton Pump? *J. Bioenergetics Biomembr.* **18** (1986) 1–20.

[4] BOYER, P.: The Unusual Enzymology of ATP Synthase; *Biochemistry* **26** (1987) 8503–8507.

[5] BRETON, J.; NABEDRYK, E.: Pigment and Protein Organization in Reaction Center and Antenna Complexes. In: BARBER, J., Ed.; *The Light Reactions*, Elsevier Science Publ., Amsterdam, Biomedical Division, (1987); pp 159–195.

[6] BRETON, J.; MARTIN, J.-L.; MIGUS, A.; ANTONETTI, A.; ORSZAG, A.: Femtosecond Spectroscopy of Excitation Energy Transfer and Initial Charge Separation in the Reaction Center of the Photosynthetic Bacterium *Rhodopseudomonas viridis*; *PNAS USA* **83** (1986) 5121–5125.

[7] BRUNE, D. C.; KING, G. H.; BLANKENSHIP, R. E.: Interactions between Bacteriochlorophyll *c* Molecules in Oligomers and in Chlorosomes of Green Photosynthetic Bacteria. In: SCHEER, H.; SCHNEIDER, S., Eds.; *Photosynthetic Light-Harvesting Systems, Organization and Function*, Walter de Gruyter, Berlin, New York (1988); pp 141–151.

[8] BYLINA, E. J.; ISMAIL, S.; YOUVAN, D. C.: Site-Specific Mutagenesis of Bacterio-Chlorophyll-Binding Sites Affects Biogenesis of the Photosynthetic Apparatus. In: YOUVAN, D. C.; DALDAL, F., Eds; *Current Communications in Molecular Biology: Microbial Energy Transduction*, Cold Spring Harbor Laboratory (1986); pp 63–70.

[9] BYLINA, E. J.; YOUVAN, D. C.: Directed Mutations Affecting Spectroscopic and Electron Transfer Properties of the Primary Donor in the Photosynthetic Reaction Center; *PNAS USA* **85** (1988) 7226–7230.

[10] CROSS, R. L.: The Number of Functional Catalytic Sites on F_1-ATPases and the Effects of Quaternary Structural Asymmetry on Their Properties; *J. Bioenergetics Biomembr.* **20** (1988) 395–405.

[11] DALDAL, F.; DAVIDSON, F.; CHENG, S.: Isolation of the Structural Genes for the Rieske-FeS Protein Cytochrome b_1; *J. Mol. Biol.* **195** (1987) 1–12.

[12] DAVIDSON, E.; DALDAL, F.: Primary Structure of the bc_1 Complex of *Rhodopseudomonas capsulata*; *J. Mol. Biol.* **195** (1987) 13–24.

[13] DEISENHOFER, J.; EPP, O.; MIKI, K.; HUBER, R.; MICHEL, H.: Structure of the Protein Subunits in the Photosynthetic Reaction Center of *Rhodopseudomonas viridis* at 3 Å Resolution; *Nature (London)* **318** (1985) 618–624.

[14] DONOHUE, T. J.; KAPLAN, S.: Synthesis and Assembly of Bacterial Photosynthetic Membranes. In: STAEHELIN, L. A.; ARNTZEN, C. J., Eds.; *Photosynthesis III, Encyclopedia of Plant Physiol.*, **Vol. 19**, Springer Verlag, Heidelberg, Berlin, New York (1986); pp 632–639.

[15] DRACHEVY, S. M.; DRACHEV, L. A.; KONSTANTINOV, A. A.; SEMENOV, A. Y.; SKULACHEV, U. P.; ARUTJUNJAN, A. M.; SHUVALOV, V. A.; ZABEREZHNAYA, S. M.: Electrogenic Steps in the Redox Reactions Catalyzed by Photosynthetic Reaction-Center Complex from *Rhodopseudomonas viridis*; *Eur. J. Biochem.* **171** (1988) 253–264.

[16] DREWS, G.; DAWES, E. A., Eds.: *Molecular Biology of Membrane Bound Complexes in Phototrophic Bacteria*, Plenum Press, New York (1990).

[17] DREWS, G.: Organization and Assembly of Bacterial Antenna Complexes in Photosynthetic Light Harvesting Systems. In: SCHEER, H.; SCHNEIDER, S., Eds.; *Photosynthetic Light-Harvesting Systems, Organization and Function*, Walter de Gruyter, Berlin (1988); pp 233–246.

[18] DREWS, G.; OELZE, J.: Organization and Differentiation of Membranes of Phototrophic Bacteria; *Adv. Microb. Physiol.*, **22** (1981) 1–92.

[19] FALK, G.; HAMPE, A.; WALKER, J. E.: Nucleotide Sequence of the *Rhodospirillum rubrum atp* Operon; *Biochem. J.* **228** (1985) 391–407.

[20] FERGUSON, S. J.; JACKSON, J. B.; MCEWAN, A. G.: Anaerobic Respiration in the Rhodospirillaceae; *FEMS Microbiol. Rev.* (1987) 117–143.

[21] FUTAI, M.; NOUMI, T.; MAEDA, M.: Mechanism of F_1-ATPase Studied by the Genetic Approach; *J. Bioenergetics Biomembr.* **20** (1988) 469–480.

[22] GABELLINI, N.: Organization and Structure of the Genes for Cytochrome b/c_1 Complex in Purple Photosynthetic Bacteria. A Phylogenetic Study Describing the Homology of the b/c_1 Subunits between Prokaryotes, Mitochondria and Chloroplasts; *J. Bioenergetics Biomembr.* **20** (1988) 59–83.

[23] GABELLINI, N.; HARNISCH, U.; MCCARTHY, J. E. G.; HAUSKA, G.; SEBALD, W.: Cloning and Expression of the *fbc* Operon Encoding the FeS Protein, Cytochrome b and Cytochrome c_1 from the *Rhdopseudomonas sphaeroides* b/c_1 Complex; *EMBO Journal* **4** (1985) 549–553.

[24] GABELLINI, N.; SEBALD, W.: Nucleotide Sequence and Transcription of the *fbc* Operon from *Rhodopseudomonas sphaeroides*; *Eur. J. Biochem.* **154** (1986) 569–579.

[25] GARCÍA, A. F.; VENTUROLI, G.; GAD'ON, N.; FERNNDEZ-VALASCO, J. G.; MELANDRI, B. A.; DREWS, G.: The Adaptation of the Electron Transfer Chain of *Rhodopseudomonas capsulata* to Different Light Intensities; *Biochim. Biophys. Acta* **890** (1987) 335–345.

[26] HAUSKA, G.; NITSCHKE, W.; HERRMANN, R. G.: Amino Acid Identities in the Three Redox Center-Carrying Polypeptides of Cytochrome bc_1/b_6f Complexes; *J. Bioenergetics Biomembr.* **20** (1988) 211–228.

[26a] HOLZAPFEL, W.; FINKELE, U.; KAISER, W.; OESTERHELT, D.; SCHEER, H.; STILZ, H. V.; ZINTH, W.: Observation of a Bacteriochlorophyll Anion Radical During the Primary Charge Separation in a Reaction Center; *Chem. Phys. Lett.* **160** (1989) 1–7.

[27] HUNTER, C. N.; VAN GRONDELLE, R.; OLSEN, J. D.: Photosynthetic Antenna Proteins: 100 ps before Photochemistry Starts; *TIBS* **14** (1989) 72–76.

[28] KILEY, P. J.; KAPLAN, S.: Molecular Genetics of Photosynthetic Membrane Biosynthesis in *Rhodobacter sphaeroides*; *Microbiol. Rev.* **52** (1988) 50–69.

[29] KIRMAIER, C.; HOLTEN, D.; BYLINA, E. J.; YOUVAN, D. C.: Electron Transfer in a Genetically Modified Bacterial Reaction Center Containing a Heterodimer; *PNAS USA* **85** (1988) 7562–7566.

[30] KNOX, R. S.: Theory and Modeling of Excitation Delocalization and Trapping. In: STAEHELIN, L. A.; ARNTZEN, C. J., Eds.; *Photosynthesis III, Encyclopedia of Plant Physiol.*, **Vol. 19**, Springer Verlag, Heidelberg, Berlin, New York (1986); pp 286–298.

[31] LJUNGDAHL, P. O.; PENNOYER, J. D.; ROBERTSON, D. E.; TRUMPOWER, B. L.: Purification of Highly Active Cytochrome bc_1 Complexes from Phylogenetically Diverse Species by a Single Chromatographic Procedure; *Biochim. Biophys. Acta* **891** (1987) 227–241.

[32] MELONEY, P. C.: Coupling to an Energized Membrane. In: NEIDHARDT, F. C., Ed.; *Escherichia coli and Salmonella typhimurium*, **Vol. 1**, Am. Soc. Microbiol., Washington (1987); pp 222–243.

[33] MÄNTELE, W. G.; WOLLENWEBER, A. M.; NABEDRYK, E.; BRETON, J.: Infrared Spectroelectrochemistry of Bacterio-Chlorophylls and Bacteriopheophytins; *PNAS USA* **85** (1988) 8468–8472.

[34] MATSUNO-YAGI, A.; HATEFI, Y.: Role of Energy in Oxidative Phosphorylation; *J. Bioenergetics Biomembr.* **20** (1988) 481–502.

[35] MCCARTY, R. E.; NALIN, C. M.: Structure, Mechanism and Regulation of the Chloroplast H^+-ATPase. In: STAEHELIN, L. A.; ARNTZEN, C. J., Eds.; *Photosynthesis III, Encyclopedia of Plant Physiol.*, **Vol. 19**, Springer Verlag, Heidelberg, Berlin, New York (1986); pp 576–583.

[36] MICHEL, H.; WEYER, K. A.; GRUENBERG, H.; LOTTSPEICH, F.: The Heavy Subunit of the Photosynthetic Reaction Center from *Rhodopseudomonas viridis*: Isolation of the Gene, Nucleotide and Amino Acid Sequence; *EMBO Journal* **4** (1985) 1667–1672.

[37] MICHEL, H.; EPP, O.; DEISENHOFER, J.: Pigment-Protein Interactions in the Photosynthetic Reaction Center from *Rhodopseudomonas viridis*; *EMBO Journal* **5** (1986) 2445–2457.

[38] MICHEL, H.; DEISENHOFER, J.: Relevance of the Photosynthetic Reaction Center from Purple Bacteria to the Structure of Photosystem II; *Biochemistry* **27** (1988) 1–7.

[39] MICHEL, H.; WEYER, K. A.; GRUENBERG, H.; DUNGER, T.; OESTERHELL, D.; LOTTSPEICH, F.: The Light and Medium Subunits of the Photosynthetic Reaction Center from *Rhodopseudomonas viridis*: Isolation of the Genes, Nucleotide and Amino Acid Sequence; *EMBO Journal* **5** (1986) 1149–1158.

[40] MICHEL-BEYERLE, M. E.; PLATO, M.; DEISENHOFER, J.; MICHEL, H.; BIXON, M.; JORTNER, J.: Unidirectionality of Charge Separation in Reaction Centers of Photosynthetic Bacteria; *Biochim. Biophys. Acta* **932** (1988) 52–70.

[41] OESTERHELT, D.: Light-Driven Proton Pumping in Halobacteria; *Bio Science* **35** (1985) 18–21.

[42] PARLAR, H.; SCHUHMANN, W.: Photosynthese analoge Reduktion von Wasser; *Nachr. Chem. Techn. Lab.* **36** (1988) 1101–1109.

[43] PRINCE, R. C.: Light Driven Electron and Proton Transfer in the Cytochrome bc_1 comples. In: STAEHELIN, L. A.; ARNTZEN, C. J., Eds.; *Photosynthesis III, Encyclopedia of Plant Physiol.*, **Vol. 19**, Springer Verlag, Heidelberg, Berlin, New York (1986); pp 539–546.

[44] ROBERT, B.; LUTZ, M.: Structures of Antenna Complexes of Several Rhodospirillales from Their Resonance Raman Spectra; *Biochim. Biophys. Acta* **807** (1985) 10–23.

[45] ROBERTSON, D. E.; DUTTON, P. L.: The Nature and Magnitude of the Charge-Separation Reactions of Ubiquinol Cytochrome c_2 Oxidoreductase; *Biochim. Biophys. Acta* **935** (1988) 273–291.

[46] ROBERTSON, D. E.; DAVIDSON, E.; PRINCE, R. C.; VAN DEN BERG, W. H.; MARRS, B. L.; DUTTON, P. L.: Discrete Catalytic Sites for Quinone in the Ubiquinal-Cytochrome c_2 Oxidoreductase of *Rps. capsulata*; *J. Biol. Chem.* **261** (1986) 584–591.

[47] SCOLNIK, P. A.; MARRS, B. L.: Genetic Research with Photosynthetic Bacteria; *Ann. Rev. Microbiol.* **41** (1987) 703–726.

[48] SNOZZI, M.; CROFTS, A. R.: Electron Transport in Chromatophores from *Rhodopseudomonas sphaeroides* GA Fused with Liposomes; *Biochim. Biophys. Acta* **766** (1984) 451–463.

[49] STAEHELIN, L. A.; ARNTZEN, C. J., Eds.; *Photosynthesis III, Encyclopedia of Plant Physiol.*, **Vol. 19**, Springer Verlag, Heidelberg, Berlin, New York (1986); pp 1–725.

[50] STAEHELIN, L. A.; GOLECKI, J. R.; FULLER, R. C.; DREWS, G.: Visualization of the

Supramolecular Architecture of Chlorosomes in Freeze Fractured Cells of *Chloroflexus aurantiacus*; *Arch. Microbiol.* **119** (1987) 269–277.

[51] Staehelin, L. A.; Golecki, J. R.; Drews, G.: Supramolecular Organization of Chlorosomes and of Their Membrane Attachment Sites in *Chlorobium limicola*; *Biochim. Biophys. Acta* **589** (1980) 30–45.

[52] Strotmann, H.: Evaluation of Results on Nucleotide-Binding Sites of the ATPase Complex. In: Staehelin, L. A.; Arntzen, C. J., Eds.; *Photosynthesis III, Encyclopedia of Plant Physiol.*, **Vol. 19**, Springer Verlag, Heidelberg, Berlin, New York (1986); pp 584–594.

[53] Tadros, M. H.; Frank, R.; Dörge, B.; Gad'on, N.; Takemoto, J. Y.; Drews, G.: Orientation of the B800–850, B870 and Reaction Center Polypeptides on the Cytoplasmic and Periplasmic Surfaces of *Rhodobacter capsulatus* membranes; *Biochemistry* **26** (1987) 7680–7687.

[54] van Grondelle, R.; Bergström, H.; Sundström, V.; Gillbro, T.: Energy Transfer within the Bacteriochlorophyll Antenna of Purple Bacteria at 11 K, Studied by Picosecond Absorption Recovery; *Biochim. Biophys. Acta* **894** (1987) 313–326.

[55] Venturoli, G.; Fernndez-Velasco, J. G.; Crofts, A. R.; Melandri, B. A.: The Effect of the Size of the Quinone Pool on the Electrogenic Reactions in the Ubiquinol-Cytochrome c_2 Oxidoreductase of *Rhodobacter capsulatus*. Pool Behavior at the Quinone Reductase Site; *Biochim. Biophys. Acta* **935** (1988) 258–272.

[56] Vos, M.; Nuijs, A. M.; van Grondelle, R.; van Dorssen, R. J.; Gerola, P. D.; Amesz, J.: Excitation Transfer in Chlorosomes of Green Photosynthetic Bacteria; *Biochim. Biophys. Acta* **891** (1987) 275–285.

[57] Vos, M.; van Dorssen, R. J.; Amesz, J.; van Grondelle, R.; Hunter, C. H.: The Organization of the Photosynthetic Apparatus of *Rhodobacter sphaeroides*: Studies of Antenna Mutants Using Singlet-Singlet Quenching; *Biochim. Biophys. Acta* **933** (1988) 132–140.

[58] Wacker, T.; Gad'on, N.; Steck, K.; Welte, W.; Drews, G.: Isolation of Reaction Center and Antenna Complexes from the Halophilic Purple Bacterium *Rhodospirillum salexigens*. Crystallization and Spectroscopic Investigations of the B800–850 Complex; *Biochim. Biophys. Acta* **933** (1988) 299305.

[59] Wacker, T.; Gad'on, N.; Becker, A.; Mäntele, W.; Kreutz, W.; Drews, G.; Welte, W.: Crstallization and Spectroscopic Investigation with Polarized Light of the Reaction Center – B875 Light-Harvesting Complex of *Rhodopseudomonas palustris*; *FEBS Lett.* **197** (1986) 267–273.

[60] Wechsler, T.; Suter, F.; Fuller, R. C.; Zuber, H.: The Complete Amino Acid Sequence of the Bacteriochlorophyll *c* Binding Polypeptide from Chlorosomes of *Chloroflexus aurantiacus*; *FEBS Lett.* **181** (1984) 173–178.

[61] Weyer, K. A.; Schäfer, W.; Lottspeich, F.; Michel, H.: The Cytochrome Subunit of the Photosynthetic Reaction Center from *Rhodopseudomonas viridis* is a Lipoprotein; *Biochemistry* **26** (1987) 2909–2914.

[62] Witt, H. T.: Examples for the Cooperation of Photons, Excitons, Electrons, Electric Fields and Protons in the Photosynthesis Membrane; *New J. Chem.* **11** (1987) 91–101.

[63] Yeales, T. O.; Komiya, H.; Rees, D. C.; Allen, J. P.; Feher, G.: Structure of the Reaction Center from *Rhodobacter sphaeroides* R-26: Membrane Protein Interactions; *PNAS USA* **84** (1987) 6438–6442.

[64] Ysern, X.; Amzel, L. M.; Pedersen, P. L.: ATP Synthases – Structure of the F_1 Moiety and its Relationship to Function and Mechanism; *J. Bioenergetics Biomembr.* **20** (1988) 423–450.

Biological Actions and Induction Mechanisms of the Interferons

by J. BODE and H. HAUSER

Contents

Prof. Dr. Jürgen BODE and Dr. Hansjörg HAUSER,
GBF – Gesellschaft für Biotechnologische Forschung mbH,
Genetik von Eukaryonten,
Mascheroder Weg 1,
D-3300 Braunschweig-Stöckheim, Fed. Rep. Germany

[1] This subject is covered by an abundance of primary literature. Therefore only a few recent articles will be mentioned at the end of a paragraph, and where review articles are available, they will be given preference. Reviews of a more general nature are at the beginning of the reference list at the end of the chapter.

1 Biological actions of the interferons

1.1 Introduction

Interferons (IFNs) were discovered because they can prevent the spread of viral infections. This *antiviral* activity has been extensively investigated, and the molecular basis of some of its aspects is now known in considerable detail. Only when cloning procedures had been devised and the pure proteins had become available was it established with certainty that a homogeneous interferon species has several different biological activities. These additional activities have been termed *antiproliferative* and *immunoregulatory*, although this description is only roughly accurate (Figure 1).

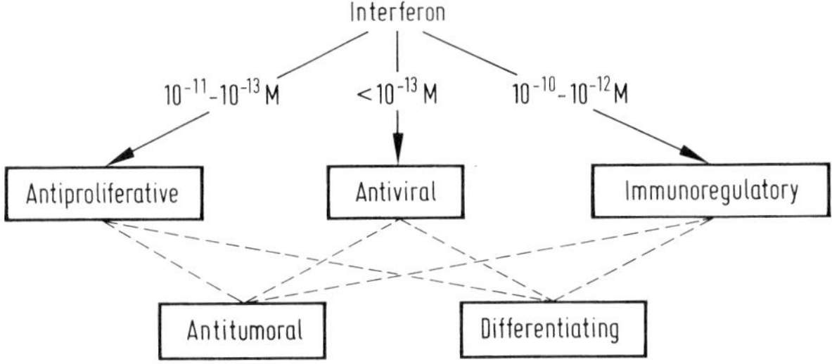

Fig. 1 Biological activities of the interferons
The three main activities (antiviral, antiproliferative, immunoregulatory) are observed at different interferon concentrations, and are characterized by different values of K_d. The secondary activities are due to components of these systems

Interferons slow down the cell cycle, lead to an increase in the cell volume, and cause alterations within the cytoskeleton and the cell membrane. They are inducers for the expression of certain cell surface antigens, and they affect the immune system. Due to a combination of these effects, they participate in differentiation processes and exhibit antitumor activity.

In some respects, interferons are related to the hormones, or, more accurately, to the prostaglandins. Like the latter, they are produced close to the site of their action, where they function in an autokrine mode (the cell acts upon itself) or in a parakrine mode (the cell acts on its nearest neighbors). Interferons show antiviral activity at concentrations of 3×10^{-14} M, which is orders of magnitude below the concentration at which typical peptide hormones are active (10^{-10} M-10^{-12} M), but most of interferon's other activities require higher concentrations (Figure 1). We therefore expect that there may be several, possibly overlapping, pathways of signal transduction for these mediator molecules.

Traditionally, interferons are classified as belonging to one of two classes according to their physicochemical and antigenic properties, their inducers, and the type of cell from which they originate. This division into classes I (IFN-α and IFN-β) and II (IFN-γ) is supported by recent evidence concerning gene structure and chromosomal localization. All class I interferon genes are localized on human chromosome 9 and lack an intron – an unusual situation in higher eukaryotes. IFN-γ is encoded by a gene on human chromosome 12 and comprises several introns.

In this chapter we attempt to trace the signal transduction path to the nucleus, beginning with a functional analysis of individual interferon domains. At present, this scheme must be considered incomplete; we have therefore tried to combine information available for class I and II interferons, although this may not be justified for particular aspects.

The second part of this chapter deals with the induction mechanisms of the interferon genes themselves. Here too we have a gap that has not yet been bridged, between the known composition of a number of inducers and the final process of induction, for which considerable information is available. We therefore emphasize the structural basis of gene activity. Several more years of intensive research are needed before we can link the triggering signals and the ultimate effects.

1.2 Structure-function relationships

Class I interferons constitute a multigene family with evolutionary interrelations. Hence, early hypotheses about the relevance of certain protein domains, proposed on the basis of sequence homologies, could subsequently be tested by *in vitro* mutagenesis.

Class I consists of at least 15 nonallelic members of the IFN-α family and of IFN-β. Of the 166 amino acid residues, 37 are invariant between these species; most of the invariant amino acids are located in segments 29–33, 47–50 and 136–150. Only the first and last of these proved essential in the more common tests for biological activity (Figure 2).

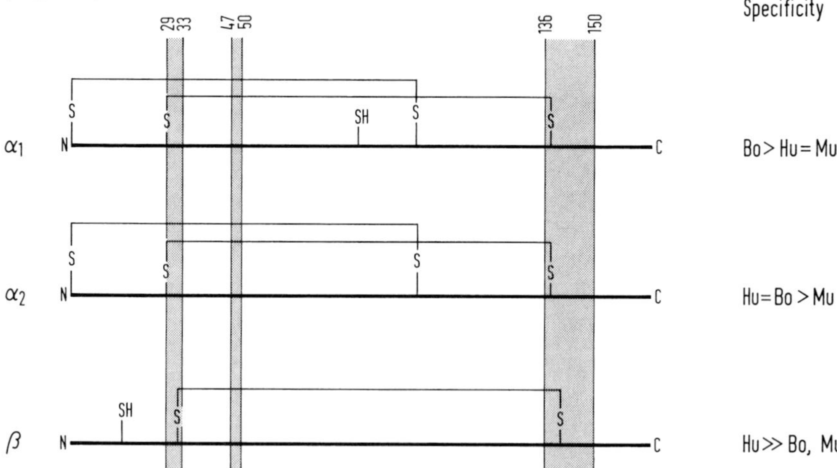

Fig. 2 Class I interferons are members of a gene family that has conserved elements
The position of isolated and bridged cysteine residues is shown in relation to the highly conserved regions (hatched areas). The corresponding antiviral activities on bovine (*Bo*), human (*Hu*), and murine (*Mu*) cell systems are listed on the right

Observations on species specificity have been vital to deepening our understanding. Strictly speaking, the specificity is a property only of β- and γ-IFNs; in contrast, the family of α-IFNs also triggers an antiviral state in murine and bovine cells. For example, Hu-IFN-α1 is highly active in bovine cells but less so in human and murine cells, while Hu-IFN-α2 provides better antiviral protection for human and bovine than for murine cells. With observations such as these, we were able to establish which segments are responsible for the biological effects in certain species.

Most secretory proteins are stabilized by disulfide bridges that are the result of the protein's biogenetic history. The positions of these bridges give us clues to a rough outline of the structure-function relationships in the protein (Figure 2). Human α-interferons have four Cys residues in homologous positions (1, 29, 99, 139), forming two disulfide bridges (1–99 and 29–139). IFN-β comprises three cysteines (17, 31, 141) and one disulfide bridge (31–141). A mutation eliminating the 1–99 bridge results in an IFN-α derivative with full biological activity; when the single Cys-17 is replaced by a Ser, the result is an IFN-β resistant to oxidation.

From this standpoint, the most complicated interferon is α1, with a fifth singular Cys at position 86, as is the case with all murine interferons and the single rat interferon sequenced to date. Replacing this residue with Tyr, Ser, or Gly has no effect on the activity in human cells, but the preferential activity in bovine cells is depressed significantly, and the activity in murine cells is simultaneously abolished.

Data on interferons soon led us to conclude that evolutionary invariance is not sufficient for us to identify the regions relevant to biological action. All functional amino acid residues have to remain constant, but conservation of the coding sequence may instead guarantee the operation of transcription or translation. This may explain the rather unexpected finding that the constant residues Phe-48, Gln-49, and Glu-62 could be replaced by residues of a fundamentally different nature without noticeably affecting either activity or specificity.

In what follows we will concentrate on Hu-IFN-α1 and Hu-IFN-α2. For the purpose of this discussion these species will be divided into three domains, the amino terminus (1–64), the central domain (65–120), and the carboxy terminus (121–166).

At the amino terminus, we find that exchanging the α1 and α2 termini raises the previously negligible murine and human activities of α1 and lowers the corresponding activities of α2.

In the central domain, among 10 variable residues, only Thr-80 (in α2) and Cys-86 (in α1) proved important for antiviral action in the mouse system.

At the carboxy terminus, replacing the variable α1 residues Lys-121, Arg-125, and Thr-132 by the corresponding α2 residues Arg, Gln, and Lys resulted in a nearly 200-fold decrease in activity in the murine but not in the human system. It may be significant that these residues lie on the hydrophilic side of a postulated amphipathic fourth helix in the series of helices, so that a simultaneous interaction with cellular structures is possible.

The action of interferons is mediated by receptors on the plasma membrane. It is therefore tempting to speculate that the amino terminus governs the interaction with the respective human or murine receptor. Moreover, the murine receptor seems to enable additional interactions with the central domain and (especially) the carboxy terminus. Within the complex of Hu-IFN-α1 with its receptor, the carboxy terminal segment is unconstrained, since a synthetic peptide comprising the terminal 16 amino acids can prevent the binding of antibodies to the interferon-receptor complex. Not unexpectedly, truncating the carboxy terminus by 10 amino acids is without effect; a similar result is obtained for Hu-IFN-γ [76, 80, 84].

1.3 Interferon receptors

Information on insulin, growth factors, and LDL/cholesterol has been sufficient for us to localize some basic functions of receptors within the plasma membrane (Figure 3). Data available for the interferon system are still fragmentary, but the suggestion is strong that the same principles apply with regard to internalization, processing, and signal termination.

Most information about the function of interferon receptors was derived from the human and mouse systems. Once pure iodinated interferon species were available, insights into the identity and structural transitions thereof were possible. We found in competition experiments that the signal of type I interferons is mediated by the same α/β receptor, whereas the type II or γ receptor is a separate species. This conclusion was supported by the synergistic actions of the interferons, by the results of mapping their genetic loci, and by the segregation of interferon specificities in somatic cell hybrids.

1.3.1 Receptor of IFN-α/β (type I)

The α/β receptor is encoded by a gene on human chromosome 21 or murine chromosome 16, respectively. The fact that fibroblasts from patients with Down's syndrome (trisomy 21) respond exceptionally strongly to type I interferons is thus explained.

The receptors isolated from lymphoblastoid cells (Daudi and Namalva cell lines) and cross-linked with labeled interferon migrated in SDS gels with mobilities corresponding to molecular masses of 95 and 100 kd. When they were isolated using milder detergents, they appeared to be in aggregates that might have been multiples of these values.

Most cells express a rather low number (200–10,000) of α/β-receptors with a high affinity ($K_d = 10^{-9}$–10^{-11}), the number depending on the cell type and degree of differentiation. Regarding these data, we note that it is very difficult to establish and maintain true equilibrium conditions throughout a binding measurement because of the phenomenon of *down regulation*. Steady-state conditions are also difficult to approach. Most binding curves do not show true saturation. Experimenters using nonlinear Scatchard plots have deduced the existence of additional low-affinity receptors, nonspecific binding, or both. It is true, however, that there is a measurable biological response even at low levels of saturation. Under those circumstances, the antiviral effect is usually observed before the antiproliferative actions [70].

1.3.2 Receptor of IFN-γ (type II)

Both the affinity and number of IFN-γ receptors can be compared to those of IFN-α/β receptors. A singular affinity between $K_d = 10^{-10}$ and 10^{-11} accords with the results of most binding studies.

Surprisingly, increases in the level of IP_3 are not always accompanied by corresponding increases in intracellular calcium. On the other hand, expression of MHC II genes resulting from the action of IFN-γ in the malignant monocytotic HL-60 cell line can be mimicked by a calcium ionophore. In this case, available evidence suggests an activation of the calcium-calmodulin pathway following Ca^{2+} transfer from the extracellular space.

1.4.3 Pathways dependent upon cAMP

Although some pathways triggered by the interferons are modulated by cAMP, cAMP does not control their central activities. Increases in intracellular cAMP occur slowly in these systems and could be interpreted as a means of terminating the initial signal; this interpretation is supported by the fact that the rapid actions of the interferons, which are mediated by IP_3, are inhibited by activators of adenylate cyclase (Figure 3). Temporal correlations between increases in cAMP and antiproliferative actions of the interferons may be indicative of the relevance of pathway I in this context. This is also true for an increase in phagocytotic activity [72].

1.4.4 Other participating elements

Pathways I and II both seem to terminate in kinases, which are believed to participate in activating *transcription factors* (TF). The mechanism may involve either a direct activation of transcription factors or a dissociation of an inhibitory subunit (see Section 1.6). Activation in these cases necessarily means the capacity to bind DNA. Thus, proof for such a mechanism is usually established by synthetic DNA fragments that comprise critical promoter sequences.

In this context, we should mention the appearance of a factor termed "E", released from a cryptic form in the cytosol within 30 seconds after fibroblasts are treated with interferon. A few minutes later, the cytosolic variant (which is extractable at low ionic strength) disappears; this has been ascribed to its translocation into the nucleus, followed by its binding to chromatin.

Factor E seems to be identical with a factor ISGF3 that was earlier discovered by other investigators. ISGF3 arises in the absence of new protein synthesis, and its appearance correlates with transcriptional activation in HeLa cells and fibroblasts. Two other factors, ISGF2 and ISGF1, recognizing the same sequences, have also been detected. ISGF2 can be detected 90 minutes after interferon treatment; it requires new protein synthesis and correlates with the cessation of transcriptional activities. Finally, ISGF1 is believed to be a constitutive factor that covers the control elements during phases when there is no transcription. It may thus serve to prevent the packaging of the promoter in inactive chromatin. A more detailed discussion of the interplay of such factors requires knowledge of their DNA recognition sites, that is, of the interferon-stimulated response element (ISRE), and will be postponed until later (see Section 1.5.1; also [25, 54]).

1.5 Interferon-induced genes

Activities initiated by interferons require transcription. This observation initiated an intensive search for interferon-induced proteins. Conventional techniques were used first, and helped to identify in chicken fibroblasts at least twelve proteins inducible by IFN-α/β that were not present in control cells. IFN-γ induces the same twelve plus another twelve; four of these were completely absent before treatment. The induction of these sets of proteins follows different kinetics depending on whether the interferon is of type I or II: IFN-α/β induces a transient synthesis, but syntheses induced by IFN-γ continue for at least 24 hours. Adding certain inhibitors of RNA synthesis 4–8 hours after the onset of interferon treatment may lead to the phenomenon of superinduction. As postulated for the induction of interferons, these treatments appear to deactivate a shutoff mechanism, which may be based on the expression of a repressor molecule (Section 2.2.6).

Human interferon-inducible genes have been localized on human chromosomes 1, 4, 6, 10–13, 15, 16, and 21; murine counterparts are found on chromosomes 1–3, 12, 16, and 17. In some cases loci comprise families consisting of homologous genes that may have arisen from a series of gene duplications. Examples are the human 1–8 family, with its members 9–27 and 6–16, and the murine 203 family; these have been identified on the basis of their homology with IFN-induced mRNAs [66, 85].

1.5.1 The mechanism of induction: ISRE and IRS

The regions flanking the 5′ terminus of many IFN-inducible genes contain homologous sequences called *interferon-stimulated response elements* or ISREs. Researchers now agree that a 13–14 bp consensus A/GGGA/GAAN(N)GAAACT exists. Corroboration of this sequence in various genes is summarized in Table 1; clearly, ISREs can vary with respect to their orientation and the distance from the start of transcription, as can enhancers. Several examples lead us to suspect that ISREs are part of a larger functional unit, the so-called *interferon response sequence* (IRS). Within the IRS, the ISRE controls the potency of another element that itself acts as a constitutive enhancer of transcription. These two elements may overlap or be separate.

The ISRE of the human gene for 2′,5′-oligoadenylate synthetase E is part of a 72-base-pair fragment between positions 87 and 159. In its 5′ part this sequence has a constitutive enhancer that is reminiscent of the enhancer of the immunoglobulin heavy chain. The terminal 16 base pairs constitute the ISRE control element. Under artificial conditions, this ISRE can be used to render the constitutive thymidine kinase promoter inducible by herpes simplex interferon [11, 22].

The interferon-stimulated genes (ISG) 15 and 54 have sequence homologies within segments 114 to 92, and 89 to 113, respectively (note that the orientations differ). Both these gene segments include an ISRE (Table 1) and each is able, as part of an extended sequence, to control the TK promoter from a position upstream or downstream of the gene.

The ISRE mediates interferon induction irrespective of the type of interferon. This is exemplified by two members of the human 1–8 gene family, 9–27 and 6–16, which feature comparable ISRE sequences but respond differentially to interferons of classes

Table 1 Interferon-stimulated response elements (ISRE) of various genes

Consensus	$\genfrac{}{}{0pt}{}{A}{G}$ G G $\genfrac{}{}{0pt}{}{A}{G}$ A A N (N) G A A A C T				
Gene				Orientation	Induced by
Hu-OASE (E)	−100	A G G A A A C − G A A A C c	− 88	→	IFN-α/β
Mu-OASE	− 72	G G G A A A T G G A A A C T	− 59	→	IFN-α/β
Hu-ISG15	−108	G G G A A A C C G A A A C T	− 95	→	IFN-α/β
Hu-ISG54	− 90	G G G A A A G T G A A A C T	−103	←	IFN-α/β
Hu-9−27	−172	A G G A A A T A G A A A C T	−159	→	IFN-α/β, -γ
Hu-6−16	−112	G G G A A A A T G A A A C T	− 99	→	IFN-α/β
Hu-IP10	−222	t G G A A A G T G A A A C c	−209	→	IFN-γ
Mu-H-2D[d]	−138	G c A G A A G T G A A A C T	−151	←	IFN-α/β, -γ
Mu-H-2K[b]	−137	G c A G A A G T G A A A C T	−150	←	IFN-α/β, -γ
Mu-*β2m*	−146	A a G A A c A T G A A A C T	−133	→	IFN-α/β, -γ
Hu-IFN-β (PRD I)	− 78	G G a G A A G T G A A A g T	− 65	→	Poly(rI):(rC) IFN-β
Hu-IFI54	− 85	G G G A A A G T G A A A C T	− 98	←	Poly(rI):(rC) IFN-β
Mu-Mx	−119	c a G A A A C − G A A A C T	−131	←	NDV, IFN-α

I and II (Table 1). Under different conditions, the ISRE derived from 6–16 was found to be capable of mediating responses to IFN-γ. It is clear that different flanking sequences have an effect on the specificity as well as the magnitude of the response.

The action of promoter sequences is mediated mainly by binding proteins, the so-called transcription factors. Novel techniques of probing DNA-protein interactions permit insights not only into the specificity and affinity of the interaction, but also into its dynamics during the induction process. The following is an outline of three widely used standard procedures, which have been applied to almost all the systems referred to in Table 1. The procedures are also integral to the experiments to be described in Section 2.2. Some principles that appear to hold for a number of cases will then be explained using a typical example.

Band-shift or gel retardation assays. Synthetic DNA fragments corresponding to sequences that are potentially important for gene regulation are radioactively labeled and are incubated with protein extracts from cells. In this way, binding proteins are labeled in accordance with their affinity for a particular sequence, so that they can be visualized on a polyacrylamide gel. The mass of the protein can be estimated from its mobility. The specificity of the interaction can be deduced using unlabeled competitor DNA of the same or a homologous sequence.

DNase I footprinting. The complex reconstituted from a labeled DNA segment and a binding protein is analyzed. This complex can be recovered from a scaled-up variant of the previous technique. If the conditions for limited degradation are established, for example, with nucleolytic enzymes, such as DNase I, or reagents, such as MPE-Fe(II) (methidiumpropyl-EDTA iron) or copper *o*-phenanthroline, ideally a single cut is introduced per DNA molecule. This is sterically possible only in regions not protected by protein. If the fragments are sorted by length on a sequencing gel, "windows" appear in those places where proteins were associated, where the DNA has been shielded from the reagent.

In some variations of this procedure, guanine residues are methylated in the regions of DNA that lack proteins (*methylation protection*), or the effect of a previous limited methylation upon complexing with the protein is investigated (*methylation interference*). Since DNA can be cleaved at alkylated guanosines, the previous standard procedure can be repeated for analysis.

Analysis of sites hypersensitive to DNase I. The aforementioned methods have been optimized for analyses of DNA-protein interactions *in vitro*. The accessibility to DNase I, particularly at times when active transcription is occurring, serves as a qualitative probe of structural alterations *in vivo*. So-called *DNase I-hypersensitive sites* may occur within sensitive regions, and have been ascribed to torsionally stressed conformations and a reduced protection by proteins. This procedure also relies on sorting DNA fragments by length on a gel. However, in this instance terminal labels are provided by short complementary DNA probes.

The band-shift technique was used to demonstrate that ISRE sequences may bind both the factors modulated by IFN-α/β and those modulated by IFN-γ. An elaborate interplay of at least three factors, forming distinct complexes with DNA, was observed in all systems. We have introduced these factors already as ISGF1 to ISGF3, names given by DARNELL's group. According to detailed analyses, these factors bind within the ISRE to different, though overlapping, sequences on both strands. Appearance and disappearance of these factors in a band-shift assay follows the same kinetics as a corresponding modulation within the control region of interferon-induced genes. Thus it can be shown that the ISRE of the ISG15 gene has protein protection that ends on both sides at DNase I-hypersensitive sites. One of the flanking regions is extended upon induction by 60 bp, signifying the presence of a protein with an extended binding site. Some of this information will again be useful when we discuss more fully the function of some of the well-characterized interferon-induced gene products [24, 53, 63].

1.5.2 The ISRE has more functions

Induction of the IFN-β gene can be increased significantly for certain cells if they have been treated with IFN-α or -β prior to induction. This phenomenon will be discussed in some detail in Section 2.2.6. Moreover, the IFN-α gene in mouse cells is slightly but reproducibly inducible by IFN-α itself. It may be relevant in this context that the PRD I element (63 to 77) of the IFN-β gene comprises a nearly ideal ISRE sequence (Table 1 and Figure 6). The footprint binding by a factor that binds PRD I is indistinguishable from the corresponding factor binding to ISRE elements. We therefore suspect that there are crossover points not only between the signaling pathways of particular interferon types but also to viral induction pathways. We are probably dealing with a general phenomenon, as several genes that can be induced both by interferon and by viruses have been reported recently. Sequences derived from the human IFI54 gene and the murine Mx gene correspond closely enough that it seems likely that an ISRE element may also be active in these cases (Table 1).

In some cases (Hu-ISG15, Hu-IFI54), the ISRE sequence can be interpreted as being a dimer of an element with the consensus base sequence GAAA(G/C)T, and some other examples resemble this situation, at least approximately. In fact, it has been reported that synthetic oligomers of GAAAGT mediate induction by interferon, but oligomers of GAAACT permit induction by interferon and viruses. Obviously, variants of the ISRE have the unprecedented capability of processing three signals from quite separate sources: those originally from interferons of classes I and II and those of double-stranded RNA [31, 40, 83].

1.6 The antiviral state: a complex phenomenon

Interferons bring about in their target cells an antiviral state, which thereafter is no longer limited to the type of virus that triggered it. In this state, the cell is capable of inhibiting virus replication at several stages between the association of the infective particle with the cell membrane and the maturation and release of the progeny. The specific targets depend to some extent on the type of infecting virus. The most general and efficient block is at the level of translation. Two proteins arise upon induction that remain cryptic until they become activated by *double-stranded RNA* (dsRNA), which acts as a positive effector. These two activities act in concert upon the protein synthesis apparatus (Figure 4a). In *in vitro* studies, the double strand formed from synthetic poly(rI) and poly(rC) is used as a model of dsRNA; in what follows, it will be called poly(rI):(rC).

There is general agreement that double-stranded RNA becomes a mediator of interferon action as it arises during the replication cycle of several virus types. However, this must be demonstrated for each individual virus type.

Reovirus has a dsRNA genome, which replicates in the cytoplasm of infected cells. Viruses with a single-stranded RNA genome (EMC, VSV) are known to be particularly efficient inducers. In experiments with intercalators and cross-linking reagents, workers have demonstrated that double-stranded segments of sufficient length arise from these viruses during replication. Upon transcription, HIV proviruses develop a hairpin structure in the 5′ TAR segment, which is not transcribed. For dsDNA viruses (vaccinia, SV40) dsRNA could arise by symmetric transcription, but this is only a guess at this time.

The dsRNA plays a dual role as an interferon-inducing agent and as a mediator of the antiviral state, making possible the coordinated strategy to prevent viral infections shown in Figures 4a and 4b. The infected cells first synthesize and secrete interferon. This early parakrine signal induces a dsRNA-dependent antiviral potential in adjacent cells. The induced enzymes do not affect cellular metabolism as long as there is no acute infection. On the other hand, the antiviral activity is spontaneously available after the second signal (dsRNA) appears. We will first discuss these dsRNA-dependent antiviral mechanisms, before presenting several less clearly defined factors contributing to the antiviral state.

a)

dsRNA
IFN
OASE
Kinase (dsI)
PDE
ppp(A2'p5')nA
ATP
RNase L
Nt
mRNA
eIF-2α (P)
Pase
Inhibition of viral protein synthesis

b)

Immune response
Induction of MHC
IFN
Induction of dsRNA-dependent enzymes
Pathway independent of dsRNA
OASE
Kinase
Viral mRNA (Degradation)
eIF-2 (Inactivation)
tRNA (Degradation)
Cap methylation
Protein glycosylation
Membrane function
Translation blocked
Infection blocked

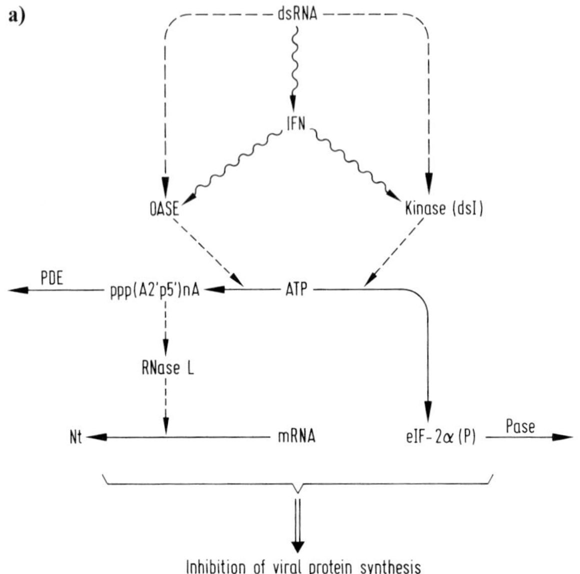

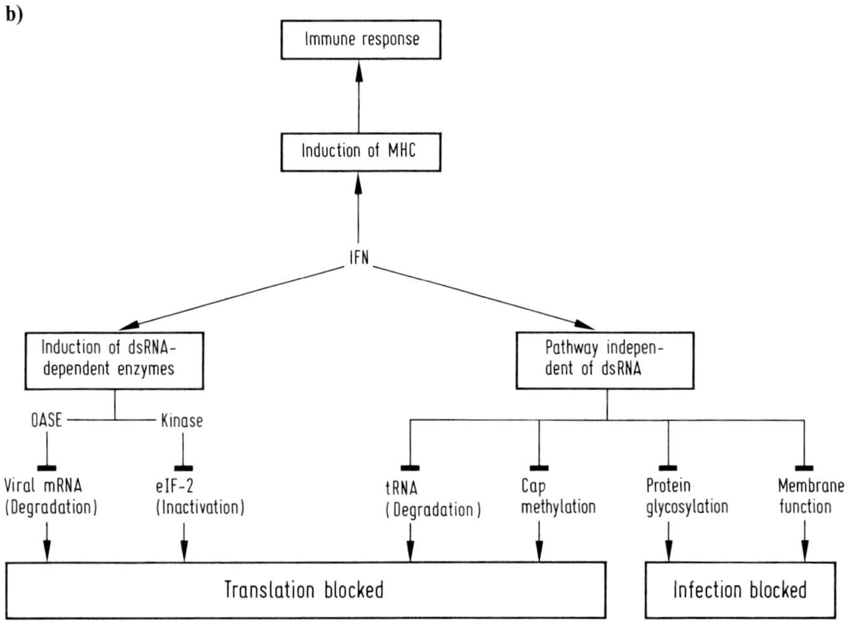

1.6.1 Proteins activated by double-stranded RNA

eIF-2 kinase (dsI). The first step in elucidating the molecular details of the mechanism was the discovery of the so-called *heme-regulated inhibitor* (HRI). Heme depletion leads to a translational block by phosphorylating the 38-kd α subunit of *eukaryotic initiation factor 2* (eIF-2α). Heme was found to be a highly efficient inhibitor of a novel type of kinase specific to a limited number of substrates.

Phosphorylated eIF-2 blocks translation because it does not interact normally with the *reversing factor* (RF). RF binds so strongly to the eIF-2α(P)/GDP complex that it is incapable of catalyzing the GDP-GTP exchange that would serve to prepare eIF-2 for its function in the preinitiation complex. The recycling of the RF is thus interrupted.

The next discovery after HRI was the *dsRNA-dependent kinase* (dsI), which causes the same effect but arises from a different signal. For their activation, both kinases require an autophosphorylation, and both inactivate eIF-2 by phosphorylating the same serine residue. This type of kinase is known to be inhibited by 2-aminopurine, although side reactions have recently been reported. A functional difference between these enzymes may arise from the fact that dsI but not HRI has the ability to associate directly with the ribosome; that may cause its action to become restricted to its immediate environment, making possible a discrimination between cellular and viral mRNAs. There is evidence, obtained recently, that dsI may have other substrates as well: it may participate in activating latent transcription factors (see Section 1.4.4; also [28, 42, 55]).

2′,5′-Oligoadenylate synthetase (OASE). OASE is expressed by extracellular stimuli, most notably the interferons. This dsRNA-dependent enzyme can be simultaneously purified and activated on dsRNA affinity columns. That is exactly what was done in 1978 to convert sizable amounts of ATP into the corresponding oligomers, for which the following structure could be deduced:

$$pppA^{2'}p^{5'}A^{2'}p^{5'}A$$

While the trimer is the predominant product *in vitro*, higher oligomers (up to 15-mers) are synthesized by the cellular biosynthetic apparatus.

The 2′,5′ bond of the oligoisoadenylate is unusual, and apparently unique for nucleotides of biological origin. It therefore resists the more common nucleases. When the turnover of this class of compounds was studied, a *2′,5′-phosphodiesterase* (PDE) that degrades starting from the 3′ end was discovered. Depending on cell type, this enzyme may be constitutive or inducible by interferons.

◀ *Fig. 4* Elements contributing to the antiviral state.

a) IFN-induced enzyme activities, activatable by double-stranded RNA (dsRNA) or poly(rI):(rC). Chemical conversions are marked with solid arrows, activations with dashed arrows, and transcriptional induction events with wavy lines. Abbreviations not explained in the text: Pase, protein phosphatase; Nt, nucleotide(s)

b) The antiviral syndrome. Contributions besides the dsRNA-dependent factors have been included. Inhibitory activities are denoted by a solid bar; where applicable, the level at which the inhibition acts is stated in parentheses

To date the only known function of a 2′,5′-adenylate oligomer is the activation of a ribonuclease, RNase L, that is otherwise latent. This enzyme preferentially cleaves oligo(U) segments, and *in vitro* both viral and cellular DNAs are degraded. However, *in vivo* a preferential degradation of viral mRNAs was observed, at least in particular cellular systems. This selectivity has been ascribed to the fact that 2′,5′-adenylate oligomers are formed at the site where OASE is activated, and are bound to partially double-stranded replicative viral intermediates. The 2′,5′-phosphodiesterase maintains their concentration gradient. It seems possible that OASE and RNase L may form a functional complex, as has been observed for other enzymes that act in a coordinated manner.

Results for cloning the OASE enzyme(s) have already been published (this is in contrast to the situation for dsI). According to the data, human cells produce two mRNAs, differing in their 3′ ends, which arise from differential splicing of the product of the E gene on chromosome 12. The resulting proteins have molecular masses of 41.5 and 46 kd. Mice appear to have two nonallelic forms of an OASE gene comprising interferon-stimulated response elements (ISREs), as shown in Table 1. These ISREs appear to control a repressor-binding sequence, which causes the quiescent state of the gene in cells not treated with interferons [11, 22].

1.6.2 Other antiviral activities

The antiviral state is caused mainly by dsRNA-dependent steps acting at the level of protein synthesis. This aspect is supplemented by dsRNA-independent activities. In sum, the collective property known as an antiviral state comprises alterations at the cell surface that act in concert with the immune system to control virus infections.

Modifications of tRNA. If extracts from IFN-treated cells are used in *in vitro* translation experiments, the activity is low but can be enhanced by the addition of certain tRNAs. The Leu-, Lys-, and Ser-tRNAs were the reasons for the low activity: they failed to act as acceptors in an aminoacylation reaction. In mouse L cells, a similar effect could be ascribed to the degradation of the C – C – A terminus, caused by an interferon-induced phosphodiesterase that acted mainly on 2′, 5′ bonds.

This phosphodiesterase may have the primary function of controlling the local concentrations of the 2′,5′-adenylate oligomer, which is both the OASE product and the RNase L activator, thereby preventing a global degradation of cellular RNA. This primary effect would then be followed by depletion of particular tRNAs, resulting in a general blocking of elongation. The blocking would continue until the tRNAs are resynthesized or repaired by nucleotidyl transferases. The enzyme might also be responsible for an inhibition of translation independent of dsRNA [71].

Methylation of mRNA. Almost all eukaryotic and viral mRNAs carry at their 5′ end a 7-methylguanosine residue attached by a triphosphate bridge in a 5′,5′ bond, a condition called *capping*. This cap may be modified by methylation of a few terminal nucleotide residues and it acts to stabilize mRNAs. It also promotes the formation of the 48S preinitiation complex in concert with *cap-binding proteins* (CBPs). Extracts from interferon-treated cells were found to contain an inhibitor of cap methylation, which appeared to compete with the methyl transferases for binding sites on mRNA. Conceivably, this inhibitor interferes with the process of attachment of ribosomes to mRNA.

Mx protein. IFN induces a 72-kd protein in the cells of mice resistant to influenza viruses, called *Mx^+ mice*. How this *Mx protein* affects viral replication is still unknown. Analogs have also been identified in humans.

The murine Mx 1 gene comprises 14 exons over a total length of 55 kb. When we discussed the data in Table 1, we mentioned that interferons are not the only inducers of this protein; viruses and dsRNA are alternative inducers. Thus, an immediate antiviral response is triggered in virus-infected cells; uninfected cells can establish an antiviral state as the result of the interferon signal [40].

MHC complexes. Proteins encoded by the *major histocompatibility complex* (MHC) are membrane proteins with multiple functions. Dealing with them in a section on the antiviral state means considering the events on the cell surface that justify their inclusion in the discussion.

The MHC complex consists of three coupled gene clusters. In humans, class I comprises several loci on chromosome 6, called HLA-A, HLA-B, HLA-C and HLA-D (HLA stands for *human leucocyte antigen*). The analogous loci in the mouse are H-2K and H-2D on chromosome 17. HLA- and H-2 genes code for the long chains of dimeric glycoproteins localized in the cellular membrane. In all cases, the second chain is β_2-microglobulin (β_2-m). Class I antigens are expressed by nearly all nucleated cells if they are at the appropriate state of differentiation. The antigens mediate the cellular immune response exerted by macrophages, cytotoxic *T*-lymphocytes (CTTL), and granulocytes: the cells of the immune system recognize viral and tumor antigens only if these are presented together with the corresponding class I antigen on the cells that are to be attacked. In most tissues or cell lines with a low level of expression of MHC I, this can be enhanced by IFN-γ and to some extent by IFN-α and IFN-β [33].

Class II gene products, which are also dimeric glycoproteins, are usually found only in macrophages and B lymphocytes. The human gene family is called DR, and the murine one, Ia. MHC II proteins participate in the activation of helper *T* lymphocytes (T_H cells). This takes place as follows: antigens that have been processed by macrophages are presented to T_H cells as part of a complex with MHC II proteins; the T_H cells are activated and become capable of increasing the production of antibodies in those B cells that recognize the same antigen. MHC class II genes are specifically regulated by IFN-γ. Stimulation involves characteristic sequences other than the ISRE; these have been called the *X box* and the *Y box*; the Y box comprises a CCAAT sequence in reverse orientation. Regulatory features appear complex; they follow different kinetics, and include steps dependent on protein synthesis, as well as the activation of preexisting factors [16, 20, 27].

In humans, the MHC III region lies in the 1000-kb DNA segment that separates the MCH I and MHC II complexes. Here, various apparently unrelated components are encoded. Examples include the C2, B, and C4 complement factors, P-450 hydroxylase, and the cytokines TNF-α and TNF-β. Several of these components are regulated by IFN-γ [69, 78].

The following discussion will be confined to the MHC class I genes whose regulation has been documented best. Proteins of this class interact with viral antigens on the cell surface; they are then capable of triggering the lysis of cells by cytotoxic *T* cells. Thus IFN-induced expression of MHC I antigens contributes to the antiviral action of interferons at the level of cell-cell interaction.

The MHC I genes, as well as the β_2-microglobulin (β_2-*m*) genes, are regulated by the interferons. During translation, the MHC I heavy chain is inserted into the endoplasmic reticulum, where it combines with β_2-*m* after glycosylation. Mutants such as the Daudi cell line, which are characterized by deficient β_2-*m* expression, accumulate the HLA heavy chain in its high-mannose form, indicating that the association with β_2-*m* is important for further processing in the Golgi apparatus. Moreover, the association of HLA and β_2-*m* must be catalyzed by an additional factor, which also responds to interferon [44].

For some reason the syntheses of the heavy and light chains do not occur simultaneously. However, both genes contain a standard ISRE sequence (Table 1), which makes them responsive to the same transcription factor. Workers who analyzed sites around the H2-D locus that were hypersensitive to DNase I found that the ISRE sequence can be part of four alternative chromatin structures, depending on whether it is noninducible, inducible during differentiation, inducible by interferon, or constitutive. The conformations of these structures appear to determine whether transcription factors have access to the ISRE, other promoter elements, or both [41, 57].

1.7 Antiproliferative activities

At present, the best understood of the factors that have antiproliferative action are the interferons themselves. Although the discovery of the antiproliferative effect came shortly after that of the antiviral potency, definite proof that this effect was intrinsic in interferons had to wait until pure preparations produced by cloning technology became available. We have already mentioned that the antiproliferative actions are triggered at higher interferon concentrations than the antiviral ones; different signal transducing mechanisms may be operating. This does not preclude, however, the participation of some antiviral activities along with the antiproliferative effects.

dsT. At certain phases of their growth and differentiation, 3T3 preadipocytes spontaneously secrete interferon. This interferon in turn induces and activates dsI, apparently in an autokrine mode. The activation, which we deduce to have occurred because dsI is autophosphorylated, appears to occur in the absence of dsRNA, and can be mimicked by experimental treatment with IFN-β. It is still unclear whether there are different mechanisms for activating the enzyme, or if mechanisms exist which could lead to a synthesis of endogenous RNAs [62].

OASE. The 2′,5′-oligoadenylate metabolism is controlled by an interferon-dependent activation of OASE, as we have already described, but other contributions have also been postulated. Different cell types have basal OASE activities ranging over at least three orders of magnitude. Broadly speaking, there is an inverse correlation between OASE and cell growth, which leads us to suspect that OASE has a more general role in growth regulation.

An extension of the analyses to all enzymes participating in the 2′,5′-oligoadenylate system showed high OASE activities throughout the S phase, while the activity of 2′-phosphodiesterase increased in G1. The third relevant enzyme, RNase L, showed little variation over the cell cycle.

OASE may have additional functions besides those in the actual oligo(A) system. Evidence for this comes from a side reaction during which OASE inactivates NAD^+ by 2′-adenylation. This reaction could affect energy metabolism as well as other NAD^+-dependent processes such as ADP ribosylation. Similar influences have been ascribed to alterations in the metabolism of tryptophan; another interferon-inducible enzyme, indolamine-2,3-dioxygenase (IDO), causes the observed changes.

IDO. (indolamine : oxygen-2,3-oxidoreductase (decyclizing)) is a cytosolic, monomeric, heme-containing enzyme. As opposed to its tetrameric counterpart in liver, it is widespread in normal and malignant tissues. The enzyme, once induced, accelerates the catabolism of tryptophan, the first step of which is oxidative decyclization of tryptophan to *N*-formylkynurenine. Pleiotropic effects may arise from decreased levels of tryptophan as well as from the modulation of Trp-dependent pathways. The decarboxylation to serotonin (a neurotransmitter and immunoregulator) may be repressed, while metabolites from the kynurenine pathway accumulate. These events may be responsible for some part of the antiproliferative and antitumor effects of the interferons. Some of the side effects of interferon treatment have also been ascribed to processes mediated by IDO [18].

1.8 Interferons as members of the cytokine network

Interferons are the prototypical *cytokines*, which are secreted glycoproteins that exert parakrine actions. The conventional term "interferon effects" seems inadequate, since the complete spectrum of actions in which the interferons participate cannot be explained without reference to their contribution within the network of cytokines.

Within the cytokine network, the interferons act in concert with particular agents that display proliferative activity, such as the four *colony-stimulating factors*. These CSF activities initiate the proliferation of certain stem cells, which differentiate into granulocytes, macrophages, both granulocytes and macrophages, or general leukocytes. Accordingly, these activities have been termed G-CSF, M-CSF (or CSF-1), GM-CSF and multi-CSF (or IL-3). A second group of partners for the interferons is composed of the *interleukins* (IL-2 to IL-8), which analogously promote the differentiation of *T* and B cells. Finally, this class of agents comprises the *growth factors*, with PDGF (platelet-derived growth factor) as a prototype competence factor and EGF (epidermal growth factor), IGF (insulin-like growth factor), and TGF-α/β (transforming growth factor) as typical progression factors.

a)

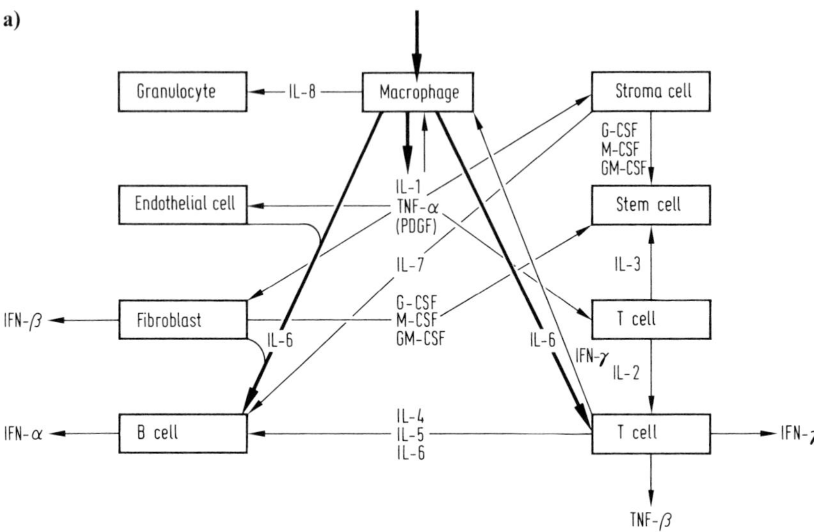

b)

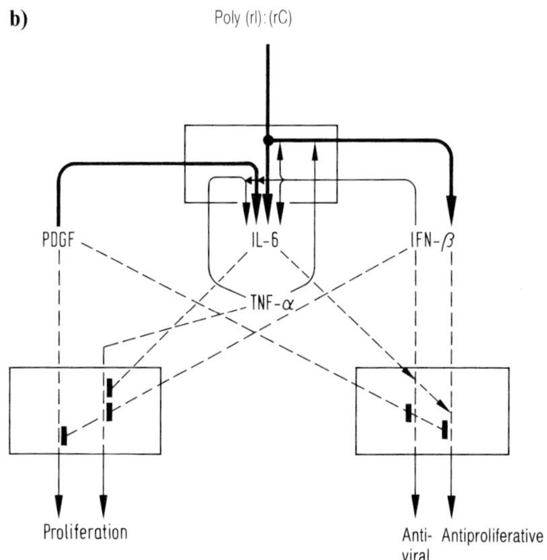

Fig. 5 The cytokine network

a) Cytokines as intercellular signals. The scheme comprises alternative induction mechanisms for the interferons (except for IFN-*γ*) but not their actions.

b) The cytokine network as derived for fibroblasts. The autokrine and parakrine activities of IFN-*β* are enclosed in the three boxes (representing three cells), and interactions with the other cytokines have been added. Primary and secondary induction events are shown as solid arrows. Modulating activities are shown as dashed lines; their result may be activation (the lines end in arrowheads) or inhibition (bars). poly(rI):(rc) simulates aspects of the viral infection. In certain respects, PDGF corresponds to IL-1 (as shown in Figure 5a)

A schematic view of the cytokine network is summarized in Figure 5a. When stimulated by infectious agents, macrophages secrete IL-1 and *tumor necrosis factor* α (TNF-α); this stimulus can be mimicked *in vitro* by lipopolysaccharides. The cells, different types of which are shown in the boxes to either side, are then triggered by the central mediators, and as a result they secrete further cytokines. Among these latter, IFN-γ acts to enhance the primary response after it has been secreted by *T* cells.

There is general agreement that IL-6 is the central mediator in the acute phase of inflammation. In the standard fibroblast system (Figure 5b), IL-6 is (co)induced with IFN-β. An alternative term for IL-6 is IFN-β_2 because of its apparent antiviral activity (which is not generally accepted). Another property IL-6 shares with IFN-β (now frequently called IFN-β_1) is an antiproliferative effect. However, the primary function of IL-6 seems to be its stimulatory action on B-cell differentiation (Figure 5a), which led to its being classified as *B-cell stimulating factor-2* (BSF-2). Other typical activities are observed upon stimulating *T* cells and stem cells. The gene encoding IL-6 has been located on chromosome 7; it has similarities to the gene of G-CSF but not of the other interferons.

Within the cytokine network, induction of the interferons is mediated by IL-1, IL-6, TNF-α and, in an analogous manner, by PDGF. This then explains the indirect antiviral effect of these cytokines. As the IFNs are capable of terminating the stimulus initiated by the growth factors, this mechanism may act as a feedback control, from a functional point of view. This hints at an intricate coordination of proliferative and antiproliferative contributions, which again is best understood for fibroblasts. The fibroblast system may serve as a model for the events occurring in the stroma cells, which are another cell population of this type (Figure 5b).

Poly(rI):(rC) induces IFN-β (IFN-β_1) and IL-6 (IFN-β_2) at the same time. Afterwards, IFN-β stimulates transcription of its own gene (*priming*) and that of IL-6. The result is a local mechanism for amplifying the antiproliferative response.

Between positions 169 and 124, near the IL-6 gene, is located a control element that is 70% homologous with the c-*fos* enhancer. This may provide growth factors like PDGF with a means of limiting the proliferative response after the c-*fos* gene has been activated. The action of IL-6 in turn is restricted because its mRNA has a short half-life, the result of an AU-rich nontranslated 3′-OH terminus that promotes degradation of the mRNA.

The fibroblast system is therefore valuable as a tool for investigating the antagonistic effects of cytokines, as a first step towards learning about cytokine properties *in vivo*. Thus, the proliferative activity of PDGF (and, by analogy, of TNF-α) is limited by IFN-β and IL-6, while the antiviral activity of IFN-β is restricted by PDGF. In contrast to TNF-α, PDGF (and IL-1) will induce IL-6 without developing an antiviral state: apparently, these two have evolved to compensate for the specific effects of the various polyfunctional proteins [43, 45, 61, 82].

2 Induction of interferons

2.1 Introduction

Because interferons have such dramatic biological effects on practically all cells, their permanent presence is not compatible with a functional organism. Indeed, all interferons are inducible proteins that are secreted by vertebrate cells as a response to different inducers. When cell culture lines from mice or humans, not infected by viruses, are analyzed by very sensitive methods that can detect small amounts of specific interferon mRNA, these types of mRNA have not been found.

Type I interferons (IFN-α/β) are efficiently secreted in response to all viral infections. Usually the virus infection can be simulated by certain types of double-stranded RNA (for example, poly(rI):(rC)). Other substances, regarded as physiological or synthetic inducers, that lead to the synthesis of type I interferon have been found.

Whereas type I interferons are produced by nearly all cell types, IFN-γ (type II interferon) is secreted only by *T* lymphocytes and macrophages, when immunocompetent cells are antigenically stimulated. Treating cultivated *T* cells with mitogens simulates this effect *in vitro*. The activation of IFN-γ therefore appears very similar to the induction of lymphokines. This subject will be discussed in Section 2.3.

There is additional evidence to confirm the fundamental differences between type I and type II interferons. The homology of the regulation of type I interferons in mammalian species is due to significant homologies at the level of the structural genes, namely the promoters. In contrast, the IFN-γ regulatory sequences show no homology to promoters of type I interferon. Obviously the antiviral properties of IFN-γ are not due to a common evolutionary development with type I interferons. Properties other than the antiviral ones differ in important ways from those of type I interferons and are the basis for distinguishing the two [7, 8, 30].

2.2 Activation of type I interferons

Because interferons have numerous functions, and in particular have antiproliferative properties, their synthesis is strictly controlled. As mentioned above, in healthy organisms no interferon activity is detectable. Contrariwise, if one finds interferons in blood or other tissue fluids one can deduce a previous virus infection.

Note, however, that there are circumstances in which the localized presence of small amounts of interferons is not indicative of a pathological state. Which interferons function as regulators of cell growth (autokrine interferon) has become clear as a result of recent work on the limited induction of interferons. Physiological components, such as growth factors and differentiation factors, seem to be the inducers. Therefore a distinction must be made between the viral and the nonviral inducers of the synthesis of type I interferon (compare Figure 5).

2.2.1 Viral induction of interferon

It has been shown in numerous investigations that virtually all viruses can induce interferon, but successful induction depends on the type of cell. A particular virus may be a very potent inducer in one cell line but may induce no interferon production on

Table 2 Inducers of interferon genes

Inducers of type I interferon
Viridae
Double-stranded RNA (dsRNA)
Lipopolysaccharides (LPS)
Mycoplasmas
Interleukin-1 (IL-1)
Platelet-derived growth factor (PDGF)
Colony-stimulating factor (M-CSF)
10-Carboxymethyl-9-acridanone (CMA)
Viral transactivating proteins:
– SV40 T antigen
– Adenovirus type 12 E1a/b proteins
– Hepatitis B virus X protein
Inducers of type II interferon
Antigens
Mitogens
Enterotoxins

contact with a second cell line. On the other hand, this second cell line may be able to produce interferon on contact with a different virus type.

The viral component responsible for triggering the induction signal has in many cases not been defined. Apart from its function in establishing the antiviral state of the cell, dsRNA by itself is a strong inducer of interferon (Section 1.6.1); dsRNA is a component of many viruses or an intermediate of their replication. There are recent reports that the virus-encoded proteins of some virus species act as inducers of interferon: they act as transactivators of interferon transcription (Table 2).

Isolation of virus components usually leads to weak induction of interferons; we conclude that complexes of viral proteins and nucleic acids are responsible for efficient induction of interferon. Alternatively, several viral components may act synergistically.

2.2.2 Nonviral inducers of interferon

A number of chemical and physiological substances act as inducers of interferon (Table 2). Bacterial LPS is thought to be an important natural inducer of interferon, and endogenous cytokines are also able to induce interferons.

A group of endogenous substances or processes, some known and some not known, can induce the secretion of very small amounts of IFN-γ. This kind of interferon production is called "physiologic", "spontaneous", "autokrine", or "endogenous". Certain cell growth factors or inducers of cellular differentiation are believed to be the causative agents. This seems to be a specific property of IFN-γ and has not been recognized in the case of other type I interferons (Figure 5).

2.2.3 Differential regulation of IFNs-α/β

The synthesis of individual species of type I interferons is differentially regulated. Which of the interferon species is induced depends on the type of cell and the type of inducer.

This is interesting because identical inducers mediate the synthesis of all type I interferons.

Lymphoid cells induced by Newcastle disease virus (NDV) secrete mainly IFN-α, but fibroblasts induced with the same virus secrete mainly IFN-β. The composition of the different IFN-α species depends on the inducer and the cell type. As previously mentioned, certain inducers cause the synthesis of only IFN-β. This holds not only for physiological inducers but also for substances like dsRNA (for fibroblasts) or 10-carboxymethyl-9-acridanone (for macrophages).

It is generally true, however, that viral induction leads to the concomitant induction of IFN-α and -β. A common mechanism of signal transduction seems to be operative [14, 39, 56].

2.2.4 Signal transduction

Because of the variety of interferon-inducing agents, besides the virus-induced interferon activation mechanism, we believe there are additional signal transmission pathways. We summarize the details of what is known in the following.

Activation of the secretion of interferon protein is controlled primarily at the transcriptional level. Virus-induced transcription of interferon mRNA does not require ongoing protein biosynthesis: apparently, proteins already on hand are sufficient for these processes. This leads us to conclude that proteins modified after translation mediate the initiation of interferon-specific transcription. Evidence in support of this comes from experiments with 2-aminopurine, an inhibitor of protein kinases: 2-aminopurine specifically inhibits interferon induction in several cell types [89].

2.2.5 Transcriptional control of interferon genes

In efforts to elucidate the mechanism of signal transmission for the past decade or so, investigations have been directed specifically to the DNA sequences responsible for transcriptional activation.

In these experiments, reverse genetics is used to determine the sequences of the regulatory DNA. Molecularly cloned interferon genes are transferred into heterologous cells. For example, if a human gene for interferon is introduced into mouse cells, its regulation can be followed against the background of mouse-specific interferon induction. The DNA sequences responsible for transcriptional activation can be pinpointed when reporter genes are used in combination with techniques for mutagenesis of regulatory DNA and determination of its regulatory capacity in heterologous cells.

DNA regions with regulatory capacity. DNA sequences flanking the 5′ terminus have been shown to be responsible for the virus-induced expression of several genes encoding human and murine type I interferon. When the techniques just described were used, the virus-responsive region was found to be a sequence of fewer than 100 base pairs. This region mediates the virus inducibility of gene expression by heterologous promoters.

Figure 6 is a schematic depiction of the highly homologous DNA sequences of the human and murine interferon promoters. Single sequence elements within the *interferon regulatory element* (IRE) have been resolved by detailed analysis. Within the region,

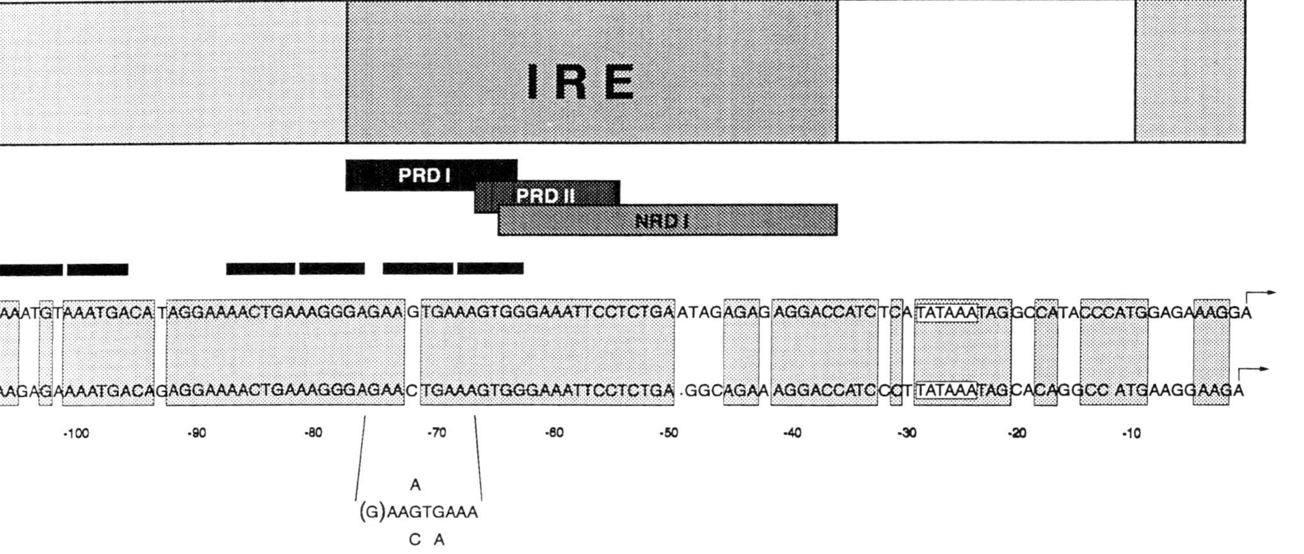

Fig. 6 Regulatory sequences of the genes of human and murine IFN-β

The DNA sequences (human, above, and murine, below) of genes coding IFN-β that react to viral stimuli are contained in a region of 100 bp flanking the 5′ terminus of the gene. Homologous regions of the two promoter sequences are located particularly within the essential domains (shaded). The minimal DNA element that can confer virus inducibility on heterologous promoters is called the *IRE*. The shaded areas to the left and right of this element make positive contributions to virus-induced regulation. *PRD I* is one of the essential positive regulatory elements. The horizontal black bars are hexanucleotide sequences with high homology to the central AAGTGA sequence of PRD I. IRF-1 and IRF-2 can bind to these hexanucleotide sequences. *PRD II* is the second virus-inducible element with positive regulatory activity. In some cell types, PRD II acts as a constitutive enhancer. A protein factor similar or identical to NF-kappaB can bind to PRD II. *NRD I* is a domain that confers negative activity only in the uninduced state of the gene. The *TATA box* is the site to which RNA polymerase II binds. The arrows mark the site of the start of transcription, and the numbering refers to the position in the sequence relative to the start of transcription *positive* and *negative regulatory domains* (PRD, NRD) have been identified: the action of PRD II is constitutive; those of both PRD I and PRD II are positive; and that of NRD I, negative. Before any induction, the activity of the constitutive element is compensated for by that of the negative domain (NRD I). Induction by a virus initiates the release of repressing properties of NRD I and also activates the inducible elements PRD I and PRD II. In addition to the DNA sequences within the IRE, DNA elements further upstream (and further downstream), located close to the site where transcription is initiated, contribute cooperatively to the process of transcriptional activation [31, 35, 36, 38, 46, 68].

DNA-binding proteins. Recent experiments, using the band-shift assays and DNase I footprinting described in Section 1.5.1, resulted in the identification of proteins that bind specifically to DNA elements within the human and murine IFN-β promoter. Two of these factors, IRF-1 and IRF-2, were further characterized by molecular cloning of their genes. Both bind specifically to a hexanucleotide sequence also within PRD I. If IRF-1 is overexpressed in heterologous cells, Type I interferons are transiently induced. One possibility is that IRF-1 activates transcription of IFN-β, and IRF-2 opposes its activity by competing for the same binding site. When IRF-1 is overexpressed for a long time in the presence of neutralizing antibodies against type I interferons, the cells die. This is confirmation of the regulatory function of the DNA-binding protein, with the added suggestion that it may have further regulatory activities.

IRF-1 binds to a sequence that is found in several interferon-inducible promoters (Table 1). We mentioned in Section 1.4.4 that three distinct proteins bind to the interferon stimulated regulatory region (ISRE). In a recent publication [63], Pine *et al.* have shown that IRF-1 is one of these factors.

A protein that binds specifically to PRD II is identical or at least very similar to *nuclear factor kappaB* (NF-kappaB), a cytoplasmic protein that is released from an inhibitor as a consequence of the action of different inducers, and that is able to bind specifically to target sequences on chromosomal DNA. The target sequences are found in immunoglobulin promoters as well as in the HIV-1 promoter. Alternatively, these promoters can also be induced by tumor promoters like TPA. The TPA-mediated induction of the IFN-β gene in certain cell types could be mediated by the same mechanism.

Not enough proteins have yet been characterized to give us a good explanation of the inducible transcription of the promoters of type I interferon. We surmise that further factors participate directly or indirectly in the binding of interferon promoter elements, and suspect that these proteins undergo additional modifications after translation. This problem will be the target of future investigations [34, 52, 58].

The studies reported above did not teach us much about the path over which signals for interferon inducers are transmitted. However, we can now guess how interferon genes could be differentially regulated. The modular components of promoters from these genes react because specific proteins bind as a result of various stimuli. The composition of the DNA-binding and signal-mediating proteins is responsible for the differential activation of an individual interferon gene. Current work is directed at closing the gap in our understanding of primary induction signals on the one hand and activation of the genes on the other; efforts are concentrated on identifying and understanding the molecular mechanisms of transcriptional activation by these DNA-binding proteins and their regulation.

Chromatin structure. Changes in the sites that are hypersensitive to DNase I (see Section 1.5.1) within the regulatory region of the human and murine IFN-β genes correlated closely in hypersensitivity with the phases of transcriptional activation. Further experiments with nucleases established that changes within the nucleosomal structure of the chromatin were responsible for a measurable increase in the susceptibility to viral induction.

When the IFN-β gene is in the uninduced state, six nucleosomes at defined positions within the region flanking the 5′ terminus of the gene display *nucleosomal phasing*. Some of these nucleosomes are lost on induction, the result being a significant change in the topology of the DNA in a 1.7-kb region upstream of the start of transcription [13, 88].

When a region of DNA 1.7 kb upstream of the site where transcription starts was examined, it became possible to guess how such local activating processes might work on the molecular level. This region, which is close to a site hypersensitive to DNase I, is protected against DNase I degradation by a bulwark of proteins. The sequence can be characterized as a *scaffold-attached region* (SAR), a DNA element responsible for anchoring chromosomal DNA to the protein matrix of the nucleus and regarded as the basis for the domain structure into which the genomes of eukaryotic cells are organized. The human IFN-β gene is part of a 14-kb domain of DNA between two SAR elements, and its murine counterpart is located within a 20-kb domain. As a result of recent findings, we are able to propose a functional definition of the SAR elements; we find that a gene in such a domain is shielded from influences from its chromosomal surroundings and displays an increased potential for transcriptional activity [12].

2.2.6 *Characteristics of interferon induction*

Shut off. In virus-treated cells, transcription is activated and interferon-specific mRNA begins to be produced. This mRNA disappears within a few hours, even in the presence of the inducers. A second induction stimulus a few hours after the first one does not lead to any increase in the synthesis of interferon-specific mRNA. The regulation of interferon-specific mRNA thus is clearly different from that of genes induced by steroid hormones; in the latter case, whenever the inducer is present, the target genes are transcriptionally active.

This *shut off regulation* of interferon synthesis is believed to be related to the induction-mediated shortening of the half-life of the mRNA: in fact, the half-life of interferon mRNA in fibroblast cells is usually on the order of 30 minutes. Alternatively, we might explain the phenomenon as a blocking of transcriptional activation, a secondary event in response to induction. The mechanism of this effect will be the subject of future research [19, 74].

Superinduction. The phenomenon of *superinduction* was observed when production of IFN-β in various mammalian cell types was being optimized: if inhibitors of translation are present during induction, the induction of mRNA is potentiated for several hours, and the mRNA does not disappear for a long time. If the inhibitors are withdrawn some hours after virus induction, the secretion of interferon protein correlates with the presence of intracellular mRNA. The inhibitors alone cannot induce interferon production.

In some cell types, such as human foreskin fibroblasts, following the protocol for superinduction yields a hundredfold more interferon mRNA, compared with the amount produced when virus alone is used as inducer. On the other hand, some cell types show no superinduction at all. The superinduction phenomenon is not restricted to IFN-β: at least some species of IFN-α can also be superinduced.

The phenomenon of superinduction may act at either or both of two possible levels, and current experiments have been designed to distinguish between the two molecular mechanisms: the rate of transcription may increase, or the half-life of the mRNA may be longer. In both cases we should be able to find repressors or degrading enzymes with extremely short half-lives, and these repressing agents would disappear immediately if translation were blocked. To date, experimenters have found evidence for the existence of both mechanisms. On the one hand, one group has reported that the half-life of interferon-specific mRNA in superinduced cells is lengthened, and, on the other, our group found that promoter activities from IFN-β and -α may be potentiated when the protocol for superinduction is followed [26, 75].

Priming. In some cell types the extent of interferon production can be stimulated by *priming*. Priming involves treating the cells with very low amounts of homologous interferon before inducing them. In living organisms, priming appears to be very important for antiviral protection. A cell with a primary virus infection first secretes interferon and subsequently produces new viruses. If the cell is pretreated with low concentrations of interferon, the production of interferon is potentiated and hyperreaction to virus infection is induced.

The effect of priming can also be detected in the absence of ongoing protein synthesis. Therefore it seems likely that no interferon-induced proteins are involved in this mechanism. Priming results in an increased transcription of interferon genes, but the target sequences for priming have not yet been verified. Since there are sequence homologies between virus and interferon-regulated promoter sequences (see Section 1.2.7 and Table 1), it seems reasonable to postulate that interferon treatment may share a molecular mechanism with virus induction [29, 79].

2.2.7 Genes co-induced with interferon

When cells are treated with viruses, a series of genes is induced, only a few of which are the genes for the interferons. Some of these genes have been identified and isolated by molecular cloning. Surprisingly enough, several of them are inducible not only by viruses and dsRNA but also by cytokines and by substances that stimulate growth and differentiation. The function of most of these genes co-induced with interferon is unknown. One guess is that they are part of the regulatory network of cytokines or substances that regulate growth and differentiation [50].

2.3 Biogenesis of type II interferons

INF-γ differs from type I interferons in being secreted only by *T* cells and macrophages. The two classes of interferons are also regulated differently. For example, when immunocompetent *T* cells are stimulated by antigen, they produce IFN-γ exclusively, but when the same cells are treated with virus, they produce only type I interferons. Our knowledge of the regulation of IFN-γ is still incomplete, particularly when compared with the research done on the regulation of genes for type I interferon. This is partly because the low efficiency of gene transfer into T cells restricts the use of reverse genetics with them. It has been successfully demonstrated, however, that transcriptional activation brings about most of the secretion of IFN-γ. Posttranscriptional events, such as the superinduction of IFN-γ, are also significant in this system [30, 37, 51, 77, 87].

2.4 Interferon induction during embryonal development

Any cell type from an adult organism can produce interferon after contact with a suitable virus, but cells from the early embryonic state cannot. This was investigated using the mouse model. All workers report that before the sixth day of embryonal development (trophoblast stage), cells are not capable of producing interferon. Cells from the inner cell mass and cells of the embryonal ectoderm subsequently acquire this ability.

It is interesting that embryonal cells after interferon treatment are not able to develop an antiviral state, despite the fact that they express interferon receptors and can activate one of the interferon-activatable enzymes (OASE). Whether this coupling is accidental or not is not known. However, it is plausible that the coupling comes about because the related regulatory sequences from genes regulated by viruses or by interferon are unable to respond to both signals.

Undifferentiated embryonic carcinoma cell lines can be compared to embryonal cells. The carcinoma cells can neither produce interferon nor establish an interferon-mediated antiviral state. If these cells in culture are treated with agents such as retinoic acid, they differentiate and simultaneously acquire both capabilities. When DNase I footprinting (Section 1.5.1) was used to study murine embryonal carcinoma cells, a protein was found that binds specifically to a DNA region upstream of the virus-inducible element of the human IFN-β promoter in embryonal cells. Subsequent experiments used DNase I mapping: DNase sensitivity in the region surrounding the murine IFN-β gene increases during embryonal development. Apparently there is a domain that becomes more susceptible during differentiation, and until this happens the gene it surrounds cannot be induced [10, 23, 49].

3 Future prospects

This chapter, like its forerunners, is a brief review of current knowledge and concepts in the molecular biology of interferons. At this writing, the techniques of reverse genetics, which are best at elucidating the terminal portion of the signal transduction pathway, are giving the most satisfactory results. Nevertheless, we did not want to limit the discussion in this chapter to the final events of gene activation. For this reason we have tried to present a variety of aspects in the framework of a more comprehensive signal transmission pathway, although at the time being the picture is incomplete. The same holds for the events occurring between the viral infection of a cell or the reception of specific parakrine signals and the activation of interferon genes. We expect that research activities in the next few years will be directed towards closing these gaps.

4 Acknowledgements

We thank W. DIRKS for participating in informative discussions, and both W. DIRKS and C. SCHARNHORST for communicating unpublished results. We appreciate the critical reading of the manuscript by H. D. HALLER (Ithaca, NY), also thank I. DORTMUND for her help with the preparation of the manuscript and H. SCHÄDLICH for her linguistic assistance.

5 References

5.1 Review articles

[1] HAUSER, H.: Interferons. In: HABENICHT, A., Ed.; *Growth Factors, Differentiation Factors and Cytokines*, Springer Verlag, Berlin (1989); pp 243–253.

[2] JACOBSEN, H.: Struktur und Wirkung von Interferonen. In: NIEDERLE, N.; VON RUSSOW, R., Eds.; *Interferone – präklinische und klinische Befunde*, Springer Verlag, Berlin (1990).

[3] LENGYEL, P.: Biochemistry of Interferons and Their Actions; *Ann. Rev. Biochem.* **51** (1982) 251–282.

[4] PESTKA, S.; LANGER, J.; ZOON, K. C.; SAMUEL, C. E.: Interferons and Their Actions; *Ann. Rev. Biochem.* **56** (1987) 727–777.

[5] ROMEO, G.; FIORUCCI, G.; ROSSI, G. B.: Interferons in Cell Growth and Development; *Trends Genet.* **5** (1989) 19–24.

[6] SEN, G. C.: Biochemical Pathways in Interferon Action; *Pharmac. Ther.* **24** (1984) 235–257.

[7] STEWART, II, W. E.: *The Interferon System*, Springer, Vienna, New York (1979).

[8] WEISSMANN, C.; WEBER, H.: The Interferon Genes; *Progr. Nucl. Acids Res. Mol. Biol.* **33** (1986) 251–300.

5.2 Specialized literature

[9] AGUET, M.; DEMBIC, Z.; MERLIN, G.: Molecular Cloning and Expression of the Interferon γ Receptor; *Cell* **55** (1988) 273–280.

[10] BARLOW, D. P.; RANDLE, B. J.; BURKE, D. C.: Interferon Synthesis in the Early Post-Implantation Mouse Embryo; *Differentiation* **27** (1984) 229–235.

[11] BENECH, P.; VIGNERON, M.; PEREZ, D.; REVEL, M.; CHEBATH, J.: Interferon-Responsive Regulatory Elements in the Promoters of the Human 2′-5′-Oligo(A)synthetase Gene.; *Mol. Cell. Biol.* **7** (1987) 4498–4504.

[12] BODE, J.; MAASS, K.: Chromatin Domain Surrounding the Human Interferon-Beta Gene as Defined by Scaffold-Attached Regions; *Biochemistry* **27** (1988) 4706–4711.

[13] BODE, J.; PUCHER, H. J.; MAASS, K: Chromatin Structure and Induction-Dependent Conformational Changes of Human Interferon-β Genes in a Mouse Host Cell; *Eur. J. Biochem.* **158** (1986) 393–401.

[14] BREHM, G.; KIRCHNER, H.: Analysis of the Interferons Induced in Mice *in Vivo* and in Macrophages *in Vitro* by Newcastle Disease Virus and by Polyinosinic-Polycytidylic Acid; *J. Interferon Res.* **6** (1986) 21–28.

[15] BURWIN, S. J.; JONES, A. L.: The Association of Polypeptide Hormones and Growth Factors with the Nuclei of Target Cells; *Trends Biochem. Sci.* **12**(4) (1987) 159–162.

[16] BUUS, S.; SETTE, A.; COLON, S. M.; MILES, C.; GREY, H. M.: The Relation between Major Histocompatibility Complex (MHC) Restriction and the Capacity of Ia To Bind Immunogenic Peptides; *Science* **235** (1987) 1353–1358.

[17] CALDERON, J.; SHEEHAN, K. C. F.; CHANCE, C.; THOMAS, M. L.; SCHREIBER, R. D.: Purification and Characterization of the Human Interferon-γ Receptor from Placenta; *PNAS USA* **85** (1988) 4837–4841.

[18] CARLIN, J. M.; OZAKI, Y.; BYRNE, G. I.; BROWN, R. R.; BORDEN, E. C.: Interferons and Indoleamine-2,3-dioxygenase: Role in Antimicrobial and Antitumor Effects; *Experientia* **45**(6) (1989) 535–541.

[19] CAVALIERI, R. L.; HAVELL, E. A.; VILCEK, J.; PESTKA, S.: Induction and Decay of Human Fibroblast Interferon mRNA; *PNAS USA* **74** (1977) 4415–4419.

[20] CELADA, A.; KLEMSZ, M. J.; MAKI, R. A.: Interferon-Gamma Activates Multiple Pathways To Regulate the Expression of the Genes for Major Histocompatibility Class II I-ABeta, Tumor Necrosis Factor and Complement Component C3 in Mouse Macrophages; *Eur. J. Immunol.* **19**(6) (1989) 1103–1109.

[21] CERNESCU, C.; CONSTANTINESCU, S. N.; BALTA, F.; POPESCU, L. M.; CAJAL, N.: Interferon-Induced Antiviral State Is Inhibited by Neomycin and Mimicked by Diacylglycerols; *Biochem. Biophys. Res. Comm.* **151** (1988) 402–407.

[22] COHEN, B.; PERETZ, D.; VAIMAN, D.; BENECH, P.; CHEBATH, J.: Enhancer-Like Interferon Responsive Sequences of the Human and Murine (2′-5′)Oligoadenylate Synthetase Gene Promoters; *EMBO J.* **7** (1988) 1411–1419.

[23] COVENEY, J.; SCOTT, G.; KING, R.; BURKE, D. C.; SKUP, D.: Changes in the Conformation of the Interferon Beta Gene during Differentiation and Induction; *Biochem. Biophys. Res. Comm.* **121**(1) (1984) 290–296.

[24] DALE, T. C.; ROSEN, J. M.; GUILLE, M. J.; LEWIN, A. R.; PORTER, A. C. G.; KERR, I. M.; STARK, G. R.: Overlapping Sites for Constitutive and Induced DNA Binding Factors Involved in Interferon-Stimulated Transcription; *EMBO J.* **8**(3) (1989) 831–839.

[25] DALE, T. C.; IMAM, A. M. A.; KERR, I. M.; STARK, G. R.: Rapid Activation by Interferon α of a Latent DNA-Binding Protein Present in the Cytoplasm of Untreated Cells; *PNAS USA* **86** (1989) 1203–1207.

[26] Dinter, H.; Hauser, H.: Superinduction of the Human Interferon-Beta Promoter; *EMBO J.* **6**(3) (1987) 599–604.

[27] Dorn, A.; Durand, B.; Marfing, C.; Le Meur, M.; Benoist, C.; Mathis, D.: Conserved Major Histocompatibility Complex Class II Boxes – X and Y – Are Transcriptional Control Elements and Specifically Bind Nuclear Proteins; *PNAS USA* **84** (1987) 6249–6253.

[28] Edery, I.; Petryshyn, R.; Sonenberg, N.: Activation of Double-Stranded RNA-Dependent Kinase (dsI) by the TAR Region of HIV-1 mRNA: A Novel Translational Control Mechanism; *Cell* **56** (1989) 303–312.

[29] Enoch, T.; Zinn, K.; Maniatis, T.: Activation of the Human Beta-Interferon Gene Requires an Interferon-Inducible Factor; *Mol. Cell Biol.* **6**(3) (1986) 801–810.

[30] Epstein, L. B.: The Special Significance of Interferon-Gamma. In: Vilcek, J.; De-Mayer, E., Eds.; *Interferons and the Immune System*, Elsevier, Amsterdam (1984); Vol. 2, pp 185–219.

[31] Fan, C. M.; Maniatis, T.: Two Different Virus-Inducible Elements Are Required for Human β-Interferon Gene Regulation; *EMBO J.* **8**(1) (1989) 101–110.

[32] Fan, X.-D.; Goldbern, M.; Bloom, B. R.: Interferon-γ-Induced Transcriptional Activation Is Mediated by Protein Kinase C; *PNAS USA* **95** (1988) 5122–5125.

[33] Flavell, R. A.; Allen, H.; Burkly, L. C.; Sherman, D. H.; Waneck, G. L.; Widera, G.: Molecular Biology of the H-2 Histocompatibility Complex; *Science* **233** (1986) 437–443.

[34] Fujita, T.; Kumura, Y.; Miyamoto, M.; Barsoumian, E. L.; Taniguchi, T.: Induction of Endogenous IFN-α and IFN-β Genes by a Regulatory Transcription Factor, IRF-1; *Nature (London)* **337** (1989) 270–272.

[35] Goodbourn, S.; Zinn, K.; Maniatis, T.: Human Beta-Interferon Gene Expression Is Regulated by an Inducible Enhancer Element; *Cell* **41**(2) (1985) 509–520.

[36] Goodbourn, S.; Burstein, H.; Maniatis, T.: The Human Beta-Interferon Gene Enhancer Is Under Negative Control; *Cell* **45**(4) (1986) 601–610.

[37] Hardy, K. J.; Pterlin, B. M.; Atchison, R. E.; Stobo, J. D.: Regulation of Expression of the Human Interferon Gamma Gene; *PNAS USA* **82**(23) (1985) 8173–8177.

[38] Hauser, H.; Gross, G.; Bruns, W.; Hochkeppel, H. K.; Mayr, U.; Collins, J.: Inducibility of Human β-Interferon Gene in Mouse L-Cell Clones; *Nature (London)* **297** (1982) 650–654.

[39] Hiscott, J.; Cantell, K.; Weissmann, C.: Differential Expression of Human Interferon Genes; *Nucleic Acids Res.* **12**(9) (1984) 3727–3746.

[40] Hug, H.; Costas, M.; Staeheli, P.; Aebi, M.; Weissmann, C.: Organization of the Murine Mx Gene and Characterization of Its Interferon- and Virus-Inducible Promoter; *Mol. Cell. Biol.* **8** (1988) 3065–3079.

[41] Israel, A.; Kimura, A.; Kieran, M.; Yano, O.; Kanellopoulos, J.; Bail, O. L.; Kourilski, P.: A Common Positive Trans-Acting Factor Binds to Enhancer Sequences in the Promoters of Mouse H-2 and β2-Microglobulin Genes; *PNAS USA* **84** (1987) 2653–2657.

[42] Kaufman, R. J.; Murtha, P.: Translational Control Mediated by Eukaryotic Initiation Factor-2 Is Restricted to Specific mRNAs in Transfected Cells; *Mol. Cell. Biology* **1987**(4) (1987) 1568–1571.

[43] Kelso, A.; Gough, N. M.: Coexpression of Granulocyte-Macrophage Colony-Stimulating Factor, Gamma Interferon, and Interleukins 3 and 4 Is Random in Murine Alloreactive T-Lymphocyte Clones; *PNAS USA* **85**(23) (1988) 9189–9193.

[44] Klar, D.; Hämmerling, G. J.: Induction of Assembly of MHC Class I Heavy Chains with β2-Microglobulin by Interferon-γ; *EMBO J.* **8**(2) (1989) 475–481.

[45] KOHASE, M.; MAY, L. T.; TAMM, I.; VILCEK, J.; SEHGAL, P. B.: Cytokine Network in Human Diploid Fibroblasts: Interactions of Beta-Interferons, Tumor Necrosis Factor, Platelet-Derived Growth Factor, and Interleukin-1; *Mol. Cell. Biol.* **7** (1987) 273–280.

[46] KUHL, D.; DE LA FUENTE, J.; CHATURVEDI, M.; PARIMOO, S.; RYALS, J.; MEYER, F.; WEISSMANN, C.: Reversible Silencing of Enhancers by Sequences Derived from the Human IFN-Alpha Promoter; *Cell* **50** (1987) 1057–1069.

[47] KUSHNARYOV, V. M.; MACDONALD, H. S.; LEMENSE, G. P.; DEBRUIN, J.; SEDMAK, J. J.; GROSSBERG, S. E.: Quantitative Analysis of Mouse Interferon-Beta Receptor-Mediated Endocytosis and Nuclear Entry; *Cytobios* **53** (1988) 185–197.

[48] KUSHNARYOV, V. M.; MACDONALD, H. S.; SEDMAK, J. J.; GROSSBERG, S. E.: The Cellular Internalization of Recombinant Gamma Interferon Differs from That of Natural Interferon Gamma; *Biochem. Biophys. Res. Comm.* **157** (1988) 109–114.

[49] LAMBROPOULOS, A. F.; KOLIAIS, S. I.: Induction of 2′,5′ Oligo(A) Synthetase in Vero Cells; *Microbiologica* **12** (1989) 105–109.

[50] LAMMERS, R.; GROSS, G.; MAYR, U.; COLLINS, J.: Alternative Mechanisms for Gene Activation Induced by Poly(rI):poly(rC) and Newcastle Disease Virus; *Eur. J. Biochem.* **178**(1) (1988) 93–99.

[51] LEBENDIKER, M. A.; TAL, C.; SAYAR, D.; PILO, S.; EILON, A.; BANAI, Y.; KAEMPFER, R.: Superinduction of the Human Gene Encoding Immune Interferon; *EMBO J.* **6**(3) (1987) 585–589.

[52] LENARDO, M. J.; FAN, C.-M.; MANIATIS, T.; BALTIMORE, D.: The Involvement of NF-kappaB in β-Interferon Gene Regulation Reveals Its Role As Widely Inducible Mediator of Signal Transduction; *Cell* **57** (1989) 287–294.

[53] LEVY, D. E.; KESSLER, D. S.; PINE, R.; REICH, N.; DARNELL, JR., J. E.: Interferon-Induced Nuclear Factors That Bind a Shared Promoter Element Correlate with Positive and Negative Transcriptional Control; *Genes and Dev.* **2** (1988) 383–393.

[54] LEVY, D. E.; KESSLER, D. S.; PINE, R.; DARNELL, JR., J. E: Cytoplasmic Activation of ISGF3, the Positive Regulator of Interferon-Alpha-Stimulated Transcription, Reconstituted *in Vitro*; *Genes and Dev.* **3** (1989) 1362–1371.

[55] LONDON, I. M.; LEVIN, D. H.; MATTS, R. L.; THOMAS, N. S. B.; PETRYSHYN, R.; CHEN, J.-J.: Regulation of Protein Synthesis. In: BOYER, P. D.; KREBS, E. G., Eds.; *The Enzymes*, Vol. 18, Academic Press, New York (1987); Chapter 12, pp 359–380.

[56] MANIATIS, S.; GOODBOURN, T.: Overlapping Positive and Negative Regulatory Domains of the Human β-Interferon Gene; *PNAS USA* **85** (1988) 1447–1451.

[57] MASCHEK, U.; PÜLM, W.; HÄMMERLING, G. J.: Altered Regulation of MHC Class I Genes in Different Tumor Cell Lines Is Reflected by Distinct Sets of DNAse I Hypersensitive Sites; *EMBO J.* **8** (1989) 2297–2304.

[58] MIYAMOTO, M.; FUJITA, T.; KIMURA, Y.; MARUYAMA, M.; HARADA, H.; SUDO, Y.; MIYATA, T.; TANIGUCHI, T.: Regulated Expression of a Gene Encoding a Nuclear Factor, IRF-1, That Specifically Binds to the IFN-β Gene Regulatory Elements; *Cell* **54** (1988) 903–913.

[59] MURAKAMI, K.; ROUTTENBERG, A.: Direct Activation of Purified Protein Kinase C by Unsaturated Fatty Acids (Oleate and Arachidonate) in the Absence of Phospholipids and Ca^{++}; *FEBS Lett.* **192** (1985) 189–193.

[60] NISHIZUKA, Y.: Phospholipid Degradation and Signal Translation for Protein Phosphorylation; *Trends Biochem. Sci.* **1983**(1), (1983) 13–16.

[61] OPDENAKKER, G.; CABEZA-ARVELAIZ, Y.; VAN DAMME, J.: Interaction of Interferon with Other Cytokines; *Experientia* **45**(6) (1989) 513–520.

[62] PETRYSHYN, R.; CHEN, J.-J.; LONDON, I. M.: Detection of Activated Double-Stranded RNA-Dependent Protein Kinase in 3T3–F442A Cells; *PNAS USA* **85** (1988) 1427–1431.

[63] PINE, R.; DECKER, T.; KESSLER, D. S.; LEVY, D. E.; DARNELL, JR., J. E.: Purification and Cloning of Interferon-Stimulated Gene Factor 2 (ISGF2): ISGF2 (IRF-1) Can Bind to the Promoters of Both Beta Interferon- and Interferon-Stimulated Genes but Is Not a Primary Transcriptional Activator of Either; *Mol. Cell. Biol.* **10** (1990) 2448–2457.

[64] PREMECZ, G.; MARKOVITS, A.; FÖLDES, I.: Antiviral Activity of Phorbol Myristate Acetate and Possible Relationships with Interferon Action; *FEBS Lett.* **180** (1985) 300–302.

[65] PREMECZ, G.; MARKOVITS, A.; BAGI, G.; FARKAS, T.; FÖLDES, I.: Phospholipase C and Phospholipase A2 Are Involved in the Antiviral Activity of Human Interferon-Alpha; *FEBS Lett.* **249**(2) (1989) 257–260.

[66] REVEL, M.; CHEBATH, J.: Interferon-Activated Genes; *Trends Biochem. Sci.* **11** (1986) 166–170.

[67] ROCHETTE-EGLY, C.; TOVEY, M. G.: Cyclic GMP Levels in Interferon Treated Cells; *Antiviral Res.* **5** (1985) 127–135.

[68] RYALS, J.; DIERKS, P.; RAGG, H.; WEISSMANN, C.: A 46–Nucleotide Promoter Segment from an IFN-Alpha Gene Renders an Unrelated Promoter Inducible by Virus; *Cell* **41**(2) (1985) 497–507.

[69] SARGENT, C. A.; DUNHAM, I.; CAMPBELL, R. D.: Identification of Multiple HTF-Island Associated Genes in the Human Major Histocompatibility Complex Class III Region; *EMBO J.* **8**(8) (1989) 2305–2312.

[70] SARKAR, F. H.; GUPTA, S. L.: Interferon Receptor Interaction; *Eur. J. Biochem.* **140** (1984) 461–467.

[71] SCHMIDT, A.; CHERNAJOVSKY, V.; SCHULMAN, L.; FEDERMAN, P.; BERISSI, H.; REVEL, M.: An Interferon-Induced Phosphodiesterase Degrading (2′–5′)Oligoadenylate and the C-C-A Terminus of tRNA; *PNAS USA* **76** (1979) 4788–4792.

[72] SCHNECK, J.; RAGER-ZISMAN, B.; ROSEN, O. M.; BLOOM, B. R.: Genetic Analysis of the Role of cAMP in Mediating Effects of Interferon; *PNAS USA* **79** (1982) 1879–1883.

[73] SCHRAMM, M.; SELINGER, Z.: Message Transmission: Receptor Controlled Adenylate Cyclase System; *Science* **225** (1984) 1350–1356.

[74] SEHGAL, P. B.; DOBBERSTEIN, B.; TAMM, I.: Interferon Messenger RNA Content of Human Fibroblasts during Induction, Shutoff and Superinduction of Interferon Production; *PNAS USA* **74** (1977) 3409–3413.

[75] SEHGAL, P. B.; LYLES, D. S.; TAMM, I.: Superinduction of Human Fibroblast Interferon Production: Further Evidence for Increased Stability of Interferon mRNA; *Virology* **89** (1978) 186–198.

[76] SPANJAARD, R. A.; VAN HIMBERGEN, J. A. J.; VAN DUIN, J.: The Cysteines in Position 1 and 86 of Rat Interferon-Alpha1 Are Indispensable for Antiviral Activity; *FEBS Lett.* **249**(2) (1989) 186–188.

[77] STORCH, E.; KIRCHNER, H.; BREHM, G.; HÜLLER, K.; MARCUCCI, F.: Production of Interferon-Beta by Murine T-Cell Lines Induced by 10-Carboxymethyl-9-acridanone; *Scand. J. Immunol.* **23** (1986) 195–199.

[78] STRUNK, R. C.; SESSIONS COLE, F.; PERLMUTTER, D. H.; COLTEN, H. R.: γ-Interferon Increases Expression of Class III Complement Genes C2 and Factor B in Human Monocytes and in Murine Fibroblasts Transfected with Human C2 and Factor B Genes; *J. Biol. Chem.* **260** (1985) 15280–15285.

[79] STUART, II, W. E.; GOSSER, L. B.; LOCKART, JR., R. Z.: Priming: A Nonantiviral Function of Interferon; *J. Virol.* **7** (1971) 792–801.

[80] VALENZUELA, D.; WEBER, H.; WEISSMANN, C.: Is Sequence Conservation in Interferons Due to Selection for Functional Proteins; *Nature (London)* **313** (1985) 698–700.

[81] WAGNER, K. G.: Der Inositstoffwechsel und seine Bedeutung für die zelluläre Regulation. In: PRÄVE, P.; SCHLINGMANN, M.; CRUEGER, W.; ESSER, K.; THAUER, R.; WAGNER,

F., Eds.; *Jahrbuch Biotechnologie Band 2*, Carl Hanser Verlag, München, Wien (1988/89); pp 5–28 (in German). WAGNER, K.G.: The Metabolism of Inositol and Its Significance in Cellular Regulation. In: FINN, R.K.; PRÄVE, P.; SCHLINGMANN, M.; CRUEGER, W.; ESSER, K.; THAUER, R.; WAGNER, F., Eds.; *Biotechnology Focus 2*, Hanser Publishers, Munich, Vienna, New York, (1988/89); pp 5–26 (in English).

[82] WALTHER, Z.; MAY, L. T.; SEHGAL, P. B.: Transcriptional Regulation of the Interferon-β2/B Cell Differentiation Factor BSF-2/Hepatocyte-Stimulating Factor Gene in Human Fibroblasts by Other Cytokines; *J. Immunol.* **140** (1988) 974–977.

[83] WATHELET, M. G.; CLAUSS, I. M.; CONTENT, J.; HUEZ, G.: Regulation of Two Interferon Inducible Human Genes by Interferon, Poly(rI):(rC) and Viruses; *Eur. J. Biochem.* **174** (1988) 323–329.

[84] WEBER, H.; VALENZUELA, D.; LUJBER, G.; GUBLER, M.; WEISSMANN, C.: Single Amino Acid Changes That Render Human IFN-Alpha 2 Biologically Active on Mouse Cells; *EMBO J.* **6** (1987) 591–598.

[85] WEIL, J.; EPSTEIN, C. J.; EPSTEIN, L. B.: A Unique Set of Polypeptides Is Induced by γ-Interferon in Addition to Those Induced in Common with α- and β-Interferons; *Nature (London)* **301** (1983) 437–439.

[86] YAP, W. H.; TEO, T. S.; TAN, Y. H.: An Early Event in the Interferon-Induced Transmembrane Signaling Process; *Science* **234** (1986) 355–358.

[87] YOUNG, H. A.; VARESIO, L.; HWU, P.: Posttranscriptional Control of Human Gamma Interferon Gene Expression in Transfected Mouse Fibroblast; *Mol. Cell. Biol.* **6** (1986) 2253–2256.

[88] ZINN, K.; MANIATIS, T.: Detection of Factors That Interact with Human Beta-Interferon Regulatory Region *in Vivo* by DNAse I Footprinting; *Cell* **45**(4) (1986) 611–618.

[89] ZINN, K.; KELLER, A.; WHITTEMORE, L. A.; MANIATIS, T.: 2-Aminopurine Selectively Inhibits the Induction of Beta-Interferon, c-fos, and c-myc Gene Expression; *Science* **240** (1988) 210–213.

Enzymes in Carbohydrate Synthesis

by R. Stiller and J. Thiem

Contents

Dipl.-Chem. Regine Stiller and Prof. Dr. Joachim Thiem,
Institut für Organische Chemie und Biochemie der Universität Hamburg,
Martin-Luther-King-Platz 6,
D-2000 Hamburg 13, Fed. Rep. Germany

1 Introduction

For many years we have thought of carbohydrates in terms of their principal functions in nature, as structural materials (cellulose) or energy stores (glycogen, starch), and have restricted classical carbohydrate chemistry predominantly to the chemistry of monosaccharides. Over the past 30 years we have discovered the biological relevance of heterooligosaccharides in glycoproteins [1] and glycolipids [2], and consequently have accorded the synthesis of these structures increasing significance.

Oligosaccharides are an intimate part of the membrane-bound proteins on cell surfaces, and constitute the recognition sites for hormones, antibodies, toxins, viruses, and bacteria [3–5]. They form the molecular basis for blood group properties [6] and have key functions in cell recognition, cell growth, and cell differentiation [7]. Furthermore, certain heterooligosaccharides are found only in tumor tissues and not in normal cells [8,9].

All these different functions of heterooligosaccharides are ascribable to the enormous structural variation possible because of the polyfunctionality of sugars. Thus, biological information is inherent not only in the sequence but also in the type of connection, configuration, and branching. This very variety makes the straightforward synthesis of a defined oligosaccharide a difficult project.

Even though in recent years some elegant methods for glycosylation have been developed and older methods have been improved [10–13], the task of synthesizing molecules with regio- and stereospecific interglycosidic linkages remains a demanding one if classical chemical reactions are used.

In sum, making each specific molecule involves special chemical techniques incorporating many reaction steps and protective groups; very seldom can an approach that works with monosaccharides be transferred to analogous syntheses of oligosaccharides. The synthesis of a trisaccharide may be a superlative achievement, and there are impressive examples in which oligosaccharides with even more than three sugars have been made, but all these preparations are time-consuming and the overall yields are usually low. As most of the heterooligosaccharides of interest are produced in nature by enzymes, it is entirely plausible to use the synthetic potential of enzymes *in vitro* too.

2 Use of enzymes in organic chemistry

Enzymes accelerate chemical reactions without influencing the thermodynamic equilibrium. Since most reactions involving enzymes are reversible, by selecting appropriate conditions one can drive the reaction in the desired direction. Enzymes are active under mild conditions, in aqueous solution at temperatures near 37 °C and physiological values of pH, so even labile substances can be reacted. Enzymic reactions are usually highly selective with regard to functional groups (*chemoselectivity*), reaction sites (*regioselectivity*), and stereochemistry (*stereoselectivity*).

Therefore, enzymes are ideal candidates for syntheses, whether the reactions be of, at, or with polyfunctional substances such as carbohydrates. For instance, because enzymes are regio- and stereoselective, the experimenter can dispense with time-consuming procedures involving protective group chemistry. Enzymes have also been used with great success in other areas of preparative organic chemistry; witness the numerous reviews [14–21]. Nevertheless, numerous difficulties forestall the widespread use of enzymes. Because they are highly substrate-specific, many enzymes catalyze only one reaction. Unfortunately, only about 300 enzymes are commercially available. Some of them are cheap (250,000 units of lipase from *Candida cylindracea* from Sigma costs US $8.60 in 1991), but others are extremely expensive (0.1 unit of 2,6-α-sialyltransferase from rat liver, also from Sigma, costs US $124.85 in 1991). Therefore, many enzymes have to be isolated by the synthetic chemists themselves, who must be familiar with many biochemical techniques [22].

Enzymes are complex biomolecules, often unstable under the usual chemical reaction conditions and generally markedly sensitive to variations in the pH, temperature, or ionic strength of solutions. Unwanted contamination by bacteria or by proteases can be held in check to only a limited extent by antibacterial agents or protease inhibitors. All the factors that affect the three-dimensional structure of proteins, such as organic solvents or oxygen, lower the catalytic activity of enzymes.

Immobilizing enzymes can lessen the impact of many of these drawbacks [23–27]. To immobilize an enzyme requires that the amino functions of the enzyme be covalently bonded to functional groups of a solid support, such as synthetic polymers based on polyacrylamide [for example PAN, poly(acrylamide-*co*-*N*-acryloxysuccinimide)] [28], inorganic material such as silica gel [29], polysaccharides such as CNBr-activated SEPHAROSE, or a variety of other possibilities.

The technique of membrane-enclosed enzymatic catalysis (MEEC), a procedure in which the enzyme is contained in a dialysis bag, was recently suggested for chemoenzymatic reactions [30]. This approach, well known to biochemists, was successfully applied to some complex reactions [30–32]. If enzymes are immobilized, not only are they

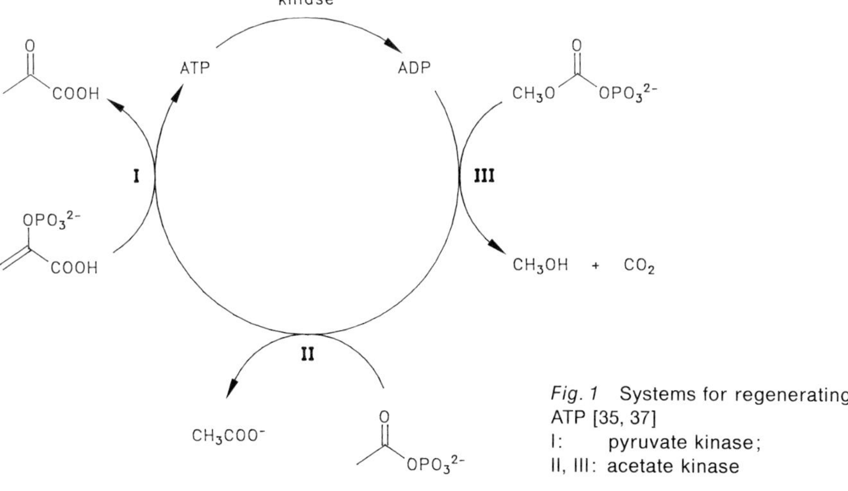

Fig. 1 Systems for regenerating ATP [35, 37]
I: pyruvate kinase;
II, III: acetate kinase

often more stable, but also they can be easily separated from reaction mixtures and can be used several times, an important consideration if the enzymes are expensive and hard to obtain. Enzymatic reactions become possible even in organic solvents [33] and aqueous two-phase systems [34].

One aggravating disadvantage is that expensive cofactors must frequently be supplied in stoichiometric quantities. In response, methods have been developed to regenerate the cofactors *in situ* so that only catalytic amounts need be added.

The regeneration of adenosine 5′-triphosphate (ATP), the ubiquitous molecule important in carbohydrate metabolism as in all biology, has been well worked out (Figure 1). The feedback inhibitor ADP is usually removed with the systems phosphoenolpyruvate (PEP)/pyruvate kinase (EC 2.7.1.40) (I); acetylphosphate (AcP)/acetate kinase (EC 2.7.2.1) (II); and methoxycarbonyl phosphate (MCP)/acetate kinase (III) [35–37].

In this chapter we describe some aspects of the use of enzymes in carbohydrate chemistry.

3 Functionalization and synthesis of monosaccharides

3.1 Acylations

Some enzymes retain their activity even in organic solvents, and reaction routes can be influenced by the choice of solvent [33]. For example, lipases in organic solvents esterify saccharides nearly quantitatively, but in aqueous solutions they hydrolyze esters. (This is discussed in detail in the chapter by Erdmann *et al.* in this book.)

Therisod and Klibanov [38] used a lipase from porcine pancreas (PPL, triacylglycerol lipase; EC 3.1.1.3) in pyridine to acylate monosaccharides selectively at the primary –OH group on C-6 with the reagent trichloroethyl butyrate (Figure 2). The reaction has 82–100% regioselectivity and yields of 90%.

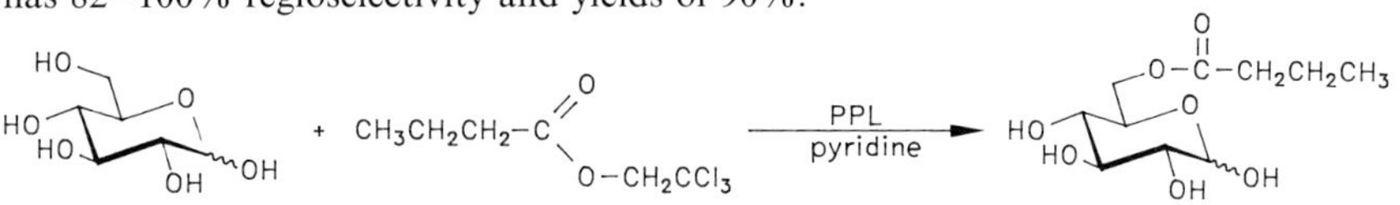

$$\text{monosaccharide} + CH_3CH_2CH_2-C(=O)-O-CH_2CCl_3 \xrightarrow[\text{pyridine}]{\text{PPL}} \text{6-}O\text{-}(C(=O)-CH_2CH_2CH_3)\text{-monosaccharide}$$

Fig. 2 Esterification of monosaccharides using porcine pancreatic lipase (PPL, triacylglycerol lipase) [38]

Monosaccharides acylated at C-6 can be converted further to diacylated products [39]. Depending on the kind of lipase and the choice of solvent, either 2,6- or 3,6-diesters were synthesized. The reagents were trichloroethyl carboxylates, and the solvent was methylene chloride or tetrahydrofuran instead of pyridine.

Kloosterman *et al.* [40] studied the regioselectivity of various lipases. The model substrate was 1,2-*O*-isopropylidene-α-D-xylofuranose, which has one primary and one secondary –OH group. They concluded that the primary hydroxy function was preferentially acylated.

The trouble with lipases is that while their activity increases with decreasing solvent polarity, the solubility of sugars decreases. Riva *et al.* (in Klibanov's laboratory) [41] reported that other hydrolases, such as proteases, an example of which is subtilisin from *Bacillus subtilis* (EC 3.4.21.14), can also be used to acylate mono- and oligosaccharides. Subtilisin, which is more active in polar solvents such as dimethylformamide than are lipases, can be used to acylate cellobiose, maltose, and sucrose.

Lipases also make possible the selective deacylation of protected monosaccharides. The back reaction is favored in aqueous solutions with their huge excess of the nucleophile water. Sweers and Wong [42] were the first to use this in the deacylation of methyl 2,3,4,6-tetra-*O*-acyl-α-D-hexopyranosides with a lipase from *Candida cylindracea*. Here, deacylation is selective for the substituent on the primary C-6, with yields of 80–90%.

Kloosterman *et al.* [43, 44] investigated the regioselectivity of different lipases for deacylation, as well as the influence of the nature and chain length of the acyl groups. They found that primary –OH groups are deacylated first. They were able to use 1,6-anhydro-β-D-glucopyranoses as model substrates to study the regioselectivity of various lipases with respect to secondary –OH groups [44].

In a recent study, lipase-catalyzed acylation and deacylation reactions were studied in furanosides also [45].

3.2 *De novo* synthesis of monosaccharides

Enzymes have turned out to be very helpful in the *de novo* synthesis of certain monosaccharides. Generally, two chiral carbonyl compounds are combined in an aldol-type reaction.

Fructose 1,6-diphosphate aldolase (fructose-diphosphate aldolase, FDP aldolase; EC 4.1.2.13) from rabbit muscle is often used. In carbohydrate metabolism, this enzyme catalyzes the condensation of dihydroxyacetone phosphate (DHAP) and glyceraldehyde 3-phosphate (GA-3-P) to fructose 1,6-diphosphate. Since this reaction is reversible, the enzyme also catalyzes the cleavage of fructose 1,6-diphosphate under appropriate reaction conditions. FDP aldolase is highly specific for DHAP [46], but it accepts many aldehydes, even unphosphorylated ones, so the aldehyde component can be varied in order to design a particular sugar.

To date more than fifty aldehydes, and even alkoxy and amino derivatives, have taken part in the enzymatic conversions [47]. On the other hand, aldehydes that are sterically hindered, aromatic, or α, β-unsaturated will not serve as substrates. Aldehydes that are phosphorylated or that have electronegative substituents in the α position are converted faster than their unphosphorylated or unsubstituted counterparts. Aldehydes for these reactions must be of highest enantiomeric purity, because otherwise mixtures of fructose and sorbose derivatives are formed, and the separation of these diastereoisomers is rather difficult. After the ketose phosphate products have the phosphate removed in an acid cleavage, the ketose can be converted to the corresponding aldose by D-xylose ketol-isomerase (EC 5.3.1.5). This enzyme accepts a broad range of substrates, but only ketoses of the fructose and not the sorbose type.

FDP aldolase is a rather stable enzyme, which can be used in free or immobilized form as well as in the MEEC technique. Moreover, it has been used in the presence of organic cosolvents such as ethanol or dimethylsulfoxide [47].

In 1983 Wong and Whitesides synthesized ^{13}C-labeled fructose and glucose in millimolar quantities using FDP aldolase [48]. GA-3-P, which was synthesized from glyceraldehyde by glycerol kinase (EC 2.7.1.30), was converted to DHAP with triose-phosphate isomerase (EC 5.3.1.1) (Figure 3). DHAP was then condensed with the remaining GA-3-P to fructose 1,6-diphosphate, with FDP aldolase. The product was hydrolyzed with acid to fructose 6-phosphate; the phosphate was removed with acid phosphatase (EC 3.1.3.2), or the fructose 6-phosphate was converted to glucose 6-phosphate by glucose-6-phosphate isomerase (EC 5.3.1.9) and the phosphate removed by acid phosphatase.

If lactaldehyde is condensed with DHAP, the products are 6-deoxyhexoses [49]. Other products such as 3-deoxyglucose or 6-*O*-methyl-D-glucose are obtained if the aldehyde component is varied [50]. Even unusual sugars with eight or nine carbon

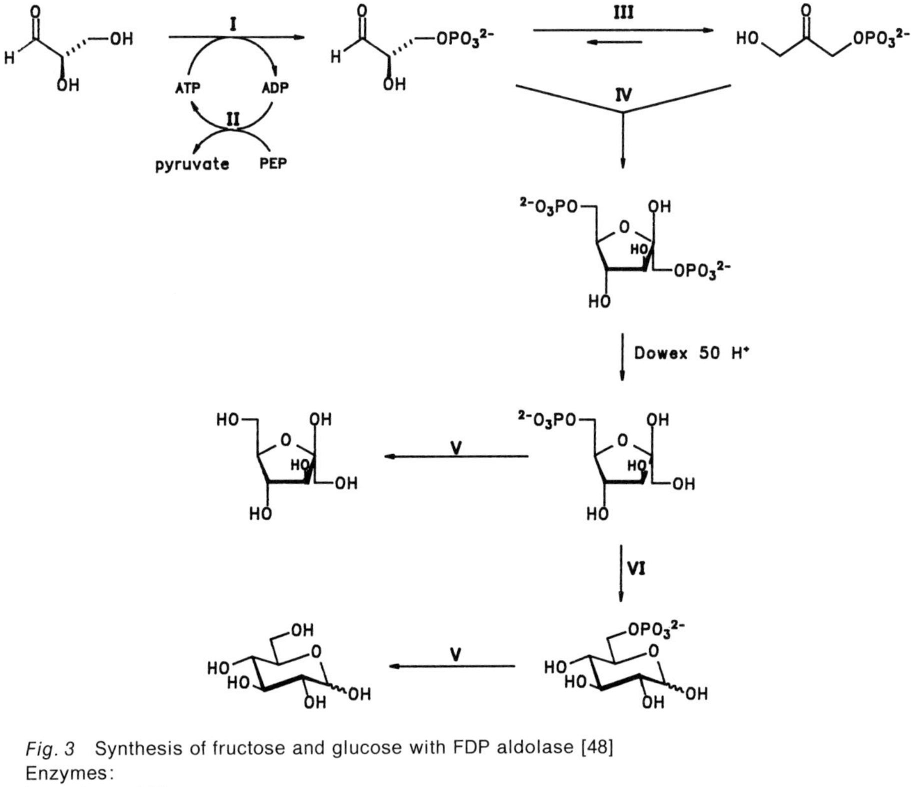

Fig. 3 Synthesis of fructose and glucose with FDP aldolase [48]
Enzymes:
I = glycerol kinase;
II = pyruvate kinase;
III = triose-phosphate isomerase;
IV = FDP aldolase;
V = acid phosphatase;
VI = glucose-6-phosphate isomerase

atoms can be synthesized if DHAP is condensed with pentose 5-phosphates or hexose 6-phosphates. For example, D-*glycero*-D-*altro*octulose 1,8-diphosphate was synthesized in 80% yield starting from ribulose 5-phosphate [51]. This is a way of making a wide variety of unusual sugars [52]. 2-Deoxy- and 2-amino-2-deoxy- derivatives of phosphorylated aldoses can also be substrates of these enzymes. Deoxyfluoro sugars, which are potential enzyme inhibitors of pharmaceutical relevance, were made by starting with fluorinated aldehydes; for instance, 3-fluoro-2-hydroxypropanal was used in the synthesis of the antitumor agent 6-deoxy-6-fluoroglucose [53].

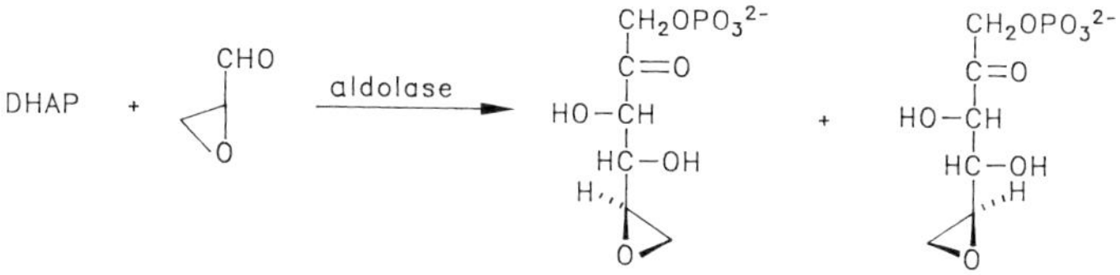

Fig. 4 Synthesis of hexose epoxides with FDP aldolase [54]

As early as 1973, O'CONNELL and ROSE [54] took advantage of the synthetic potential of FDP aldolase to synthesize hexose epoxides (Figure 4). These compounds are used as inhibitors in studies of the mechanism of isomerases.

Recently, even disaccharides have been made this way [55]. Condensing glycolaldehyde glycosides with DHAP gave glycosyl-(1→5)-D-xylulose 1-phosphate, which was dephosphorylated with acid phosphatase and isomerized with xylose isomerase [D-xylose ketol-isomerase (EC 5.3.1.5)] ; the final product was a disaccharide containing D-xylose (Figure 5).

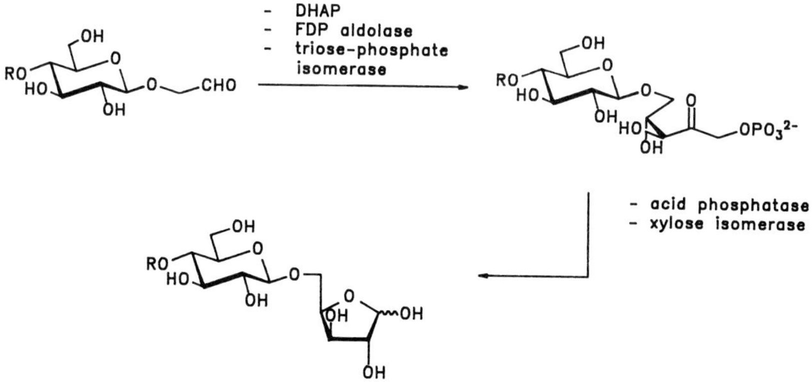

Fig. 5 Synthesis of disaccharides with FDP aldolase [55]

When DHAP was condensed with 3-azido-2-hydroxypropanal and the products catalytically hydrogenated, the glucosidase inhibitors 1-deoxynojirimycin and 1-deoxymannojirimycin were produced [56] (Figure 6).

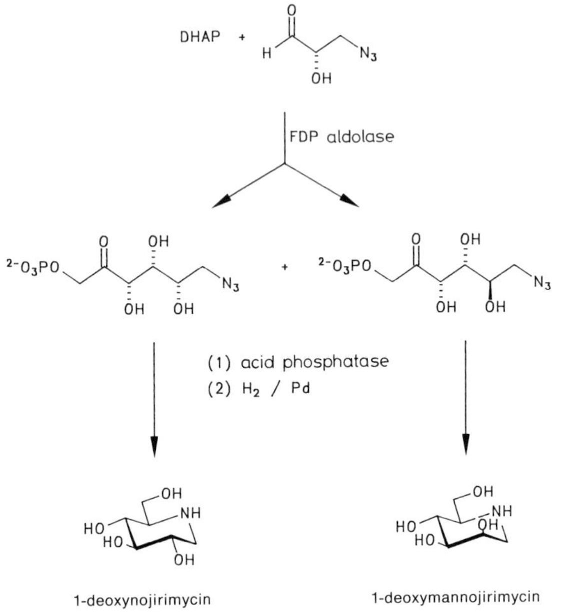

DHAP +

FDP aldolase

(1) acid phosphatase
(2) H_2 / Pd

1-deoxynojirimycin

1-deoxymannojirimycin

Fig. 6 Synthesis of 1-deoxynojirimycin and 1-deoxymannojirimycin [56]

These syntheses all have the disadvantage of requiring the expensive substrate DHAP in equimolar amounts. DHAP could be synthesized by enzymatic phosphorylation of dihydroxyacetone with glycerol kinase [48], but even if the ATP is regenerated this method will not suffice to prepare large amounts of DHAP [57]. An alternative, the *in situ* liberation of DHAP by aldol cleavage of fructose 1,6-diphosphate, is not suitable for preparative uses either [52]. In this latter method, fructose 1,6-diphosphate is cleaved enzymatically into DHAP and GA-3-P; the GA-3-P is then converted to DHAP by triose-phosphate isomerase.

Dihydroxyacetone, in the presence of inorganic arsenate, can be used instead of DHAP as a substrate for the aldolase [53]. There is also an encouraging novel chemical approach in which the stable barium salt of 2,5-bis(phosphonooxymethyl)-2,5-diethoxy-1,4-dioxane is synthesized [58]; this compound can readily be transformed to DHAP by treating it with an acid cation exchange resin (Figure 7).

Sialic acids (*N*-acetylneuraminic acids, NANs) are widespread in nature [59]. As components of many glycoconjugates they play a very important part in numerous molecular recognition processes. Originally, they had to be isolated from biological sources [60], but a recent approach makes use of the enzyme *N*-acetylneuraminate lyase from *Clostridium* (EC 4.1.3.3). Augé *et al.* in 1984 [61] were the first to use this enzyme to prepare NAN in an aldol-type reaction between *N*-acetylmannosamine and pyruvate (Figure 8).

Since then, this enzyme has been used to synthesize various derivatives of NAN. It permits a certain amount of variation in the mannose component, but is strictly specific for pyruvate. With 2-acetamido-6-*O*-acetyl-2-deoxy-D-mannose as a starting material, the cell acceptor determinant of the influenza virus, *N*-acetyl-9-*O*-acetylneuraminate, could be synthesized [62, 63]. The mannose derivative was made either chemically [62] or enzymatically [63]. In the latter case, *N*-acetylmannosamine was acylated at the C-6 position by protease N from *Bacillus subtilis*, and then condensed with pyruvate (Figure 9).

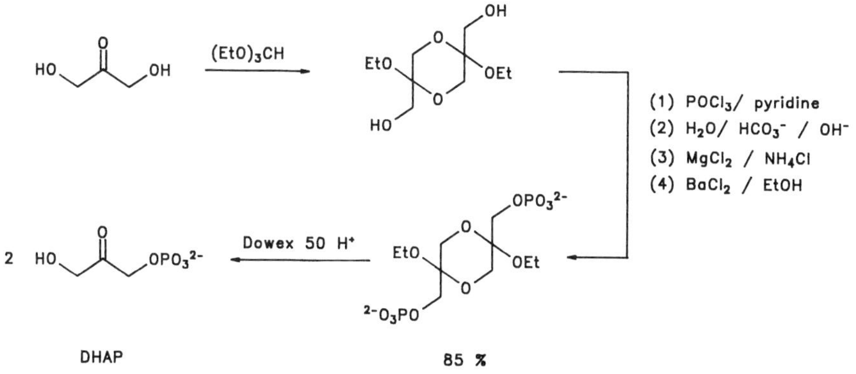

Fig. 7 Formation of dihydroxyacetone phosphate (DHAP) [58]

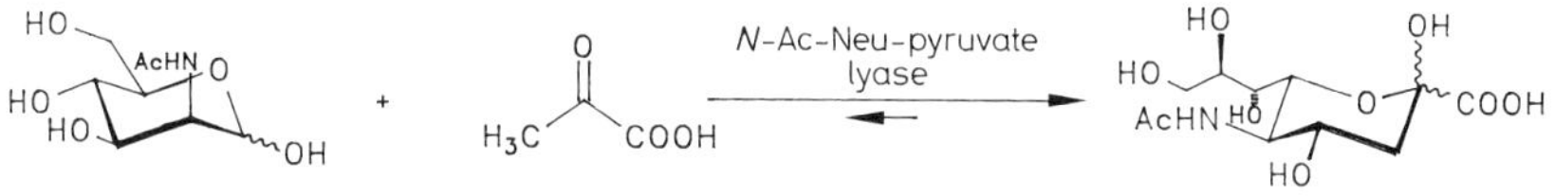

Fig. 8 Preparative synthesis of *N*-acetylneuraminic acid (NAN) [61]

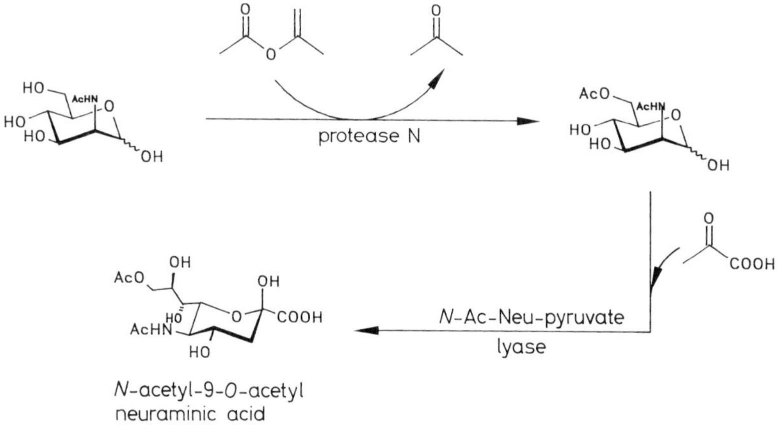

Fig. 9 Enzymatic synthesis of *N*-acetyl-9-*O*-acetylneuraminate [62]

Reaction of mannose with pyruvate gives the deaminated 3-deoxy-D-glycero-D-*galacto*nonulosonic acid in 84% yield [64]. Many neuraminate derivatives have been synthesized to date from chemically modified mannosamines [65, 66]. Because *N*-acetylmannosamine is very expensive, chemical derivatization of *N*-acetylglucosamine followed by alkaline epimerization is preferable [67, 68].

BROSSMER *et al.* [69] used an NAN synthase (EC 4.1.3.19) from *Neisseria meningitidis* to make 5-acetamido-9-azido-3,5,9-trideoxyneuraminate. This reaction differs from lyase catalysis in that phosphoenolpyruvate instead of pyruvate is condensed with the mannose derivative, with subsequent phosphate cleavage.

3-Deoxy-D-*manno*-2-octulosonate 8-phosphate (KDO-8-P) [70] is significant in the biosynthesis of the cell walls of Gram-negative bacteria. This compound was made using 3-deoxy-D-*manno*octulosonate aldolase (EC 4.1.2.23) for the condensation between arabinose 5-phosphate (Ara-5-P) and phosphoenolpyruvate [71]. The Ara-5-P substrate was produced by phosphorylating arabinose with hexokinase (EC 2.7.1.1) using the MEEC technique. The ATP consumed was regenerated with PEP and pyruvate kinase (Figure 10).

Transketolases (EC 2.2.1.1) may also be used to make monosaccharides [72–74]. In this reaction a hydroxyacetyl group is transferred to an aldose phosphate. Transketolase from yeast was used to prepare D-xylulose 5-phosphate from glyceraldehyde 3-phosphate and hydroxypyruvate [72]. These enzymes have the advantage that they also accept unphosphorylated substrates, in contrast to aldolases. BOLTE *et al.* [73] used spinach leaf transketolase to condense various aldoses with hydroxypyruvate, forming ketoses; for instance, 5-deoxy-D-xylulose was synthesized, starting from DL-lactaldehyde (Figure 11).

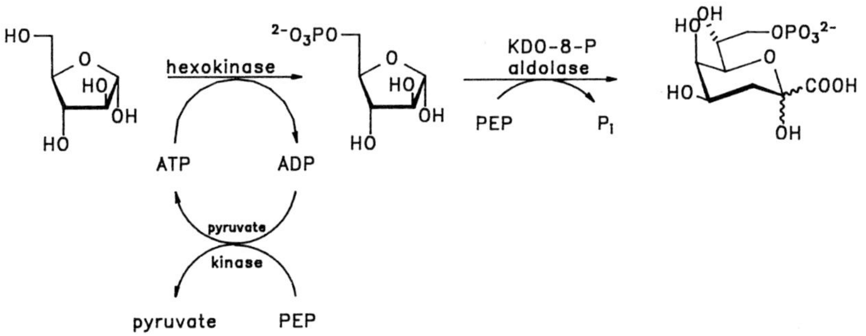

Fig. 10 Synthesis of 3-deoxy-D-*manno*-2-octulosonate 8-phosphate (KDO-8-P) [71]

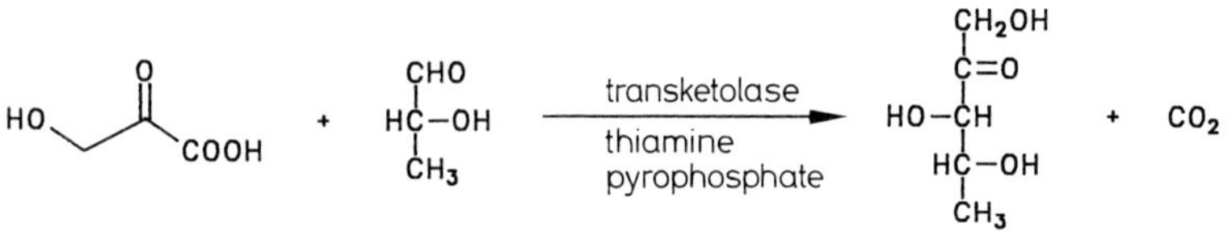

Fig. 11 Synthesis of monosaccharides using transketolases [73]

Spinach leaf transketolase was also used to synthesize unusual sugar phosphates such as D-*glycero*-D-*altro*octulose 8-phosphate [51]. This compound was prepared from D-allose 6-phosphate and hydroxypyruvate in 85% yield. The enzyme was also used in a multienzyme synthesis of 3-deoxy-D-*arabino*-heptulosonic acid 7-phosphate (DAHP), which is an important intermediate in the shikimic acid pathway for the biosynthesis of aromatic amino acids [75]. For the reaction, fructose, a large excess of PEP, and catalytic amounts of ATP were incubated together in the presence of four immobilized enzymes. Hexokinase phosphorylated the fructose to fructose 6-phosphate; the ATP consumed was regenerated with PEP/pyruvate kinase. Fructose 6-phosphate was then converted to erythrose 4-phosphate by transketolase. The erythrose 4-phosphate was further condensed with PEP to give the product DAHP, with DAHP synthase from *E. coli* as catalyst. Although the yield was 85%, the four-step reaction system is at a serious disadvantage because the enzymes are difficult and expensive to obtain.

Therefore, a four-step chemoenzymatic synthesis of DAHP was developed recently [76]. The key step, in which the required *threo* configuration was introduced, was catalyzed by FDP aldolase: *N*-acetyl-DL-aspartate-β-semialdehyde was condensed with dihydroxyacetone phosphate. After three more chemical reactions, DAHP was produced with an overall yield of 13%.

Bakers' yeast, which is used successfully in many fields of organic chemistry [14, 20], can also be used in the *de novo* synthesis of monosaccharides. For instance, the racemic α-acetoxyketone was reduced enantioselectively by oxidoreductase present in bakers' yeast (Figure 12). The blocked tetraol produced was further converted to 4-deoxy-D-mannose [77]. In an analogous reaction, 2,3-dideoxy-2-*C*-methyl-D-glucose derivatives could be synthesized [78].

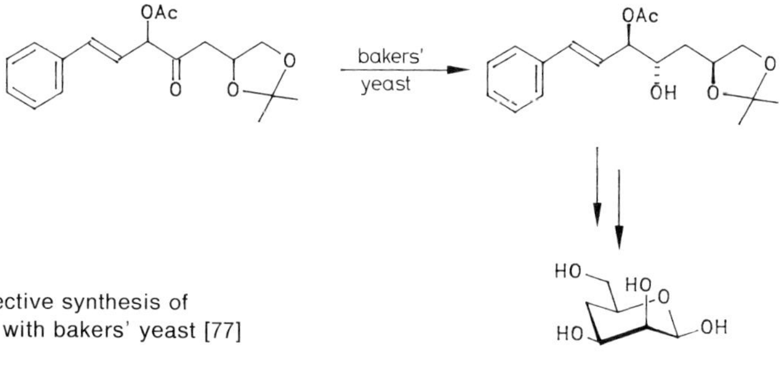

OAc
O O O
bakers' yeast
OAc
OH O O
HO HO O
HO OH

Fig. 12 Enantioselective synthesis of 4-deoxy-D-mannose with bakers' yeast [77]

3.3 Further reactions of monosaccharides

There is a growing demand for the synthesis of phosphorylated sugar compounds, which serve as intermediates or inhibitors in sugar metabolism. Various chemically synthesized deoxyfluoro sugars were phosphorylated by hexokinase, which has a broad substrate specificity [79].

Even the synthesis of 2-deoxy-α-D-glucose 1-phosphate, which is difficult to do chemically, was possible when enzymes were used [80]. 2-Deoxy-D-glucose 6-phosphate

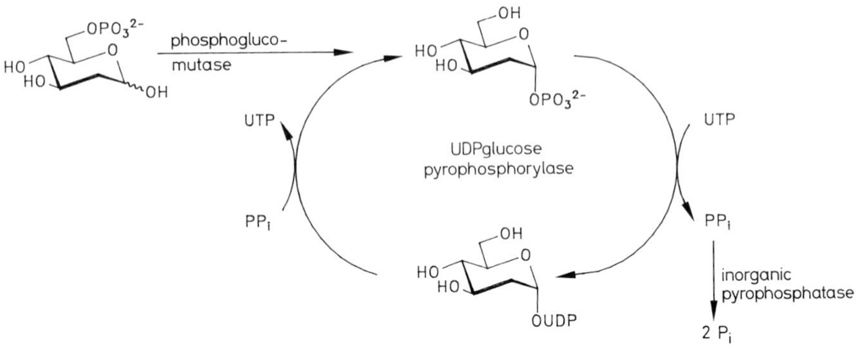

Fig. 13 Synthesis of 2-deoxy-α-D-glucose 1-phosphate [80]

was converted to 2-deoxy-α-D-glucose 1-phosphate by phosphoglucomutase (EC 5.4.2.2.) (Figure 13). Because the equilibrium in this reaction is strongly to the side of the 6-phosphate, a *UTP: glucose-1-phosphate* uridyltransferase (UDPglucose pyrophosphorylase; EC 2.7.7.9) was added to convert the 1-phosphate to the UDP-sugar compound and thus remove it from the equilibrium mixture. The nucleotide sugar was isolated and then was cleaved into UDP and 2-deoxy-α-D-glucose 1-phosphate by reverse catalysis with *UTP: glucose-1-phosphate* uridyltransferase.

Several linked enzymic reactions were used to synthesize a central intermediate in nucleotide biosynthesis, 5-phospho-D-ribose 1-pyrophosphate (PRPP) [81]. First, ribose was phosphorylated in the 5-C position by hexokinase (Figure 14). The ribose 5-phosphate was then converted to the product by PRPP synthase (ribose-phosphate pyrophosphokinase; EC 2.7.6.1). In this reaction, ATP was converted to AMP, which was regenerated to ATP in a combined reaction of adenylate kinase (EC 2.7.4.3) and pyruvate kinase.

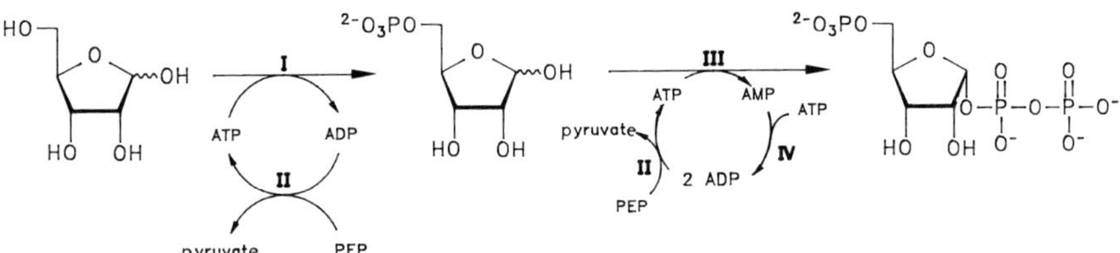

Fig. 14 Synthesis of 5-phospho-D-ribosyl 1-pyrophosphate (PRPP) [81].

Enzymes: I = hexokinase;
II = pyruvate kinase;
III = PRPP synthase
IV = adenylate kinase

A multienzyme system was used to synthesize ribulose 1,5-diphosphate, an important metabolite in CO_2 assimilation in plants [82]. Glucose was converted to glucose 6-phosphate by hexokinase, and was then oxidized to 6-phosphogluconic acid by glu-

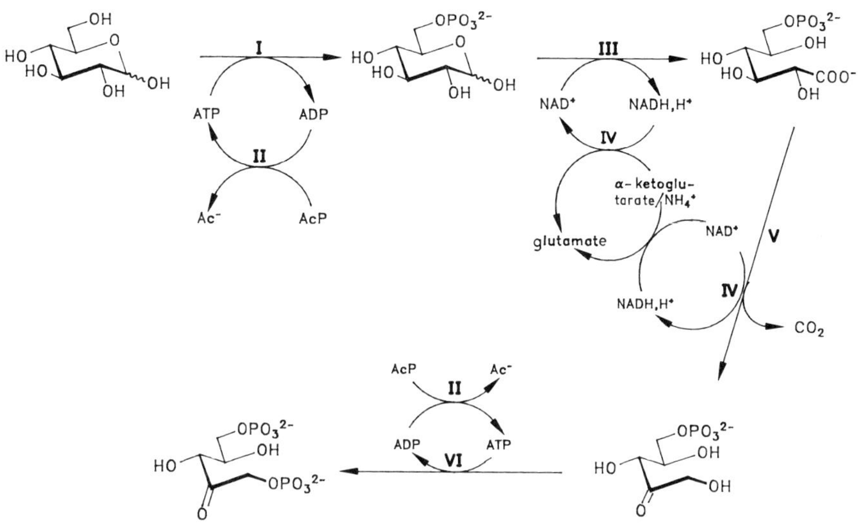

Fig. 15 Synthesis of ribulose 1,5-diphosphate [82].
Enzymes: I = hexokinase;
II = acetate kinase;
III = glucose-6–phosphate dehydrogenase;
IV = glutamate dehydrogenase;
V = phosphogluconate dehydrogenase (decarboxylating);
VI = phosphoribulokinase

cose 6-phosphate dehydrogenase (EC 1.1.1.49) (Figure 15). This compound was decarboxylated oxidatively by 6-phosphogluconate dehydrogenase (EC 1.1.1.44). The resulting ribulose 5-phosphate was converted to ribulose 1,5-diphosphate with phosphoribulokinase (EC 2.7.1.19). The ATP consumed was regenerated with the acetylphosphate/acetate kinase system, and NAD^+ was regenerated with α-ketoglutarate and glutamate dehydrogenase (EC 1.4.1.2). The product was obtained in gram amounts.

Some enzymes may even accept unphysiological substrates. For instance, sucrose phosphorylase (EC 2.4.1.7) converted glucosyl fluoride to glucose 1-phosphate in the presence of phosphate [83].

As our final example in this section, microsomal epoxide hydrolase (EC 3.3.2.3) was used to cleave 3,4-epoxytetrahydropyran to *trans* diols exclusively [84].

4 Synthesis of oligosaccharides

4.1 Reactions using glycosidases

The first studies of the mechanisms and specificities of glycosidases were done in the 1940s [85]. These hydrolases remove sugar moieties from the nonreducing ends of oligosaccharide chains. The enzymes are very specific with respect to the nature of the sugar as well as the configuration of the bond to be cleaved, but they exhibit a certain variability as regards the binding site. The reactions catalyzed by glycosidases are reversible, and the equilibrium can be shifted in the direction of bond formation by an appropriate choice of conditions of temperature, pH, solvent, or concentration.

In principle, we can take advantage of the transferase potential of glycosidases in two ways [86]. Free monosaccharides can be reacted with an acceptor, either an alcohol or a saccharide compound, in a *reverse hydrolysis*. This method usually gives only poor yields and is difficult to handle; it is therefore rarely used. Transglycosylations, in which glycosides or other activated glycosyl donors instead of free monosaccharides are reacted with alcohols, are much more straightforward. Nevertheless, the lack of regioselectivity results in mixtures of isomers with (1 → 2)-, (1 → 3)-, (1 → 4)-, and (1 → 6)-glycosidic linkages. The reaction is somewhat regioselective if certain glycosides are used as acceptors [87]. Often, the transferase activity is greater in organic solvents than in water. WILLEMOT *et al.* studied various solvent systems [88].

If α-galactosidase (EC 3.2.1.22) is used to glycosylate *N*-acetylglucosamine with galactose or methyl α-galactoside, the result is *N*-acetylmelibiosamine (Figure 16) [89–91].

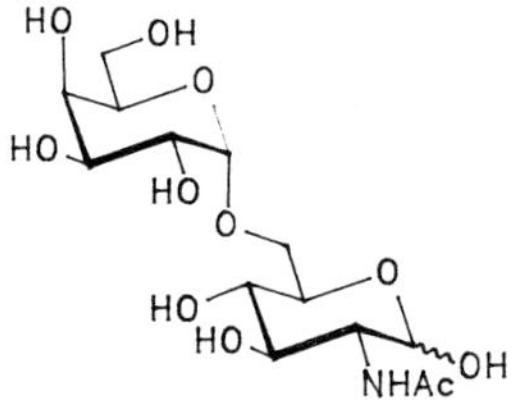

Fig. 16 *N*-Acetylmelibiosamine

With phenyl α-glucoside as donor and α-glucosidase (EC 3.2.1.20) from bakers' yeast, glucose was transferred to the following acceptors: D-glucose [92], D-mannose [93], D-xylose [94], and D-fructose [95]. Maltose, a cheap glucose donor for α-glucosidases, was used in reactions with various acceptors such as sorbose [96–103].

With β-galactosidase (EC 3.2.1.23) from *E. coli* as catalyst, transglycosylation proceeds with lactose as the galactosyl donor [86, 104]. Using immobilized enzyme and with *N*-acetylgalactosamine as acceptor, the disaccharide 6-*O*-β-D-galactopyranosyl-2-acetamido-2-deoxy-D-galactose was obtained in yields up to 20% [105]. When β-galactosidase from bovine testis was used, the product disaccharide was (1 → 3)-β-linked rather than (1 → 6) [106]. Even trisaccharides have been synthesized by transglycosylations. For example, sucrose was galactosylated by forming (1 → 6)-β-linkages. Donors were either *o*-nitrophenyl β-D-galactoside or lactose [107].

AJISAKA *et al.* developed a new method for reverse hydrolysis [108, 109]. The enzyme was β-galactosidase immobilized on SEPHAROSE 4 B in a column reactor. The mixture of galactose and *N*-acetylglucosamine was circulated through the reaction column and a column filled with activated charcoal. The charcoal adsorbed the disaccharide products and removed them from the reaction mixture, so the reaction equilibrium shifted further toward bond formation. After the reaction had gone to completion, the products were eluted from the charcoal by aqueous ethanol. Thus, *N*-acetyllactosamine and *N*-acetylmelibiosamine were prepared in a total yield of 16% [109]. This technique was proposed for the synthesis of trisaccharides also [110].

Lactose was often the β-galactosyl donor in transgalactosylations. In the presence of an alcohol, lactose was first cleaved into glucose and a galactoside, which then served as acceptor for another β-galactosyl residue formed from lactose (Figure 17). Galactosyl disaccharides were thus synthesized *in situ* by incubating lactose with various alcohols and β-galactosidase as catalyst [111].

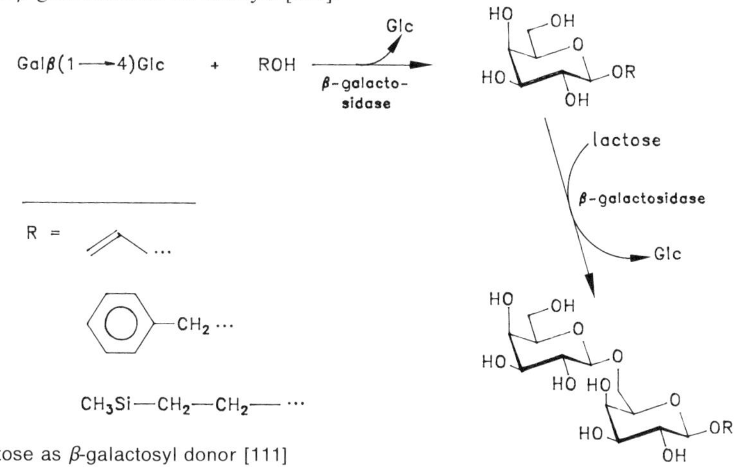

Fig. 17 Lactose as β-galactosyl donor [111]

Cellotriose was formed when a cellobiose solution was incubated with β-glucosidase (EC 3.2.1.21) from almonds [112]. Other kinds of cello-oligomers were synthesized with cellulases (EC 3.2.1.4) [113]; donors in these reactions were cellotriose, cellopentaose or *p*-nitrophenyl β-cellobioside, and the acceptor was ^{14}C-labeled glucose.

The regioisomers laminaribiose, sophorose, and gentiobiose were prepared with an overall yield of 40% when a 90% (w/v) glucose solution was incubated with β-glucosidase (EC 3.2.1.21) at 55 C [114]. An amylase from *Bacillus macerans* (cyclomaltodextrin glucanotransferase; EC 2.4.1.19) was used to synthesize 4-*O*-(α-D-glucopyranosyl)-1-deoxynojirimycin (Figure 18). This compound, which is similar to acarbose, might be effective in treating diabetes mellitus [115].

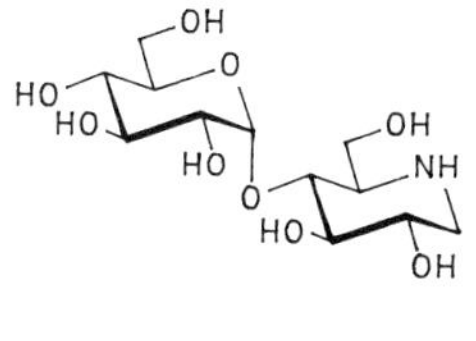

Fig. 18 4-*O*-(α-D-Glucopyranosyl)-1-deoxynojirimycin

Amylases also catalyze the synthesis of branched cyclodextrins [116] and various maltooligomers [117, 118].

Lysozyme (EC 3.2.1.17), which cleaves the polysaccharide muramic acid of bacterial cell walls, can be used in oligosaccharide synthesis. For instance, a disaccharide consisting of *N*-acetylglucosamine (GlcNAc) in (1 → 4)-*β*-linkage with *N*-acetylxylosamine was synthesized [119]. When the reducing *N*-acetylglucosamine moiety of chitin was transferred to the C-2 position of lactose using lysozyme, the product was the trisaccharide shown in Figure 19 with $n = 0$. Furthermore, the enzyme could catalyze the glycosylation at the GlcNAc moiety with more *N*-acetylglucosamine to produce the three novel oligosaccharides with $n = 1-3$ as shown in Figure 19 [120].

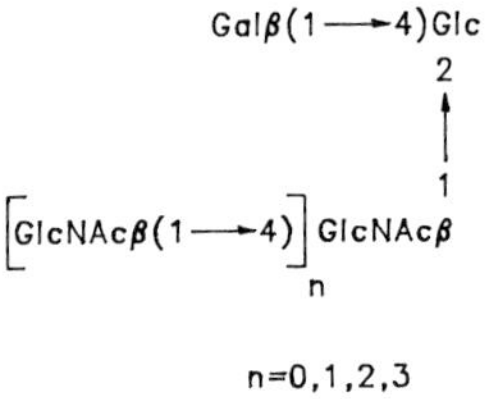

Galβ(1 → 4)Glc
2
↑
1
[GlcNAcβ(1 → 4)]$_n$ GlcNAcβ

n=0,1,2,3

Fig. 19 Products in the synthesis of oligosaccharides with lysozyme [120]

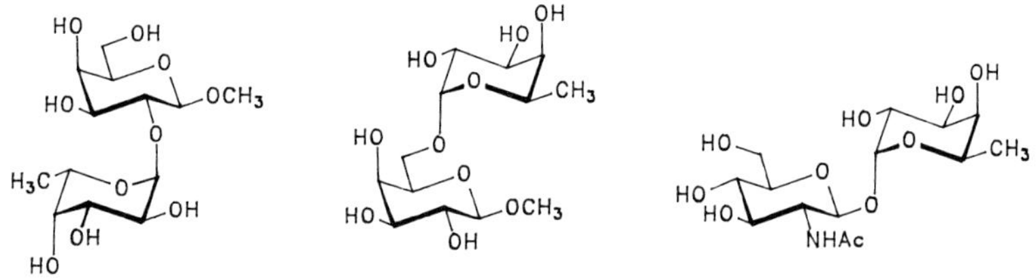

Fig. 20 Disaccharides produced by enzymatic fucosylation [122]

Our group has recently demonstrated the synthesis of fucosylated oligosaccharides with α-L-fucosidase (EC 3.2.1.51) from porcine liver [121, 122]. The glycosyl donors were *p*-nitrophenyl α-L-fucoside and α-L-fucosyl fluoride; methyl *β*-D-galactoside and *N*-acetylglucosamine were the acceptors. The products (Figure 20) were obtained in yields up to 20 %.

Even phosphorylases, which catalyze the phosphorolytic cleavage of oligosaccharides to form glycose 1-phosphates *in vivo*, can be used in oligosaccharide synthesis. Thus, cellobiose phosphorylase from *Clostridium thermocellum* (EC 2.4.1.20) transferred glucose 1-phosphate to glucose, 2-deoxyglucose, mannose, glucosamine, xylose, and arabinose to form (1 → 4)-*β*-linkages [123]. Amylose phosphorylase (EC 2.4.1.1) was used in a continuous enzyme reactor to form maltooligomers from glucose 1-phosphate [124].

Glycosidases can be used for transglycosylations to an alcohol component that is not a carbohydrate, as well as for oligosaccharide synthesis. An example is the synthesis of labile cardiac glycosides by Ooi *et al.* [125]. Gitoxigenin galactoside was made from phenyl *β*-galactoside with *β*-galactosidase as catalyst (Figure 21). Because the steroid is soluble only in organic media, the reaction was carried out in 50 % aqueous acetonitrile.

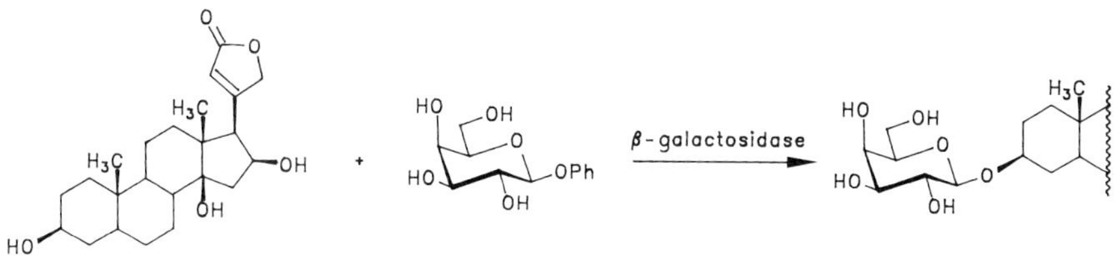

Fig. 21 Synthesis of gitoxigenin galactoside using β-galactosidase [125]

Lactase from *Kluyveromyces lactis* (EC 3.2.1.108) was used in the synthesis of various alkyl β-glucosides [126], transglucosylating phenyl β-D-glucoside with several primary and secondary alcohols. Yields were 10-70%.

KUSAMA *et al.* [127] synthesized the artificial sweetener stevioside by transglucosylation (Figure 22). The reaction was catalyzed by an enzyme system from *Streptomyces* sp., consisting of glucan endo-1,3-β-glucosidases (β-1,3-glucanases; EC 3.2.1.39) that cleave the compound curdlan, a (1 → 3)-β-linked polysaccharide, to lower oligosaccharides. Moreover, the enzyme system contained β-glucosidases that catalyzed the transfer reactions to the products A, B, and C (compare Figure 23).

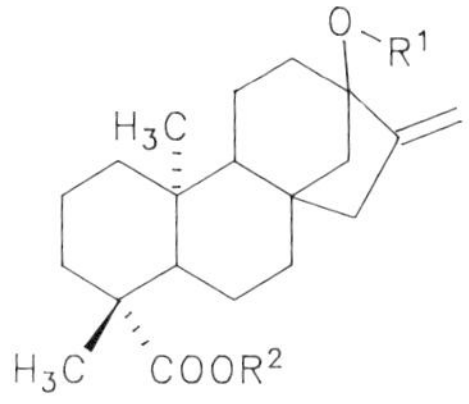

product	R^1	R^2
stevioside	Glcβ(1→2)Glcβ-	Glcβ-
A	Glcβ(1→2)Glcβ-	Glcβ(1→3)Glcβ-
B	Glcβ(1→3)Glcβ(1→2)Glcβ-	Glcβ-
C	Glcβ(1→2)Glcβ-	Glcβ(1→3)Glcβ(1→3)Glcβ-

Fig. 22 Stevioside and products of transglucosylation [127]

Some nonspecific enzymes can be used in syntheses with nonphysiological substrates. Surprisingly, glycosyl fluorides are accepted by some glycosidases as well as by transferases. BARNETT [128–131] showed that various glycosyl fluorides were hydrolyzed by certain glycosidases, with formation of hydrogen fluoride. If the hydrolysis was reversed, oligosaccharides were formed. If a concentrated solution of α-glucosyl fluoride was incubated with immobilized α-glucosidase [132], isomaltosyl fluoride was obtained in 25% yield; in a reaction analogous to the transglycosylations, isomaltose and panose were also formed, each in 6% yield (Figure 23). In a similar reaction, α-galactosyl fluoride was converted to the disaccharide (1 → 6)-α-galactobiosyl fluoride with α-galactosidase.

Many sucrases also accept α-glucosyl fluoride as a substrate. A polysaccharide resembling glycogen was formed with α-glucosyl fluoride and fructose in the presence of amylosucrase [133]. Similar reactions with maltosyl fluoride have also been studied [134–138]; lower maltooligomers with between 3 and 12 glucose units were synthesized in 80% yield by α-amylase [137]. Dextransucrase, which usually catalyzes the formation of a (1 → 6)-α-glucan (dextran) from sucrose, also converted α-glucosyl fluoride to dextrans [139–141]. The disaccharide 4-*O*-β-D-xylopyranosyl-D-xylopyranose was syn-

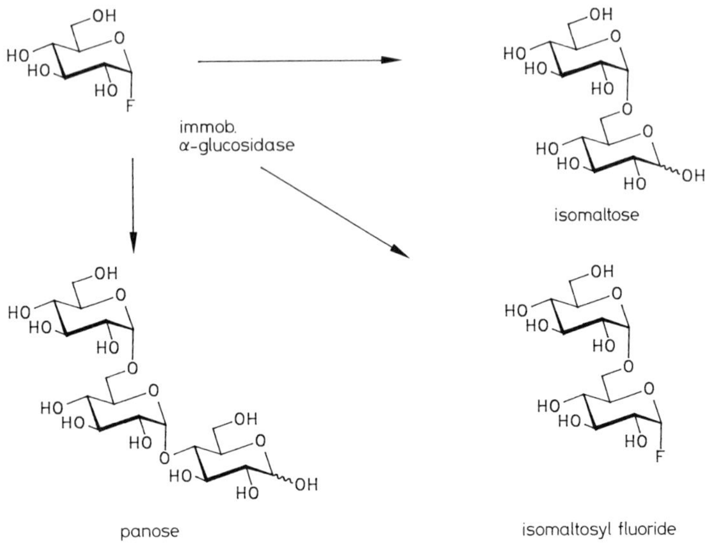

Fig. 23 Reaction of α-glucosyl fluoride with immobilized α-glucosidase [132]

thesized from xylopyranosyl fluoride with a β-xylosidase from *Bacillus pumilus* [142]. It was also possible to make α,α-trehalose [143] and the analogous α-D-glucopyranosyl-α-D-xylopyranoside [144] with the enzyme trehalase, using α-glucosyl fluoride, together with α-glucose or α-xylose, respectively.

Glycosidases were further used to make *O*-glycosidic bonds between sugars and amino acids such as serine. For instance, α-mannosidase (EC 3.2.1.24) was used in the synthesis of serine α-D-mannoside, and α-*N*-acetylgalactosaminidase (EC 3.2.1.49) formed serine *N*-acetyl-α-D-galactosaminide [145].

4.2 Reactions with transferases

Glycosidases have the advantage of converting a sugar without prior activation, so they do not need expensive precursors or cofactors. Furthermore, many of them are available at low or reasonable prices, whereas transferases are extremely expensive, or else difficult to isolate and handle. Nevertheless, glycosidases lack regioselectivity, and it is difficult to separate the product oligosaccharides from their starting materials – two enormous disadvantages. Moreover, yields are often low and reaction conditions are troublesome. Therefore, when highly specific oligosaccharides need to be synthesized, transferases should be the enzymes of choice. These enzymes are distinctly substrate-, regio-, and stereospecific [146, 147], and the consequence is that very pure oligosaccharides can be obtained. Transferases require that the sugar to be transferred be in activated form as a nucleotide compound, which can be prepared either chemically or enzymatically in preliminary reactions. There are also impressive examples of multienzyme systems in which the nucleotide sugar is made *in situ*.

A central disaccharide unit of many heterooligosaccharides is *N*-acetyllactosamine. NUNEZ and BARKER [148] were the first to describe a preparative enzymatic synthesis of this compound; they used a lactose synthase from bovine colostrum (galactosyltransferase; EC 2.4.1.22). Here the galactose moiety of UDPgalactose was transferred to *N*-acetylglucosamine. Unfortunately, the UDP that is also formed inhibits the reaction and has to be removed. Therefore, WONG *et al.* [149] worked out a reaction cycle in which UDP is converted to UTP by the PEP/pyruvate kinase system (Figure 24). UTP

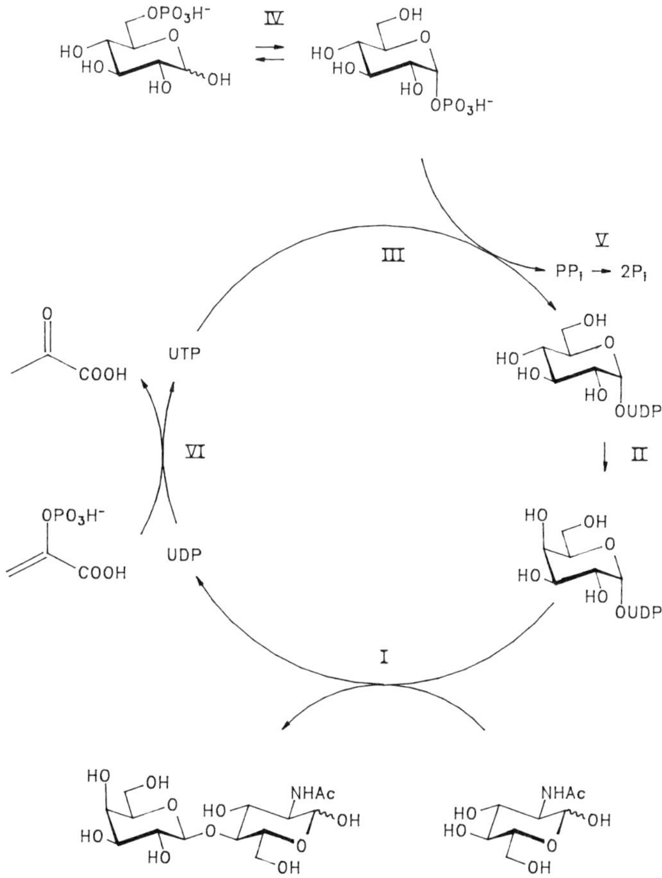

Fig. 24 Synthesis of *N*-acetyllactosamine [149].
Enzymes: I = galactosyltransferase;
II = UDPglucose 4-epimerase;
III = UDPglucose pyrophosphorylase;
IV = phosphoglucomutase;
V = inorganic pyrophosphatase;
VI = pyruvate kinase

is then immediately used for the *in situ* synthesis of UDPglucose. First, glucose 6-phosphate is isomerized to glucose 1-phosphate by phosphoglucomutase and is then converted to UDPglucose. In order to push the equilibrium in the desired direction, the pyrophosphate product is cleaved to phosphate by inorganic pyrophosphatase (EC 3.6.1.1). An epimerase (EC 5.1.3.2) then converts UDPglucose to UDPgalactose, which is then transferred to the acceptor *N*-acetylglucosamine, forming *N*-acetyllactosamine in the transferase-catalyzed step.

The enzymes were immobilized on a polyacrylamide-based carrier [28], and the yield was 70%. The cycle also worked with enzymes immobilized on silica [150, 151]. This substratum seemed to be easier to handle and much cheaper.

AUGÉ *et al.* extended this cycle to the synthesis of higher oligosaccharides by using acceptors other than *N*-acetylglucosamine for the galactosyl transferase. The enzymes were immobilized on agarose prior to the reaction. They used this system to galactosylate disaccharides [152], and could even make penta- and hexasaccharides if they used a tetrasaccharide acceptor [66, 153]. Recently, chitobiose and cellobiose were galactosylated with or without the presence of α-lactalbumin [154, 155]. Prior to the transfer reaction, commercial UDPglucose was converted to UDPgalactose. If free enzymes were used, yields were up to 60% (Figure 25).

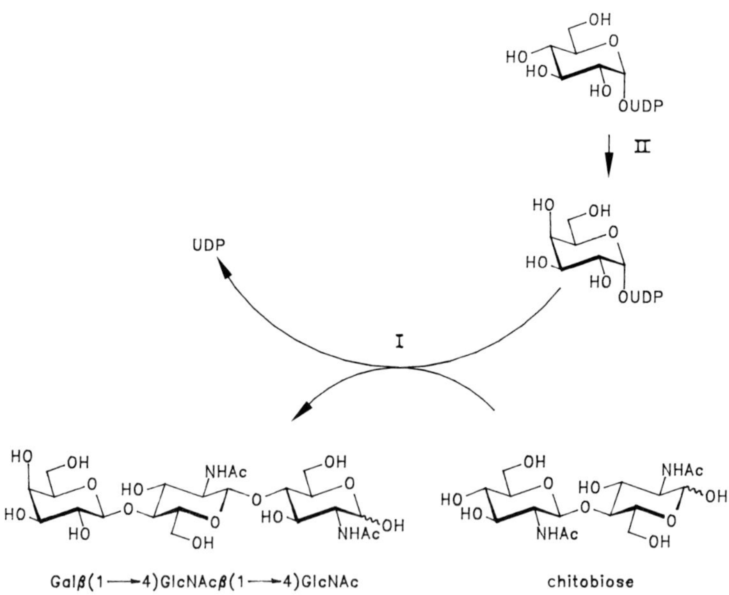

Fig. 25 Galactosylation of chitobiose [155].
Enzymes: I = galactosyltransferase;
II = UDPglucose 4-epimerase

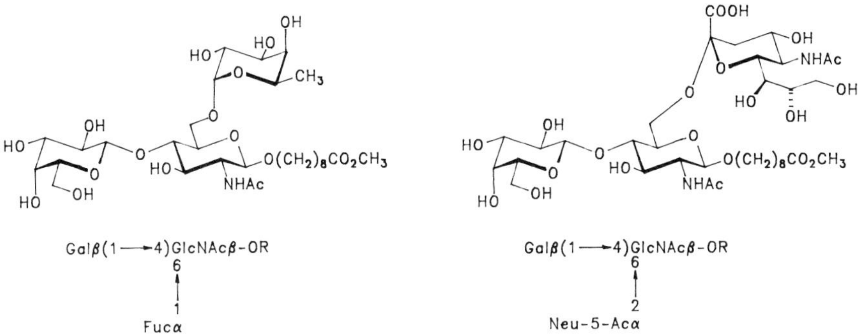

$Gal\beta(1 \rightarrow 4)GlcNAc\beta$-OR (6 ← 1 Fucα)

$Gal\beta(1 \rightarrow 4)GlcNAc\beta$-OR (6 ← 2 Neu-5-Acα)

Fig. 26 Antigen-determining trisaccharides [156]

Galactose was also transferred to the C-4 position of fucosylated or sialylated *N*-acetylglucosamine [156], forming the antigen-determining trisaccharides shown in Figure 26.

Yates *et al.* [157] isolated a 1,3-α-galactosyltransferase from *type B* human blood. This enzyme is also capable of transferring UDP-*N*-acetylgalactosamine to 2′-fucosyllactose, and was thus used in the synthesis of the tetrasaccharide that determines *blood group A*, shown in Figure 27.

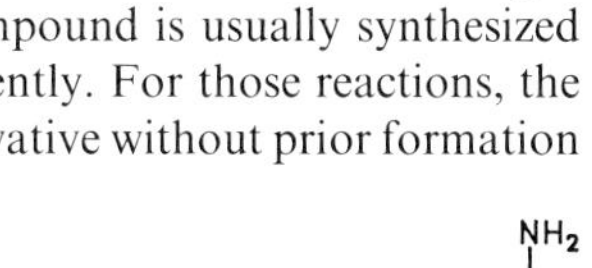

$GalNAc\alpha(1 \rightarrow 3)Gal\beta(1 \rightarrow 4)Glc$ (2 ← 1 Fucα)

Fig. 27 Tetrasaccharide determinant of blood group A

In oligosaccharide biosynthesis, transferases are used to transfer a nucleotide sugar to a growing oligosaccharide chain. This nucleotide compound is usually synthesized from a sugar phosphate, but sialylations are done differently. For those reactions, the free neuraminic acid is activated directly to the CMP derivative without prior formation of a 1-phosphate (Figure 28).

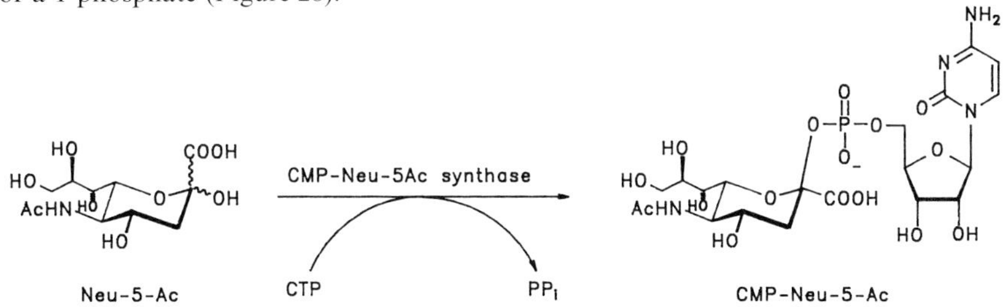

Neu-5-Ac —(CMP-Neu-5Ac synthase; CTP → PP_i)→ CMP-Neu-5-Ac

Fig. 28 Physiological formation of CMPneuraminate

Kean and Roseman in 1966 [158] were the first to report a preparative enzymatic synthesis of CMPneuraminic acid. They used an acylneuraminate cytidylyltransferase (CMPsialate synthase; EC 2.7.7.43) from porcine submaxillary glands. Later, Schauer and co-workers optimized the reaction by using various enzyme sources as well as better isolation methods [159, 160]. Nevertheless, because the very expensive CTP was needed in high concentrations, the preparative synthesis of CMPneuraminic acid did not appear very attractive. In recent work, CTP has been made enzymatically from the cheaper CMP. SIMON *et al.* in WHITESIDES's laboratory [32] used myokinase (adenylate kinase), which is available at a low price, with the MEEC technique. CMP and a catalytic amount of CTP symproportionate to CDP in the presence of myokinase. CDP was then converted to CTP by the PEP/pyruvate kinase system. The resulting solution of CTP was then used in the synthesis of CMPneuraminic acid [67]. In a preceding reaction, neuraminic acid was made from *N*-acetylmannosamine and pyruvate with an aldolase (*N*-acetylneuraminate lyase). These two crude reaction solutions were then mixed and CMPneuraminate was made, in gram quantities, with immobilized CMPneuraminate synthase (Figure 29).

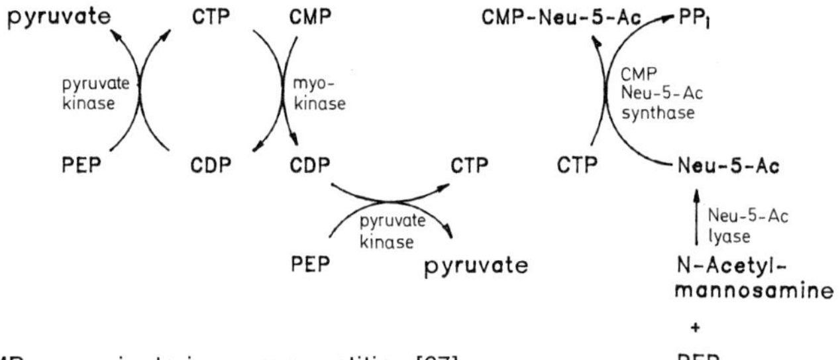

Fig. 29 Synthesis of CMPneuraminate in gram quantities [67]

CTP was also made from CMP using a nucleoside-phosphate kinase (EC 2.7.4.4.) [161, 162]. AUGÉ and GAUTHERON used CTP made in this way to synthesize various CMPneuraminate derivatives (Figure 30).

$$CMP + PEP \xrightarrow[ATP \rightleftarrows ADP]{E_1/E_2} 2\ CH_3-\overset{O}{\overset{\|}{C}}-COOH + CTP$$

$$CTP + \text{(neuraminate: } R^1O, R^2, HO, OH, COOH) \xrightarrow{E_3/E_4} \text{(CMP-neuraminate: } R^1O, R^2, HO, OCMP, COOH) + PP_i$$

R^1	R^2
H	NHAc
H	$NHCOCH_2OH$
Ac	NHAc
H	OH

E_1: nucleoside-phosphate kinase
E_2: pyruvate kinase
E_3: CMPneuraminate synthase
E_4: inorganic pyrophosphatase

Fig. 30 Formation of various CMPneuraminate derivatives [162]

The CMPneuraminic acid can be used in oligosaccharide synthesis as a substrate for various α-2,6-as well as α-2,3-sialyltransferases. Thus the trisaccharide α-D-Neu5Ac-(2 → 6)-β-Gal-(1 → 4)GlcNAc was produced in 50% yield by sialylating *N*-acetyllactosamine with a sialyltransferase from bovine colostrum (CMP-*N*-acetylneuraminate-β-galactoside α-2,6-sialyltransferase; EC 2.4.99.1) [150, 151]. Higa and Paulson [163] used glycoproteins as acceptors for sialyltransferases and made a series of (2 → 3)-α- as well as (2 → 6)-α-sialylated products. The sialyltransferases even accepted chemically synthesized disaccharides [164], forming sialylated trisaccharides. Recently, a combination of chemical and enzymatic methods was used to produce a disialotetrasaccharide in 10% yield (Figure 31) [165].

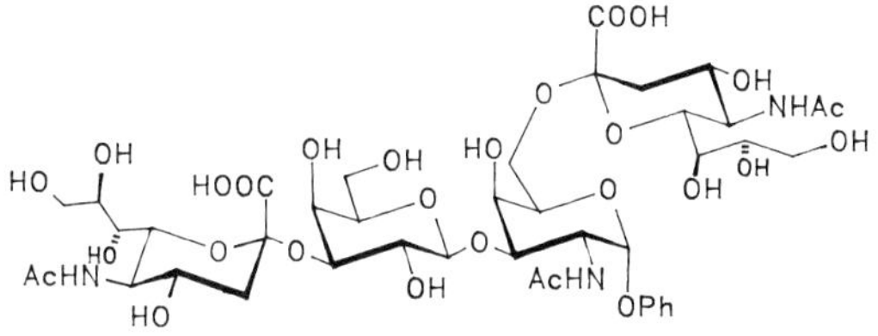

Fig. 31 Disialotetrasaccharide [165]

The glycoside α-D-Neu5Ac-(2 → 3)-β-Gal-(1 → 3)-α-GalNAc-1-OEt was synthesized with a sialyltransferase (CMP-*N*-acetylneuraminate--galactoside α-2,3-sialyltransferase; EC 2.4.99.4) from porcine submaxillary glands, immobilized on agarose [166]. Another trisaccharide synthesized by combining glycosidase and sialyltransferase reactions [167] was α-D-Neu5Ac-(2 → 3)-β-Gal-(1 → 3)-β-GalNAc-1-OEtBr (Figure 32).

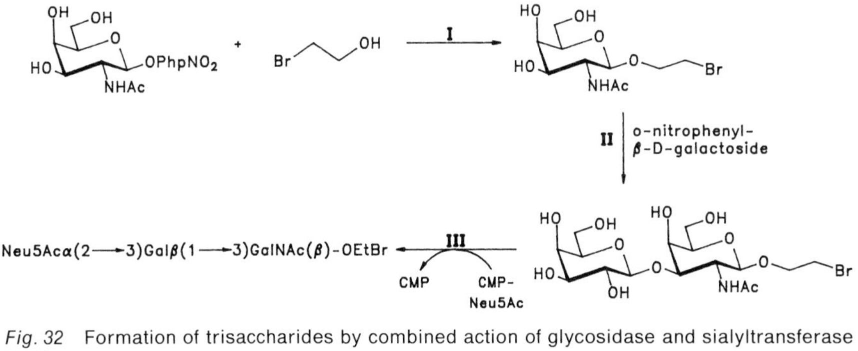

Fig. 32 Formation of trisaccharides by combined action of glycosidase and sialyltransferase [167].
Enzymes: I = *N*-acetyl-β-D-hexosaminidase
II = β-(1 → 3)-galactosyltransferase
III = α-(2 → 3)-sialyltransferase

Transferases and glycosidases will also react with oligosaccharides bound to a polymer matrix. In a preceding reaction, the oligosaccharide acceptor is chemically bound to a polymer with a reactive spacer [168, 169]. The products can be cleaved from the carrier by photolysis [170, 171] or by enzymatic hydrolysis [172], depending on the kind

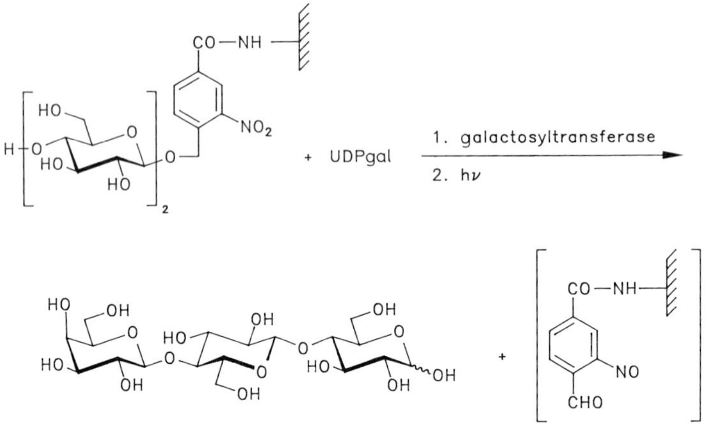

Fig. 33 Galactosylation of cellobiose on a solid support [170, 171]

of bond. For example, a photosensitive bond between a 4-carboxy-2-nitrobenzyl glycoside and the matrix was used in the galactosylations of glucose and cellobiose [170, 171] (Figure 33).

Another system involves the *N*-glycosidic linkage between the sugar and a phenylalanine-containing spacer bound to polyacrylamide. This bond can be digested with chymotrypsin [172].

In some cases, transferases can be used to synthesize nonnatural products. Immobilized glucuronosyltransferase (EC 2.4.1.17) from rat liver was used to make the cytostatic compound Laetrile® from UDPglucuronate and mandelonitrile [173] (Figure 34). Recently, the same enzyme from bovine liver was used to bind glucuronic acid to the β-blockers propafenone and 5-hydroxypropafenone [174].

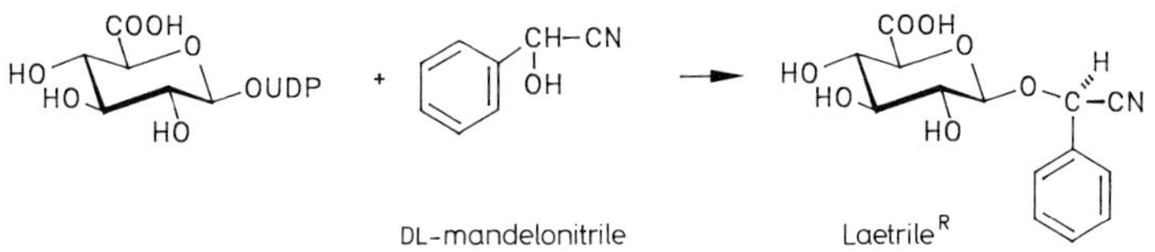

Fig. 34 Enzymatic synthesis of Laetrile® [173]

Cyclic glucooligomers with $(1 \rightarrow 4)$-α-linkages, called *cyclodextrins*, were made from α-glucosyl fluoride [175–178]. If cyclodextrin-α-$(1 \rightarrow 4)$-glucosyltransferase from *Klebsiella pneumoniae* (cyclomaltodextrin glucanotransferase, CGT; EC 2.4.1.19) was used, α- and β-cyclodextrins were obtained in 30% and 38% yield, respectively, along with 32% of linear maltooligomers (Figure 35).

This enzyme could also be used in a solid-phase reaction. Glucose or maltose was bound to a water-soluble polymer and incubated in the presence of α-glucosyl fluoride.

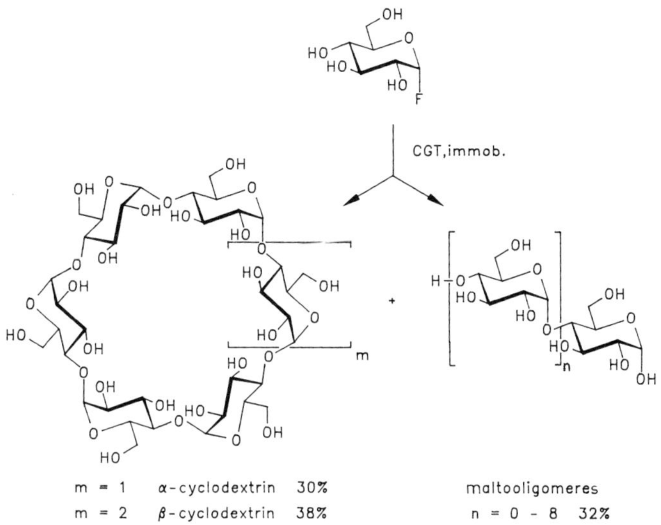

Fig. 35 Enzymatic synthesis of cyclodextrins [175, 176]

m = 1 α-cyclodextrin 30%
m = 2 β-cyclodextrin 38%
maltooligomeres n = 0 - 8 32%

The product maltooligomers, with up to 16 glucose units, were cleaved from the polymer by photolysis [179, 180].

Finally, CARD and co-workers [180, 181] reported that glucose can be transferred to 1-deoxy-1-fluorofructose or 1-azido-1-deoxyfructose with the enzyme sucrose synthase (EC 2.4.1.13).

5 Concluding remarks

Classical carbohydrate chemistry has attained an impressive stage of development, but the tried and true chemical methods have limitations. Synthetic approaches are therefore of interest, and enzymes used in preparative synthesis seem especially promising. To use them, however, we must know some basic biochemical techniques. Furthermore, the scale-up of biochemical reactions to preparative dimensions is not a trivial task. The field of enzymatic synthesis is just beginning to develop. The increasing number of research groups engaged in using enzymes particularly for carbohydrate work attests to the growing interest in this field.

6 Acknowledgements

Regine STILLER thanks the *Studienstiftung des deutschen Volkes*, and Joachim Thiem thanks the *Deutsche Forschungsgemeinschaft*, the *Bundesministerium für Wissenschaft und Forschung*, and the *Fonds der Chemischen Industrie* for supporting their work.

7 References

[1] RADEMACHER, T. W.; PAREKH, R. B.; DWEK, R. A.: Glycobiology; *Ann. Rev. Biochem.* **57** (1988) 785.

[2] HAKOMORI, S.-I.: Glycosphingolipids in Cellular Interaction, Differentiation and Oncogenesis; *Ann. Rev. Biochem.* **50** (1981) 733.

[3] SHARON, N.; LIS, H.: Glycoproteins: Research Booming on Long-Ignored, Ubiquitous Compounds; *Chem. Eng. News* **58** (13) (1981) 21.

[4] HAKOMORI, S.-I.: Glykosphingolipide; *Spektrum d. Wiss.* **1986** (7) 90.

[5] WILEY, D. C.; WILSON, J. A.; SKEHEL, J. J.: Structural Identification of the Antibody-Binding Sites of Hong Kong Influenza Hemagglutinin and Their Involvement in Antigenic Variation; *Nature (London)* **289** (1981) 373.

[6] GINSBURG, V.: Enzymatic Basis for Blood Groups in Man; *Adv. Enzymol.* **36** (1972) 131.

[7] EDELMANN, G. M.: Cell Adhesion Molecules; *Science* **219** (1983) 450.

[8] FEIZI, T.: Demonstration by Monoclonal Antibodies That Carbohydrate Structures of Glycoproteins and Glycolipids Are Oncodevelopmental Antigens; *Nature (London)* **314** (1985) 53.

[9] FEIZI, T.; CHILDS, R. A.: Carbohydrate Structures of Glycoproteins and Glycolipids as Differentiation Antigens, Tumor-Associated Antigens and Components of Receptor Systems; *Trends Biochem. Sci;* **10** (1985) 24.

[10] PAULSEN, H.: Fortschritte bei der selektiven chemischen Synthese komplexer Oligosaccharide; *Angew. Chem.* **94** (1982) 184; *Angew. Chem. Int. Ed. Engl.* **21** (1982) 155.

[11] PAULSEN, H.: Synthesis of Complex Oligosaccharide Chains of Glycoproteins; *Chem. Soc. Rev.* **13** (1984) 15.

[12] OGAWA, T.; YAMAMOTO, H.; NUKADA, T.: Synthetic Approach to Glycan Chains of a Glycoprotein and a Proteoglycan; *Pure Appl. Chem.* **56** (1984) 779.

[13] SCHMIDT, R. R.: Neue Methoden zur Glycosid- und Oligosaccharidsynthese – gibt es Alternativen zur Koenigs-Knorr-Methode? *Angew. Chem.* **98** (1986) 213; *Angew. Chem. Int. Ed. Engl.* **25** (1986) 212.

[14] JONES, J. B.: Enzymes in Organic Synthesis; *Tetrahedron* **42** (1986) 3351.

[15] JONES, J. B.: Enzymes as Chiral Catalysts; *Asym. Synth.* **5** (1985) 309.

[16] WHITESIDES, G. M.; WONG, C.-H.: Enzyme in der organischen Synthese; *Angew. Chem.* **97** (1985) 617; *Angew. Chem. Int. Ed. Engl.* **24** (1985) 617.

[17] WHITESIDES, G. M.; WONG, C.-H.: Enzyme as Catalysts in Organic Synthesis; *Aldrichimica Acta* **16** (1983) 27.

[18] WONG, C.-H.; DRUECKHAMMER, D. G.; DURRWACHTER, J. R.; LACHER, B.; CHAUVET, C. J.; WANG, Y.-F.: Enzyme-Catalyzed Synthesis of Carbohydrates. In: HORTON, D.; HAWKINS, L. D.; MCGARVEY, G. J., Eds.; *Trends in Synthetic Carbohydrate Chemistry;* ACS Symposium Series **356** (1989) 317.

[19] SUCKLING, C. J.; SUCKLING, K. E.: Enzymes in Organic Synthesis; *Chem. Soc. Rev.* **3** (1974) 387.

[20] MULZER, J.: Enzym-unterstützte Wirkstoffsynthese; *Nachr. Chem. Techn. Lab.* **32** (1984) 520, 589.

[21] ALTENBACH, H. J.: Enzymreaktionen – neue nützliche Anwendungen; *Nachr. Chem. Techn. Lab.* **36** (1988) 1114.

[22] COOPER, T. G.: *Biochemische Arbeitsmethoden*, Walter de Gruyter, Berlin (1981).

[23] CHIBATA, I., Ed.; *Immobilized Enzymes*, John Wiley and Sons, New York (1978).

[24] TREVAN, M. D.: *Immobilized Enzymes: Introduction and Application in Biotechnology*, John Wiley and Sons, New York (1980).

[25] HARTMEIER, W.: *Immobilisierte Biokatalysatoren*, Springer Verlag, Berlin (1986).
[26] KENNEDY, J. F.; CABRAL, J. M. S.: Enzyme Immobilization. In: KENNEDY, J. F., Ed.; *Biotechnology*, VCH-Verlagsgesellschaft, Weinheim (1987); Vol. 7a, p 347.
[27] KLIBANOV, A. M.: Immobilized Enzymes and Cells as Practical Catalysts; *Science* **219** (1983) 722.
[28] POLLAK, A.; BLUMENFELD, H.; WAX, M.; BAUGHN, R. L.; WHITESIDES, G. M.: Enzyme Immobilization by Condensation Copolymerization into Cross-Linked Polyacrylamide Gels; *J. Am. Chem. Soc.* **102** (1980) 6324.
[29] WEETALL, H. H.: Covalent Coupling Methods for Inorganic Support Materials; *Meth. Enzymol.* **44** (1976) 134.
[30] BEDNARSKI, M. D.; CHENAULT, H. K.; SIMON, E. S.; WHITESIDES, G. M.: Membrane-Enclosed Enzymatic Catalysis (MEEC): A Useful, Practical New Method for the Manipulation of Enzymes in Organic Synthesis; *J. Am. Chem. Soc.* **109** (1987) 1283.
[31] STILLER, R.; THIEM, J.: unpublished; STILLER, R.: *Präparativ-enzymatische Synthesen von Vorstufen zum Aufbau komplexer Oligosaccharide*. Dissertation, Univ. Münster (1988).
[32] SIMON, E. S.; BEDNARSKI, M. D.; WHITESIDES, G. M.: Generation of Cytidine 5′-Triphosphate using Adenylate Kinase; *Tetrahedron Lett.* **29** (1988) 1123.
[33] KLIBANOV, A. M.; SAMOKHIN, G. P.; MARTINEK, K.; BEREZIN, I. V.: A New Approach to Preparative Enzymic Synthesis; *Biotechnol. Bioeng.* **19** (1977) 1351.
[34] SUZUKI, H.; YAMAZAKI, J.: Adenosine 5′-Triphosphate Recycling in an Enzyme Reactor Based on Aqueous Two-Phase Systems; *Meth. Enzymol.* **136** (1987) 45.
[35] CRANS, D. C.; KASLAUSKAS, R. J.; HIRSCHBEIN, B. L.; WONG, C.-H.; ABRIL, O.; WHITESIDES, G. M.: Enzymatic Regeneration of Adenosine 5′-Triphosphate, Acetyl Phosphate, Phosphoenolpyruvate, Methoxycarbonyl Phosphate, Dihydroxyacetone Phosphate, 5′-Phospho-α-D-ribosyl Pyrophosphate, Uridine-5′-diphosphoglucose; *Meth. Enzymol.* **136** (1987) 263.
[36] CRANS, D. C.; WHITESIDES, G. M.: A Convenient Synthesis of Disodium Acetyl Phosphate for Use in *in Situ* ATP Cofactor Regeneration; *J. Org. Chem* **48** (1983) 3130.
[37] HIRSCHBEIN, B. L.; MAZENOD, F. P.; WHITESIDES, G. M.: Synthesis of Phosphoenolpyruvate and its Use in Adenosine Triphosphate Cofactor Regeneration; *J. Org. Chem.* **47** (1982) 3765.
[38] THERISOD, M.; KLIBANOV, A. M.: Facile Enzymatic Preparation of Monoacylated Sugars in Pyridine; *J. Am. Chem. Soc.* **108** (1986) 5638.
[39] THERISOD, M.; KLIBANOV, A. M.: Regioselective Acylation of Secondary Hydroxyl Groups in Sugars Catalyzed by Lipases in Organic Solvents; *J. Am. Chem. Soc.* **109** (1987) 3977.
[40] KLOOSTERMAN, M.; SCHOEMAKER, H. E.; KLOOSTERMAN-CASTRO, E. N.,; MEIJER, E. M.: Sugar-Alcohol Fatty Acid Ester: An Enzymatic Approach. In: Lichtenthaler, F. W.; Neff, K. H., Eds.; *Carbohydrates 1987, Abstracts of the 4th European Carbohydrate Symposium (Eurocarb IV)*, Darmstadt, July 1987, Gesellschaft Deutscher Chemiker (European Carbohydrate Organization); Abstr. D-17.
[41] RIVA, S.; CHOPINEAU, J.; KIEBOOM, A. P. G.; KLIBANOV, A. M.: Protease-Catalyzed Regioselective Esterification of Sugars and Related Compounds in Anhydrous Dimethylformamide; *J. Am. Chem. Soc.* **110** (1988) 584.
[42] SWEERS, H. M.; WONG, C.-H.: Enzyme-Catalyzed Regioselective Deacylation of Protected Sugars in Carbohydrate Synthesis; *J. Am. Chem. Soc.* **108** (1986) 6421.
[43] KLOOSTERMAN, M.; MOSMULLER, E. W. J.; SCHOEMAKER, H. E.; MEIJER, E. M.: Application of Lipases in the Removal of Protective Groups on Glycerides and Glycosides; *Tetrahedron Lett.* **28** (1987) 2989.
[44] KLOOSTERMAN, M.; DEVRIES, N. K.; SCHOEMAKER, H. E.; MEIJER, E. M.: Lipase-

Catalyzed Regio- and Stereoselective Hydrolysis of Acyl Esters on Carbohydrates; XIVth Int. Carbohydr. Symp., Stockholm, August 1988, Abstr. D-36.

[45] Hennen, W. J.; Sweers, H. M.; Wang, Y.-F.; Wong, C.-H.: Enzymes in Carbohydrate Synthesis: Lipase-Catalyzed Selective Acylation and Deacylation of Furanose and Pyranose Derivatives; *J. Org. Chem.* **53** (1988) 4939.

[46] Bischofberger, N.; Waldmann, H.; Saito, T.; Simon, E. S.; Lees, W.; Bednarski, M. D.; Whitesides, G. M.: Synthesis of Analogues of 1,3-Dihydroxyacetone Phosphate and Glyceraldehyde 3-Phosphate for Use in Studies of Fructose-1,6-diphosphate Aldolase; *J. Org. Chem.* **53** (1988) 3457.

[47] Bednarski, M. D.; Simon, E. S.; Bischofberger, N.; Fessner, W.-D.; Kim, M.-J.; Lees, W.; Saito, T.; Waldmann, H.; Whitesides, G. M.: Rabbit Muscle Aldolase as a Catalyst in Organic Synthesis; *J. Am. Chem. Soc.* **111** (1989) 627.

[48] Wong, C.-H.; Whitesides, G. M.: Synthesis of Sugars by Aldolase-Catalyzed Condensation Reactions; *J. Org. Chem.* **48** (1983) 3199.

[49] Wong, C.-H.; Mazenod, F. P.; Whitesides, G. M.: Chemical and Enzymatic Syntheses of 6-Deoxyhexoses; Conversion to 2,5-Dimethyl-4-hydroxy-2,3-dihydrofuran-3-one (Furaneol) and Analogues; *J. Org. Chem* **48** (1983) 3493.

[50] Durrwachter, J. R.; Sweers, H. M.; Nozaki, K.; Wong, C.-H.: Enzymatic Aldol Reaction/Isomerization as a Route to Unusual Sugars; *Tetrahedron Lett.* **27** (1986) 1261.

[51] Kapuscinski, M.; Franke, F. P.; Flanigan, I.; MacLead, J. K.; Williams, J. F.: Improved Methods for the Enzymic Preparation and Chromatography of Octulose Phosphates; *Carbohydr. Res.* **140** (1985) 69.

[52] Bednarski, M. D.; Waldmann, H. J.; Whitesides, G. M.: Aldolase Catalyzed Synthesis of Complex C_8 and C_9 Monosaccharides; *Tetrahedron Lett.* **27** (1986) 5807.

[53] Durrwachter, J. R.; Drueckhammer, D. G.; Nozaki, K.; Sweers, H. M.; Wong, C.-H.: Enzymatic Aldol Condensation/Isomerization as a Route to Unusual Sugar Derivatives; *J. Am. Chem. Soc.* **108** (1986) 7812.

[54] O'Connell, E. L.; Rose, I. A.: Affinity Labeling of Phosphoglucose Isomerase by 1,2-Anhydrohexitol-6-phosphates; *J. Biol. Chem.* **248** (1973) 2225.

[55] Fessner, W.-D.; Whitesides, G. M.: Enzymatic Aldol Condensation – Key Step to Unusual Oligosaccharides. In: Lichtenthaler, F. W.; Neff, K. H., Eds.; *Carbohydrates 1987, Abstracts of the 4th European Carbohydrate Symposium (Eurocarb IV)*, Darmstadt, July 1987, Gesellschaft Deutscher Chemiker (European Carbohydrate Organization); Abstr. A-141.

[56] Pederson, R. L.; Kim, M.-J.; Wong, C.-H.: A Combined Chemical and Enzymatic Procedure for the Synthesis of 1-Deoxynojirimycin and 1-Deoxymannojirimycin; *Tetrahedron Lett.* **29** (1988) 4645.

[57] Crans, D. C.; Whitesides, G. M.: Glycerol Kinase: Synthesis of Dihydroxyacetone Phosphate, *sn*-Glycerol-3-phosphate, and Chiral Analogues; *J. Am. Chem. Soc.* **107** (1985) 7019.

[58] Effenberger, F.; Straub, A.: A Novel Convenient Preparation of Dihydroxyacetone Phosphate and its Use in Enzymatic Aldol Reactions; *Tetrahedron Lett.* **28** (1987) 1641.

[59] Schauer, R.: Chemistry, Metabolism, and Biological Functions of Sialic Acids; *Adv. Carbohydr. Chem. Biochem.* **40** (1982) 131.

[60] Pozsgay, V.: Jennings, H.; Kasper, D. L.: 4,8-Anhydro-*N*-acetylneuraminic Acid – Isolation from Edible Bird's Nest and Structure Determination; *Eur. J. Biochem.* **162** (1987) 445.

[61] Augé, C.; David, S.; Gautheron, C.: Synthesis with Immobilized Enzyme of the Most Important Sialic Acid; *Tetrahedron Lett.* **25** (1984) 4663.

[62] Augé, C.; David, S.; Gautheron, C.; Veyrieres, A.: Synthesis with an Immobilized Enzyme of *N*-Acetyl-9-*O*-acetylneuraminic Acid, a Sugar Reported as a Component of Embryonic and Tumor Antigens; *Tetrahedron Lett.* **26** (1985) 2439.

[63] Kim, M.-J.; Hennen, W. J.; Sweers, H. M.; Wong, C.-H.: Enzymes in Carbohydrate Synthesis: *N*-Acetylneuraminic Acid Aldolase Catalyzed Reactions and Preparation of *N*-Acetyl-2deoxy-D-neuraminic Acid Derivatives; *J. Am. Chem. Soc.* **110** (1988) 6481.

[64] Augé, C.; Gautheron, C.: The Use of an Immobilized Aldolase in the First Synthesis of a Natural Deaminated Neuraminic Acid; *J. Chem. Soc., Chem. Commun.* **(1987)** 859.

[65] Augé, C.; David, S.; Gautheron, C.; Malleron, A.; Cavayé, B.: Preparation of Six Naturally Occurring Sialic Acids with Immobilized Acylneuraminate Pyruvate Lyase; *Nouv. J. Chem.* **12** (1988) 733.

[66] David, S.; Augé, C.: Immobilized Enzymes in Preparation Carbohydrate Chemistry; *Pure Appl. Chem.* **59** (1987) 1501.

[67] Simon, E. S.; Bednarski, M. D.; Whitesides, G. M.: Synthesis of CMP-NeuAc from *N*-Acetylglucosamine: Generation of CTP from CMP Using Adenylate Kinase; *J. Am. Chem. Soc.* **110** (1988) 7159.

[68] Spivak, C. T.; Roseman, S.: *N*-Acetyl-D-mannosamine (2-Acetamido-2-deoxy-D-mannose) and D-mannosamine hydrochloride; *J. Am. Chem. Soc.* **81** (1959) 2403.

[69] Brossmer, R.; Rose, U.; Kasper, D.; Smith, T. L.; Grasmuk, H.; Unger, F. M.: Enzymic Synthesis of 5-Acetamido-9-azido-3,5,9-trideoxy-D-*glycero*-D-*galacto*-2-nonulosonic Acid, a 9-Azido-9-deoxy Derivative of *N*-Acetylneuraminic acid; *Biochem. Biophys. Res. Commun.* **96** (1980) 1282.

[70] Unger, F. M.: The Chemistry and Biological Significance of 3-Deoxy-D-manno-2-octulosonic Acid (KDO); *Adv. Carbohydr. Chem. Biochem.* **38** (1981) 323.

[71] Bednarski, M. D.; Crans, D. C.; DiCosimo, R.; Simon, E. S.; Stein, P. D.; Whitesides, G. M.; Schneider, M. J.: Synthesis of 3-Deoxy-D-manno-2-octulosonate-8-phosphate (KDO-8-P) from D-Arabinose: Generation of D-Arabinose-5-Phosphate Using Hexokinase; *Tetrahedron Lett.* **29** (1988) 427.

[72] Mocali, A.; Aldinucci, D.; Paoletti, F.: Preparative Enzymic Synthesis and Isolation of D-Threo-2-pentulose 5-Phosphate (D-Xylose 5-Phosphate); *Carbohydr. Res.* **143** (1985) 288.

[73] Bolte, J.; Demuynck, C.; Samaki, H.: Utilization of Enzymes in Organic Chemistry: Transketolase Catalyzed Synthesis of Ketoses; *Tetrahedron Lett.* **28** (1987) 5525.

[74] Yokota, A.; Sasajima, K.: Formation of 1-Deoxyketoses by Pyruvate Dehydrogenase and Acetoin Dehydrogenase; *Agric. Biol. Chem.* **50** (1986) 2517.

[75] Reimer, L. M.; Conley, D. L.; Pompliano, D. L.; Frost, J. W.: Construction of An Enzyme-Targeted Organophosphonate Using Immobilized Enzyme and Whole Cell Synthesis; *J. Am. Chem. Soc.* **108** (1986) 8010.

[76] Turner, N. J.; Whitesides, G. M.: A Combined Chemical-Enzymatic Synthesis of 3-Deoxy-D-arabinoheptulosonic Acid 7-Phosphate; *J. Am. Chem. Soc.* **111** (1989) 624.

[77] Fronza, G.; Fuganti, C.; Grasselli, P.; Servi, S.: Baker's Yeast Mediated Synthesis of 4-Deoxy-D-lyxohexopyranose (4-Deoxy-D-mannose); *J. Org. Chem.* **52** (1987) 2086.

[78] Fronza, G.; Fuganti, C.; Grasselli, P.; Servi, S.: Baker's Yeast Mediated Synthesis of Epimeric 2,3-Dideoxy-2-*C*-methyl D-Glucose Derivatives; *Tetrahedron Lett.* **27** (1986) 4363.

[79] Drueckhammer, D. G.; Wong, C.-H.: Chemoenzymatic Syntheses of Fluorosugar Phosphates and Analogues; *J. Org. Chem.* **50** (1985) 5912.

[80] Percival, M. D.; Withers, D. G.: Applications of Enzymes in the Syntheses and Hydrolytic Study of 2-Deoxy-α-D-glucopyranosyl Phosphate; *Can. J. Chem.* **66** (1988) 1970.

[81] GROSS, A.; ABRIL, O.; LEWIS, J. M.; GERESH, S.; WHITESIDES, G. M.: Practical Synthesis of 5-Phospho-D-ribosyl α-1-pyrophosphate (PRPP): Enzymatic Routes from Ribose 5-Phosphate or Ribose; *J. Am. Chem. Soc.* **105** (1983) 7428.

[82] WONG, C.-H.; POLLACK, A.; MCCURRY, S. D.; SUE, L. M.; KNOWLES, J. R.; WHITESIDES, G. M.: Synthesis of Ribulose 1,5-Bisphosphate: Routes from Glucose 6-Phosphate (Via 6-Phosphogluconate) and from Adenosine Monophosphate (Via Ribose 5-Phosphate); *Meth. Enzymol.* **89** (1982) 108.

[83] GOLD, A. M.; OSBER, M. P.: α-D-Glucopyranosyl Fluoride: Substrate of Sucrose Phosphorylase; *Biochem. Biophys. Res. Commun.* **42** (1971) 469.

[84] BARILI, P. L.; BERTI, G.; CATELANI, C.; COLONNA, F.; MASTRORILLI, E.: Kinetic Resolutions of Racemic 3,4-Anhydropyranosides with Microsomal Epoxide Hydrolase: A Totally Synthetic Approach to Both Enantiomeric Forms of α- and β-Boivinopyranosides and to β-D-Olivopyranosides. In: Lichtenthaler, F. W.; Neff, K. H., Eds.; *Carbohydrates 1987, Abstracts of the 4th European Carbohydrate Symposium (Eurocarb IV)*, Darmstadt, July 1987, Gesellschaft Deutscher Chemiker (European Carbohydrate Organization); Abstr. A-102.

[85] PIGMAN, W. W.: Specificity, Classification, and Mechanism of Action of the Glycosidases; *Adv. Enzymol.* **5** (1944) 41.

[86] WALLENFELS, K.; WEIL, R.: β-Galactosidase. In: BOYER, P., Ed.; *The Enzymes*, Academic Press, New York (1972); 3rd edition, **Vol. 7**, p 617.

[87] NILSSON, K. G. I.: A Simple Strategy for Changing the Regioselectivity of Glycosidase-Catalysed Formation of Disaccharides; *Carbohydr. Res.* **167** (1987) 95.

[88] WILLEMOT, R. M.; LARONTE, V.; DURAND, G.: Use of Glycosidases in Water-Miscible Organic Solvents; XIVth Int. Carbohydr. Symp., Stockholm, August 1988, Abstr. B-82.

[89] CLANCY, M. J.; WHEELAN, W. M.: Enzymic Polymerisation of Monosaccharides II; *Arch. Biochem. Biophys.* **118** (1967) 730.

[90] WATKINS, W. M.: Enzymic Synthesis of Nitrogen-Containing Disaccharides by α-Galactosyl Transfer; *Nature (London)* **181** (1958) 117.

[91] TARENTINO, A. L.; MALEY, F.: The Enzymic Synthesis of *O*-α-D-Glucopyranosyl(1–6)-2-amino-2-deoxy-D-glucopyranose and its *N*-Acetyl Derivative; *Arch. Biochem. Biophys.* **130** (1969) 80.

[92] CHIBA, S.; SUGAWARA, S.; SHIMOMURA, T.; NAKAMURA, Y.: Comparative Biochemical Studies on α-Glucosidases. I. Substrate Specificity and Transglucosidation Action of an α-Glucosidase of Brewers' Yeast; *Agric. Biol. Chem.* **26** (1962) 787.

[93] CHIBA, S.; SHIMOMURA, T.: Comparative Biochemical Studies on α-Glucosidases. IV. Two α-D-Glucopyranosyl-D-mannoses Synthesized by Brewers' Yeast α-Glucosidase; *Agric. Biol. Chem.* **33** (1969) 807.

[94] CHIBA, S.; TAKAHASHI, N.; SHIMOMURA, T.: Comparative Biochemical Studies on α-Glycosidases. V. Two α-D-Glucopyranosyl-D-xyloses Synthesized by Brewers' Yeast α-Glucosidase; *Agric. Biol. Chem.* **33** (1969) 813.

[95] CHIBA, S.; SHIMOMURA, T.: Enzymic Synthesis of Oligosaccharides. I. α-D-Glucopyranosyl-D-fructoses Synthesized by the Transglucosidation Reaction of Brewers' Yeast α-Glucosidase; *Agric. Biol. Chem.* **35** (1971) 1292.

[96] CHIBA, S.; SHIMOMURA, T.: Enzymic Synthesis of Oligosaccharides. II. α-D-Glucopyranosyl-L-sorboses Synthesized by the Transglucosidation Reaction of Brewers' Yeast α-Glucosidase; *Agric. Biol. Chem.* **35** (1971) 1363.

[97] MATSUSAKA, K.; CHIBA, S.; SHIMOMURA, T.: Enzymic Synthesis of Oligosaccharides. IV. Enzymic Synthesis of 1,4-Di-α-D-glucosyl-L-sorbose and 1-α-Isomaltosyl-L-sorbose; *Agric. Biol. Chem.* **39** (1975) 725.

[98] MANNERS, D. J.; PENNIE, I. R.; STARK, J. R.: Studies on Carbohydrate-Metabolising

Enzymes. XVII. The Enzymic Synthesis of Some Disaccharides Containing α-D-Glycosyl Residues; *Carbohydr. Res.* **7** (1968) 291.

[99] MANNERS, D. J.; STARK, J. R.; TAYLOR, D. C.: Studies on Carbohydrate-Metabolising Enzymes. XXIII. *Trans* β-D-Glucosylation by Extracts of *Tetrahymena pyriformis* and *Ochromonas malhamensis*; *Carbohydr. Res.* **16** (1971) 123.

[100] CHIBA, S.; SHIMOMURA, T.; HATAKEYAMA, K.: Enzymic Synthesis of Oligosaccharides. III. Enzymic Synthesis of α-D-Glucopyranosyl-2-deoxy-D-glucose by Transglucosidation Reaction of Buckwheat α-Glucosidase; *Agric. Biol. Chem.* **39** (1975) 591.

[101] NISHI, K.; CHIBA, S.; SHIMOMURA, T.: Enzymic Synthesis of Oligosaccharides. V. Enzymic Synthesis of a Branched Trisaccharide, 2,4-Di-α-glucosylglucose; *Agric. Biol. Chem.* **39** (1975) 737.

[102] CHIBA, S.; ASADA-KOMATSU, J.; KIMURA, A.; KAWASHIMA, K.: A New Trisaccharide, 3-α-D-Glucosylsucrose, Synthesized by the Transglucosidase Action of Immobilized Buckwheat α-Glucosidase; *Agric. Biol. Chem.* **48** (1984) 1173.

[103] CHIBA, S.; YAMANA, O.: Enzymic Synthesis of a New Disaccharide, 6-*O*-α-Glucopyranosyl-2-deoxy-D-glucose, by Transglycosylation of *Schizosaccharomyces pombe* α-Glucosidase; *Agric. Biol. Chem.* **44** (1980) 549.

[104] HUBER, R. E.; KURZ, G.; WALLENFELS, K.: A Quantitation of the Factors Which Affect the Hydrolase and Transgalactosylase Activities of β-Galactosidase (*E. coli*) on Lactose; *Biochemistry* **15** (1976) 1994.

[105] HEDBYS, L.; LARSSON, P.-O.; MOSBACK, K.; SVENSSON, S.: Synthesis of the Disaccharide 6-*O*-β-D-Galactopyranosyl-2acetamido-2-deoxy-D-galactose Using Immobilized β-Galactosidase; *Biochem. Biophys. Res. Commun.* **123** (1984) 8.

[106] HEDBYS, L.; JOHANNSON, E.; MOSBACH, K.; LARSSON, P.-O.; GUNNARSON, A.; SVENSSON, S.: Glycosidases in Carbohydrate Synthesis 1: Synthesis of Galβ-(1–3)GalNAc by a new Enzymic Method Comprising the Sequential use of β-Galactosidase from Bovine Testes and *Escherichia coli*; XIVth Int. Carbohydr. Symp., Stockholm, August 1988, Abstr. B-124.

[107] SUYAMA, K.; ADACHI, S.; TOBA, S.; SOHMA, T.; HWANG, C.-J.; ITOH, T.: Isoraffinose (6-*O*-Galactosylsucrose) Synthesized by the Intermolecular Transgalactosylation Reaction of *Escherichia coli* β-Galactosidase; *Agric. Biol. Chem.* **50** (1986) 2069.

[108] AJISAKA, K.; FUJIMOTO, H.; NISHIDA, H.: Enzymic Synthesis of Disaccharides by Use of the Reversed Hydrolysis Activity of β-D-Galactosidases; *Carbohydr. Res.* **180** (1988) 35.

[109] AJISAKA, K.; FUJIMOTO, H.; NISHIDA, H.: Enzymatic Synthesis of Oligosaccharides by Use of a Reversed Hydrolysis Activity of β-Galactosidase. In: Lichtenthaler, F. W.; Neff, K. H., Eds.; *Carbohydrates 1987, Abstracts of the 4th European Carbohydrate Symposium (Eurocarb IV)*, Darmstadt, July 1987, Gesellschaft Deutscher Chemiker (European Carbohydrate Organization); Abstr. A-140.

[110] AJISAKA, K.; FUJIMOTO, H.: Enzymic Synthesis of Trisaccharides by Use of β-Galactosidase; XIVth Int. Carbohydr. Symp., Stockholm, August 1988, Abst. B-81.

[111] NILSSON, K. G. K.: A Simple Strategy for Changing the Regioselectivity of Glycosidase-Catalysed Formation of Disaccharides II. Enzymic Synthesis *in Situ* of Various Acceptor Glycosides; *Carbohydr. Res.* **180** (1988) 53.

[112] TANAKA, T.; OI, S.: Cellotriose Synthesis using D-Glucose as a Starting Material by Two Step Reactions of Immobilized β-Glucosidases; *Agric. Biol. Chem.* **49** (1985) 1267.

[113] KANDA, T.; NODA, I.; WAKABAYASHI, K.; NISIZAWA, K.: Transglycosylation Activities of Exo- and Endo-type Cellulases from *Irpex lacteus* (*Polyporus tulipiferae*); *J. Biochem.* **93** (1983) 787.

[114] AJISAKA, K.; NISHIDA, H.; FUJIMOTO, H.: The Synthesis of Oligosaccharides by the Reversed Hydrolysis Reaction of β-Glucosidase at High Substrate Concentration and at High Temperature; *Biotechn. Lett.* **9** (1987) 243.

[115] Ezure, Y.: Enzymic Synthesis of 4-*O*-α-D-Glucopyranosylmoranoline; *Agric. Biol. Chem.* **49** (1985) 2159.

[116] Abe, J.-I.; Mizowaki, N.; Hizukuri, S.; Koizumi, K.; Utamura, T.: Synthesis of Branched Cyclomaltooligosaccharides using *Pseudomonas* isoamylase; *Carbohydr. Res.* **154** (1986) 81.

[117] Takasaki, Y.: Studies on Enzymic Production of Oligosaccharide I. Production of Maltohexaose by α-Amylase from *Bacillus circulans* G-6; *Agric. Biol. Chem.* **46** (1982) 1539.

[118] Takasaki, Y.: Maltohexaose-Producing Amylase from *Bacillus* species; *Deupun Kagaku* **29** (1982) 145; *J. Jpn. Soc. Starch Sci.* **29** (1982) 145.

[119] Van Eikeren, P.; White, W. A.; Chipman, D. M.: Synthesis of Oligosaccharides Containing 2-Acetamido-2-deoxyxylose by Chemical and Enzymic Methods; *J. Org. Chem.* **38** (1973) 1831.

[120] Deya, E.; Nojiri, K.; Igarashi, S.: New Oligosaccharides Synthesized by Transglycosylation with Lysozyme; *Nippon Nogei Kagaku Kaishi* **58** (1984) 273; *Chem. Abstr.* **101** (1984) 38754r.

[121] Siriwardena, A.; Thiem, J.; Treder, W.: Preparative Enzymatic Synthesis of Fucosyl Disaccharides. In: Lichtenthaler, F. W.; Neff, K. H., Eds.; *Carbohydrates 1987, Abstracts of the 4th European Carbohydrate Symposium (Eurocarb IV)*, Darmstadt, July 1987, Gesellschaft Deutscher Chemiker (European Carbohydrate Organization); Abstr. B-50.

[122] Svensson, S. C. T.; Thiem, J.: Enzymatic Synthesis of Fucosyl Disaccharides; XIVth Int. Carbohydr. Symp., Stockholm, August 1988, Abstr. B-50.
Purification of α-L-Fucosidase by C-Glycoside Affinity Chromatography and the Enzymic Synthesis of α-L-Fucosyl Disaccharides; *Carbohydr. Res.* **200** (1990) 391.

[123] Alexander, J. K.: Disaccharide Synthesis by Cellobiose Phosphorylase; *Arch. Biochem. Biophys.* **123** (1968) 240.

[124] Nakatani, H.; Tanaka, A.; Hiromi, K.: Continuous Production of Maltooligosaccharides from Glucose 1-Phosphate by Amylase-Phosphorylase Reactor; *J. Appl. Biochem.* **5** (1983) 371.

[125] Ooi, Y.; Hashimoto, T.; Mitsuo, N.; Satoh, T.: Enzymic Synthesis of Chemically Unstable Cardiac Glycosides by β-Galactosidase from *Aspergillus oryzae*; *Tetrahedron Lett.* **25** (1984) 2241.

[126] Mitsuo, N.; Takeishi, H.; Satoh, T.: Synthesis of -Alkyl Glucosides by Enzymic Transglucosylation; *Chem. Pharm. Bull.* **32** (1984) 1183.

[127] Kusama, S.; Kusakabe, J.; Nakamura, Y.; Eda, S.; Murakami, K.: Studies on the β-1,3-Glucanase System of *Streptomyces*. V. Transglucosylation into Stevioside by the Enzyme System from *Streptomyces* sp; *Agric. Biol. Chem.* **50** (1986) 2445.

[128] Barnett, J. E. G.; Jarvis, W. T. S.; Munday, K. A.: Enzymic Hydrolysis of the Carbon-Fluorine Bond of α-D-Glucosyl Fluoride by Rat Intestinal Mucosa. Localization of Intestinal Maltase; *Biochem. J.* **103** (1967) 699.

[129] Barnett, J. E. G.; Jarvis, W. T. S.; Munday, K. A.: The Hydrolysis of Glycosyl Fluorides by Glycosidases; *Biochem. J.* **105** (1967) 669.

[130] Barnett, J. E. G.: The Hydrolysis of Glycosyl Fluorides by Glycosidases. Determination of the Anomeric Configuration of the Products of Glycosidase Action; *Biochem. J.* **123** (1971) 607.

[131] Barnett, J. E. G.: Unusual Substrates for Enzymes Which Hydrolyse Glycosidic Bonds; *Int. Biochem. J.* **6** (1975) 321.

[132] Thiem, J.; Treder, W.; Schlingmann, M.; Keller, R.; Wiesner, M.: Disaccharidfluoride; Verfahren zur enzymatischen Herstellung mit α-Glycosyfluoriden als Substraten; Patent application P 37 22 812.9 (1987).

[133] Okada, G.; Hehre, E. J.: *De novo* Synthesis of Glycosidic Linkages by Glycosylases: Utilization of α-D-Glucopyranosyl Fluoride by Amylosucrase; *Carbohydr. Res.* **26** (1973) 240.

[134] Figures, W. R.; Edwards, J. R.: α-D-Glucopyranosyl Fluoride as a D-Glucopyranosyl Donor for a Glycosyltransferase Complex from *Streptococcus mutans* FA1; *Carbohydr. Res.* **48** (1976) 245.

[135] Hehre, E. J.; Brewer, C. F.; Genghof, D. S.: Scope and Mechanism of Carbohydrase Action. Hydrolytic and Nonhydrolytic Action of β-Amylase on α- and β-Maltosyl Fluoride; *J. Biol. Chem.* **245** (1979) 5942.

[136] Okada, G.; Genghof, D. S.; Hehre, E. J.: The Predominantly Nonhydrolytic Action of Alpha Amylase on α-Maltosyl Fluoride; *Carbohydr. Res.* **71** (1979) 287.

[137] Kitahata, S.; Brewer, C. F.; Genghof, D. S.; Sawai, T.; Hehre, E. J.: Scope and Mechanism of Carbohydrase Action. Stereocomplementary Hydrolytic and Glucosyl-Transferring Actions of Glucoamylase and Glucodextranase with α- and β-D-Glucosyl Fluoride; *J. Biol. Chem.* **256** (1981) 6017.

[138] Palm, D.; Blumenauer, G.; Klein, H. W.; Blanc-Muesser, M.: α-Glucan Phosphorylases Catalyze the Glucosyl Transfer from α-D-Glucosyl Fluoride to Oligosaccharides; *Biochem. Biophys. Res. Commun.* **111** (1983) 530.

[139] Genghof, D. S.; Hehre, E. J.: *De novo* Glycosidic Linkage Synthesis by Glycosylases. α-D-Glucosyl Fluoride Polymerization by Dextransucrase; *Proc. Soc. Exp. Biol. Med.* **140** (1972) 1298.

[140] Grier, D. J.; Mayer, R. M.: Dextransucrase: Studies on Donor Substrate Specificity; *Arch. Biochem. Biophys.* **212** (1981) 651.

[141] Mae Jung, S.; Mayer, R. M.: Dextransucrase: Donor Substrate Reactions; *Arch. Biochem. Biophys.* **208** (1981) 288.

[142] Kasumi, T.; Tsumuraya, Y.; Brewer, C. F.; Kersters-Hilderson, H.; Claeyssens, M.; Hehre, E. J.: Catalytic Versatility of *Bacillus pumilus* β-Xylosidase: Glycosyl Transfer and Hydrolysis Promoted with α- and β-D-Xylosyl Fluoride; *Biochemistry* **26** (1987) 3010.

[143] Hehre, E. J.; Sawai, T.; Brewer, C. F.; Nakamo, M.; Kanda, T.: Trehalase: Stereocomplementary Hydrolytic and Glycosyl Transfer Reactions with α- and β-D-Glucosyl Fluoride; *Biochemistry* **21** (1982) 3090.

[144] Kasumi, T.; Brewer, C. F.; Reese, E. T.; Hehre, E. J.: Catalytic Versatility of Trehalase: Synthesis of α-D-Glucopyranosyl-α-D-xylopyranoside from β-D-Glucosyl Fluoride and α-D-Xylose; *Carbohydr. Res.* **146** (1986) 39.

[145] Johannson, E.; Hedbys, L.; Mosbach, K.; Larsson, P.-O.; Gunnarson, A.; Svensson, S.: Glycosidases in Carbohydrate Synthesis 2: Applications of Glycosidase-Catalyzed Reversion Reactions; XIVth Int. Carbohydr. Symp., Stockholm, August 1988, Abstr. B-125.

[146] Beyer, T. A.; Sadler, J. E.; Rearick, J. I.; Paulson, J. C.; Hill, R. L.: Glycosyltransferases and Their use in Assessing Oligosaccharide Structure and Structure Function Relationships; *Adv. Enzymol.* **52** (1981) 23.

[147] Watkins, W. M.: Glycosyltransferases. Early History, Development and Future Prospects; *Carbohydr. Res.* **149** (1986) 1.

[148] Nunez, H. A.; Barker, R.: Enzymatic Synthesis and Carbon-13 Nuclear Magnetic Resonance Conformational Studies of Disaccharides Containing β-D-[1-^{13}C] Galactopyranosyl Residues; *Biochemistry* **19** (1980) 489.

[149] Wong, C.-H.; Haynie, S. L.; Whitesides, G. M.: Enzyme-Catalyzed Synthesis of *N*-Acetyllactosamine with *in Situ* Regeneration of Uridine 5′-Diphosphate Galactose; *J. Org. Chem.* **47** (1982) 5416.

[150] Thiem, J.; Treder, W.: Synthesis of the Trisaccharide Neu-5-Ac-α(2–6)Gal-

β(1–4)GlcNAc by the Use of Immobilized Enzymes; *Angew. Chem.* **98** (1986) 1100; *Angew. Chem. Int. Ed. Engl.* **25** (1986) 1096.

[151] Thiem, J.; Treder, W.; Keller, R.; Schlingmann, M.: Verfahren zur enzymatischen Synthese eines *N*-Acetyl-neuraminsäure-haltigen Trisaccharids; *Ger. Offen* DE 3 626 915 (8.8.1986; 11.2.1988); *Chem. Abstr.* **109** (1988) 72046m.

[152] Augé, C.; David, S.; Mathieu, C.; Gautheron, C.: Synthesis with Immobilized Enzymes of Two Trisaccharides, One of Them Active as the Determinant of a Stage Antigen; *Tetrahedron Lett.* **25** (1984) 1467.

[153] Augé, C.; Mathieu, C.; Mérienne, C.: The Use of an Immobilized Cyclic Multi-Enzyme System to Synthesize Branched Penta- and Hexasaccharides Associated with Blood-Group I Epitopes; *Carbohydr. Res.* **151** (1986) 147.

[154] Thiem, J.; Treder, W.; Wiemann, T.: Formation of Oligosaccharides by Preparative Enzymatic Glycosylations; DECHEMA Biotechnology Conferences, VCH-Verlagsgesellschaft, Weinheim (1988); Vol. 2, p 189.

[155] Thiem, J.; Wiemann, T.: Kombinierte chemoenzymatische Synthese von *N*-Glycoproteinbausteinen; *Angew. Chem.* **102** (1990) 78; *Angew. Chem. Int. Ed. Engl.* **29** (1990) 82.

[156] Palcic, M. M.; Srivastava, O. P.; Hindsgaul, O.: Transfer of D-Galactosyl Groups to 6-*O*-Substituted 2-Acetamido-2-deoxy-D-glucose Residues by Use of Bovine D-Galactosyltransferase; *Carbohydr. Res.* **159** (1987) 315.

[157] Yates, A. D.; Feeney, J.; Donald, A. S. R.; Watkins, W. M.: Characterisation of a Blood-Group A-Active Tetrasaccharide Synthesised by a Blood-Group *B* Gene-Specified Glycosyltransferase; *Carbohydr. Res.* **130** (1984) 251.

[158] Kean, E. L.; Roseman, S.: CMP-Sialic Acid Synthethase; *Meth. Enzymol.* **8** (1966) 208.

[159] Schauer, R.; Wember, M.; Ferreira Do Amaral, C.: Synthesis of CMP-Glycosides of Radioactive *N*-Acetyl-, *N*-Glycoloyl-, *N*-Acetyl-7-*O*-acetyl- and *N*-Acetyl-8-*O*-acetyl Neuraminic Acids by CMP-Sialate Synthase from Bovine Submaxillary Glands; *Hoppe Seyler's Z. Physiol. Chem.* **353** (1972) 883.

[160] Haverkamp, J.; Beau, J.-M.; Schauer, R.: Improved Synthesis of CMP-Sialates Using Enzymes from Frog Liver and Equine Submandibular Glands; *Hoppe Seyler's Z. Physiol. Chem.* **360** (1979) 159.

[161] Augé, C.; Gautheron, C.: An Efficient Synthesis of Cytidine Monophospho-*N*-acetyl-neuraminic Acid with Four Immobilized Enzymes. In: Lichtenthaler, F. W.; Neff, K. H., Eds.; *Carbohydrates 1987, Abstracts of the 4th European Carbohydrate Symposium (Eurocarb IV)*, Darmstadt, July 1987, Gesellschaft Deutscher Chemiker (European Carbohydrate Organization); Abstr. A-170.

[162] Augé, C.; Gautheron, C.: An Efficient Synthesis of Cytidine Monophosphosialic Acids with Four Immobilized Enzymes; *Tetrahedron Lett.* **29** (1988) 789.

[163] Higa, H. H.; Paulson, J. C.: Sialylation of Glycoprotein Oligosaccharides with *N*-Acetyl-, *N*-Glycoloyl-, and *N,O*-Diacetylneuraminic Acids; *J. Biol. Chem.* **260** (1985) 8838.

[164] Sabesan, S.; Paulson, J. C.: Combined Chemical and Enzymatic Synthesis of Sialyloligosaccharides and Characterization by 500-MHz ^{1}H and ^{13}C NMR Spectroscopy; *J. Am. Chem. Soc.* **108** (1986) 2068.

[165] De Heij, H. T.; Kloosterman, M.; Koppen, P. L.; Van Boom, J. H.; Van Den Eijnden, D. H.: Combined Chemical and Enzymatic Synthesis of a Disialylated Tetrasaccharide Analogue to M and N Bloodgroup Determinants of Glycophorin A; *J. Carbohydr. Chem.* **7** (1988) 209.

[166] Nilsson, K. G. I.; Gudmundsson, B.-M. E.: Synthesis of CMP-Neu5Ac and Neu5Acα(2–3)Galβ(1–3)GalNAcα-OEt with Immobilised Cytidine 5′-Monophos-

phosphosialate Synthase and β-D-Galactoside α(2–3)Sialyltransferase. In: *Sialic Acids 1988, Proceedings of the Japanese-German Symposium on Sialic Acids*, Berlin, May 1988, Abstr. p 30.

[167] NILSSON, K. G. I.: Glycosidase- and Sialyltransferase-Catalysed Synthesis of Sialic Acid Containing Trisaccharides. In: *Sialic Acids 1988, Proceedings of the Japanese-German Symposium on Sialic Acids*, Berlin, May 1988, Abstr. p 28.

[168] ZEHAVI, U.: Polymers as Supports for Enzymic Oligo- and Polysaccharide Synthesis; *Reactive Polymers* **6** (1987) 189.

[169] ZEHAVI, U.: Applications of Photosensitive Protecting Groups in Carbohydrate Chemistry; *Adv. Carbohydr. Chem. Biochem.* **46** (1988) 179.

[170] ZEHAVI, U.; HERCHMAN, M.: Enzymic Synthesis of Oligosaccharides on a Polymer-Support, Light Sensitive, Water Soluble Substituted Poly(vinyl Alcohol); *Carbohydr. Res.* **128** (1984) 160.

[171] ZEHAVI, U.; SADEH, S.; HERCHMAN, M.: Enzymic Synthesis of Oligosaccharides on a Polymer-Support. Light Sensitive, Substituted Polyacrylamide Beads; *Carbohydr. Res.* **124** (1983) 23.

[172] ZEHAVI, U.; HERCHMAN, M.: Enzymic Synthesis of Oligosaccharides on an α-Chymotrypsin-Sensitive Polymer; *Carbohydr. Res.* **133** (1984) 339.

[173] FENSELAU, D.; PALLANTE, S.; BATZINGEN, R. B.; BENSON, W. R.; BARRON, R. P.; SHEININ, F. B.; MAIENTAL, M.: Mandelonitrile -Glucuronide; Synthesis and Characterization; *Science* **198** (1977) 625.

[174] NEIDLEIN, R.; WU, M.; HEGE, H. G.: Synthesis of Glucuronides of Propafenone and 5-Hydroxypropafenone by Sepharose-Bound Uridine 5′-Diphosphoglucuronyltransferase; *Arzneim.-Forsch./Drug Res.* **38** (1988) 1257.

[175] TREDER, W.; THIEM, J.; SCHLINGMANN, M.: Enzymatic Synthesis of Cyclodextrins with α-Glucosylfluoride as a Substrate for Cyclodextrin-α(1 → 4)glucosyltransferase; *Tetrahedron Lett.* **27** (1986) 5605.

[176] THIEM, J.; TREDER, W.; KELLER, R.; SCHLINGMANN, M.: Enzymatische Synthese von Cyclodextrinen mit α-Glucosyl-fluorid als Substrat für Cyclodextrin-α-(1 → 4)-glucosyltransferase; *Ger. Offen* DE 3,626,231 (2.8.1986; 4.2.1988); Synthesis of α- and β-Cyclodextrins from α-Glucosyl Fluoride Using Cyclodextrin α-(1 → 4)Glucosyltransferase; *Chem. Abstr.* **109** (1988) 228758h.

[177] TREDER, W.; THIEM, J.; KELLER, R.; SCHLINGMANN, M.: Preparative Enzymic Syntheses of Oligosaccharides with α-Glycosyl Fluorides as Substrates. In: Lichtenthaler, F. W.; Neff, K. H., Eds.; *Carbohydrates 1987, Abstracts of the 4th European Carbohydrate Symposium (Eurocarb IV)*, Darmstadt, July 1987, Gesellschaft Deutscher Chemiker (European Carbohydrate Organization); Abstr. A-101.

[178] HEHRE, E. J.; MIZUKAMI, K.; KITAHATA, S.: Cyclodextringlycosyltransferase. Catalysis of Irreversible C-F-Glycosylic Bond Cleavage and *de Novo* Maltosidic Linkage Synthesis; *Denpun Kagaku* **30** (1983) 76; *J. Jpn. Soc. Starch Sci.* **30** (1983) 76.

[179] TREDER, W.; ZEHAVI, U.; THIEM, J.; HERCHMAN, M.: Disproportionation by Immobilized Cyclodextrin-α,1–4-Glucosyltransferase Reaction with Soluble Polymeric Acceptors; *Biotechn. Appl. Biochem.* **11** (1989) 362.

[180] CARD, P. J.; HITZ, W. D.: Synthesis of 1′-Deoxy-1′-fluorosucrose via Sucrose Synthetase Mediated Coupling of 1-Deoxy-1-fluorofructose with Uridine Diphosphate Glucose; *J. Am. Chem. Soc.* **106** (1984) 5348.

[181] CARD, P. J.; HITZ, W. D.; RIPP, K. G.: Chemoenzymatic Syntheses of Fructose-Modified Sucroses via Multienzyme Systems. Some Topographical Aspects of the Binding of Sucrose to a Sucrose Carrier Protein; *J. Am. Chem. Soc.* **108** (1986) 158.

Techniques in Biotechnology

Fluorescence in Biotechnology

by W. Schmidt and H. Schneckenburger

Contents

Dr. habil. W. Schmidt,
Universität Konstanz,
D-7750 Konstanz,
and
Fachhochschule Aalen,
D-7080 Aalen, Fed. Rep. Germany

Prof. Dr. H. Schneckenburger,
Fachhochschule Aalen,
D-7080 Aalen,
and
Institut für Lasertechnologien in der Medizin an der Universität Ulm,
D-7900 Ulm, Fed. Rep. Germany

1 Introduction

The history of natural science and technology documents the close ties between basic science and the associated science known as technology. This applies especially to biochemistry and its practical derivative, biotechnology, which serve each other in exemplary fashion. Light spectroscopy in particular is a classic, noninvasive analytical technique in that the sample under study remains essentially intact. On-site *in vivo* measurements permit a continual monitoring of metabolic processes possible in no other analytical technique. From the optical standpoint, light spectroscopy can be subdivided into the rough categories of *absorption*, *reflected light* (or *reflection*), *scattering* or *dispersion*, and *luminescence spectroscopy*, which are further subdivided (Fig. 1, modified after SCHMIDT [1]).

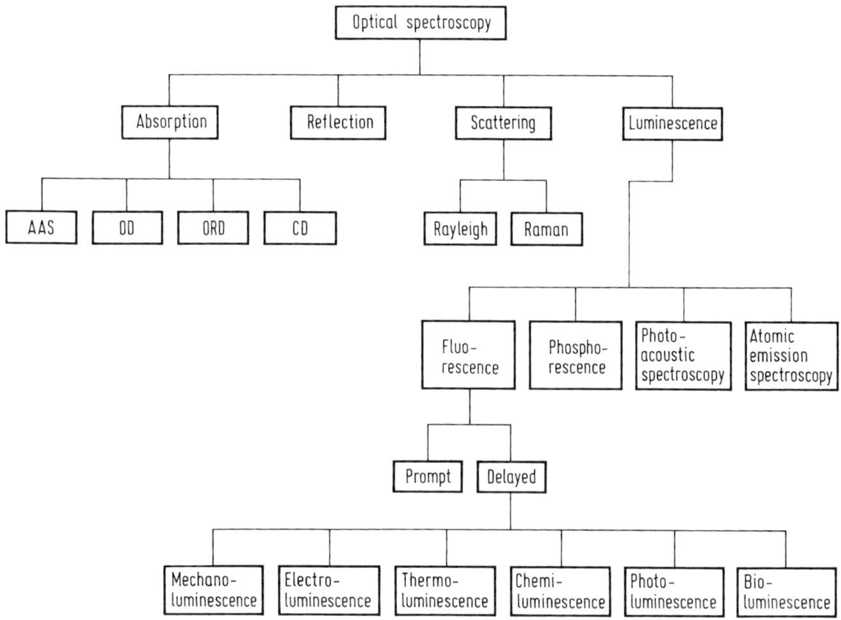

Fig. 1 Schematic diagram of the types of optical spectroscopy, with special reference to luminescence

In this chapter we restrict ourselves to one area of light spectroscopy, *luminescence spectroscopy*, of special importance to biotechnology. We discuss a variety of applications, although space limitations preclude an in-depth treatment of the subject. We recommend the textbooks of LAKOWICSZ [2] and DEMAS [3] for further study on theoretical and practical fluorimetric principles. DERENIAK and CROWE [4] treat the subject of radiation detectors in detail, and KEITZ deals with light diffraction and light measurements [5]. The *Handbook of optics* [6] provides answers to nearly all light-spectroscopic and instrumental questions. This chapter is intended to give a balanced discussion of theory and instrumental technology, supplemented by a series of practical examples.

Initially, we should consider two different aspects of luminescence analysis. First, the technique permits the identification of certain luminescing components and, concomitantly, in a limited manner, the determination of their concentration. Second, the technique also involves a probe that, like a transmitter, sends light signals detailing minute functional changes within a microscopic unit such as a cell. Deciphering these messages is the task of the analyst. In addition, fluorescence microscopy, being an image-rendering technique, permits a direct and very specific view of microstructures present within a cell (Section 5).

A basic, frequently overlooked difference between absorption and luminescence spectroscopy is the method of measuring. The absorption, $A(\lambda)$, of a given species is defined as

$$A(\lambda) = \log I_0/I = \varepsilon(\lambda) \cdot c \cdot l \qquad (1)$$

where

I_0 is the intensity of the light beam incident upon the sample,
I is the intensity of the transmitted light beam,
$\varepsilon(\lambda)$ is the extinction coefficient,
c is the concentration, and
l is the path length.

However, the nonabsorbed light I is in fact what is measured, and A is therefore determined indirectly. This provides, on the one hand, an elegant means of eliminating the instrumental dispersion of an absorption photometer by comparing a sample with a reference sample, thus arriving at an absolute absorption measurement A, sometimes also called *OD* (for *optical density*). The measurement of absorption with this method, however, is many orders of magnitude less sensitive than a measurement by means of luminescence, because of the required difference ($A = \log I_0 - \log I$). With weakly absorbing samples a (small) difference between two large signals must be detected; if absorption is strong, the intensity I of the transmitted light will be small; thus, in both cases the *signal-to-noise ratio*, *SNR*, is seriously affected.

In contrast to absorption, luminescence measurements are relative, yet of a direct nature. Light emitted by a given species is measured directly, which means that light has interacted with the species. However, the correction of a luminescence spectrum with regard to instrumental dispersion is more difficult because of the absence of a reference: an objective measurement procedure analogous to that for absorption spectroscopy is not available. Fluorescence data, therefore, intrinsically carry an absolute error of about $\pm 10\%$. This technique is fundamentally more informative than is absorption spectroscopy, however. While in absorption spectroscopy the absorption coefficient, $\varepsilon(\lambda)$, remains the sole parameter that can be determined, a whole series of independent parameters can be determined with fluorescence spectroscopy: parameters such as quantum yield, ϕ, distance for nonradiative energy transfer, R_c, lifetime, τ, polarization, p, and many others, each of which provides specific information about a given system.

2 Fluorescence spectra

2.1 Fundamentals

Fluorescence spectroscopy is simply the reverse of absorption spectroscopy. This statement is based upon KIRCHHOFF's (1859) famous principle, which postulates that each substance absorbing at a given wavelength (color) can also emit quanta at the same wavelength. EINSTEIN clarified the theory behind this principle. In essence, the same information can be obtained with either absorption spectroscopy or fluorescence. Thus, it is not surprising that a fluorescence spectrum can be directly determined, using thermodynamic principles, from the first excited state of the absorption spectrum, according to the following equation:

$$f(\lambda) = 8\pi\varepsilon(\lambda)\,\lambda^{-4} \cdot \exp(hc/kT\lambda) \qquad (2)$$

where

λ is the wavelength in nm,
the fluorescence $f(\lambda)$ is a function of the absorption $\varepsilon(\lambda)$,
h is Planck's constant, $6.626 \cdot 10^{-34}$ J · s,
c is the speed of light, $2.998 \cdot 10^{8}$ m · s^{-1},
k is the Boltzmann constant, $1.38062 \cdot 10^{-23}$ J · K^{-1}, and
T is the absolute temperature in K.

An example is shown in Fig. 2, where the fluorescence bands are calculated from the α band of chlorophyll *a* and compared to the measured fluorescence bands.

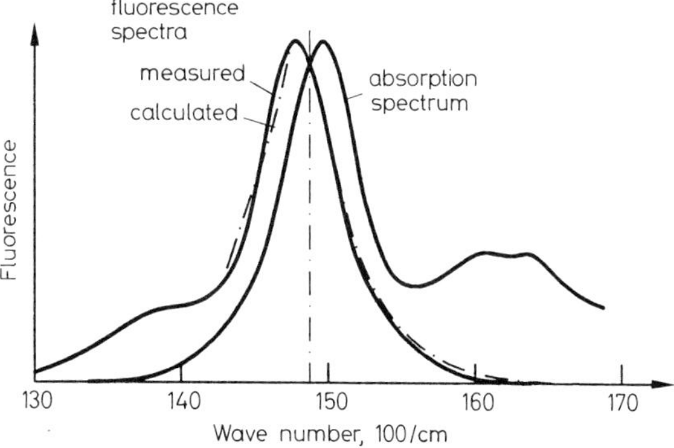

Fig. 2 Comparison of the absorption spectrum (α bands) and fluorescence spectrum of chlorophyll *a* (solid lines). The fluorescence spectrum can also be calculated from the absorption spectrum on thermodynamic grounds (dashed line). The *wave number*, k, is defined as $k = 1/\lambda$, and has the dimension of energy (units of 1/cm, where 1 cm^{-1} = 1.99×10^{-23} J)

As we mentioned in the introduction, there are *practical* differences between fluorescence and absorption spectroscopy that are particularly pertinent to molecular spectroscopy. In order to understand these, and to fully avail ourselves of the advantages of fluorescence spectroscopy, we must first explore some of the theory of fluorescence.

2.2 Potential diagram

Generally, the light emission of *electronically* excited molecules that is produced upon their return to the ground state is termed *luminescence*. This can be further subdivided into *fluorescence* and *phosphorescence*, based upon the difference in the electronic nature of the emission states of the emitting molecules (whether the emission is from the first excited singlet, S_1, or triplet state, T_1, respectively).

With *potential diagrams* we can use luminescence to learn more about the different photochemical and photophysical characteristics of molecules and particularly of complex systems. Figure 3 is a potential diagram for a simple hypothetical diatomic molecule; the potential energy of the ground state and the first singlet and triplet states is depicted as a function of nuclear distance. In the plots that result, one finds all possible molecular energy levels in the form of discrete vibrational levels upon which are overlaid a series of rotational states with even smaller energy intervals. This is usually called a *potential well*.

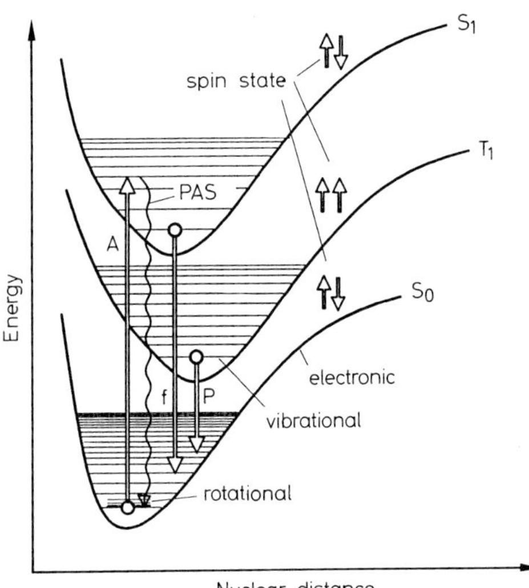

Fig. 3 Typical potential diagram for a hypothetical diatomic molecule. Excited states, characterized by their electronic configurations (singlet, S_0, S_1; triplet, T_1), are at a greater nuclear distance. The total energy is the sum of the electronic, vibrational, and rotational contributions, which result in the typical broad band spectra. The transition states for absorption, *A*, fluorescence, *f*, phosphorescence, *P*, and photoacoustic spectroscopy, *PAS*, are illustrated

The superposition of all these individual states produces the broad band spectra characteristic of molecules, in contrast to the sharp line spectra of atoms that can be excited only electronically. Electron transfer always occurs "vertically" (Franck-Condon principle), that is, the nuclear distance does not change during the transfer and neither does the corresponding absorption, *A*, fluorescence, *F*, or phosphorescence, *P*. The transfer of electrons, as illustrated in Figure 3, has a significant influence upon the shape and position of molecular luminescence spectra when it occurs primarily in vibrationally excited states.

In a broader context we should also designate *photoacoustic spectroscopy* (*PAS*) as emission spectroscopy; this technique is of increasing importance, but we will not discuss it in depth here. The energy released by electron transfer is in the form of *heat* rather than light [7] (shown in Figure 3 as the wavy line). *Atomic emission spectroscopy* (*AES*), important for the determination of microelements in biotechnology, is also beyond the scope of this discussion [8].

2.3 Jablonski diagram

If the potential curves are plotted without regard to the variable nuclear distances, then the different molecular excited states can be succinctly illustrated in a *Jablonski diagram*, which depicts only the energy relationships. Figure 4 is the Jablonski diagram of chlorophyll *a* and the observed absorption and emission spectra from which it is derived.

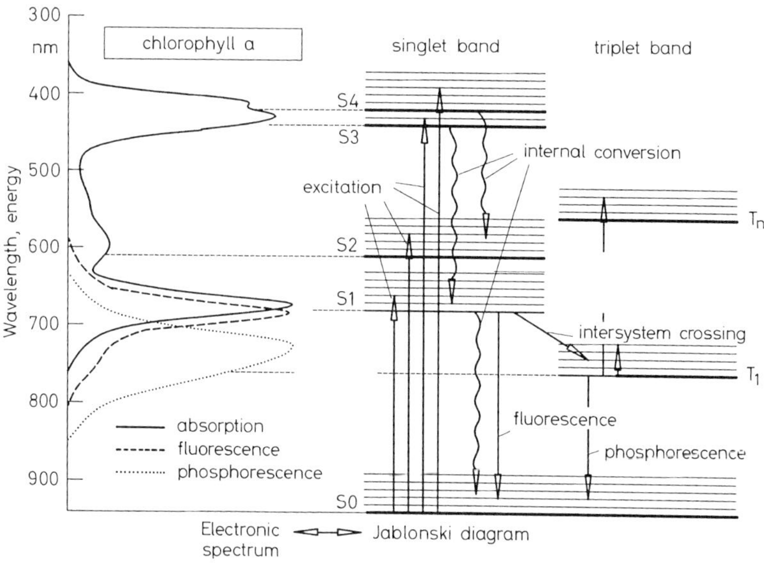

Fig. 4 Comparison of the Jablonski diagram and the corresponding chlorophyll *a* spectrum. The ordinate is only approximate because workers in the biological sciences use wavelength, λ, rather than wave number, k; the Jablonski diagram includes a representation of energy states. The various transitions are shown

2.4 Modification of fluorescence spectra

Fluorescence spectra, spectral behavior, and fluorescence intensity (quantum yield, ϕ) are all sensitive to external influences such as the temperature and the affinity or viscosity of the solvent. The solvent and the microenvironment strongly influence the fluorescence spectra [2]. With increasing polarity of the microenvironment the emission becomes weaker and the spectra are generally red-shifted (Figure 5 is an example of this with flavin), an effect that has analytical applications and that we will discuss in further

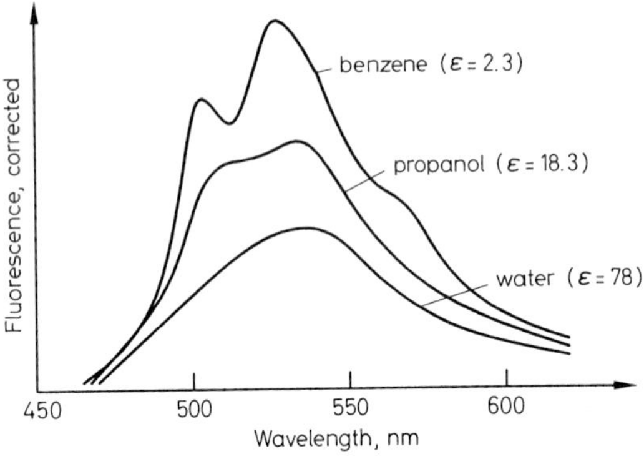

Fig. 5 Dependence of the fluorescence emission spectrum (corrected) of flavin on solvent polarity (represented by the dielectric constant, ϵ). The more hydrophilic microenvironment resulting from hydrogen bonding leads to a reduced fluorescence and a slight spectral red shift (excitation wavelength, 380 nm; spectral resolution, 5 nm)

detail. Membrane-bound flavin is known as a biochemical switch for changing from 1- to 2-electron transfer [9]. Dipalmitoyl lecithin undergoes a phase change at 41 °C from crystalline to liquid crystalline, and the flavin chromophore passes into the more hydrophobic portion of the membrane [10]. Figure 6 is a demonstration of how elegantly fluorescence monitors the kinetics of this process. Interestingly, this is an example of the phenomenon in solid-state physics called *hysteresis*, in which the means of arriving at a condition takes on special importance. Here, the fluorescence emission is a function not only of the present condition (temperature, in this example) but also of the way the present condition was reached (whether the present temperature was reached from a higher or lower one). The system "knows" its previous history.

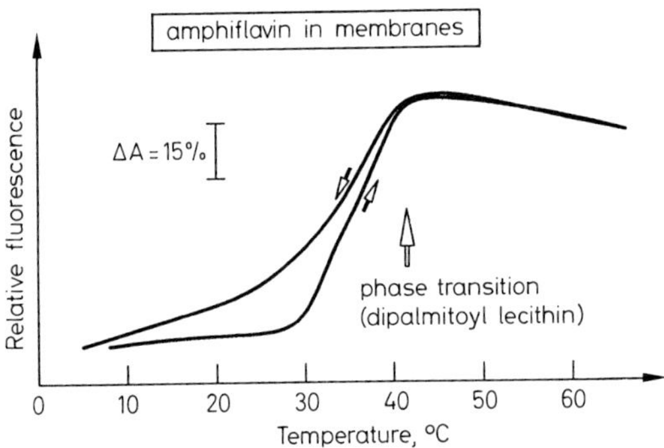

Fig. 6 Temperature dependence of the fluorescence of amphiflavin anchored in a single-layered dipalmitoyl lecithin vesicle membrane by a long aliphatic hydrocarbon chain; excitation wavelength, 370 nm; emission wavelength, 540 nm. The hysteresis is distinct and reproduces well with cyclic temperature changes

2.5 Measurement of fluorescence spectra

Figure 7 is a sketch of the basic design of a fluorimeter for determining fluorescence spectra, distinguishing between the paths of the irradiating (excitation) and emitted beams (fluorimeter design by Jobin Yvon, model JY3, modified). Grating 1 produces monochromatic radiation, a portion of which is diverted to a quantum counter (see

Section 2.6) by a half-silvered mirror; the remainder is used to excite the sample. The scattered fluorescence emission is detected at a right angle to the incident beam, thus eliminating the relatively large amounts of stray and transmitted light. Accordingly, depending on whether the excitation wavelength (grating 1) or the emission wavelength (grating 2) is varied, one observes fluorescence excitation or emission spectra, respectively.

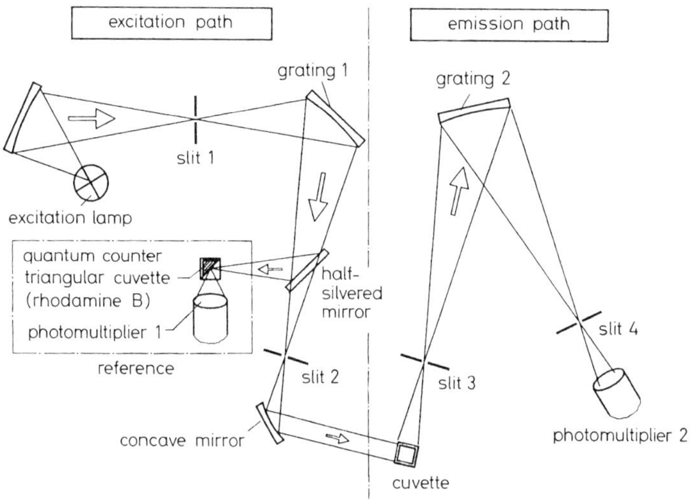

Fig. 7 Basic design of a fluorimeter, showing excitation and emission beam paths with two monochromators (gratings 1 & 2). Part of the excitation light is deflected to correct for the dispersion of the lamp and the energy of the excitation beam (see text). The four slits are usually adjustable. Nowadays the photomultiplier signal is generally processed by an on- or off-line computerized data logger

2.6 Correction of fluorescence spectra

As mentioned earlier, the recorded spectra are distorted by instrumental dispersion, the wavelength dependence of the lamp, and reflections from both the monochromators and the detector. There are a number of ways to eliminate these influences. Because theoretically the excitation and absorption spectra are identical, a correction curve (correction spectrum), $A(\lambda)$, for the excitation beam path (lamp and grating 1) can be determined; the excitation spectrum of a molecule is compared with an (independently determined) absorption spectrum (*quotient formation*). Furthermore, the correction curve $A(\lambda)$ of the excitation beam (lamp and grating 1) can also be determined with a *quantum counter*. A highly concentrated solution of rhodamine B in ethylene glycol (3 g/l) is a reliable quantum counter. Such a concentrated sample absorbs *all* incident photons regardless of their wavelength, λ. Since the emission of a specific wavelength is observed, the dispersion of the emission beam path is irrelevant and, characteristically for a quantum counter, the measured signal is proportional to the incident number of quanta. Because of the high sample absorption, the correction spectrum is measured only in the reflection mode with a triangular cuvette. The variable scattered light from the irradiation beam is minimized by irradiating the triangular cuvette at an angle other than 45° and by using appropriate cutoff filters. The rhodamine B quantum counter is useful for a wide band of wavelengths from 200 to 600 nm. Unfortunately, there is no corresponding quantum counter able to correct wavelengths above 600 nm [11].

The correction curve, $E(\lambda)$, and the efficiency of the emission path (grating 2 and photomultiplier) are required to calculate a corrected emission spectrum. Because the direct measurement of $E(\lambda)$ is difficult and time-consuming, the publication of uncorrected fluorescence spectra – in contrast to absorption spectra – is still the rule. However, a computerized method involving the comparison of the *measured* emission spectrum of a known standard with the experimental emission spectrum gives satisfactory results. A published series of standard spectra for quinine sulfate, β-naphthol, and 3-aminophthalamide, among others, covers the spectral range from 300 to 600 nm [12]. One can also employ calibrated standard lamps having documented emission spectra. The measured signal divided by the expected signal produces the correction curve $E(\lambda)$ for the emission path.

Alternatively, $E(\lambda)$ can be determined without external standards by using a quantum counter and a matte white nonfluorescent surface, such as a magnesium oxide coating on glass, placed in the cuvette holder. First, the irradiation correction, $A(\lambda)$, is measured using a quantum counter. Subsequently, the total sensitivity of the fluorimeter is determined as a function of the wavelength, λ, in a mode in which the irradiation and emission monochromators are synchronously driven (an option most modern fluorimeters have). The correction spectrum for the emission path of the monochromator, $E(\lambda)$, is given by dividing the total sensitivity distribution, $T(\lambda)$, into the correction curve of the excitation path, $A(\lambda)$.

In Figure 8 are plotted both the measured and the corrected fluorescence excitation and emission spectra of flavin in aqueous solution. The correction of the excitation spectrum is clearly apparent. The emission spectra are not as dramatically altered, but they have the typical spectrometer-dependent red shift of 10 to 20 nm that makes the detailed comparison of uncorrected spectra problematic.

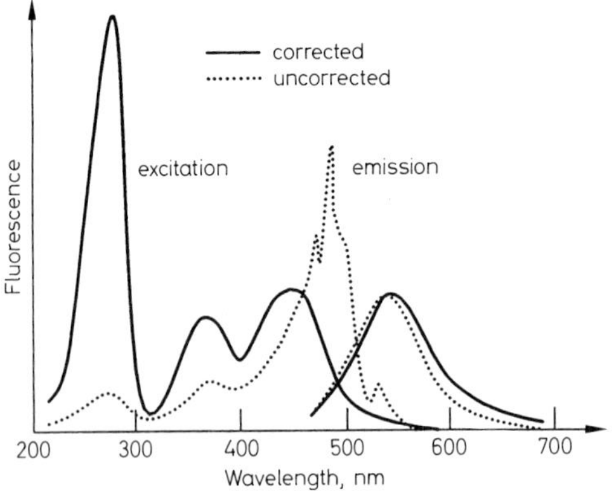

Fig. 8 Fluorescence excitation and emission spectra of flavin in aqueous solvents, both uncorrected and corrected. Excitation spectra are strongly modified by the system, and definitely require correction, while the emission spectra are only slightly distorted and shifted

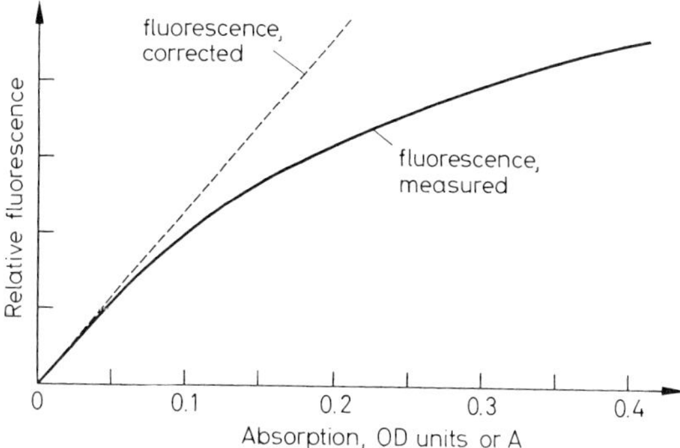

Fig. 9 Dependence of fluorescence emission upon the intrinsic absorption of a homogeneous sample. All fluorescing substances follow this general scheme, with only minor variations. Fluorescence spectra and kinetics can be reliably determined only with highly diluted samples ($A < 0.05$). The average distance between individual fluorophores should be significantly larger than R_c (see Section 3.5)

A practical, highly relevant, but often-neglected fact requires mentioning. For all fluorescent substances, without exception, fluorescence and absorption are strongly nonlinear above an optical density of $A = 0.05$ (Figure 9). Therefore, fluorescence measurements should be made with *appropriately diluted* samples. Particularly difficult are fluorescence measurements on turbid samples such as thylakoid membranes. Even when chloroplast suspensions are highly dilute, the *local* concentration of chlorophyll is many moles per liter and the optical density, A, is extremely high.

2.7 Sources of error

Even after fluorescence spectra are corrected, unanticipated peaks and shoulders are often found. These are due to a number of causes.

(a) Generally, in fluorescence spectroscopy no filters are used to block the higher orders of dispersion from the grating; and
(b) with increasingly dilute samples of aqueous solutions the Raman bands become more pronounced.

These sources of error are usually present when altered emission spectra are observed upon varying the excitation wavelength. Emission spectra should always be examined carefully for the possibility that higher-order dispersions are overlaid on the true emission peak, especially if "emission peaks" are unusually sharp and are superimposed on a broad background fluorescence.

Fluorescence is system-specific, technically determined, isotropic, and not collimated (in some contrast to the radiation path of the typical absorption photometer). Only very little emission (see Figures 7 and 16) falls on the entrance slit of the emission monochromator and subsequently from the exit slit onto the detector. Therefore, one chooses an emission slit as wide as possible and as narrow as necessary. Slit widths of

5 and 20 mm have proved useful (without the loss of significant data), because emission spectra are, by nature, very wide and have little structure. The slit widths on the excitation side can be narrower ($\Delta\lambda \approx$ 2-5 nm) because of the collimated excitation beam, but the distinct structure of the luminous discharge lamp introduces a photometric uncertainty of about $\pm 10\%$ even after the excitation spectrum is corrected and smoothed. In contrast to the classical double-beam absorption spectrophotometer ($d\lambda/dt \approx 100\ \text{nm} \cdot \text{min}^{-1}$), the scanning speed of a fluorimeter – adjusted to the light conditions – can be raised at will.

Of particular importance is low-temperature fluorescence at $-196\,°C$ (the temperature of liquid air). Because fluorescence intensity is strongly temperature-dependent (radiationless transitions $S_1 \rightarrow S_0$ are reduced with decreasing temperature, see Section 3.1), the emission of some substances has been detected only in this manner [13]. Forked optical fibers in the incident beam (reflection configuration) have proved useful in measuring frozen solid samples that are optically thick. One branch of the optical fiber brings the irradiation beam to the 0.2-ml sample held in a brass-Plexiglas cuvette that is immersed in liquid nitrogen outside the fluorimeter [14]. The fluorescence emission is captured by the second branch of the fiber and delivered to the emission monochromator.

If one is interested in the kinetic behavior of a spectral band of a particular fluorophore, both monochromators can be replaced by more light-efficient filters (interference, glass, or plastic) whose transmission maxima are at the excitation and emission maxima. This permits the photomultiplier to be placed very close to the sample, thus increasing the sampling area and consequently increasing the detection efficiency by several orders of magnitude (see Figures 7 and 16).

As previously mentioned, nothing equivalent to the Beer-Lambert law so useful in absorption spectrophotometry exists for determining the concentration of a fluorophore. A calibration curve, empirically determined from the fluorescence intensities of different concentrations ($A < 0.05$) of the specific substance and measured under the same conditions, has proved useful. It should again be mentioned that particular care should be taken with systems having high local concentrations (such as red blood cells or chloroplasts).

2.8 Chlorophyll fluorescence

If one allows green, photosynthetically active chloroplasts to rest 20 minutes in the dark, the coordinated (synchronized) fluorescence of photosystems PS II and PS I is lost (the individual components are shown in the upper part of Figure 18, in the simplified diagram of the photosynthetic electron transport chain, the path of which can be determined either directly or indirectly by spectrophotometry). When the primary electron acceptors Q_a and Q_b are completely oxidized, the reaction center of PS II is *open* and the system is in *state I*. If photosynthesis is subsequently activated by irradiation (induction), the photosystems are again synchronized, assuring a more or less optimal linear electron transport through both systems, from water on the donor side to NADP on the acceptor side. The strongly varying fluorescence emission (Kautsky effect, Figure 10) seen during the induction phase is a generally useful analytical tool in photosyn-

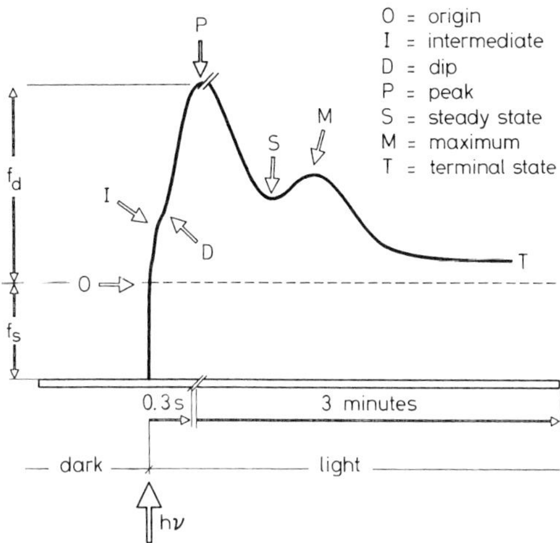

Fig. 10 Typical time course of the red fluorescence of dark-adapted, photosynthetically active samples after switching to white light (Kautsky kinetics). These kinetics reflect the current status of the electron transport chain in the thylakoid membranes. Peak fluorescence is reached approximately 0.3 s after the light comes on. After an additional 3 min, the system has again adapted to the terminal state, *T*. The baseline fluorescence, f_s, is distinct from the variable fluorescence, f_d

thesis research [15]. We can distinguish between the background fluorescence, f_s, and the variable fluorescence, f_d, which is found only in photosynthetically active tissues [16]. In the light, the single-electron acceptor Q_A is reduced and transfers its electron to the two-electron acceptor Q_B (state I). In the maximum fluorescent state, *P*, both Q_A and Q_B and the plastoquinone pool, *PQ*, are completely reduced. With the subsequent increase in the pH gradient across the thylakoid membrane, the fluorescence falls within a few minutes to an equilibrium level, *T*, with the system now in *state II*. This decrease in fluorescence (quenching) from *P* to *T* is composed of two parts, the photochemical, *Q*, and the nonphotochemical quenching, *E*, that can be distinguished by a specially designed chlorophyll fluorescence technique [17].

2.9 Phosphorescence

In principle, phosphorescence and fluorescence provide the same analytical information. Because the spin-forbidden transition of T_1 to S_0 is very unlikely, the intrinsic lifetime (that is, the natural lifetime that is not shortened by external influences such as quenchers) of the T_1 state lasts on the order of microseconds. Biologically interesting phosphorescent molecules generally do not emit at room temperature; rather, they prematurely lose their excitation energy without emission and are observed only at low temperatures (such as −196 °C), which limits their analytical value [18].

2.10 Electronic data processing of spectra

The use of microcomputers has become essential for fluorescence, not only for on-line data acquisition, but also for control of the spectrometer (monochromator slits, optical filters [rotating polarizing filters, see Figure 16], and shutter) as well as for the subsequent off-line evaluation and storage of spectra. Before computerization, the point-by-point determination of a crude polarization spectrum took several days (if sample stability permitted; see Section 3.6), while now it takes only minutes and unstable samples can be measured; by obtaining quantity we are achieving quality! Because of space limitations we can only mention further applications, such as normalizing and correcting spectra (see Section 2.6), difference spectroscopy, derivative spectroscopy, (de)convolution, selfcorrelations, and integration over a wave number scale for the determination of quantum yield [19].

We mention a point that needs particular attention in the future: spectra are maintained as files on magnetic storage devices for later use (printing or data transfer). Recorded along with the spectrum itself are additional data such as the wavelength region and the fluorescence intensity, as well as identifying parameters as headers. Unfortunately, there is no standardized system available yet for optical spectroscopy, such as that for word processing today, that would permit modem-mediated exchange of spectral data between different systems; however, this transfer capability can be expected in the not-too-distant future [20].

3 Fluorescence lifetime

3.1 Fluorescence lifetime and spectral shape

Remarkably, only the *singlet state* S_1 is *mirror-imaged* in the truest sense of the word (Figures 4 and 8); fluorescent light is emitted exclusively from the vibrational ground state of the first excited singlet, S_1. This is because the higher electronic excited states decay via a radiationless *internal conversion* (*IC*) to the vibrational ground state of the S_1 in about 10^{-14} s, whereas the S_1 has a longer lifetime of 10^{-9} s (see below). From here it returns to the electronic ground state, S_0, via further channels such as *intersystem crossing* (*ISC*) and internal conversion (IC), as well as by fluorescence. Since the sequences of the lower energy levels of S_0, S_1, and T_1 are very similar, but are reached in the case of S_1 from "below" and in the case of T_1 from "above", it necessarily follows that the fluorescence and phosphorescence spectra mirror the absorption spectra. The *apparent lifetime*, τ, of the first excited singlet state, S_1, which can be determined by direct experiments, is always less than the *intrinsic* (*natural*) *lifetime*, τ_0, because of quenching processes. The loss of excited-state molecules with time can be calculated with a simple differential equation,

$$dn = -kn\,dt \tag{3}$$

where n is the number of excited molecules and t the time to return to the ground state. The decay constant k is the sum of the reaction constants for fluorescence (F), photochemical processes (ET, energy transfer), radiationless transfer (IC, internal conversion), and intersystem crossing (ISC):

$$k = k_F + k_{ET} + k_{IC} + k_{ISC} \tag{4}$$

We integrate to obtain the solution:

$$n = n_0 \cdot e^{-kt} \tag{5}$$

The time τ, where $k = 1/\tau$, is called the *lifetime of the excited state*, and is that time when $1/e$ (about 37%) of the molecules are left in the excited state. The so-called *half-life*, $t_{1/2}$, is defined as that time when half the molecules have decayed to the ground state, S_0. It is calculated as follows:

$$t_{1/2} = \ln 2/k = \tau \cdot \ln 2 = \tau \cdot 0.693 \tag{6}$$

3.2 Determination of fluorescence lifetime

It might be surprising that in addition to the fluorescence spectrum (see above), the fluorescence lifetime can also be determined from the corresponding absorption band by a thermodynamically based formula,

$$1/\tau_0 = 3 \cdot 10^{-9}\, k_m^2 \cdot \int_{k_1}^{k_2} \varepsilon_k \, dk \tag{7}$$

where τ_0 is the intrinsic lifetime and k_m is the wave number of the absorption peak (Figure 11). For chlorophyll *a* as a classical biological fluorophore we can integrate over the α band, using the following values for the parameters:

$\varepsilon_k = 78{,}000\ \text{cm}^2 \cdot \text{mol}^{-1}$ ($78{,}000\ \text{M}^{-1}\ \text{cm}^{-1}$),
$k_1 = 14{,}200\ \text{cm}^{-1}$,
$k_2 = 15{,}700\ \text{cm}^{-1}$,
$\phi = 26\%$ (quantum yield, see Section 3.3).

From this we compute the value of the apparent lifetime,

$$\tau = \tau_0 \cdot 0.26 = 5.0\ \text{ns}.$$

The value obtained from direct measurements and reported in the literature is 6.7 ns; this is exceptionally good agreement.

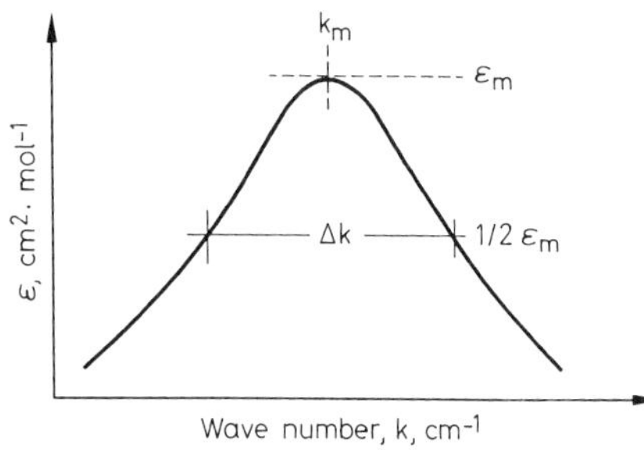

Fig. 11 The intrinsic fluorescence lifetime, τ_0, of a molecule can be estimated quite precisely using three parameters (k_m, Δk, and ϵ_m) of the absorption spectrum, as shown

Two independent methods, one *direct* and one *indirect*, can be used to determine τ experimentally [2, 3]; the direct method includes several subclasses. In the simplest case, the direct *pulse method*, a chromophore is excited with a single, short flash of intense light, and the (exponential) decay is followed; a semilogarithmic plot of the kinetic course gives the lifetime, τ, directly as the slope. With *pulse sampling* one irradiates the sample with a rapid succession of flashes; for each flash a single decay point is measured at a precise time after the flash with the aid of a "sample-and-hold" amplifier. The plot of the reaction kinetics falls off slowly. The pulse of the flash lamp occurs over a time comparable to the value of the lifetime, τ, that is being measured. In that the measured kinetics of the decay are combined with the kinetics of the flash, they are erroneous. Deconvolution (see Section 2.10) is used to separate out the kinetics of the chromophore, since the lamp kinetics are known.

The *single-photon counting method* is based on the fact that fluorescence emission is a random process. The molecules of the sample are excited by a flash and a timer is simultaneously started. The first photon emitted and detected stops the timer. The time between the flash and the detected emission is measured and the event is recorded. Most of the photons are emitted immediately after excitation by the intermittent flash, and only a few after a longer time. As a rule, there are 10^5 excitations per measured event. The experimenter selects a sufficiently weak flash intensity so that only one detectable photon is produced per 20 flashes.

The indirect *harmonic method* is based upon the amplitude and the modulated (sinusoidal) phase relationship between the exciting and emitted light. The heart of a *phase fluorimeter* is a gas-filled discharge lamp that emits a fast series of square-wave light pulses (with fast rise and decay times) that last about a nanosecond and are repeated each millisecond. Alternatively, continuously emitting light sources are used, and the pulses are generated directly by electro-optical modulators. These latter signals are more reproducible in shape and intensity than are individual flashes. Each flash excites a chromophore and, because of the finite lifetime of the first excited singlet, the emitted light appears with a constant time lag, expressed as the lifetime, τ. The *degree of modulation* of the emitted light is reduced because not all of the excited molecules have returned to the ground state between the two individual flashes (Figure 12). Therefore, the lifetime, τ, can be determined directly and independently either from the time lag given by the phase angle, ϕ, or from the demodulation factor, m:

$$\tau(\phi) = 1/\omega \cdot \tan\phi \tag{8}$$

where

$$\omega = 2\pi f \tag{9}$$

or:

$$\tau(m) = 1/\omega\{(1/m^2) - 1\}^{1/2} \tag{10}$$

While with the direct method the kinetics of the whole process are followed, the indirect method measures only a single value (m or ϕ). However, we have a reliable test of whether complex decay kinetics (superimposition of several fluorophores) is involved: if we are dealing with a single fluorescent species, then $\tau(m) \equiv \tau(\phi)$. If the decay is more complex, that is, if more species are present, then $\tau(m) > \tau(\phi)$.

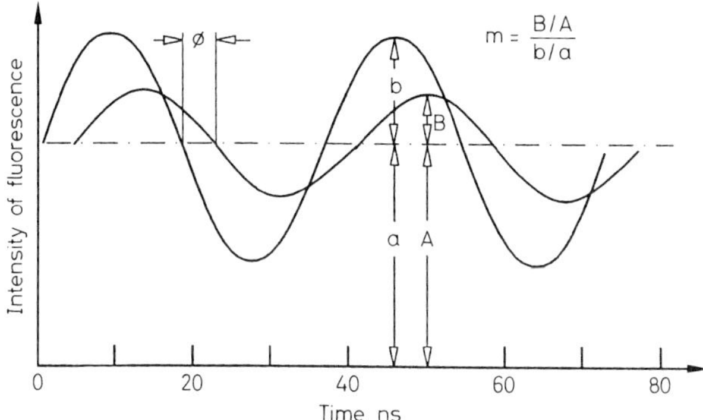

Fig. 12 Determination of the fluorescence lifetime, τ, with modulated excitation light (characterized by the large amplitude, *b*, and offset, *a*). Based upon the intrinsic lifetime, τ_0, of an excited state, the emission has a phase shift, ϕ, a smaller amplitude, *B*, and offset, *A*. The lifetime, τ, can be determined independently from either the degree of demodulation, *m*, or the phase, ϕ

Time-resolved fluorescence spectroscopy permits the selective *spectral* detection of different components in a mixture, because of different fluorescence lifetimes. The power of this time-discriminating analytical method is shown in Figure 13 with a model mixture of anthracene and perylene (*peri*-dinaphthalene), two compounds with only slightly different lifetimes.

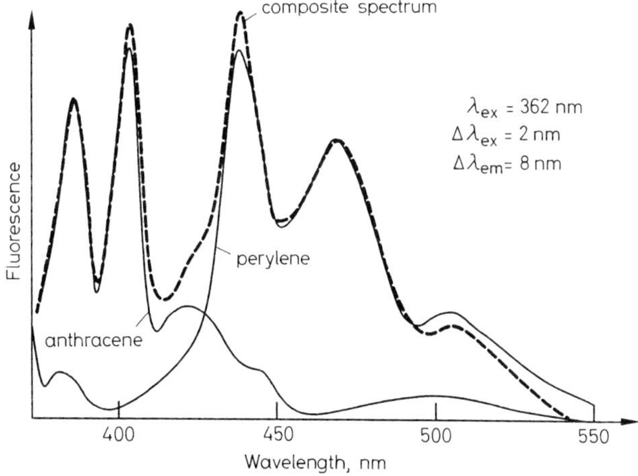

Fig. 13 Time-resolved fluorescence spectroscopy permits the selective spectral identification of various components with different fluorescence lifetimes in a mixture. A 50 : 50 mixture of anthracene and perylene (*peri*-dinaphthalene) in methanol was measured by the modulation method. The lifetimes are 4.5 ns and 5.1 ns, respectively. A difference of 600 ps permits the clean separation of two components with the appropriately adjusted phase shift of the lock-in amplifier

3.3 Fluorescence quantum yield

The quantum yield, ϕ, (not to be confused with the phase angle, ϕ) of fluorescence is a measure of the probability that an absorbed quantum will be emitted (even at a different energy level, *i.e.*, wavelength). It is defined as:

$$\phi = k_F/(k_F + k_{ET} + k_{IC} + k_{ISC}) = \tau/\tau_0 = k_F \cdot \tau \tag{11}$$

When ϕ is 1, the probability is 100%, and no relaxation channels are present. The fluorescence quantum yield, ϕ, can be experimentally determined in a number of different ways:

a) In principle, the incident and emitted quanta can be determined directly and absolutely. However, the experimental procedures and equipment required are not practical for the average laboratory.
b) The second possibility is based upon the comparison between the intrinsic lifetime, τ_0, and the apparent lifetime, τ
   ($\phi = \tau/\tau_0$).
c) The simplest and therefore the most popular method is based upon the *relative* comparison of the emission spectrum (corrected and plotted on a wavelength = energy scale) of the sample molecule X and a reference molecule, such as quinine sulfate (QS), whose absolute quantum yield, ϕ, is known from the literature [12]. The point of intersection of the two absorption bands is then determined. Irradiation at this wavelength ensures the same absorption by both species, so the two samples need not have the same concentration. The relationship of the areas under the measured emission spectra gives the quantum yield, ϕ.

3.4 Fluorescence quenching

Different mechanisms and components foster the *radiationless* transfer of the S_1 and also the T_1 states to the ground state. This is known as *quenching*, that is, the loss of fluorescence by a fluorophore. This intensively studied phenomenon, which involves direct contact between the fluorophore and the quencher, has also proven to be a valuable analytical tool in biotechnology. Various fluorophores and quenchers can be used quite specifically in particular situations, because not all fluorophores are influenced in the same way by quenchers: they react specifically in a manner determined by their particular physical-chemical properties. Therefore, the accessibility of a fluorophore can be determined by a quencher:

- Is the fluorophore embedded on the inside of a macromolecule?
- How is it localized in a membrane [21]?
- How permeable is the membrane to the quencher [10]?
- How large is the diffusion coefficient, D, of the quencher or fluorophore?

The fluorescence of molecules is generally quenched by molecular oxygen, molecular xenon, iodine, and heavy atoms that facilitate an $S_1 \rightarrow T_1$ transition (k_{ISC}, see above). The fluorescence quantum yield, ϕ, also controlled by quenching processes, serves, for example, as an index of the photochemical activity (k_{ET}) of chlorophyll (see PS I and PS II after injury to a leaf, Figure 14).

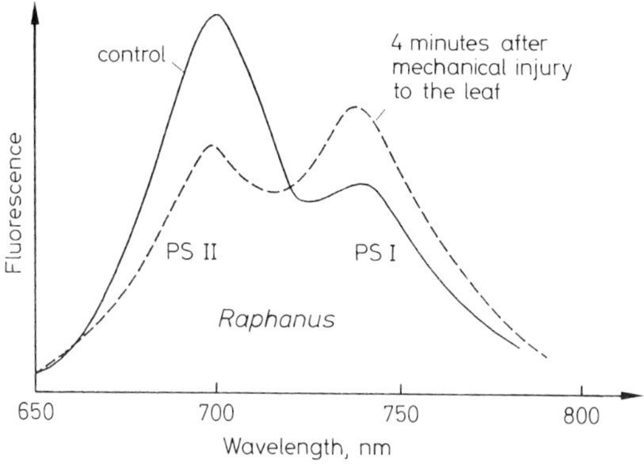

Fig. 14 The fluorescence emission ratio (690/740 nm) of photosystems I and II depends strongly on the physical condition of photosynthetically active organisms, shown here for radish leaf; the comparison is before and after mechanical injury (modified from LICHTENTHALER and RINDERLE [16])

3.5 Energy transfer

The most valuable analytical techniques in luminescence spectroscopy are the different energy transfer mechanisms between an excited molecule and its neighbor in the lowest electronic excited state, S_1 or T_1. We can distinguish three important reactions:

(a) In *electron exchange mechanisms*, the orbitals of the excited donor molecule and the acceptor molecule must overlap significantly in the ground state. Upon reciprocal electron exchange the excitation energy is transferred. The distance over which this exchange can act is small, so the process requires direct contact between donor and acceptor molecules; the distance between them cannot be more than about 10 Å. The spectral characteristics of both molecules are irrelevant.

(b) A second process, perhaps more trivial, is *energy transfer* via light quanta. The excited molecule emits a photon that is captured by an absorbing molecule, which then itself becomes excited. This presupposes a transparent solvent. This process is of less analytical interest, since it can cause erroneous measurements.

(c) The third mechanism, called the *Förster transfer mechanism* (or resonance transfer, or dipole-dipole transfer), is of significant analytical importance and is based upon the *radiationless* transfer of excitation energy over a considerable distance without the involvement of light quanta. This energy transfer is quantified by R_c, which is a measure of the molecular distance at which energy transfer as well as fluorescence emission have a probability of taking place. By this technique the membrane-bound flavin molecule was determined to have a value of R_c of 17 Å [21]. If R_c is known, the molecular dimensions can be elegantly measured. Thus, WU and STRYER in 1972 [22] fluorimetrically determined for the first time the shape and dimensions of the rhodopsin molecule by binding retinal, the natural chromophore, along with three other fluorophores, at specific locations along the protein (Figure 15).

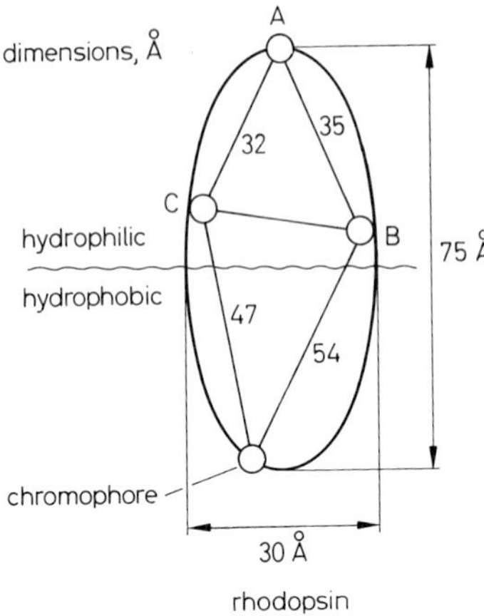

Fig. 15 Fluorimetric size measurement of the rhodopsin molecule, by use of three specifically bound fluorophores that show different spectral behavior (A: iodoacetamide; B: fluorescent disulfides; C: acridine). The distances can be determined using the known excitation and emission spectra and the known quantity R_c (after Wu and Stryer [22])

3.6 Polarization

Fluorescence produces, from a physical standpoint, a transverse *dipole emission*, which has a defined polarity with respect to the emitting molecule (transition dipole moment). If one irradiates a sample with linearly polarized light in which the electrical field vector E is all in a given direction, then the emitted fluorescent light will also be polarized. The theoretical definition of the *degree of polarization*[1] is

$$p(\lambda) = (I_{vv} - I_{vh})/(I_{vv} + I_{vh}) \tag{12}$$

in which $1/3 < p < 1/2$.

The subscripts v (vertical) and h (horizontal) refer to the relative positions of the polarizer and analyzer, respectively (Figure 16). If one is to use this formula in a realistically practical manner, the relatively strong polarization of the dispersion grating must be eliminated. This can be accomplished by a somewhat complicated procedure by which the four individual spectra are obtained one at a time by positioning the polarization filters differently. With help from a computer it is readily accomplished [19], and computed as

$$p(\lambda) = \{I_{vv} - I_{vh}(I_{hv}/I_{hh})\}/\{I_{vv} + I_{vh}(I_{hv}/I_{hh})\} \tag{13}$$

In theory, one measures the intrinsic molecular polarization $p_0(\lambda)$. There are two reasons, however, that in practice a smaller value is obtained, that is, a gradual *depolarization* occurs.

(a) First, the excitation energy may be transferred, by the mechanism described in Section 3.5, from the fluorophore to an identical neighboring molecule, which subsequently emits light.

[1] The equivalent term *anisotropy* is defined by $r = 2p/(3 - p)$ and acts additively, in contrast to p; this is an advantage if several parallel components are present

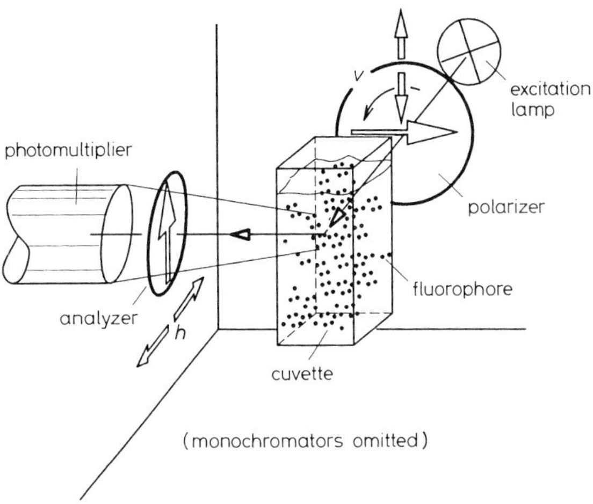

Fig. 16 Schematic diagram of apparatus for the determination of fluorescence polarization. By using all four possible positions of the polarizer and analyzer (*vv*, *hh*, *vh*, *hv*), the intrinsic background polarization can be eliminated. In conventional fluorimeters the excitation and emission beam paths are arranged vertically to minimize the Rayleigh component (see Figure 7)

(b) Secondly, the fluorophore may rotate, due to diffusion processes, between excitation and emission.

Thus R_c may be determined (see above). The rotational relaxation time, ϱ, can be determined statically by measuring the depolarization, Δp, or dynamically by following the time course of anisotropic decay. With this technique, the value of ϱ for free amphiflavin (which has an effective diameter of about 5 Å) in a fluid-crystalline dipalmitoyl lecithin membrane was found to be 10 nanoseconds [10], and ϱ for rhodopsin, a large membrane protein (with a diameter of about 15 Å, see Figure 15 [22]) was found to be 20 microseconds [23].

4 Delayed luminescence

4.1 Fundamentals

Delayed luminescence is emission from excited molecules at times considerably longer than the lifetime of prompt fluorescence, τ_0. It can be observed over many minutes or even hours. Three criteria, which we will clarify with the aid of a two-dimensional potential energy diagram, must be fulfilled (Figure 17):

(a) ΔG must be equal to or greater than hc/λ_m, where ΔG is the delivered free energy and λ_m the long wavelength limit for the excitation of the product molecule;

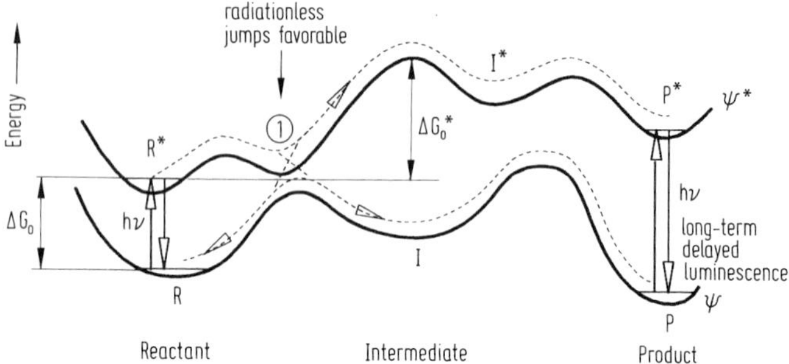

Fig. 17 Schematic potential diagram of a complex biological system depicting, two-dimensionally, the ground-state and excited-state surface as described by the wave functions ψ and ψ^* and with two stable states, *R* (reactant) and *P* (product) which can be mutually interconverted by light. Light energy supplies the required reaction enthalpy (ΔG_0) inducing the transition from the ground state *R* to the first excited state R^*. Further uptake of activation energy (ΔG_0^*) (from the environment) will bring the system – via various intermediates I^* – towards the excited state P^* of the product and finally to the ground state *P* upon emission of "normal" fluorescence light. The delay time of light emission is determined by the Boltzmann equation, taking ΔG_0^* into account; the quantum efficiency of luminescence is controlled by concomitant radiationless transitions that take place predominantly when the ground state and excited state surfaces come close to each other (position (1)).

(b) an efficient chemical reaction path must exist between the originally excited molecule, *R*, and the molecule *P*, that is, the potential curves of the ground state and the excited state must be in close proximity;
(c) the fluorescence quantum yield, ϕ, of the product, *P*, must be adequate, that is, the distance between the two potential curves must be sufficiently large.

If the last criterion is not fulfilled, then the possibility of generating luminescence by energy transfer to a highly fluorescent molecule (enhancer) exists. Luciferin, 6-hydroxybenzothiazole, and *p*-iodophenol have often been used as enhancers. We most frequently deal with delayed fluorescence, in which light from the first excited singlet state, S_1, of the product molecule *P* is emitted, even when phosphorescence (emission from state T_1) occurs.

4.2 Types of delayed fluorescence (DL)

The photochemist distinguishes two types of delayed fluorescence (DL). In *type E* DL, the energy from the excited T_1 state flows "upward" back to the S_1; in *type P* DL, a bimolecular reaction between two molecules in state T_1 produces an S_0 (ground state) and an S_1 (excited state), from which the fluorescent emission produces the expected spectrum. A complex type of DL is particularly interesting to biotechnology: the normal direction of electron flow in the electron transport chain of photosynthesis, driven by light absorption in the coupled photosystems PS II and PS I, is from the electropositive (H_2O) to the electronegative (NADP) side (Figure 18, upper). However, the elec-

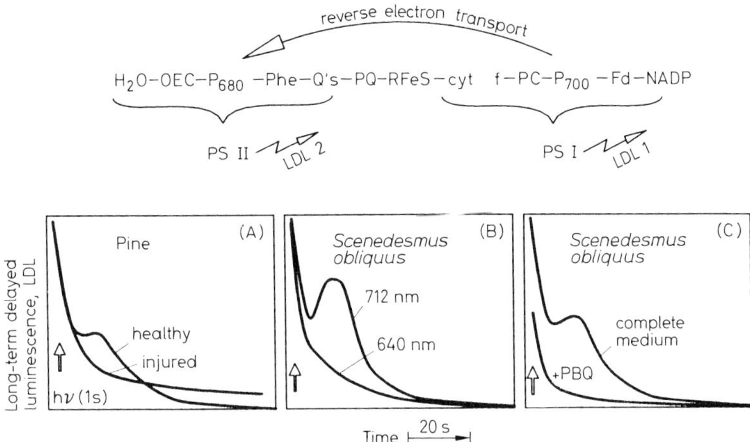

Fig. 18 *Upper*: Schematic representation of the electron transport chain in photosynthesis, including photosystems II and I (PS II, I) with their reaction centers P_{680} and P_{700}, respectively; *OEC* = the oxygen-producing complex; *Phe* = pheophytin; *Q* = quinone; *PQ* = plastoquinone; *RFeS* = the Rieske iron-sulfur protein; *PC* = plastocyanin; *Fd* = ferredoxin.
Lower: Kinetics of delayed luminescence (A) in injured and healthy pine; (B) in the green alga *Scenedesmus obliquus* in state I (640 nm) and state II (721 nm); (C) without and with addition of the herbicide parabenzoquinone (PBQ)

tron transport chain of photosynthesis has a small reverse leak current that travels from higher (negative) to lower potential (*recombinational delayed luminescence*) and coincides with a weak light emission. This process is clarified in the potential diagram (Figure 17).

4.3 Methods of excitation

Electronic excitation of molecules to produce luminescence can be obtained with the most diverse methods; by radioactive radiation, by electric fields or discharges, thermally, chemically/biochemically, photophysically/photochemically, or also simply by mechanical injury (necrosis) of tissue. Correspondingly one speaks of *electro-*, *thermo-*, *chemi-*, or *photoluminescence*, or in general of *bioluminescence* (see Figure 1).

4.4 Instrumentation requirements

One reason for the limited popularity of delayed luminescence in analytical procedures is the low level of light emitted. Light quanta production rates of 10^{-16} E/cm^2 · s or less are typical. This requires special instrumentation; standard fluorimeters are not sensitive enough to detect 10^7 quanta/cm^2 · s (sunlight has 10^{16} quanta/cm^2 · s). We can specify two types of spectrometers for the measurement of delayed luminescence that have the necessary detection sensitivity, but differ significantly in cost and specifications.

Bio- and chemiluminescence can be determined with simple, economical, commercially available spectrophotometers specially made to record direct-current measurements (detecting about 10^5 quanta/s). Even though these do not actually count *single photons* they are completely adequate without further compensation for *background radiation*. The experimental sample is placed close to the photocathode of a photomultiplier, and the photocurrent is measured (after appropriate amplification). If one wishes to measure as a function of wavelength, then a wide band filter with good transmitting properties is placed between the sample and photomultiplier. The use of adjustable monochromators is practical only when sufficient light is present; most luminescence spectra have very broad bands, so too much light is needlessly lost.

If the light sources are weak the signal-to-noise ratio (S/N) with direct current measurements is often too large (what is termed the *flicker-noise*, a mode of noise whose origin is largely unknown, depends strongly on $1/\nu$, the reciprocal of the frequency). To eliminate this excess noise with a $1/f$ power spectrum, one uses a sophisticated method to count single photons: when a light quantum falls on the photocathode of a photomultiplier, a specific pulse of current is measured on the anode. Statistically, the pulses vary in size and shape. They are amplified and then selected by defining a threshold that lies sufficiently above the background noise level and making a judgement as to whether or not each pulse is significant. Finally, all selected pulses are standardized and quantified by an appropriate counter. The digital signal is then converted to an analog signal and printed (Figure 19).

Inherent background radiation not emanating from the sample interferes in the measurement of so-called *ultraweak* photon emissions from biological or medical sam-

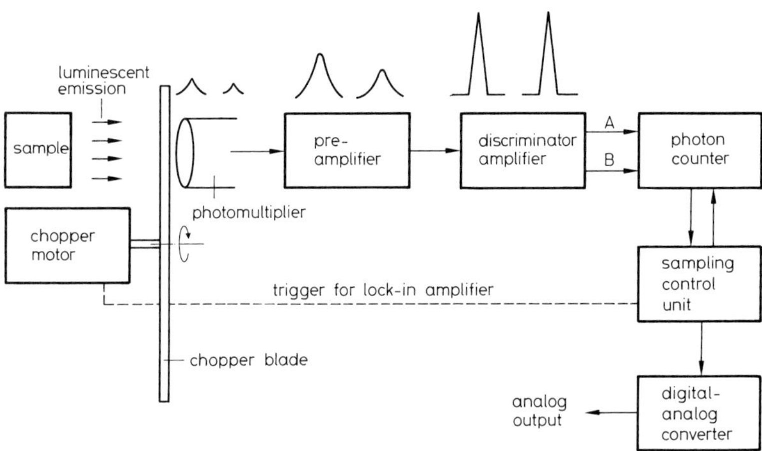

Fig. 19 Apparatus for measuring single photons. Photomultipliers produce a range of statistically variable pulses when individual electrons impinge. The pulses are initially amplified, adjusted by a discriminator to a standard shape and amplitude, and counted (channel A, up to about 10^7 photons/s). A more precise chopper, along with a lock-in amplifier (the reference signal produced by the chopper blade permits discrimination) is used to correct for the ever-present background radiation in systems that are used for the extremely sensitive detection of ultraweak radiation (a few photons per second)

ples. Most of it can be eliminated by an electronically controlled rotating shutter (*chopper*) that intermittently blocks the incident beam, thereby permitting measurement of the dark signal, which is then subtracted from the total measured signal. This technique permits the measurement of only a few photons per second (RUTH [24]).

4.5 Delayed luminescence in photosynthesis

4.5.1 Mechanisms

Delayed luminescence in photosynthetically active organisms has until now been measured in the range of microseconds to milliseconds (*SDL, short-term delayed luminescence*) [25]. One reason for the surprisingly meager interest in applications of *LDL* (*long-term delayed luminescence*) is that the kinetics and spectral data are suggestive of a trivial mechanism: the SDL decay kinetics are semilogarithmic; in limited studies of delayed luminescence spectra, they appear to be identical to the spectra of prompt fluorescence. Consequently delayed luminescence had been interpreted as a *type E luminescence* (Section 4.2).

This interpretation of SDL received support initially from the spectral behavior of thermoluminescence [26], which we are not discussing further; however, the SDL appears to be more complicated [27].

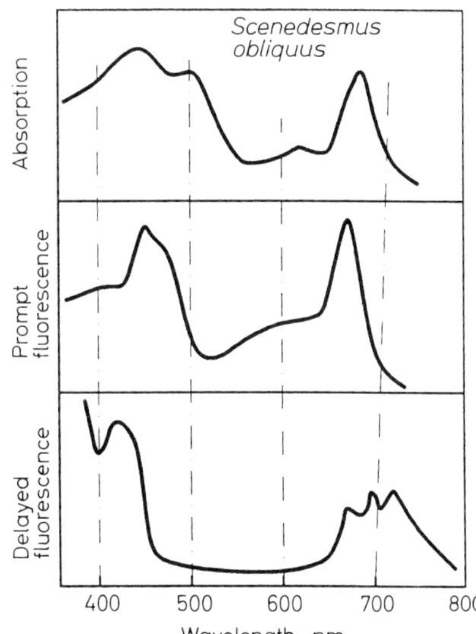

Fig. 20 Comparison of absorption and excitation spectra of prompt and long-term delayed luminescence *in vivo* of *Scenedesmus obliquus*. Prompt and delayed luminescence have quite different spectra, which means that their mechanisms differ

The LDL has completely different spectral properties from prompt fluorescence (Figure 20, action spectra); the assumption of a simple (delayed) fluorescence from chlorophyll *a* is no longer tenable. The decay kinetics are complex and not first-order. In contrast to SDL, a *functional* chloroplast membrane is essential for the LDL because membrane-active reagents have a strong influence on LDL. While the prompt fluorescence (Kautsky effect; see Figure 10 and Section 2.8) unequivocally has its origin in the antenna systems of PS I and PS II, the origin of the LDL is at the reaction center, so the LDL can be used as an indication of the state of the reaction center. Antenna mutants (LHC, with fewer antenna molecules) have an unchanged LDL; the LDL is probably based upon a reverse electron flow from PS I to PS II (Figure 18, upper). This is a physiologically irrelevant effect, an undesired dissipating reaction, which is, however, ideally useful in studying the energetics (electron transport) of photosynthesis and its sensitivity to environmental factors.

4.5.2 Ecological applications

Plant leaves and needles are difficult objects for optical spectroscopy studies, and initial results could not be interpreted because of the lack of well-defined optical characteristics [28]. Therefore, further biophysical [29] and ecological [30] studies were conducted with a more manageable organism, the green alga *Scenedesmus obliquus*, as the object of study. First, the optical properties of a microorganism make it easier to observe and, second, the external conditions, such as concentration of added reagents, pH, temperature, and redox potential, are simpler to control. Various ecological stresses have been shown to cause a significant and highly specific change in the LDL kinetics. These changes can be correlated with the effects of toxins (at the statutory concentration limits for municipal wastewater [31]) (Figure 18, lower; and [32]). Various known mutants can be used for specific studies, and the ecological significance of the changes in LDL can be investigated in the course of circadian (daily) rhythms.

The LDL makes possible different analytical approaches that virtually no other *noninvasive* technique can offer, and promises to be widely used in the future. Algae are of significant ecological importance because they are photoautotrophic organisms, that is, they photosynthetically reduce CO_2 from air, split water, and synthesize organic matter, thereby constituting the foundation of the food chain. The advantages over the conventional prompt fluorescence analyses are many.

(a) Only live algae are monitored.
(b) Scattered light from turbid samples has no effect because of the difference in time.
(c) The procedure is linear over at least 3 powers of 10. The area under the decay curve of LDL is proportional to the *living* biomass.
(d) The spectral scope of LDL enables different algal types, such as blue-green, brown, red, green, or diatoms, to be distinguished. The proportion of each algal type can be determined mathematically from the composite spectrum. With flow cytometry (see Section 7), combined with a fast fluorimeter, the continuous detection of different algal populations is possible over a longer time period if the detector is coupled to the appropriate microcomputer and control systems.
(e) The method is extraordinarily sensitive, depending only on the instrumentation used for single-photon counting.

The results from *Scenedesmus* are generally applicable to higher plants, such as coniferous and deciduous trees, according to initial experiments (Figure 18, lower). We plan a special project to pursue this further.

4.6 Chemiluminescence

In chemiluminescence, electronically excited molecules that are simultaneously good fluorophores are usually generated via strongly exothermic chemical reactions (see Section 4.3). Three mechanisms can be distinguished:

(a) oxidation of a substrate with molecular oxygen;
(b) electron transfer; and
(c) fragmentation.

The most efficient (brightest) chemiluminescence reactions are those involving the decomposition of peroxides. The reaction of luminol with hydrogen peroxide has a practical analytical application (in fluorescence immunoassay, FIA; [33]):

$$\text{Luminol} + 2\,H_2O_2 + OH^- \xrightarrow[\text{(microperoxidase)}]{\text{catalyst}} \text{aminophthalate} + N_2 + 3\,H_2O + h\nu .$$

Either luminol (in a chemically modified form) or peroxidase can serve as the tracer.

4.6.1 *Fragmentation*

Many of these processes involve the exothermal decomposition of intermediate dioxetanes by the mechanism suggested by McCapra [33a] (the dioxetane molecule is a four-membered ring comprising two carbons and two oxygens). Tetramethyl-1,2-dioxetane (TMD) is weakly fluorescent upon heating, but in the presence of a fluorophore (enhancer) it fluoresces strongly via energy transfer. There is a series of well-researched and analytically important examples of chemiluminescence in which dioxetanes are *intermediaries*. A model reaction is the thermolysis of TMD; it decomposes with an efficiency of 0.5 % into singlet acetone and with an efficiency of 50 % into triplet acetone, and thus fluoresces as well as phosphoresces:

```
     CH3
      \
H3C—C—O
      |   |
H3C—C—O
      /
    CH3
```

$$\text{TMD} \rightarrow 2\,(CH_3)_2CO + \text{phosphorescence (430 nm)} + \text{fluorescence (400 nm)}.$$

If the quantitative determination of enzymes is the goal, one can couple a specific fluorophore (fluorescence marker), such as 4-methylumbelliferone (4-MU), to a substrate upon which it does not fluoresce; the fluorescence of the marker is quenched when it is bound. The quenching is removed when the 4-MU is enzymatically uncoupled, resulting in a strong fluorescence that is directly proportional to the concentration of the enzyme. Enzyme marker kits for a large number of important enzymes are commercially available [34].

4.7 Bioluminescence

Mechanistically, bioluminescence is the same as chemiluminescence; the natural bioluminescence of certain luminescing insects or bacteria is used today for analyses in biochemistry, pharmacy, and, above all, biotechnology. We will discuss one of the most important applications. It deals with the luciferase-catalyzed oxidation of luciferin by molecular oxygen (luciferase is the enzyme and luciferin the substrate in this energy- and light-producing reaction). Different luciferin/luciferase systems are used in a variety of analytical applications [35].

4.7.1 Measurement of ATP

ATP/ATPase can routinely be determined at concentrations of 10^{-6} to 10^{-13} mol/l. NASA has reported a sensitivity of 2×10^{-17} mol/l for the detection of extraterrestrial life; at that sensitivity, bacterial contamination and the most minimal traces of life can be detected. Their analysis is based on the luciferin/luciferase system of the American firefly *Photinus pyralis*. Light is emitted in the reaction with ATP.

$$\text{luciferin} + \text{ATP} \xrightarrow[O_2,\ Mg^{2+}]{\text{luciferase}} \text{product} + h\nu .$$

The interpretation of the kinetics is difficult because the half-life of the reaction is only a few seconds. A major analytical approach for these reactions is to avoid the rapid decay of light emission and obtain constant emission.

4.7.2 Measurement of NAD(P)H

Concentrations of NAD(P)H between 10^{-9} to 10^{-7} mol/l can be measured, which is approximately 100 times more sensitive than with absorption spectroscopy. The method used is the bacterial luciferin/luciferase system with flavin mononucleotide (FMN).

$$\text{NAD(P)H} + \text{FMN} \xrightarrow{\text{FMN reductase}} \text{NAD(P)} + \text{FMNH}_2$$

$$\text{FMNH}_2 + \text{RCHO} + O_2 \rightarrow \text{FMN} + \text{RCOOH} + H_2O + h\nu\ (490\ \text{nm}).$$

These reactions can be used to detect rancidity in fats, since rancid fats produce long-chain aldehydes, which are required stoichiometrically in the above reaction.

4.7.3 Measurement of glucose

Glucose oxidase catalyzes the reaction

$$\text{glucose} + O_2 \rightarrow \text{gluconate} + H_2O_2$$

and H_2O_2 can be measured as indicated earlier. The detectable levels are between 0.7 and 10 µg per test.

4.7.4 Phagocytosis

Bacteria that enter the bloodstream are taken up by white blood cells called *granulocytes* and an intracellular *phagosome* is formed. This fuses with lysosomes, which contain an assortment of enzymes and other compounds that can oxidatively degrade the bacteria. The destruction of the bacteria is due primarily to the presence of superoxide, hydroxyl radicals, singlet oxygen, and hydrogen peroxide in the phago-lysosome. The dioxyethanedione produced is very energy-rich and upon decomposition can transfer its energy to a fluorophore. It is therefore possible to follow the kinetics of phagocytosis. There is very little intrinsic luminescence, but it can be increased by several orders of magnitude by adding an enhancer such as luminol. This procedure, among others, has found widespread clinical application. A newly developed method that relies on bioluminescence has interesting applications in tumor diagnosis: it uses single-photon counting to produce a microscopic image of various metabolites [36].

4.8 Ultraweak luminescence

Ultraweak luminescence is defined operationally. Chemi- and bioluminescence are called ultraweak luminescences when the usual direct current techniques are inadequate for light detection. In those circumstances one uses highly sensitive single-photon measurements that can detect a few photons per second (single-photon counting, see Section 4.4). It is not surprising that with the significantly high sensitivity, practically *all* organic samples, such as cells, tissues, whole blood, etc., are "very weak" photoemitters. The interpretation is intrinsically very difficult, even when the signals may in principle yield important information [24] (Figure 21).

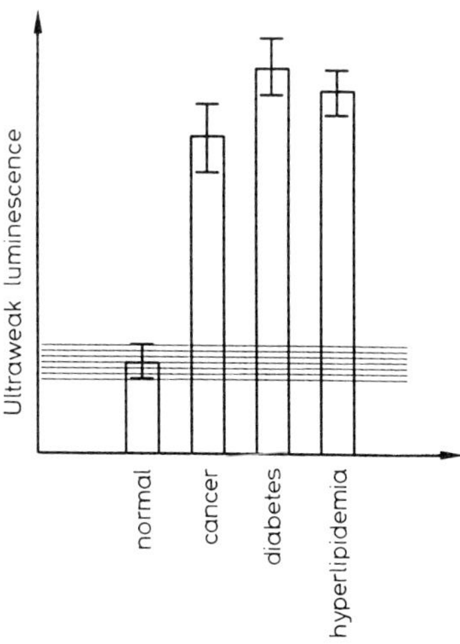

Fig. 21 Photon emission from most living tissues is detectable if sufficiently sensitive spectrometers are used. Minor tissue injury (necrosis) results in a proportionately greater emission of light, and diseased tissues have higher light emission, as is illustrated here for cancer, diabetes, and a lipid metabolism dysfunction (hyperlipidemia). Interpreting such data is naturally difficult

5 Fluorescence microscopy

Analyses that rely on optical spectroscopic analysis have become increasingly common in the fields of biology and medicine. The subfield of fluorescence microscopy appears to hold particular promise for the future. The limit of resolution for fluorescence signals from cells and microorganisms is less than a micrometer, so the source of the signal can be examined visually. High-resolution video images of fluorescence spectra can be observed while (depending on the instrumentation) the excitation or emission wavelength or the fluorescent lifetime is continuously varied. Only a few of the numerous applications of this method can be described in the sections that follow.

5.1 Principles and overview

Fluorescence microscopy is generally performed under incident illumination (Figure 22). Light from a high-pressure mercury or xenon (XBO) lamp or a laser is spectrally filtered and focused on the sample by a suitable dichroic mirror and objective lens. Fluorescent light passes through the same objective, and, due to its longer wavelength, passes through the dichroic mirror and additional spectral filters; it then – often after passing through a monochromator – falls on a highly sensitive detector, such as a

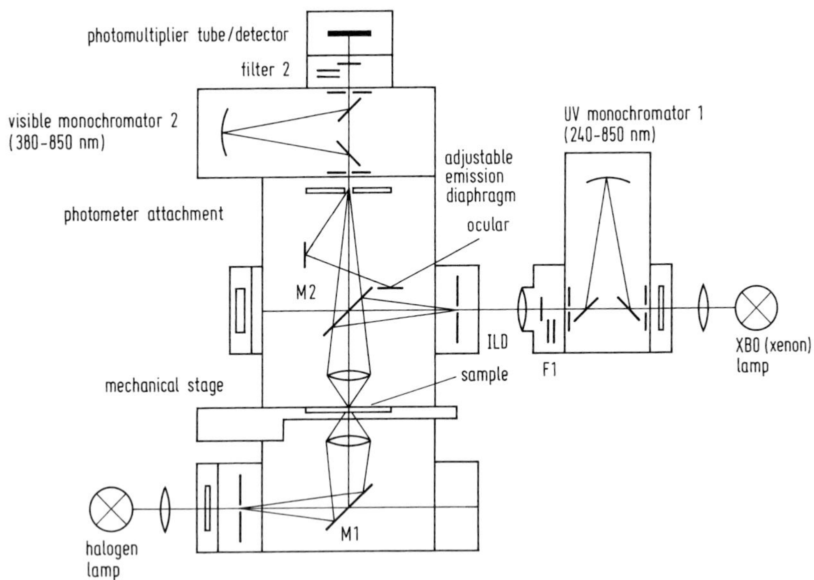

Fig. 22 Microspectrofluorimeter with adjustable excitation and emission wavelengths (with permission of Carl Zeiss AG, Oberkochen, Fed. Rep. Germany). ILD = light diaphragm; F = filter; M1 = mirror, M2 = dichroic mirror

photomultiplier or an image-amplifying camera. The sample area to be measured is often selected by a shutter located at the intermediate image plane. The minimum object size, d_{min}, that can be resolved is determined by the wavelength of the light, λ, and the numerical aperture, A, of the objective, according to the formula $d_{min} = \lambda/A$, with the maximum value of A about 1.30 for maximum-aperture oil-immersion objectives. The spectral resolution of suitable monochromators can usually be adjusted, depending on the intensity of the fluorescent light, to between 1 and 25 nm by varying the slit width.

The native fluorescence (*autofluorescence*) of cells and tissues is largely caused by nucleic acids and proteins (upon excitation by UV), additional metabolites such as porphyrins and carbohydrates, and various pigments and coenzymes [37]. The latter are of particular importance because they participate in very specific metabolic processes and thus provide information about their functions. For example, the autofluorescence of many cells upon excitation in the blue or near ultraviolet spectral range is largely due to nicotinamide adenine dinucleotide (NADH) and flavins and depends on their oxidation states [21, 38, 39]. Metabolic irregularities such as defects within the mitochondrial respiratory chain may cause a reduced oxidation of NADH and a resultant increased fluorescence intensity. All anaerobic bacteria with methane metabolism have the fluorescing coenzyme F_{420}, by which they can be detected in fermentor samples and biogas facilities [40]. A fluorescence microscopic photograph of *Methanobacterium thermoautotrophicum* is shown in Figure 23. The time-dependent bleaching upon irradiation with blue light, as well as the intensity of fluorescence, is a measure of bacterial activity [41].

The autofluorescence of many organisms is often weak and nonspecific. Therefore, *fluorescence markers*, which selectively stain certain cellular areas and organelles, are commonly introduced. These include acridine orange derivatives, which stain cell nuclei and liposomes [37]; rhodamine 123 and doxycycline for mitochondria [42, 43], 1-anilinonaphthalene-8-sulfonate (ANS) for membrane studies [44], and suitable markers for Ca^{2+} metabolism [45]. Figure 24 is the autofluorescence and emission spectra of rhodamine 123 in single, differentiated neuroblastoma cells at various excitation wavelengths. Rhodamine fluorescence clearly dominates the native fluorescence of NADH (Figure 24 A) and flavins (Figure 24 B).

Fluorescein and rhodamine stains (fluorescein isothiocyanate, FITC, and tetramethylrhodamine isothiocyanate, TRITC) are preferentially used in immunofluorescence; they couple covalently to antibodies or antigens and aid in visualizing the antigen-antibody reactions directly [46]. The ELISA technique (*enzyme-linked immunosorbent assay* [47]) is an especially sensitive immunological assay, in which antigens or antibodies are coupled to an enzyme that induces a nonfluorescing substrate to produce around 10^4 to 10^5 fluorescing molecules per minute. Characteristic end products of these reactions are substances such as fluorescein, rhodamine, naphthol, or methylumbelliferone, which under especially favorable conditions can be detected in amounts as low as 10^{-18} mol [48]. ELISA has been used successfully even with single cells [49]; in addition to diagnostic applications in immunology and prenatal testing [50], it also looks promising as a probe for numerous proteins in biotechnology [51].

Many reports of fluorescence diagnostics and photodynamic tumor therapy that use photosensitizing dyes have been appearing recently [52, 53]. Porphyrins and related compounds are often used, because they accumulate preferentially in tumors and a cytotoxic reaction occurs when light is absorbed. A fraction of the light energy absorbed is reemitted as fluorescence and can be exploited in diagnostics with sensitive detection methods. Fluorescence microscopy is particularly useful in the detection of small tumors as well as in basic studies on the uptake and distribution of photosensitizing compounds in cells and tissues.

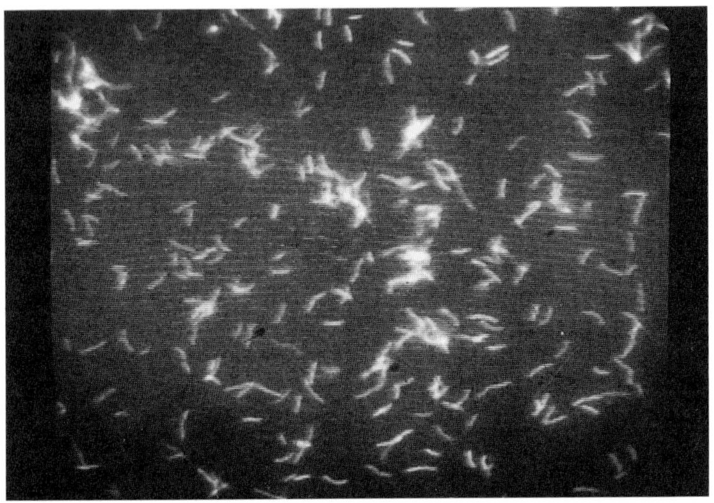

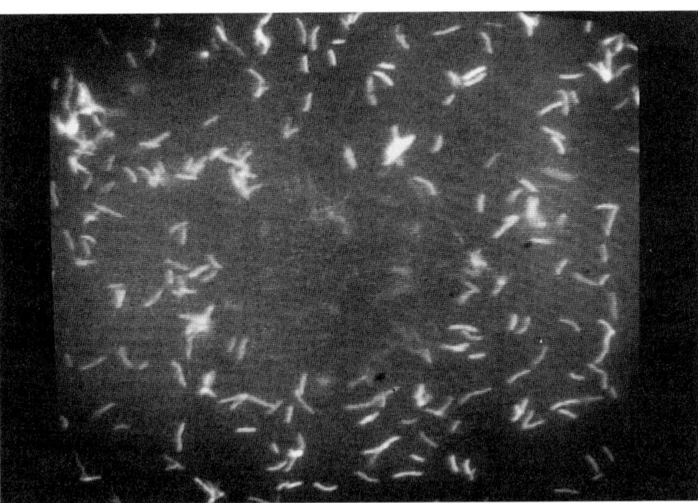

Fig. 23 Methanobacterium thermoautotrophicum as seen under fluorescence microscopy; (A) before and (B) after 8-minute bleaching with 500 mW/cm^2 (at an excitation wavelength of 405 nm; detection at 460 to 570 nm; image area 130 × 170 μm^2; from [41])

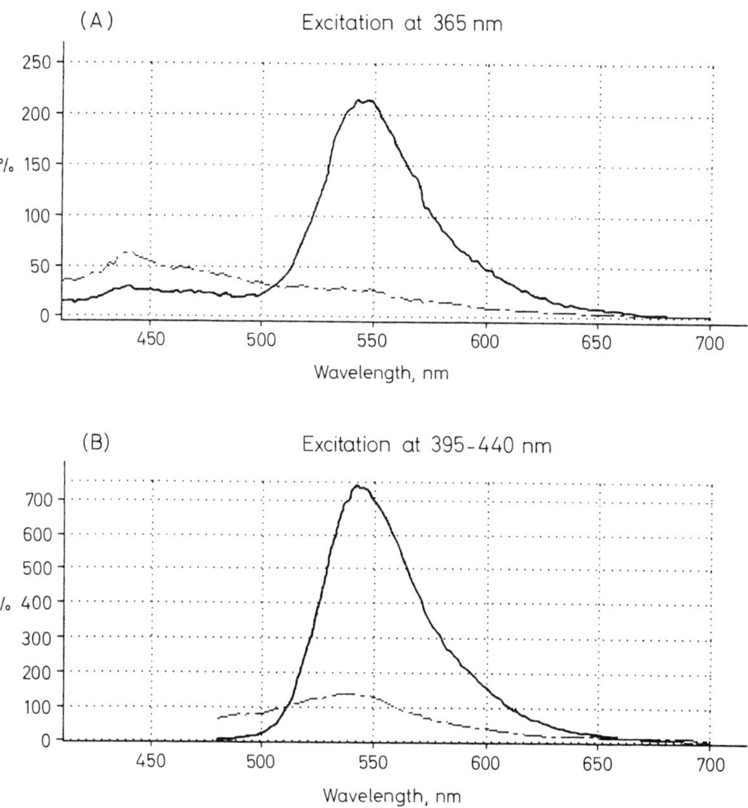

Fig. 24 Emission spectra of individual differentiated neuroblastoma cells at two different excitation wavelengths; dashed lines = autofluorescence; solid lines = treated with rhodamine 123

5.2 Time-resolved fluorescence microscopy

The single-photon counting technique is the method of choice for measuring fluorescence lifetimes in microscopic samples because of its sensitivity (Figure 25). The fluorescence photons from the area of the sample being examined are detected with a photomultiplier. After interfacing suitable amplifiers and discriminators (see Section 4.4) the time delay relative to the excitation light pulse is measured for each photon and converted to a corresponding voltage signal. The total of all voltage impulses as measured with a multichannel analyzer yields the desired fluorescence decay curve, which is subsequently resolved and corrected with a suitable computer program (for instance, to analyze decays involving multiple time constants).

Because only one fluorescence photon can be detected for each excitation pulse (see Section 3.2), the source of light impulses must have a high rate of repetition. This calls

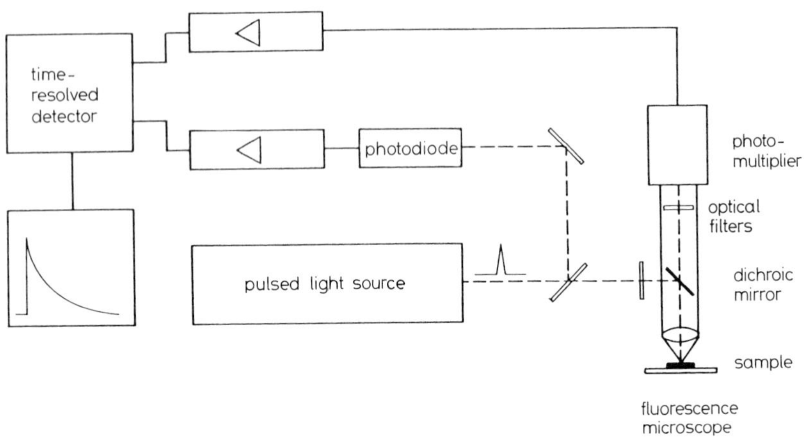

Fig. 25 Schematic representation of an instrument for time-resolved single-photon counting

for hydrogen and nitrogen gas-discharge lamps with impulse durations of about 1 ns and repetition rates of 20 kHz [54]. Shorter impulses ($\leqslant$ 10 ps), higher repetition frequencies (50-100 MHz), and improved stability can be achieved with an acousto-optical modulated laser system. A synchronously pumped laser system is optimal for many applications. In this case, an acousto-optical modulated gas or solid-state laser is used to pump a variable-dye laser that can be tuned for wavelength as well as for impulse frequency [55, 56]. Systems like this are naturally very complex (for example, the lengths of the lasers must be matched with micrometer accuracy) and costly. It remains to be seen whether they can be replaced by picosecond laser diodes of sufficient energy.

In complex samples, time-resolved fluorescence measurements permit the analysis of different fluorescing components with overlapping spectral bands, or the separation of a weak fluorescence signal from the superimposed scattered light (which has a greater intensity). If we use these methods we see a significant increase in the sensitivity of immunofluorescence and ELISA tests [57, 58]. The fluorescence decay times of many molecules depend heavily on the intracellular binding of the molecules, to certain proteins or nucleic acids, for example. In the case of intracellular flavins, the autofluorescence consists of at least three components with different lifetimes (Figure 26). DNA-coupled acridine dyes have significantly different decay times depending on whether they bridge two adenine-thymidine (AT) base pairs, two guanine-cytosine (GC) base pairs, or an AT and a GC base pair [59, 60]. Experimental results can thus yield important information about the frequency and sequence of DNA base pairs.

Time-resolved fluorescence measurements also give clues to the state of aggregation of certain molecules; the number of molecules that aggregate within a given species will increase the probability of a radiationless decay and reduce the fluorescence quantum yields and lifetimes. This is important for understanding photodynamic tumor therapy with porphyrin compounds. Aggregates of porphyrins appear to be the species that usually accumulates specifically in tumors, but monomers and dimers cause photosen-

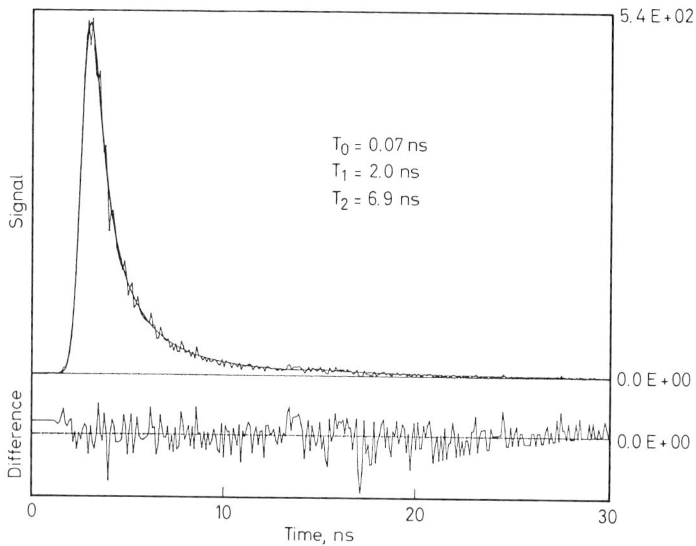

Fig. 26 Time course of autofluorescence decay in a single osteosarcoma cell after picosecond laser excitation at 420 nm. Upper curve: measured data and a curve fitted with three time constants; Lower curve: deviations. Detector range: 510 to 570 nm

sitization; the various compounds can be distinguished *in vivo* by their decay times. We have demonstrated intracellular disaggregation, essential to the photodynamic process [61].

Special techniques for time-resolved spectroscopy include the *gated single-photon counting technique* [62], which can measure, within a given time frame, fluorescence spectra of components that have variable decay rates, and a *multiple-photon counting technique* [63] that at present can be used only for decay times longer than 10 to 100 ns.

5.3 Videomicroscopy and image analysis

Video cameras of varying sensitivity are increasingly employed in diagnostics and in all situations where the imaging of information is essential. Devices used include

- conventional vidicon tubes and CCD (charge-coupled device) cameras (with modern semiconductor elements), with radiation sensitivities of 10^{-2} to 10 mW/m^2;
- SIT cameras with an amplifying silicon target (10^{-4} to 10^{-2} mW/m^2); and
- image-amplifying cameras based on photon counting technology (10^{-12} to 10^{-4} mW/m^2).

The detector is generally mounted at the intermediate image plane of the microscope. The cameras are often coupled to one or more image-storing devices where images are collected and further processed (to improve the signal-to-noise ratio, SNR). Simple computing operations include calculating the differences between two images and plotting curves and histograms. Some applications of videomicroscopy with multi-image storage devices are:

- the detection of fluorescence images at different wavelengths. In tumor diagnostics, for example, the difference between the autofluorescence of the tissue and the fluorescence of the tumor-specific dyes is computed, to improve the contrast of the tumor images (Figure 27) [64].
- the absorption of fluorescence light with varying degrees of polarization. This distinguishes fluorophores with varying degrees of polarization or rotational relaxation times (see Section 3.6), especially molecules of the same species that exist in various states of aggregation or with various intracellular associations.
- energy-transfer microscopy (see Section 3.5). With intermolecular energy transfer, the fluorescence images of the donor and the acceptor can be measured at different excitation or emission wavelengths and compared directly.

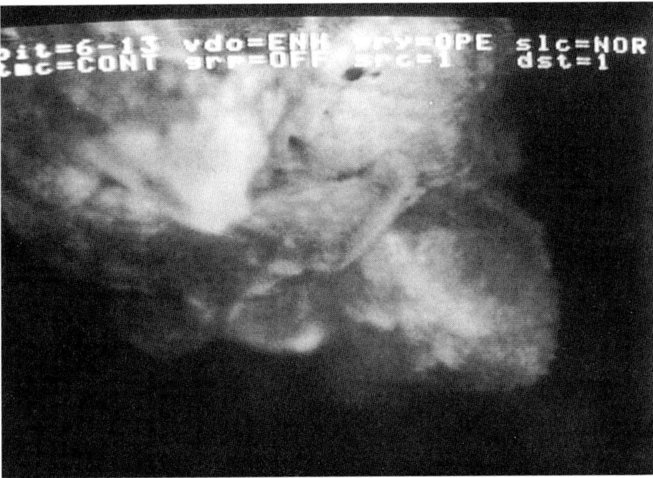

Fig. 27 Microfluorimetric detection of a squamous epithelial cell carcinoma after porphyrin enrichment and correction for autofluorescence (from [64]). Area of photograph = 4×5 mm^2

5.4 Energy-transfer microscopy

The radiationless transfer of energy from donor to acceptor molecules (see Section 3.5) occurs frequently in biological systems. Energy transfer, however, can also be induced artificially by implanting *molecular probes* selectively into specific cell areas. Energy-transfer microscopy can thus provide information about membrane architecture and intermolecular distances [65, 66], the structure and assembly of actin filaments [67], enzymatic binding sites [68], and metabolites in mitochondria [69].

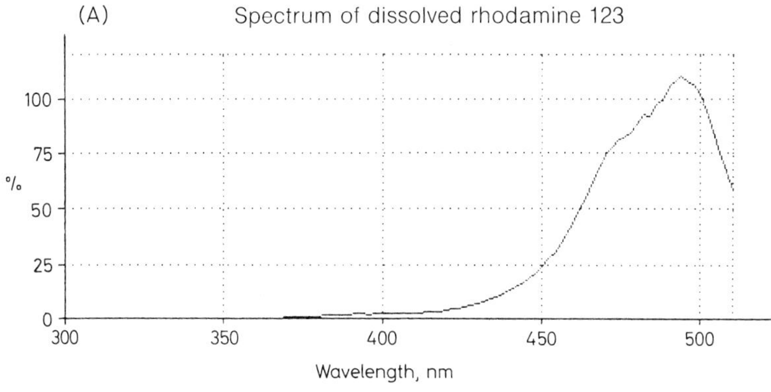

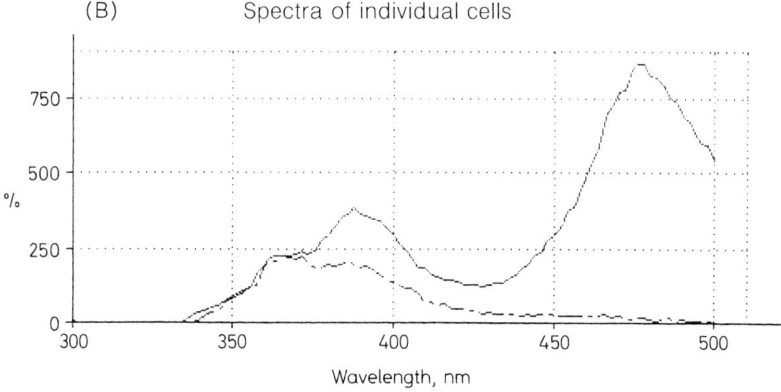

Fig. 28 Fluorescence excitation spectra of (A) rhodamine 123 in solution; (B), upper curve: intracellular rhodamine 123; (B), lower curve: cellular autofluorescence. Detector range 530 ± 10 nm (from [69])

Energy transfer can be detected spectroscopically. We show the fluorescence excitation spectra of a rhodamine 123 solution (Figure 28 A), the native fluorescence of yeast cells (Figure 28 B, lower curve), and the rhodamine 123 fluorescence of incubated yeast cells (Figure 28 B, upper curve). The rhodamine 123 solution has excitation maxima at 475 and 490 nm, and the native fluorescence maxima are at 365 and 385 nm, but the excitation spectrum of the incubated yeast cells has a distinct peak at 385 nm, in addition to the direct excitation of rhodamine at 475 nm. The peak at 385 nm may be due to absorption by mitochondrial flavins with a consequent energy transfer to rhodamine 123; its existence enables the use of rhodamine 123 in studics of respiratory metabolism.

Suitable bases for the quantitative determination of energy transfer include:

- the intensity of the relationship between acceptor and donor fluorescence under constant light excitation. This measurement, however, requires that the donor as well as the acceptor fluoresce, that their spectra do not overlap significantly, and that the fluorescence quantum yields, ϕ, are adequately known.

- energy transfer from lifetime measurements with pulse excitations. According to Section 3.1, the apparent lifetime, τ, of excited donor molecules becomes shorter with increasing energy transfer. Excitation of the acceptor via energy transfer is delayed compared to direct excitation. These measurements require that the donor *or* the acceptor fluoresce adequately. The quantum yields need not be determined, but the very rapid transfer process requires a high degree of time resolution (in the sub-nanosecond range). Studies of photosynthesis constitute an important application of this method [70, 71], since in photosynthetic processes the transfer of energy can be disrupted by environmental factors [72].

5.5 Laser-scanning microscopy

In contrast to conventional microscopy, where a sizable area is uniformly exposed to light, the sample area in laser-scanning microscopy is scanned point by point. This is done with a two-dimensional scan by a laser beam involving highly sensitive mirror galvanometers (Figure 29, [73]). The signal measured at each individual point is stored in an image memory bank and can be further processed electronically, if necessary. Typical laser-scanning microscopes are designed for both transmitted and reflected light measurements (such as for fluorescence microscopy) and can also be used as conventional microscopes (for visually examining interesting sample areas). Laser-scanning microscopy has several advantages.

- Random scattering from the sample can be avoided because of the point-by-point construction of the image. This results in a significant increase in contrast, whereas the resolution is, in principle, the same as is found in classical microscopy.

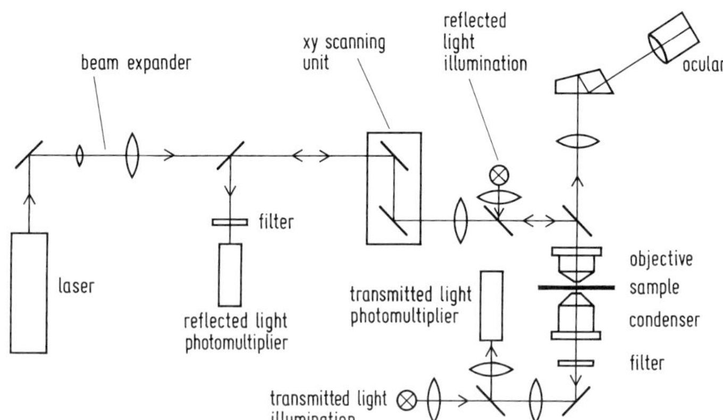

Fig. 29 Laser-scanning microscope (with permission of Carl Zeiss AG, Oberkochen, Fed. Rep. Germany)

- The detection sensitivity is greatly increased, because a highly sensitive point detecter (photomultiplier) is used instead of an area-imaging video camera.
- The relationship between the objective field and the image field can be varied with the scanning amplitude. This results in an *electronic zoom*.
- Each sampling area is only very briefly exposed to light. This reduces the often unavoidable *photobleaching*.
- If a small diaphragm is inserted in the detection beam (to the left of the scanning unit in Figure 29), the sampling areas that are located exactly in the focal plane of the microscope can be selectively monitored. Not all sampling areas in high-resolution microscopy are imaged equally sharply at the same time. If the microscope is sharply focused on a point, the cell layers above and below are not in focus. These areas can, however, be eliminated with a confocal method that makes the selective study of different layers in biological samples possible and permits, if necessary, a 3-D reproduction of the microscopic sample. This method has been used to identify individual chromosomes and genomes, to image mitochondria and endoplasmic reticulum with high contrast, and to observe cell fusion (Figure 30).

A

B

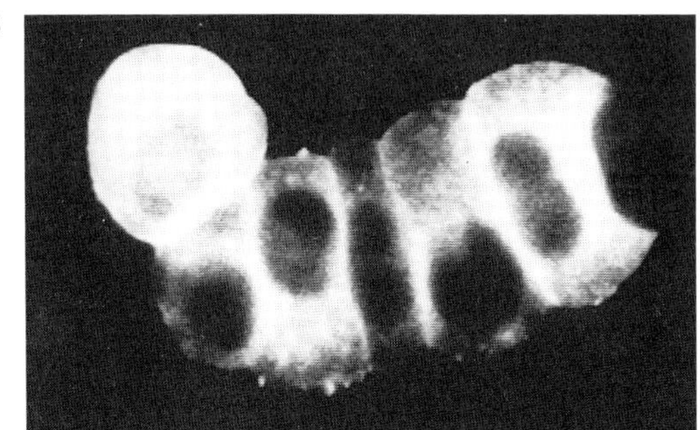

Fig. 30 Cell fusion and degranulation of rat pheochromocytoma cells as observed after extracellular stimulation with the laser-scanning microscope (area of photograph = $75 \times 100\ \mu m^2$).
(A) Difference-interference contrast image.
(B) Fluorescence image after labeling membrane with rhodamine derivative (excitation wavelength, 488 nm; emission range, 520-800 nm).
(Photographs courtesy of Dr. I. MELZNER and Dr. H. FREY)

6 Fiber-optic systems

6.1 Fundamentals of fiber optics

Classic fluorescence spectrometers and microscopic techniques suffer from limited flexibility. These methods are well-suited for measurements in cuvettes, cell cultures, or biological preparations, but they are inappropriate for the study of samples that are difficult of access, such as tissues or the insides of bioreactors. Fiber-optic systems, on the other hand, have many unique advantages.

- They are thin and flexible, with radii of curvature measured in millimeters.
- They have little attenuation and carry light signals over a range of several hundred meters.
- They can be installed directly at the sampling site.
- They can be coupled to suitable light sources, detectors, splitters, etc., with commercially available connectors.

Light transfer through glass, quartz, or plastic fibers is based on the principle of *total internal reflection* (Figure 31). A fiber core of refractive index n_1 is clad by a coating with refractive index $n_2 < n_1$. If light impinges on the core-cladding interface at an angle ε_1 that is greater than the limit angle of the fiber core relative to the cladding, then all light will be reflected. The process of reflection is repeated many times along the fiber. A minimum angle ε_1 for total reflection corresponds to a maximum angle ε_0 by which external light (refractive index n_0) can enter the fiber. This limit angle, ε_{0L}, at a sufficiently large radius is described according to the law of diffraction,

$$A = n_0 \cdot \sin \varepsilon_{0L} = (n_1^2 - n_2^2)^{0.5} \tag{14}$$

where A is the numerical aperture. If light from air ($n_0 = 1$) enters a typical glass fiber with refractive index, n_1, of 1.63 and n_2 of 1.515, the limit angle, ε_{0L}, is 37°. The calculated numerical aperture, A, is 0.60. This value can be attained in practice only with fibers of core diameter $d_c > 600$ µm. Fibers with $d_c \leqslant 50$ µm have values of A of 0.15 to 0.20. Sufficiently thin fibers can be used in biotechnology for measurements inside tissues or for time-resolved studies (in the latter case, good time resolution is obtained if the incident angle, ε_1, and the associated light paths of additional beams that impinge under different angles ε_1 differ only slightly). If the diameter of the fiber core is less than 5 to 10 µm, there is only one angle of incidence and the fibers are called *monomode*. For these fibers the time for light passage is defined exactly and the phase is retained.

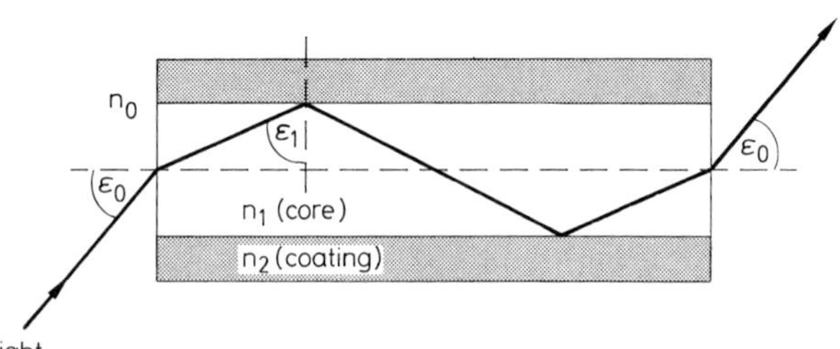

Fig. 31 Light path in an optical fiber

Monomode fiber sensors take advantage of these characteristics in making highly sensitive measurements of pressure, temperature, rotational speed, etc. [74]. The constant phase relationship is not important for fluorescence sensors because they measure intensities only.

The attenuation of commercial glass fibers is least in the red and near infrared spectral range. Here, the intensity of the light is reduced by only one-half after 0.5 to 5 km traveled. In the blue spectral range, the corresponding distance is only 20 to 100 m, because of the significantly greater light dispersion (increases with the fourth power of the wave number). The fibers used in biotechnology are usually shorter, so the attenuation is generally smaller than the losses that occur during the entering and exiting of the light.

6.2 Fiber-optic fluorescence sensors

Fluorescence measurements with glass fibers are designed with several criteria in mind.

- The excitation light should enter the fiber efficiently. Lasers are more appropriate than spectral lamps because fibers have limited entrance area and aperture. The fiber end is adjusted with an optical coupler just outside the focal point of the excitation lens (to avoid excess radiation intensity). Laser diodes can be attached directly to the fiber with plug-type connectors; however, because laser diodes have wavelengths that are usually beyond 670 nm, they are generally inappropriate for most fluorescence excitations.
- The solid angle of incidence of fluorescent light should be sufficiently wide. Thus, fibers with large core diameters are better for detecting.
- Excitation or fluorescence light must be resolvable into spectral bands. This is often more problematic in fiber-optic systems than in conventional instrumentation, because frequently there is more reflection and many spectro-optical components cannot be used. Often two separate fibers are used, one for excitation and one for the fluorescence measurement (an arrangement that reduces spectral resolution); or, alternatively, the spectral band resolution is done outside the fiber-optic system [75, 76]. This makes the design more complex and less flexible, however.

The fluorescence sensor illustrated in Figure 32 [77] represents a good practical and compact design. The spectral separation of excitation and fluorescence light takes place on a micro-optical bench (3×3 cm area). The excitation light of an argon ion laser is converted by a *self-focusing lens* into a parallel beam and is transferred to the sample by a dichroic mirror, an additional self-focusing lens, and glass fiber. The fluorescence light is accepted by the same fiber, passes through the dichroic mirror and several other filters and impinges on a sensitive detector. A fiber-coupled avalanche photodiode with a typical sensitivity of 1 V/μW is adequate. Photomultipliers are more sensitive but their integration with the measuring system is more involved. The system shown in Figure 32 has been used for *in vivo* measurements of chlorophyll fluorescence in forest trees, using Kautsky kinetics (Section 2.8 and Figures 10 and 33). The ratio of fluorescence decay, f_d, to background fluorescence, f_s, appeared suitable for determining the efficiency of photosynthesis, and correlated with plant damage due to environmental stress [78].

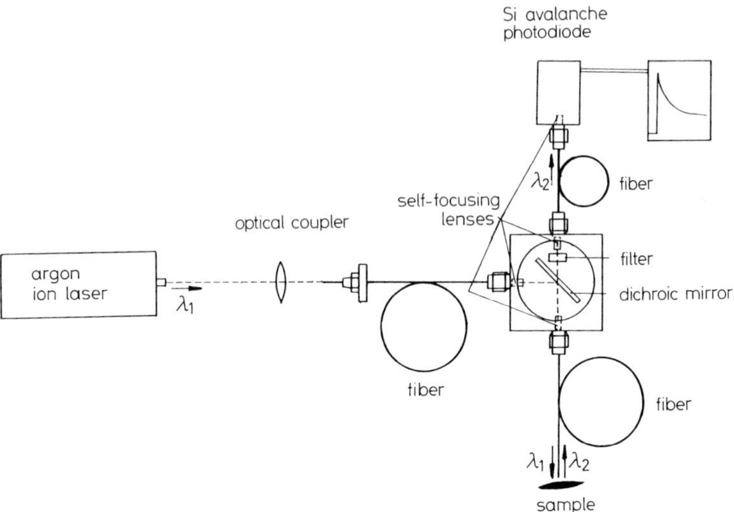

Fig. 32 Fiber-optic fluorescence sensor (from [77])

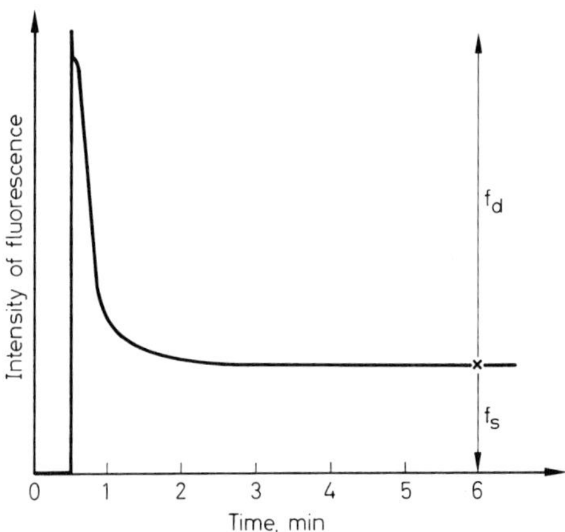

Fig. 33 *In vivo* decay of fluorescence in a dark-adapted spruce needle (Kautsky curve; see Fig. 10)

Another area of application of slightly modified fiber-optic sensors is in tumor diagnostics involving photosensitizing dyes [64].

It is important to measure the irradiation intensity, P/A, at the sample site and the solid angle, Ω, through which the fluorescence light enters the fiber (Figure 34). The solid angle in Figure 34 is calculated by the relations

$$\Omega = A_D/l^2 = \pi d^2/4l^2 \tag{15}$$

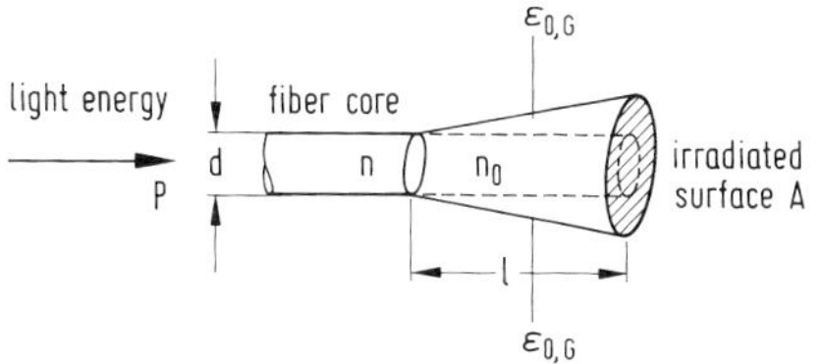

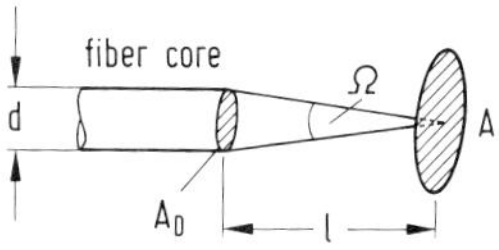

left: Irradiation with a glass fiber
numerical aperture: $A_N = n_0 \cdot \sin \epsilon_{0,G}$
beam diameter (over distance *l*): $D \approx d + 2A_N \cdot l$
irradiation intensity: $P/A = 4P/\pi D^2$

right: Light detection with a glass fiber
detection area: $A_D = \pi d^2/4$
solid angle: $\Omega = A_D/l^2 = \pi d^2/4l^2$
$(l \gg d, D \leqslant l)$

Fig. 34 Irradiation and light capture by a glass fiber

if the punctiform sample is at a sufficiently large distance ($l \gg d_c$) from the fiber. If the sample is located immediately in front of the fiber face, the solid angle is calculated from the aperture angle ε_{0L} of the fiber; the approximate relations are

$$\Omega = \pi \cdot \sin^2 \varepsilon_{0L} = \pi A^2/n_0^2. \qquad (16)$$

6.3 Biosensors

Glass-fiber probes, illustrated in Figure 32, and similar systems are used as biosensors for the detection of metabolites, dyes, and pigments that fluoresce. A common application is the determination of NADH fluorescence in microorganisms [75]. A much broader use of fiber-optic sensors is their coupling to an indicator or directly to a biochemical reaction. Thus, the pH and the partial pressure of oxygen or carbon dioxide can be determined by coupling these sensors to fluorescing dyes in solution or in a solid matrix [76, 79] (Section 4.6). The oxygen determination is generally based on measurements of the fluorescence quenching of an aromatic hydrocarbon, and occasionally also on changes in the fluorescence decay time of an oxygen indicator [80]. The most advanced technique involves a multisensor consisting of three 100-μm-thick glass fibers, with their ends coated, that detect O_2, CO_2, and pH, and an additional thermoelement for temperature measurements [81, 82]. Among the capabilities of this sensor are *in vivo* measurements inside blood vessels. Oxygen and pH sensors are used to determine glucose, ethanol, and penicillin, since oxygen consumption or pH changes are involved in the chemical reactions [83]. Finally, enzymes or fluorogenic enzymatic reactions can be identified.

A special technique is used in the *immunosensors*. They rely on total reflection of light from the surface of a glass plate or fiber coated with an optically thinner substrate. The light penetrates a few micrometers into the coating, where it is partly absorbed [84]. When the coating of an optically thin substrate, such as a fluorescence-labeled antigen-antibody complex, is immersed in a sample of the same antigen of unknown concentration, the antigen in the sample will partially replace the fluorescence-labeled antigen of the sensor. The loss of fluorescence at the sensor surface is, therefore, a measure of the concentration of the antigen in the sample. Additional techniques involving optoelectronic immunosensors are described in a review article by PLACE *et al.* [85].

6.4 Image transfer by fiber bundles

The magnitude of a signal is identified with the presence of a sensor at a specific sample site. Image-producing transfer systems are often of minor importance in biotechnology, but are essential in medical diagnostics, especially in endoscopy. Image transfer systems with as many as 30.000 single fibers, having a minimum core diameter of 8 μm, were developed for endoscopy. Because there are spaces between individual fibers, samples between 20 and 200 μm can be resolved with this system. The development of miniaturized CCD cameras makes possible the integration of the CCD chip directly into the tip of the endoscope, with the result that glass-fiber transfer systems can be dispensed with.

7 Flow cytometry

7.1 Principles and advantages

Studying and characterizing cells in complex samples such as blood with fluorescence microscopy is often too involved, and fiber technology does not give adequate resolution. Fluorescence microscopy has these particular drawbacks:

- fluorescence, scattered light, and the fluorescence of various dyes cannot be measured simultaneously;
- bleaching of dyes occurs during lengthy examination of cells, resulting in incorrect intensity measurements.

Flow cytometry, on the other hand, offers decided advantages, among which are a high flow rate of individual cells and the simultaneous measurement of up to two scattered light intensities and three fluorescence intensities. Measurements of forward-scattered light and light scattered at an angle of 90° to the incident light can provide information about cell size and degree of granulation; it is thus possible to distinguish leukocytes among lymphocytes, granulocytes, and monocytes. If the diameter, d, of the light-scattering particles is small compared to the wavelength ($d < 0.05\,\lambda$), then the scattering characteristic (which depends on the degree of polarization) is largely isotropic (*Rayleigh scattering*); if, on the other hand, the diameter, d, is of the same order of magnitude as the wavelength, then forward scattering (*Mie scattering*) occurs preferentially; the latter is more informative than the former. If leukocytes are additionally labeled with two fluorescence dyes that are measured simultaneously at different emission wavelengths, B and T lymphocytes can also be differentiated. In practice, the values of light scattered at 0° and 90° are first correlated, the intensities typical of lymphocytes are filtered out, and the lymphocytes are further analyzed for their (computer-stored) fluorescence behavior. The principle of flow cytometry is illustrated in Figure 35. A suspension of fluorescence-labeled cells is introduced into the cytometer and hydrodynamically focused, so that cells flow one after another through a narrow laser beam. Measurements of scattered light and fluorescence are done simultaneously with up to five separate detectors, whose intensity-dependent signals are correlated by a computer.

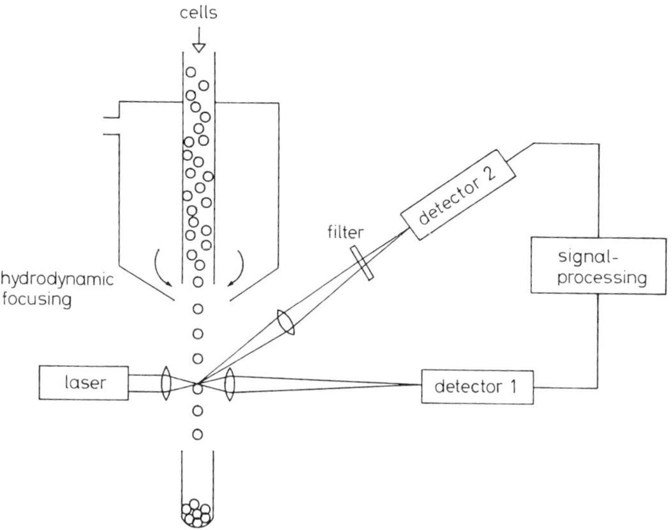

Fig. 35 Principles in the design of a flow cytometer

In some systems, the cells are sorted additionally (by size, for example). If this is the case, droplets below the measuring site containing single cells are charged positively or negatively, according to parameters stored in the computer program; they are then separated in an electric field according to their charge and collected in different vessels.

7.2 Applications

The applications of flow cytometry include those already discussed in Section 5 (see also the reviews of Kruth [86] and Quirke and Dyson [87]). Fluorescence-labeled cells are distinguished according to their content of DNA, RNA, or specific proteins. Cell permeability can be studied with the uptake of fluorescing dyes, in studies similar to those on phagocytosis with the uptake of microorganisms (Section 4.7). The possibility of detecting DNA aneuploidy is of interest in medical diagnostics; since this phenomenon is frequently associated with the early stages of cancer and other diseases, timely therapy could be initiated. Concentrating fetal cells from maternal blood by fluorescence-activated cell sorting [88] expands the possibilities of prenatal diagnostics.

The detection of fluorescence-labeled antibodies on a cellular level is just as possible as the determination of enzymatic reaction kinetics with fluorogenic substances. Not only can the overall behavior of a bioreactor be determined, but also it can be correlated [88] with the biochemistry of individual cells or particles (differing in location and enzymatic content). Highly developed experimental systems contain both laser and glass-fiber fluorescence detection systems [89], or a simultaneous multilaser excitation arrangement [90].

8 Spectral lamps and lasers

8.1 Gas-discharge lamps

Electrical discharges play a key role in many types of lamps [91]. If, for example, one puts a large voltage on an inert gas such as neon that contains traces of mercury vapor, the few ever-present scattered electrons are accelerated to a high-energy state adequate to ionize the gas. The subsequent recombination produces excited atoms of inert gas. These can then transfer their energy by various means to mercury atoms which are in turn stimulated to emit light. This is the basic principle of mercury vapor pressure lamps.

The spectral properties are largely dependent upon pressure; more precisely, they depend on the mechanisms of recombination. We can distinguish three types:

- low-pressure lamps, with a mercury partial pressure of about 10^{-3} atm, that emit lines predominantly at 253.7 nm;
- medium-pressure lamps with a pressure of about 1 atm; these require rather high currents and are rich in emitted lines; and, lastly,
- high-pressure lamps with a 30-atm partial pressure and a very small electrode distance; their spectra have broad bands that overlap from the ultraviolet to the red spectral range. The emission intensity, however, is relatively low in the red and the blue-green spectral ranges, so xenon high-pressure lamps are preferred for those wavelengths; they have a uniform spectral distribution over the entire visible light range and broad bands in the near infrared range.

8.2 Lasers

In comparison to the usual spectral lamps, the excitation of fluorescence with lasers has several advantages.

- The excitation wavelengths are exactly defined and can be adapted to the problem.
- The laser can be focused down to the submicrometer range. This means, among other things, that microscopic methods with the highest degree of resolution (such as laser scanning microscopy) can be used, and coupling to glass fibers is facilitated.
- By using suitable acousto-optical or electro-optical modulators, ultrashort laser impulses of high repetition rate and stability can be generated.

An overview of the types of lasers suitable for fluorescence measurements and the wavelengths they produce is given in Figure 36. In addition, nitrogen lasers ($\lambda = 337$ nm) and tunable dye lasers adjustable to various wavelengths ($350 \text{ nm} < \lambda < 1100 \text{ nm}$) are available. Laser diodes are not yet important in fluorescence spectroscopy; because the wavelengths available are greater than 670 nm, the photon energy is in most cases inadequate for fluorescence excitation. Most solid-state lasers also operate in the near infrared. Only the Nd : yag (neodymium : yttrium-aluminum-garnet) laser that doubles, triples, and quadruples frequency is occasionally used for fluorescence measurements.

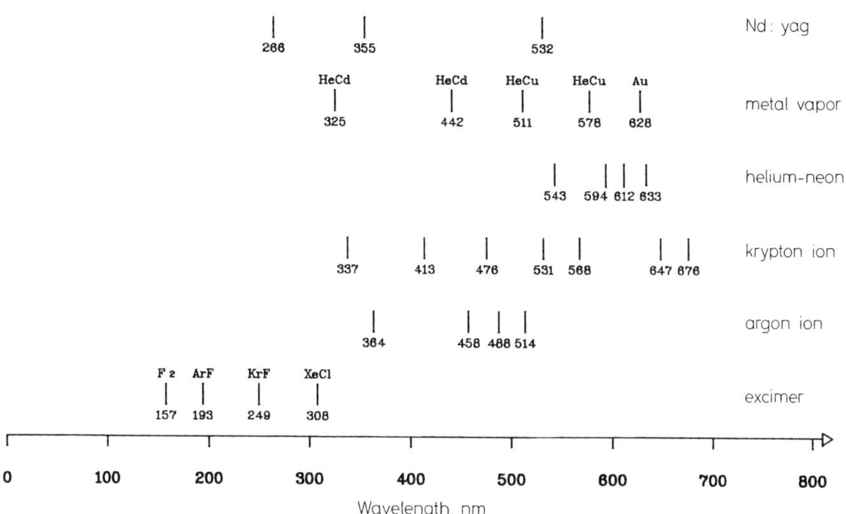

Fig. 36 Lasers useful in fluorescence spectroscopy: summary of the most important spectral lines

The average performance of commonly used types of laser ranges from about 1 mW to several W. Excimer lasers and nitrogen lasers are, in principle, impulse lasers with impulse duration times between 1 ns and 20 ns and repetition rates of up to 500 Hz. The remaining lasers listed can be operated at continuous emission of radiation.

9 Acknowledgement

We greatly appreciate the highly competent translation of the original German version of the manuscript as well as various suggestions and comments for its improvement by T. W. Bednar and E. Linsmaier-Bednar.

10 References

[1] Schmidt, W.: Luminescence of Organic Molecules. Theory and Analytical Applications in Photosynthesis. In: Lichtenthaler, H. K., Ed.; *Applications of Chlorophyll Fluorescence*, Kluwer Academic Publishers, New York, London (1988); pp 211–216.

[2] Lakowicz, J. R.: *Principles of Fluorescence Spectroscopy*, Plenum Press, New York, London (1983).

[3] Demas, J. N.: *Excited State Lifetime Measurements*, Academic Press, New York, London (1983).

[4] DERENIAK, E. L.; CROWE, D. G.: *Optical Radiation Detectors*, John Wiley & Sons, New York, Chichester, Brisbane, Toronto, Singapore (1984).

[5] KEITZ, H. A. E.: *Lichtberechnung und Lichtmessungen*, Philips Technische Bibliothek, Eindhoven, Germany (1967).

[6] DRISCOLL, W. G.; VAUGHAN, W., Eds.; *Handbook of Optics*, McGraw-Hill, New York, Toronto (1978).

[7] ROSENCWAIG, A.: Photoacoustic Spectroscopy; *Adv. Electr. Electron Phys.* **46** (1978) 207–211.

[8] BROWN, S. B., Ed.; *An Introduction to Spectroscopy for Biochemists*, Academic Press, London, New York, Toronto, Sydney, San Francisco (1980).

[9] HEMMERICH, P.: The Present Status of Flavin and Flavoenzyme Chemistry; *Prog. Chem. Org. Nat. Prod.* **33** (1976) 451–527.

[10] SCHMIDT, W.: On the Environment and Rotational Motion of Amphiphilic Flavins in Artificial Membrane Vesicles asStudied by Fluorescence; *J. Membr. Biol.* **47** (1979) 1–25.

[11] TAYLOR, D. G.; DEMAS, J. N.: Light Intensity Measurements II: Luminescent Quantum Counter Comparator and Evaluation of Some Luminescent Quantum Counters; *Analyt. Chem.* **51** (1979) 717–722.

[12] EATON, D. F.: Reference Material for Fluorescence Measurements; *J. Photochem. Photobiol.* **2** (1988) 523–531.

[13] PAPAGEORGIOU, G.: Chlorophyll Fluorescence: An Intrinsic Probe of Photosynthesis. In: GOVINDJEE *et al.*, Eds.; *Bioenergetics of Photosynthesis*, Academic Press, New York, San Francisco, London (1975); pp 319–371.

[14] KRUPINSKA, K.: Änderungen der molekularen Organisation von Thylakoidmembranen synchroner Zellen im Vergleich zu Pigmentmutanten von *Scenedesmus obliquus*; Thesis (1984), University of Marburg, Marburg, Germany.

[15] LICHTENTHALER, H. K.: *In Vivo* Chlorophyll Fluorescence as a Tool for Stress Detection in Plants. In: LICHTENTHALER, H. K., Ed.; *Applications of Chlorophyll Fluorescence*, Kluwer Academic Publishers, New York, London (1988); pp 129–142.

[16] LICHTENTHALER, H. K.; RINDERLE, U.: The Role of Chlorophyll Fluorescence in the Detection of Stress Conditions in Plants; *CRC Critical Reviews in Analytical Chemistry*, **Vol. 19**, Suppl. 1, CRC Press, Boca Raton, FL (1988); pp 30–85.

[17] SCHREIBER, U.; NEUBAUER, C.; KLUGHAMMER, C.: New Ways of Accessing Photosynthetic Activity with a Pulse Modulation Fluorometer. In: LICHTENTHALER, H. K., Ed.; *Applications of Chlorophyll Fluorescence*, Kluwer Academic Publishers, New York, London (1988); pp 63–69.

[18] WILLIAMS, A. T. R.: *An Introduction to Phosphorescence Spectroscopy*; A Perkin Elmer Production, Beaconsfield, UK (1981).

[19] SCHMIDT, W.: On-Line Computer Capability in Fluorescence Spectroscopy; *Opt. Eng.* **22** (1983) 576–582.

[20] COATES, J.: Towards a Standard Format for Spectroscopic Data; *Int. Labmate Guide* (1986) 19–22.

[21] SCHMIDT, W.: Fluorescence Properties of Isotropically and Anisotropically Embedded Flavins; *Photochem. Photobiol.* **34** (1981) 7–16.

[22] WU, C.-W.; STRYER, L.: Proximity Relationship in Rhodopsin; *PNAS USA* **69** (1972) 1104–1108.

[23] CONE, R.: Rotational Diffusion of Rhodopsin in the Light Receptor Membrane; *Nature (London)* **236** (1972) 1234–1244.

[24] RUTH, B.: *Experimenteller Nachweis ultraschwacher Photonenmission aus biologischen Systemen*; Dissertation, Marburg (1977).

[25] Arnold, W.; Sherwood, H.: Are Chloroplasts Semiconductors? *PNAS USA* **43** (1957) 105–114.

[26] Sane, P. V.; Rutherford, A. W.: Thermoluminescence from Photosynthetic Membranes. In: Govindjee; Amesz, J.; Fork, C., Eds.; *Light Emission by Plants and Bacteria*, Academic Press, Orlando, New York, San Francisco, London (1986); pp 329–360.

[27] Jursinic, P. A.: Delayed Fluorescence: Current Concepts and Status. In: Govindjee; Amesz, J.; Fork, C., Eds.; *Light Emission by Plants and Bacteria*, Academic Press, Orlando, New York, San Francisco, London (1986); pp 291–328.

[28] Schmidt, W.: Prompt and Delayed Light Emissions in Mature Tobacco Leaves: A Spectral Comparison; *Photobiochem. Photobiophys.* **9** (1985) 89–97.

[29] Schmidt, W.; Senger, H.: Long-Term Delayed Luminescence in *Scenedesmus obliquus*. I. Spectral and Kinetic Properties; *Biochim. Biophys. Acta* **890** (1987) 15–22.

[30] Schmidt, W.; Senger, H.: Long-Term Delayed Luminescence in *Scenedesmus obliquus*. II. Influence of Exogenous Factors; *Biochim. Biophys. Acta* **891** (1987) 22–27.

[31] Ökotest Magazin: Kommunale Abwassersatzungen, Grenzwerte für Schadstoffeinleitungen der Länderarbeitsgemeinschaft Wasser (LAWA) (1986); pp 36–41.

[32] Bürger, J.; Schmidt, W.: Effect of Depletion and Supplementation of Various Chemical Components on Long Term Delayed Luminescence of *Scenedesmus obliquus*; *Plant and Soil* **109** (1988) 79–83.

[33] Bracht, J.; Quick, P.: *Vom RIA zur signalverstärkten Lumineszenz*; Laborpraxis Reprint, Vogel-Verlag, Würzburg, Germany (1986); pp 1–5.

[33a] McCapra, F.: An Application of the Theory of Electrocyclic Reactions to Luminescence; *Chem. Commun.* (**1968**) 155–6.

[34] Lassak, H.-F.: *Fluorimetrie in Microplates. Applikationsbeispiele, Fluoreszenzmarker, Literaturhinweise*; Publ. Flow Laboratories GmbH, Meckenheim, Germany (1987).

[35] *Methods in Bioluminescence Analysis, "Bioluminescence"*, Product Information (1987), Handbook with Introduction and References, publ. by Boehringer, Mannheim-Biochemica, Germany.

[36] Müller-Klieser, W.; Walenta, S.; Paschen, W.; Kallinowski, F.; Vaupel, P.: Metabolic Imaging in Microregions of Tumors and Normal Tissues with Bioluminescence and Photon Counting; *J. Natl. Cancer Inst.* **80** (1988) 842–848.

[37] Udenfriend, S.: *Fluorescence Assay in Biology and Medicine*, Academic Press, New York, San Francisco, London (1962).

[38] Salmon, J.-M.; Kohen, E.; Viallet, P.; Hirschberg, J. G.; Wouters, A. W.; Kohen, C.; Thorell, B.: Microspectrofluorometric Approach to the Study of Free/Bound NAD(P)H Ratio as Metabolic Indicator in Various Cell Types; *Photochem. Photobiol.* **36** (1982) 585–593.

[39] Galland, P.; Senger, H.: The Role of Flavins as Photoreceptors; *J. Photochem. Photobiol.* **B1** (1988) 277–294.

[40] Eirich, L. D.; Vogels, G. D.; Wolfe, R. S.: Distribution of Coenzyme F_{420} and Properties of Its Hydrolytic Fragments; *J. Bacteriol.* **140** (1979) 20–27.

[41] Schneckenburger, H.; Reuter, B. W.; Schoberth, S. M.: Time-Resolved Microfluorescence for Measuring Coenzyme F_{420} in *Methanobacterium thermoautotrophicum*; *FEMS Microbiol. Lett.* **22** (1984) 205–208.

[42] Johnson, L. V.; Summerhayes, I. C.; Chen, L. B.: Decreased Uptake and Retention of Rhodamine 123 by Mitochondria in Feline Sarcoma Virus-Transformed Mink Cells; *Cell* **28** (1982) 7–14.

[43] Christopher, R. S.; Whitaker, D.; Murphy, G. F.; Hasan, T.: Ultrastructure and Dynamics of Selective Mitochondrial Injury in Carcinoma Cells after Doxycycline Photosensitization *in Vitro*; *Am. J. Pathol.* **133** (1988) 381–388.

[44] Radda, G. K.; Vanderkooi, J.: Can Fluorescence Probes Tell Us Anything about Membranes? *Biochim. Biophys. Acta* **265** (1972) 509–549.
[45] Tsien, R. Y.: Fluorescence Measurement and Photochemical Manipulation of Cytosolic Free Calcium; *TINS* **11** (1988) 419–424.
[46] Nakamura, R. M.; Dito, W. R.; Tucker, E. S., Eds.; *Immunoassays in the Clinical Laboratory*, Vol. 3, Alan R. Liss Inc., New York (1979).
[47] Engvall, E.; Perlman, P.: Enzyme-Linked Immunoassay (ELISA) Quantitative Assay of Immunoglobulin G; *Immunochem.* **8** (1971) 871–874.
[48] Sernetz, M.: Mikrofluorometrische Verfahren zur Analyse von zellulären und trägergebundenen Enzymen; *Biotech. Umschau* **3** (1979) 214–219.
[49] Wudl, L.: Paigen, K.: Enzyme Measurements on Single Cells; *Science* **184** (1974) 992–994.
[50] Hösli, P.: Quantitative Assays of Enzyme Activity in Single Cells: Early Prenatal Diagnosis of Genetic Disorders; *Clin.Chem.* **23** (1977) 1476–1484.
[51] Conway de Macario, E.; Wolin, M. J.; Macario, A. J. L.: Antibody Analysis of Relationships among Methanogenic Bacteria; *J. Bacteriol.* **149** (1982) 316–319.
[52] Dougherty, T. J.: Photosensitizers: Therapy and Detection of Malignant Tumours; *Photochem. Photobiol.* **45** (1987) 879–889.
[53] Spikes, J. D.: Phthalocyanines as Photosensitizers for the Photodynamic Therapy of Tumours; *Photochem. Photobiol.* **43** (1986) 691–699.
[54] Birch, D. L. S.; Imhof, R. E.: Coaxial Nanosecond Flashlamp; *Rev. Sci. Instrum.* **52** (1981) 1206–1212.
[55] Fleming, G. R.; Beddard, G. S.: CW Modelocked Dye Lasers for Ultrafast Spectroscopic Studies; *Optics Laser Technol.* **Oct.** (1978), 257–264.
[56] Zaal, G. J.: Lasersysteme zur Erzeugung ultrakurzer Lichtimpulse; *Laser Optoelektr.* **2** (1982) 31–37.
[57] Soini, E.; Hemmilä, I.: Fluoroimmunoassay: Present Status and Key Problems; *Clin. Chem.* **25** (1979) 353–361.
[58] Schneckenburger, H.: Time-Resolved Microfluorescence in Biomedical Diagnosis; *Opt. Eng.* **24** (1985) 1042–1044.
[59] Andreoni, A.; Sacchi, C. A.; Svelto, O.; Bottiroli, G.; Prenna, G.: Laser-Induced Fluorescence of Acridine-DNA Complexes; *Sov. J. Quantum Electron.* **8** (1978) 1255–1259.
[60] Arndt-Jovin, D.; Latt, S. A.; Striker, G.; Jovin, T. M.: Fluorescence Decay Analysis in Solution and in a Microscope of DNA and Chromosomes Stained with Quinacrine; *J.Histochem. Cytochem.* **27** (1979) 87–95.
[61] Schneckenburger, H.; Seidlitz, H. K.; Eberz, J.: Time-Resolved Fluorescence in Photobiology; *J. Photochem. Photobiol.* **B2** (1988) 1–19.
[62] Röder, B.; Wabnitz, H. B.: Time-Resolved Fluorescence Spectroscopy of Hematoporphyrin, Mesoporphyrin, Pheophorbide *a* and Chlorin e_6 in Ethanol and Aqueous Solution; *J. Photochem. Photobiol., B: Biology* **1** (1987) 103–113.
[63] Pauker, F.; Schneckenburger, H.; Unsöld, E.: Time-Resolved Multiphoton Counting; *J. Phys. E* **19** (1986) 240–241.
[64] Schneckenburger, H.; Lang, M.; Köllner, T.; Rück, A.; Herzog, M.; Hörauf, H.; Steiner, R.: Fluorescence Spectra and Microscopic Imaging of Porphyrins in Single Cells and Tissues; *Lasers Med. Sci.* **4** (1989) 159–166.
[65] Uster, P. S.; Pagano, R. E.: Resonance Energy Transfer Microscopy: Observations of Membrane-Bound Fluorescent Probes in Model Membranes and in Living Cells; *J. Cell Biol.* **103** (1986) 1221–1234.
[66] Szöllösi, J.; Damjanovich, S.; Mulhern, S. A.; Tron, L.: Fluorescence Energy Trans-

fer and Membrane Potential Measurements Monitor Dynamic Properties of Cell Membranes: A Critical Review; *Prog. Biophys. Molec. Biol.* **49** (1987) 65–87.
[67] Taylor, D. L.; Reidler, J.; Spudich, A.; Stryer, L.: Detection of Actin Assembly by Fluorescence Energy Transfer; *J. Cell Biol.* **89** (1981) 362–367.
[68] Squier, T. C.; Bigelow, D. J.; de Ancos, J. G.; Inesi, G.: Localization of Site-Specific Probes on the Ca-ATPase of Sarcoplasmic Reticulum Using Fluorescence Energy Transfer; *J. Biol. Chem.* **262** (1987) 4748–4754.
[69] Schneckenburger, H.; Rück, A.; Haferkamp, O.: Energy Transfer Microscopy for Probing Mitochondrial Deficiencies; *Analyt. Chim. Acta* **227** (1989) 227–233.
[70] Holzwarth, A. R.; Wendler, J.; Haehnel, W.: Time-Resolved Fluorescence Spectra of the Antenna Chlorophylls in *Chlorella vulgaris*. Resolution of Photosystem I Fluorescence Kinetics; *Biochim. Biophys. Acta* **806** (1985) 374–388.
[71] Berens, S. J.; Scheele, J.; Butler, W. L.; Magde, D.: Time-Resolved Fluorescence Studies of Spinach Chloroplasts – Evidence for the Heterogeneous Bipartite Model; *Photochem. Photobiol.* **42** (1985) 51–57.
[72] Schneckenburger, H.; Frenz, M.: Time-Resolved Fluorescence of Conifers Exposed to Environmental Pollutants; *Radiat. Environ. Biophys.* **25** (1986) 289–295.
[73] Wilke, V.: Optische Rastermikroskopie – Das Laser-Scan-Mikroskop; *Physik in unserer Zeit* **15** (1984) 13–18.
[74] Kersten, R. T.: Faseroptische Sensoren; *Physik in unserer Zeit* **15** (1984) 139–144.
[75] Scheper, T.; Schügerl, K.: Characterization of Bioreactors by *in Situ* Fluorimetry; *J. Biotechn.* **3** (1986) 221–229.
[76] Wolfbeis, O. S.: Fluorescence Optical Sensors in Analytical Chemistry; *Trends Anal. Chem.* **4** (1985) 184–188.
[77] Schneckenburger, H.; Bader, J.: Fiber-Optic Detection of Chlorophyll Fluorescence. In: Lichtenthaler, H. K., Ed.; *Applications of Chlorophyll Fluorescence*, Kluwer Academic Publishers, New York, London (1988); pp 255–258.
[78] Lichtenthaler, H. K.; Buschmann, C.; Rinderle, U.; Schmuck, G.: Application of Chlorophyll Fluorescence in Ecophysiology; *Radiat. Environ. Biophys.* **25** (1986) 297–308.
[79] Lübbers, D. W.; Opitz, N.: Optical Fluorescence Sensors for Continuous Measurements of Chemical Concentration in Biological Systems; *Sensors and Actuators* **4** (1984) 641–647.
[80] Lippitsch, M. E.; Pusterhofer, J.; Leiner, M. J. P.; Wolfbeis, O. S.: Fiber-Optic Oxygen Sensor with the Decay Time as Information Carrier; *Anal. Chim. Acta* **205** (1988) 1–6.
[81] Gehrich, J. L.; Lübbers, D. W.; Opitz, N.; Hansmann, D. R.; Miller, W. W.; Tusa, J. K.; Yafuso, M.: Optical Fiber Fluorescence and Its Application to an Intravascular Blood Gas Monitoring System; *IEEE Trans. Biomed. Eng.* **33** (1986) 117–132.
[82] Miller, W. W.; Yafuso, M.; Yan, C. F.; Hui, H. K.; Arik, S.: Performance of an *in Vivo* Continuous Blood Gas Monitor with Disposable Probe; *Clin. Chem.* **33** (1987) 1538–1542.
[83] Wolfbeis, O. S.: Fiber-Optic Sensors in Bioprocess Control. In: Twork, J. V.; Yancynych, A. M., Eds.; *Sensors in Bioprocess Control*, Marcel Dekker, New York (1990).
[84] Hirschfeld, T.: Total Reflexion Spectrometry; *Can. J. Spect.* **10** (1965) 128–130.
[85] Place, J. F.; Sutherland, R. M.; Dähne, C.: Optoelectronic Immunosensors: A Review of Optical Immunoasssay at Continuous Surfaces; *Biosensors* **1** (1985) 321–353.
[86] Kruth, H. S.: Flow Cytometry: Rapid Biomedical Analysis of Single Cells; *Analyt. Bioch.* **125** (1982) 242–255.
[87] Quirke, P.; Dyson, J. E. D. : Flow Cytometry: Methodology and Applications in Pathology; *J. Pathol.* **149** (1986) 79–87.

[88] HERZENBERG, L. A.; BIANCHI, D. W.; SCHRÖDER, J.; CANN, H. M.; IVERSON, G. M.: Fetal Cells in the Blood of Pregnant Women: Detection and Enrichment by Fluorescence-Activated Cell Sorting; *PNAS USA* **76**(3) (1979) 1453–1455.

[89] WILLEMS, H.; SERNETZ, M.: Flow Fluorimetric Investigations on Reaction Kinetics of a Growing Analytical Bioreactor; *Analyt. Chim. Acta* **213** (1988) 245–250.

[90] STEINKAMP, J. A.; STEWART, C. C.; CRISSMAN, H. A.: Three-Colour Fluorescence Measurements on Single Cells Excited at Three Laser Wavelengths; *Cytometry* **2** (1982) 226–231.

[91] Osram-Applikationen: Licht für Kinoprojektion. Technik und Wissenschaft, Issue August (1986)/FO 302.

Mechanical Cell Disruption Processes

by J. Schwedes and F. Bunge

Contents

Prof. Dr.-Ing. J. Schwedes and Dipl.-Ing. F. Bunge,
Institut für Mechanische Verfahrenstechnik der
Technischen Universität Braunschweig,
Volkmaroder Str. 4–5,
D-3300 Braunschweig, Fed. Rep. Germany

1 Introduction

The products of modern biotechnology are used in a wide range of applications, from research and diagnostic laboratories to the varied industrial fields of chemistry, pharmaceuticals, food technology, and so forth. The starting materials for these products are cells from animal tissue, plant cells, and microorganisms. Over the past few years, concentrated efforts have been devoted to research and development in the area of genetic engineering (especially the technique of recombinant DNA). As a result, a multitude of new bio-products is being produced and used. Since many of these substances are intracellular, *cell disruption* is the first necessary step in the processing of the product.

The aim of a cell disruption process is the partial destruction of the cell wall so that the cytoplasm is released into the surrounding medium. This disruption process has to meet several requirements, listed in Table 1. The rank and importance of these requirements depend on the kind of disruption process in question. Disruption of a small volume of cell suspension in a laboratory for purposes of analysis, and continuous cell disruption in industry differ in important ways. It matters whether microorganisms or animal tissues are to be disrupted. Furthermore, if the cell walls are fragmented into a great many particles during a continuous disruption, because too high an energy input is used, the separation of cell debris becomes technically difficult as well as expensive.

Table 1 Requirements for cell disruption processes

- no denaturation
- high degree of disruption
- ability to be sterilized
- no contamination of the product
- easy separation of cell debris
- low energy consumption
- speed
- low investment

Especially important in the disruption process are the stability of the cell wall and the size of the cell itself. Microorganisms range in size from 0.5 to 10 μm, plant cells from 20 to 200 μm, and the diameters of animal cells from 10 to 100 μm. The resistance of cell walls to applied forces differs for cells with different textures and cell wall compositions. Animal tissue can be damaged by very small shear forces, but the disruption of Gram-positive bacteria requires a much greater expenditure of energy. The energy necessary for the disruption process can be supplied as pressure, translational or rotational energy, or as nonmechanical energy (in chemical or biochemical transformations). The type of energy input permits classification into *mechanical cell disruption processes* and *nonmechanical cell disruption processes* (Table 2). This chapter considers only mechanical cell disruption processes.

Agitated ball mills and pressure homogenizers are especially suitable for the continuous disruption of large quantities. The other methods described here are in general limited to laboratory applications.

Table 2 Classification of cell disruption processes

Mechanical processes		Nonmechanical processes
Animal and plant cells	Microorganisms	
Colloid mill Hammer mill Pin mill Cutting mill Shear gap setting Chaikoff Press	Agitated ball mill Vibration mill Ultrasonic probe High-pressure homogenizer French Press Hughes Press X-Press Jet mill	Chemical methods Biological methods Physical methods

2 Fundamental principles of size reduction

The effect of a mechanical load on a particle can be described in terms of stress and strain. The solid body – here the cell – consists of a material whose resistance to mechanical deformation, that is the particle strength, can be described quantitatively. If the particle is to break down, the molecular binding forces between the fragments

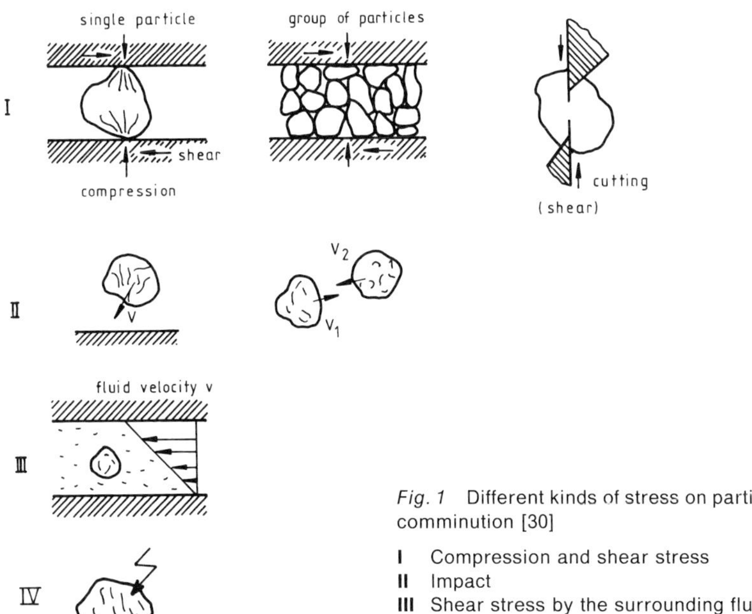

Fig. 1 Different kinds of stress on particles during comminution [30]

I Compression and shear stress
II Impact
III Shear stress by the surrounding fluid
IV Nonmechanical stress

have to be overcome. Fracture occurs if the loading energy exceeds the tear resistance of the material and if this energy exceeds the amount of energy consumed by the fracture.

The experimental determination of the strength of single particles is an expensive task. That is why comminution research often consists of investigating the phenomenological crushing behavior of real particles under ideal experimental conditions. The relationship between loading conditions and the resulting comminution is thus elucidated. One of the most important loading conditions is the type of applied stress. According to RUMPF [1] four kinds of stress can be defined (Figure 1). *Compression and shear stress* and *shear stress by the surrounding fluid* are chiefly responsible for the mechanical disruption of cells. Chemical, biological, and physical methods of disruption are the result of type IV (nonmechanical) stress.

3 Disruption of animal and plant cells

Animal tissue and plant tissue are qualitatively different from microorganisms. Various devices are available for the disruption of animal and plant cells (Table 2). Plant materials are comminuted in commercially available mills, which will not be discussed further here.

The loading energy necessary for the disruption of animal cells from organs is usually achieved with *rotor/stator* devices.

3.1 The rotor/stator principle

The widely applied rotor/stator principle is shown schematically in Figure 2. A cylindrical rotor turns with high circumferential speed (in the range of 10 to 24 m/s) concentrically inside a stator. Cell suspension fills the shear gap. Here a velocity profile is generated that leads to shear stress in the suspension. The coarsely turbulent flow pattern is overlaid with a fine structure of turbulence, which leads to the disruption of cell walls.

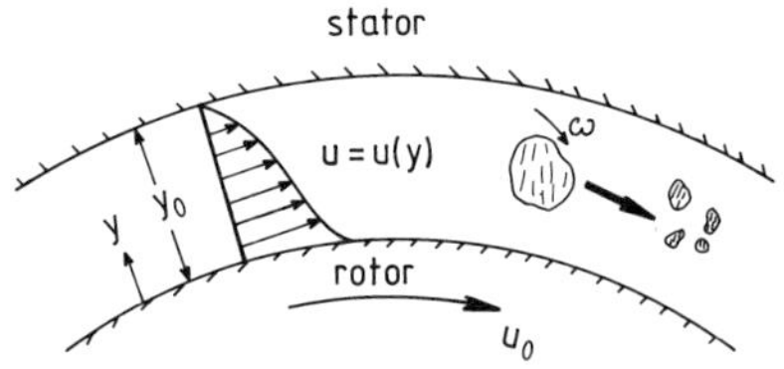

Fig. 2 Stress load on particles in a hydrodynamic shear field
Key to variables: y_0 = gap width; $u(y)$ = velocity profile; u_0 = circumferential speed of rotor; ω = angular velocity of particles

3.2 Chaikoff Press

The Chaikoff Press, described in 1957 by EMANUELL and CHAIKOFF [2], can be used for the disruption of small quantities of animal tissue. With this apparatus it is possible to destroy cells with a low particle strength while retaining most of the organelles intact.

The Chaikoff Press consists of a cylinder in which a piston with a slightly smaller diameter moves. The cell suspension is poured into the cylinder and is compressed by the piston. The suspension streams upwards through the gap between piston and cylinder wall. The resulting shear stress on the cell walls is sufficient to destroy the labile structure of animal tissue.

The gap opening and therefore the magnitude of the applied shear stress can be varied by using different pistons. The choice of gap size depends on the viscosity of the suspension and can range from 6 to 130 μm. The pressure can be as high as 20 MPa (200 bar).

4 Mechanical disruption of microorganisms

The distinction between batch and continuous flow disruption equipment is made depending on the volume of cell suspension. All the methods for disrupting microorganisms listed in Table 2 fall into the category of wet comminution, because the microorganisms in every case are suspended in medium. A further distinction among the various processes can be made by whether it is necessary to use grinding agents for the disrupting. This is the case with the vibration mill and the agitated ball mill, which do their work at atmospheric pressure. Several disruption methods are described in the following sections.

4.1 The agitated ball mill

The agitated ball mill consists of a machine housing with an attached grinding chamber into which the rotor (driven by an electric motor) projects. The configuration of mills can differ, since the grinding chamber can be arranged horizontally or vertically (Figure 3). The shape of the grinding chamber is usually cylindrical, occasionally conical.

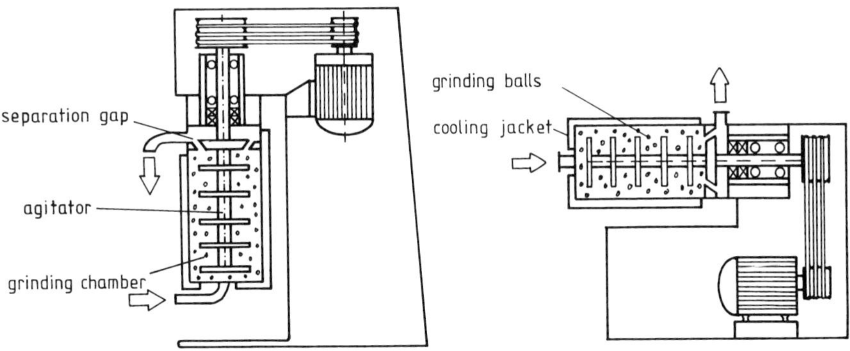

Fig. 3 Configurations of closed agitated ball mills [30]
left: vertical, right: horizontal

The rotor and the agitating elements mounted on it may differ in shape and size. The geometry of the agitating elements influences the power consumption. The most common agitator geometries are shown in Figure 4. Figure 5 consists of drawings of widely used German agitated ball mills. The volume of the grinding chambers can be 50 ml for a laboratory mill, and up to a thousand liters with throughputs of a few hundred liters per hour for industrial mills.

Besides this application as a cell disruption machine, the agitated ball mill is chiefly used for the fine grinding of crystalline material and for the disagglomeration of dyestuffs.

The grinding chamber of the agitated ball mill is filled up to 80% with beads, which are set in motion by the rotor. The mill can be operated continuously or batchwise. For batch operation the grinding chamber is filled with the cell suspension, and after a certain grinding time it is stopped and discharged. In continuous operation a steady stream of suspension is pumped through the mill and is either recirculated (multipass grinding) or passed on to the next process step. The beads are retained in the grinding chamber by a sieve or a constriction.

As with all industrial comminution processes, a large fraction of the input energy is dissipated, causing the temperature of the suspension to rise. In most agitated ball mills this energy is taken up by the coolant, which flows through the cooling channel of the

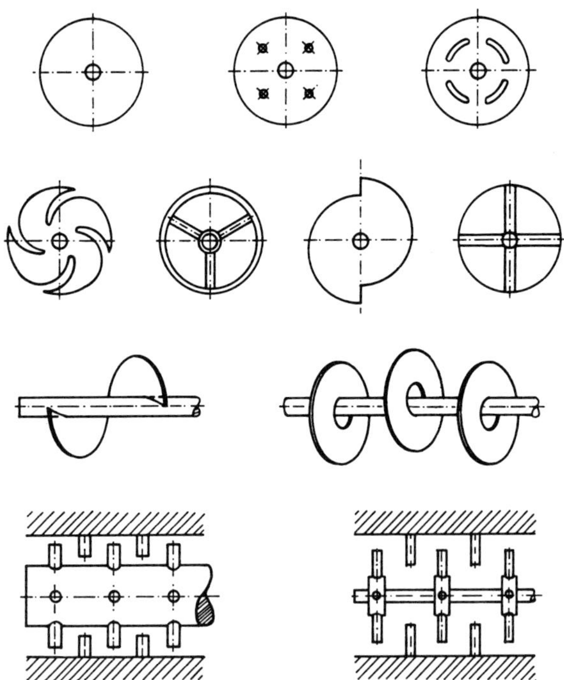

Fig. 4 Geometry of agitator discs for agitated ball mills [30]

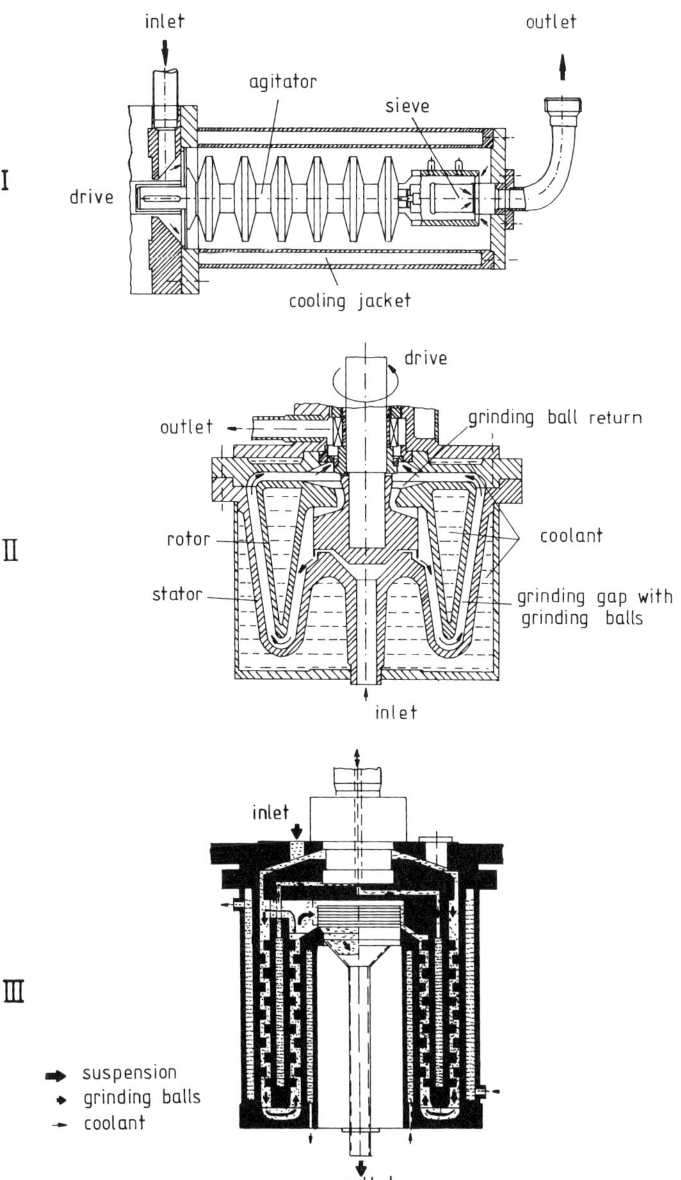

Fig. 5 Design details of different agitated ball mills

I slit mill [33]
II CoBall mill [34]
III Perl mill [35]

grinding chamber. Unacceptably high temperatures (which lead to the denaturation of proteins) are thus prevented.

In the literature, hydrodynamic shear forces and compression loading between the balls are often cited as breakdown mechanisms. However, we do not yet have any experimental foundation for statements concerning the breakdown mechanism. Various models start from the assumption that the grinding agents stream past each other in layers. Because of the difference in velocity of the layers and the rotation of the grinding agents, shear forces are induced, which lead to cell disruption. Furthermore, the grinding agents are accelerated at the agitating elements and are driven outward. This results in an exchange of momentum among the balls themselves and between the balls and the wall of the agitating chamber.

For an agitated ball mill with a given geometry, the following operational parameters can be varied:

- the size of the grinding agents,
- their density and filling ratio,
- the agitator speed,
- the throughput of suspension,
- the grinding time, and
- the cell concentration in the suspension

(see Section 6, List of symbols used).

The influence of these parameters has been examined, mostly with bakers' yeast or *Escherichia coli* [3, 4, 5]. The disruption behavior of different microorganisms differs significantly, since it depends on the particle strength, size and shape, but the way the operational parameters affect the disruption behavior is largely independent of the product.

The extent of cell disruption is assayed after the treated cell suspension has been separated in a high-speed centrifuge. The concentration of protein released into the supernatant can be compared to the activity of the intracellular enzymes (such as G-6-P dehydrogenase). The result of this determination can be compared to the maximum value, and the degree of disruption is thus calculated as a percentage. The shape and size of the cell debris can be seen and measured in electron micrographs (Figure 6). In Figure 7 the concentration of protein released is given as a function of agitator speed, for a series of batch disruption experiments. For each experimental point the mill was filled with a 10% (*w*/*v*, wet cell weight of yeast) cell suspension and was run at the selected agitator speed for the time shown as the parameter of the curves. It is apparent that a maximum concentration of released protein, hence a maximum level of cell disruption, cannot be exceeded. The maximum degree of disruption can be reached by a long grinding time at low agitator speed or by higher revolutions for a shorter (but adequate) time.

If the suspension is pumped through the mill continuously, the degree of cell disruption is lower at higher throughputs. This is shown in Figure 8.

To correlate results such as these, an experiment was performed in which the exact input of mechanical power into the grinding chamber was determined by measuring the torque and the agitator speed. In Figure 9 the concentration of protein released is given as a function of the mechanical energy input. The time of cell disruption was varied from

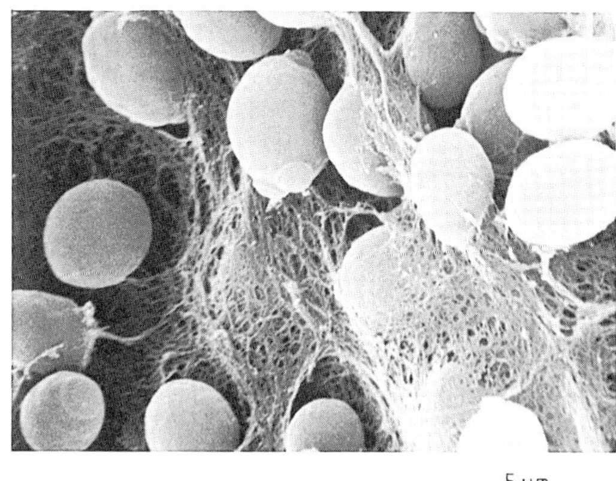

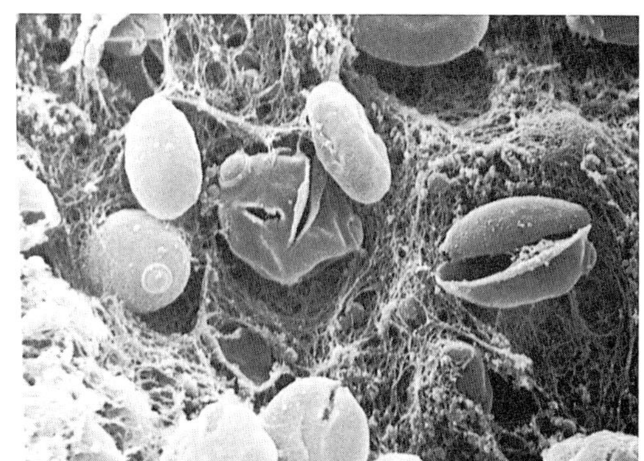

Fig. 6 *Saccharomyces cerevisiae* before (a) and after (b) disruption in an agitated ball mill

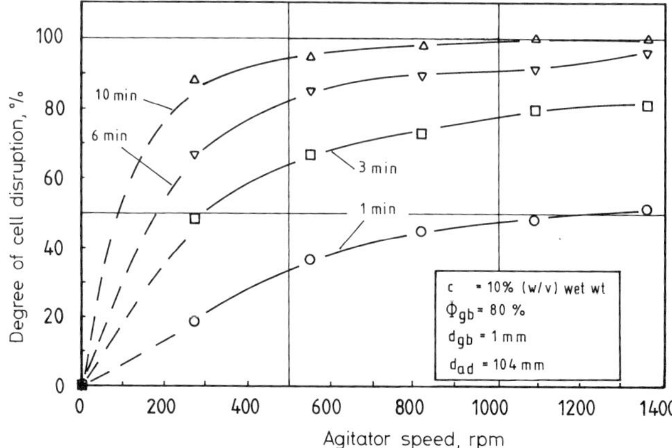

Fig. 7 Release of protein from bakers' yeast in batch experiments as a function of agitator speed

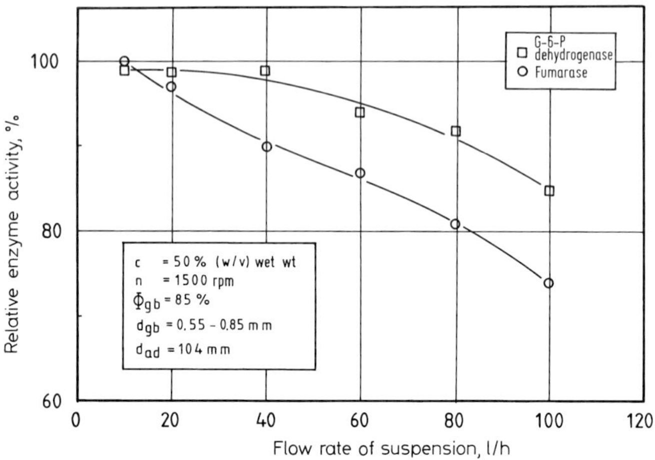

Fig. 8 Release of enzymes from bakers' yeast in a continuous process as a function of throughput [3]

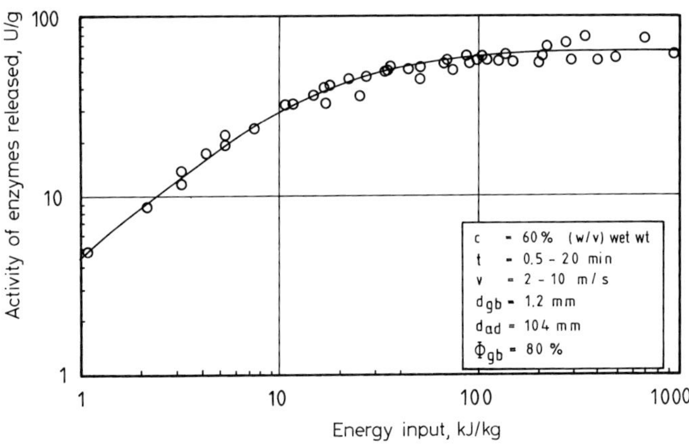

Fig. 9 Release of enzymes from bakers' yeast as a function of energy input, in batch and continuous experiments

30 sec to 20 min, and the tip speed of the agitator covered the range from 2 to 10 m/s. The same degree of cell disruption always resulted from the same energy input, independent of how the energy input was achieved. The energy necessary to attain a certain level of cell disruption for the given parameters can be read from the graph. This extensive investigation of the comminution in agitated ball mills allows the prediction of comminution results for continuous operation from only a few batch grinding experiments. Also, the scale-up of agitated ball mills has been successfully carried out [6, 7, 8].

In contrast to the grinding of crystalline material in agitated ball mills, the extent of cell disruption is almost independent of cell concentration for a wide range of parameters. Hence, the best use of the input energy is made at high cell concentrations.

The grinding agents in agitated ball mills are balls or beads of steel, ceramic, or glass. These materials differ in their density and thus in the achievable energy of momentum exchange. Wear of these materials by fracture and abrasion should not be neglected. In each instance the ball material must be compatible with the final product. The higher the viscosity and density of the cell suspension, the higher the density of the grinding agents should be. For cell disruption, hardened lead-free glass beads with high abrasion resistance are usually used. The density of the glass is 2.5 g/cm^3.

The bead diameter must be suited to the mill's geometry, which is to say it should be smaller than a quarter of the distance between the perimeter of the agitator disc and the inner wall of the cylinder. The minimum size is given by the sieve or the constriction, as the case may be for the separation of the grinding agents, and is about 200 μm. The influence of bead size on the disruption of bakers' yeast is shown in Figure 10. The degree of disruption increases, up to a maximum, with decreasing bead diameter. For the concentrated cell suspension (60% wet weight) the optimum bead diameter is about 0.8 mm. For less concentrated suspensions, with lower viscosity, the optimum bead diameter lies between 0.3 and 0.8 mm. The power consumption was the same for all points shown. Hence, the energy efficiency during cell rupture depends significantly on the diameter of the grinding agents.

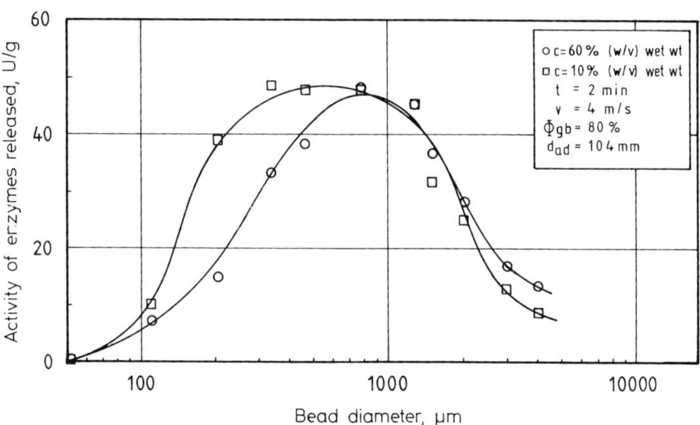

Fig. 10 Release of enzymes from bakers' yeast as a function of bead diameter

The degree of filling with grinding beads is defined as the ratio of the bulk volume of the beads to the volume of the grinding chamber. An increase in the degree of filling results in improved cell disruption [3] but only up to values of about 85%. This can be explained as follows. The volume of beads used in filling the grinding chamber has a void volume fraction of about 40%. When the beads are set in motion they flow past each other unhindered in layers, so they appear to have a cubic close-packing with a void volume fraction of 48%. The bead filling increases its volume by motion; the maximum volume the moving beads can fill is the volume of the grinding cylinder. If the maximum degree of filling is exceeded the result is high friction between the beads, which leads to a rise in temperature and to increased wear of the beads. A degree of filling of 80% is customary.

The temperature of the cell suspension should be controlled during the disruption, to prevent an unacceptable temperature rise. In principle, provision should be made to remove as heat, with a suitable coolant, all the mechanical power (as a first approximation, the electric power consumption) that is brought into a grinding chamber. For small laboratory mills with large cooling surfaces, a hookup to cooling water is sufficient; otherwise a refrigeration machine has to be connected. It is advisable to precool the suspension.

With agitated ball mills all kinds of microorganisms can be disrupted quickly and gently. Even the disruption of Gram-positive bacteria (for example, *Brevibacterium*) can be satisfactorily achieved [3].

4.2 Vibration mill

Vibation mills for cell disruption may be run continuously as well as batchwise. They consist of one or more cylindrical grinding drums. The suspension is poured into the grinding cylinder along with the grinding agents. An eccentric drive causes the cylinder to oscillate with frequencies of 100 to 6000 vibrations per minute. The grinding agents are usually lead-free hardened glass beads with diameters of 0.2 to 1 mm; the degree of filling with grinding agents should not exceed 80%.

Different vibration mills have been studied for their usefulness as cell disintegrators, by FURNESS [9], MICKLE and RAY [10], NOSSAL [11], and SCHÜTTE [12], among others. In Figure 11 a batch-operated mill with two horizontal grinding drums is shown. The empty volume in each cylinder can be filled with about 6 ml of cell suspension. Results

Fig. 11 Vibration mill with two grinding drums [32]

Table 3 Disruption of microorganisms in a vibration mill

Microorganism	Enzyme	Bead diameter, mm	Time for 50% disruption, min	Time for 90% disruption, min
Saccharomyces cerevisiae	Fumarase	0.75–1.0	10	20
Lactobacillus confusus	L-2-Hydroxy-isocaproate dehydrogenase	0.75–1.0	12	25
Brevibacterium ammoniagenes	Fumarase	0.25–0.5	15	35
Bacillus cereus	Leucine dehydrogenase	0.5–0.75	20	45

of an experiment with this mill were published by SCHÜTTE [12] and are given in Table 3. A continuously operated configuration has been investigated by ROSS [13]. A flow of several liters of suspension per hour is pumped through the vibrating cylinder. The beads are held back by a sieve. Cooling the grinding cylinder prevents an unacceptable temperature rise in the cell suspension.

The effectiveness of vibration mills as comminuting devices is explained by KURRER [14] as momentum exchange resulting from normal shock and friction in the peripheral high-energy zone of the cylinder.

4.3 Ultrasonics

According to DOULAH, the disruption of cells by ultrasonic energy is due to the development of high-energy fields of shear stress [15]. This is why ultrasonic disruption can be considered a mechanical method. The energy necessary for the disruption is brought into the cell suspension by an oscillator system in the form of acoustic waves. The probe projects into the liquid and oscillates with a resonant frequency of usually about 15–25 kHz, depending on the device.

Cavitation bubbles in the suspension collapse, releasing large amounts of mechanical energy in form of elastic compression shock waves, and eddies develop when these shock waves interact. The high-energy eddies create a field of shear stresses; microorganisms are stressed by the shear forces. Bigger eddies pass their energy on to smaller ones, which dissipate immediately in the viscous medium. In highly viscous suspensions disruption is less effective, so low cell concentrations must be chosen.

In Figure 12 an ultrasonic cell disruption apparatus is shown, of the type used mainly in laboratories for the batch treatment of small volumes. Some continuously operated devices with throughputs of about 1 liter per hour are available. The acoustic output often reaches values of a few hundred watts. Because of the small volume of fluid, the energy density that must be dissipated is high, and the temperature rise can be so large as to be a problem, making intense cooling of the sample necessary.

Fig. 12 Ultrasonic cell disruption apparatus [31]

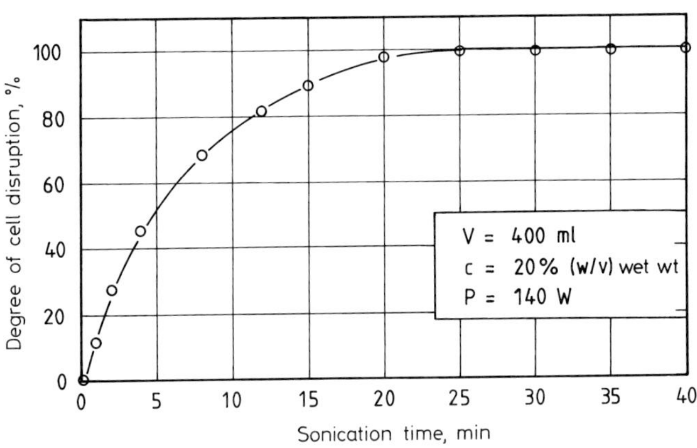

Fig. 13 Degree of disruption of bakers' yeast (20% wet wt) as a function of time in the ultrasonic field [16]

The time course for the disruption of 400 ml of a 20% yeast suspension at an acoustic power input of 140 watts is given by JAMES and co-workers [16] and is shown in Figure 13. The time for complete disruption is 21 minutes. The kinetics of the disruption can be described by a first-order reaction, as has been observed for each method of mechanical disruption. DOULAH investigated the protein release from bakers' yeast as a function of the acoustic power input. He found that a higher acoustic power input results in a markedly higher degree of disruption per unit time. Unfortunately, this goes along with a greater temperature rise.

4.4 High-pressure homogenizer

Originally, high-pressure homogenizers were used in the dairy industry to homogenize milk. Today they are not only used in food processing and the chemical industry, but also for the disruption of microorganisms. Here the high-pressure homogenizer is used as a continuously operating cell disruption apparatus.

The throughput ranges from 50 l/h for laboratory units up to a few cubic meters per hour in industrial production operations. The heart of the unit is the homogenizing valve, through which the suspension is forced at high pressure. The fluid pressure is created by a reciprocating pump and can reach values up to 100 MPa (1000 bars). A cross-sectional view of a homogenizing valve is shown in Figure 14. Its principal components are a stationary valve seat, an adjustable valve body, and an impact ring. The cell suspension impinges on the valve body; the stream is sharply deflected and is led through the homogenizing gap onto the impact ring. The gap width can be varied by adjusting the position of the valve body; thus the pressure drop and the concomitant velocity can be set. The fluid velocity in the homogenizing gap attains values up to 300 m/s, but it is strongly dependent on the valve geometry. According to TONNIUS [17], there are more than 150 patented nozzle geometries now, but the principle of construction always stays the same.

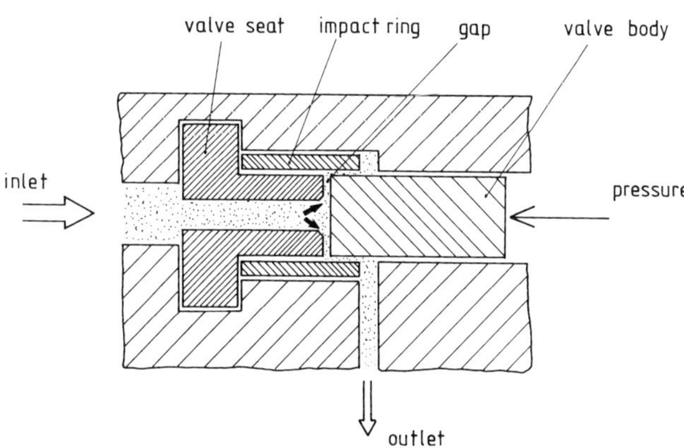

Fig. 14 Cross-sectional view of a homogenizing valve

The cell disruption takes place in the narrow homogenizing gap. The impact ring shown in Figure 14 is not critical for the homogenizing effect, but it improves the result substantially, according to TREIBER [18].

Cavitation and turbulence are the accepted mechanisms of disruption, a conclusion reached on the basis of theoretical considerations [19]. Cavitation occurs in a high-pressure homogenizer because the compressed liquid is strongly accelerated in the narrow homogenizing gap. According to the first law of thermodynamics, this leads to a significant decrease in pressure. Bubbles form when the pressure falls below the vapour pressure. Another acceleration of the gas-liquid stream gives rise to compression shock and collapse of the cavitation bubbles. The energy released thereby causes the disruption of the microorganisms in the suspension.

The stress on fluids in turbulent phenomena has been examined by KIEFER [20] and MOHR [21], among others. There was some disruption observed in the absence of cavitation. However, the contribution of turbulence to the final extent of disruption remains undetermined.

The influence of the following parameters on the disruption of microorganisms has been determined: operating pressure, number of homogenizing steps, cell concentra-

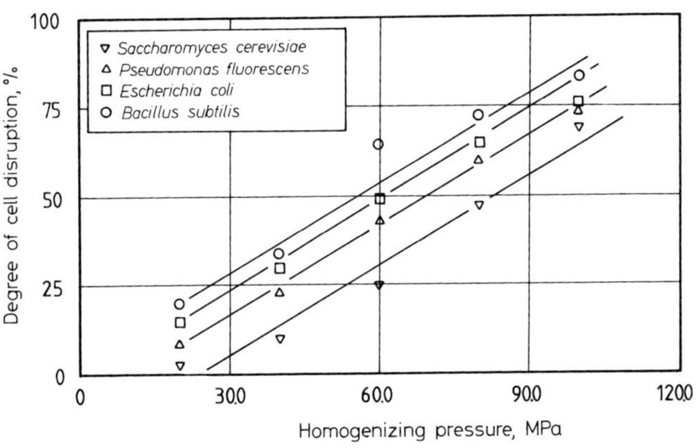

Fig. 15 Degree of disruption of different microorganisms as a function of homogenizing pressure [19]

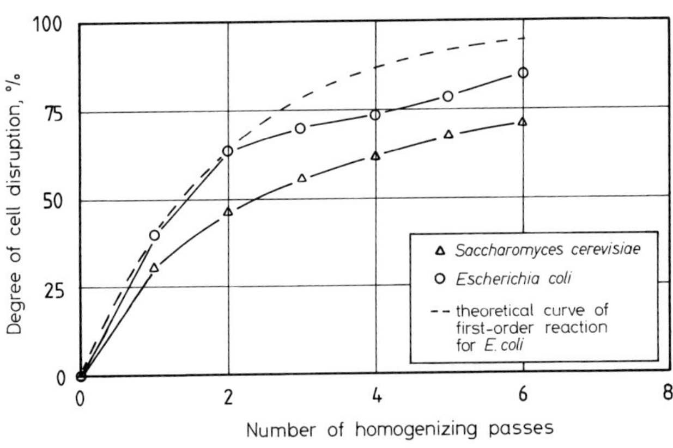

Fig. 16 Disruption of *Escherichia coli* and *Saccharomyces cerevisiae* as a function of the number homogenizing passes [19]

tion, and valve geometry. The most important of these is the operating pressure. In Figure 15 the degree of disruption is given as a function of pressure for different organisms. With higher pressure, and therefore higher applied energy, the percentage of broken cells rises. The energy necessary to achieve a certain degree of disruption differs for different organisms, such as yeast, Gram-positive, and Gram-negative bacteria. As mentioned above, this can be ascribed to their different cell wall structures. At usual pressures of 60 MPa (600 bar) bakers' yeast can be about 30% disrupted after one homogenizing pass. Obviously, several passes are necessary. The results of BÜSCHELBERGER [19] are given in Figure 16.

According to the results of the experiments of BÜSCHELBERGER and TONNIUS, the inlet temperature of the cell suspension has no influence on the final degree of disruption. HETHERINGTON [22], BROOKMAN [23] and SCHÜTTE and KULA [3] found that the extent of disruption in the high-pressure homogenizer does not depend on the wet weight, in the range of 10 to 60% cell concentration. Energetic considerations make it advisable to choose as high a concentration as possible.

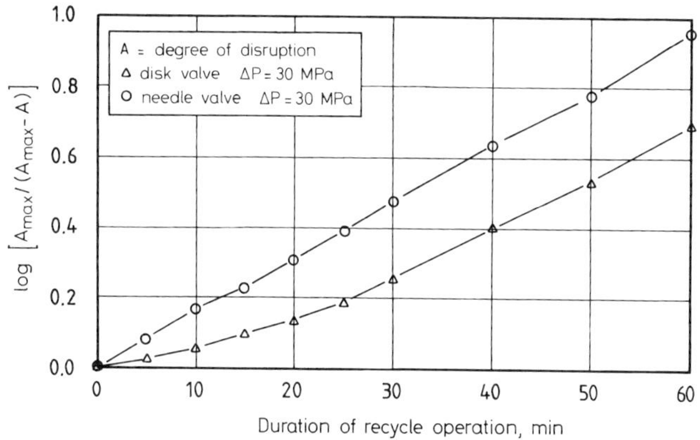

Fig. 17 Influence of valve geometry on the release of protein from bakers' yeast [3]

SCHÜTTE and KULA [3] report on work in which two homogenizing valves of different geometry were used; their results are shown in Figure 17. Data represented by triangles were obtained with a valve with a flat disk face (German *Flachventil*); those represented by circles were obtained with a needle valve (German *Kegelventil*); the pressure difference was 30 MPa (300 bar) in all experiments.

The disruption of microorganisms in a high-pressure homogenizer is very gentle because the times during which energy is applied are so short. The homogenizer can be used to break open all microorganisms; only some Gram-positive bacteria turn out to be difficult to disrupt [3].

4.5 French Press

In 1950 MILNER, LAWRENCE, and FRENCH described the French Press as a cell disruption device [24]. As shown in Figure 18, it consists of a cylinder with a piston, which is sealed against the cylinder wall. The outlet at the bottom of the cylinder is of small diameter, and can be closed off by a valve. The cell suspension, free of gas bubbles, is poured into the precooled cylinder. With the outlet closed, the piston is used to set the desired pressure (up to 200 MPa, 2000 bar), by means of a hydraulic press. Then the valve is opened so that the suspension is slowly released.

Because of its small volume of about 50 ml the French Press is a laboratory cell disruption apparatus. The disruption mechanism is comparable to that in the high-pressure homogenizer. The disruption of different microorganisms with the French Press was studied by HUGHES, WIMPENNY, and LLOYD [25]; their results are in Table 4. For most organisms, 50 % disruption was reached after the first pass. As has been found with other disruption devices, Gram-positive organisms are the most difficult to break.

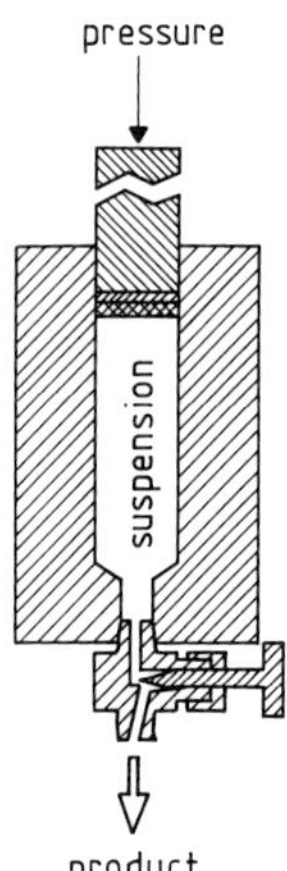

Fig. 18 Cross section of a French Press [24]

Table 4 Percent disruption of microorganisms in a French Press at different pressures [25]

Organism	Pressure, MPa			
	68	90	113	200
Escherichia coli	65	87	88	–
Bakers' yeast	28	48	56	–
Mycobacterium tuberculosis BCG	52	57	61	67
Bacillus megatherium	26	35	46	57
Chlorella pyrenoidosa	–	27	32	35
Sarcina lutea	8	9	14	25
Staphylococcus aureus	2	3	8	31

4.6 Hughes Press and X-Press

The press for cell disruption developed by HUGHES [26] in 1951 has been modified in many ways. All versions consist of a cylindrical pressure vessel and a piston. In contrast to the French or Chaikoff Press, the suspension here is in a solid, crystalline condition at temperatures below −20 °C. With a very high pressure (up to 500 MPa, 5000 bar) on the piston, the highly concentrated frozen suspension is forced through a narrow gap. Cell breakage can be attributed to a high shear stress and to the abrasive effect of ice crystals.

An apparatus developed by EDEBO [27] permits repeated operation without refilling. This apparatus is known as the X-Press and also operates as described above.

4.7 Liquid jet mill

The application of liquid jet mills for the disruption of microorganisms was not particularly successful until recently. ENGLER and ROBINSON included a high-velocity jet apparatus when they studied homogenizers [28]. Its principal components are a high-pressure chamber, a nozzle with 80 µm inside diameter, and a baffle plate. The cell suspension flows through the nozzle opening because of the differential pressure of up to 120 MPa (1200 bar) and impinges on the baffle plate below. Here, significant normal forces act on the suspension. The authors reported a degree of disruption of 80% for a suspension of *Candida utilis* with 10% (*w*/*v*) cell dry weight at 120 MPa (1200 bar) after only one pass.

Cell disruption in a highly turbulent liquid jet mill has recently been examined by KRÄMER and BOMBERG [29] in an arrangement that is still experimental. The fluid suspension is put under pressure (to 15 MPa, 150 bar) and then split. The two identical streams are introduced into a chamber where they impinge on each other as free jets with a velocity difference of 190 m/s. For a suspension of bakers' yeast the degree of cell disruption reached 40% in one pass.

5 Conclusions

This presentation and description of different mechanical cell disruption methods gives an overview of current possibilities for the disruption of microorganisms and other cell structures. Someone selecting a process suitable for a particular application should consider the points listed in Table 1. General recommendations for use of a given process are impossible to give. If Gram-positive cocci are to be disintegrated, agitated ball mills give the most satisfactory results. The improvement of mechanical disruption by chemical methods or lysis is possible and should be considered.

6 List of symbols used

Symbol	Units	Meaning
y_o	mm	gap width
$u(y)$	m/s	velocity profile
u_o	rpm = min^{-1}	circumferential speed of rotor
ω	min^{-1}	angular velocity of particles
w	kg/kg	concentration, weight fraction
ϕ_{gb}	%	degree of filling with grinding beads
d_{gb}	mm	diameter of agitator beads
d_{ad}	mm	diameter of agitator discs
n	rpm = min^{-1}	agitator rotation rate
t	min	grinding time
v	m/s	tip speed
V	ml	volume
P	W	power
c	% wet wt	cell wet weight
c	% dry wt	cell dry weight

7 References

[1] Rumpf, H.: Die Einzelkornzerkleinerung als Grundlage einer technischen Zerkleinerungswissenschaft; *Chem.-Ing.-Tech.* **37** (1965) 187–202.

[2] Emanuell, C. L.; Chaikoff, I. L.: A Hydrodynamic Homogenizer for the Controlled Release of Cellular Components from Various Tissues; *Biochim. Biophys. Acta* **24** (1957) 254–261.

[3] Schütte, H.; Kula, M.-R.: Einsatz von Rührwerkskugelmühlen und Hochdruckhomogenisatoren für den technischen Zellaufschluß; *BTF Biotech-Forum* **3** (1986) 68–79.

[4] Currie, J. A.; Dunnill, P.; Lilly, M. D.: Release of Protein from Baker's Yeast by Disruption in an Industrial Agitator Mill; *Biotech. Bioeng.* **14** (1972) 725–736.

[5] Woodrow, J. R.; Quirk, A. V.: Evaluation of the Potential of a Bead Mill for the Release of Intracellular Bacterial Enzymes; *Enzyme Microb. Technol.* **4** (1982) 385–389.

[6] Stehr, N.: Zerkleinerung und Materialtransport in einer Rührwerkskugelmühle; Dissertation, Tech. Univ. Braunschweig (1982).

[7] Weit, H: Betriebsverhalten und Maßstabsvergrößerung von Rührwerkskugelmühlen; Dissertation, Tech. Univ. Braunschweig (1987).

[8] Weit, H.; Schwedes, J.: Scale-up of Power Consumption in Agitated Ball Mills; *Chem. Eng. Technol.* **10** (1987) 398–404.

[9] Furness, G.: Some Factors Affecting the Destruction of Bacterium *E. coli* by Shaking with Glass Beads; *J. Gen. Microbiol.* **7** (1952) 335.

[10] Mickle, H.; Ray, J.: Tissue Disintegrator; *J. Microb. Soc.* **68** (1948) 10.

[11] Nossal, P. M.: A Mechanical Cell Disintegrator; *Aust. J. Exp. Biol. Med. Sci.* **31** (1953) 583–589.

[12] SCHÜTTE, H.: Aufschluß von Mikroorganismen mit der Schwingmühle MM2; presented at the Analytika-Forum, München (June 1986).
[13] ROSS, J. W.: Continuous Flow Mechanical Cell Disintegrator; *Appl. Microbiology* **11** (1963) 33–35.
[14] KURRER, K.-E.: Zur inneren Kinematik und Kinetik von Rohrschwingmühlen; Dissertation, Tech. Univ. Berlin (1986).
[15] DOULAH, M. S.: Mechanism of Disintegration of Biological Cells in Ultrasonic Cavitation; *Biotech. Bioeng.* **19** (1977) 649–660.
[16] JAMES, C. J.; COAKLEY, W. T.; HUGHES, D. E.: Kinetics of Protcin Release from Yeast Sonicated in Batch and Flow Systems at 20 kHz; *Biotech. Bioeng.* **14** (1972) 33–42.
[17] TONNIUS, F. G.: Verfahren zur Herstellung von nucleinsäurearmen Proteinkonzentraten aus Bäckerhefe; Dissertation, Tech. Univ. Karlsruhe (1982).
[18] TREIBER, A.: Zerkleinerungsmechanismen bei der Hockdruckhomogenisation von Öl-in-Wasser-Emulsionen; Dissertation, Tech. Univ. Karlsruhe (1979).
[19] BÜSCHELBERGER, H.-G.: Untersuchungen zum mechanischen Zellaufschluß von Mikroorganismen in Hochdruckhomogenisatoren; Dissertation, Tech. Univ. Karlsruhe (1987).
[20] KIEFER, P.: Der Einfluß von Scherkräften auf die Tröpfchenzerkleinerung beim Homogenisieren von Öl-in-Wasser-Emulsionen in Hochdruckhomogenisierdüsen; Dissertation, Tech. Univ. Karlsruhe (1977).
[21] MOHR, K.-H.: Zum Kavitationseinfluß auf die Dispergierung in Turbulenzfeldern hoher Energiedichte; *Int. Zeitschrift f. Lebensmitteltechnologie u. -verfahrenstechnik* **38** (1987) 563–568.
[22] HETHERINGTON, P. J.; FOLLOWS, M.; DUNNILL, P.; LILLY, M. D.: Release of Protein from Baker's Yeast by Disruption in an Industrial Homogeniser; *Trans. Inst. Chem. Eng.* **49** (1971) 142–148.
[23] BROOKMAN, J. S. G.: Mechanism of Cell Disintegration in a High Pressure Homogenizer; *Biotechnol. Bioeng.* **16** (1974) 371–383.
[24] MILNER, H. W.; LAWRENCE, N. S.; FRENCH, C. S.: Colloidal Dispersion of Chloroplast Material; *Science* **6** (1950) 633–634.
[25] HUGHES, D. E.; WIMPENNY, J. W. T.; LLOYD, D.: The Disintegration of Microorganisms. In: NORRIS, J. R.; RIBBONS, D. W. (Eds.): *Methods of Microbiology*, Academic Press, New York (1971).
[26] HUGHES, D. E.: A Press for Disrupting Bacteria and Other Microorganisms; *Br. J. Exp. Path.* **32** (1951) 97.
[27] EDEBO, L.: Disintegration of Cells. In: PERLMAN, D. (Ed.), *Fermentation Advances*, Academic Press, London (1969); pp 249–267.
[28] ENGLER, C. R.; ROBINSON, C. W.: Disintegration of *Candida utilis* Cells in High Pressure Flow Devices; *Biotech. Bioeng.* **23** (1981) 765–780.
[29] KRÄMER, P.; BOMBERG, A.: Zellaufschluß von Mikroorganismen mit Hochgeschwindigkeitsstrahlen; *Chem.-Ing.-Tech.* **60** (1988) 776–778.
[30] GOLDSCHMIDT, A.; HANTSCHKE, B.; KNAPPE, E.; VOCK, G.-F.: Glasurit-Handbuch; Vincentz Verlag, Hannover (1984); pp 154–182.
[31] SCHÖLLERSCHALL GmbH; technical bulletin.
[32] RETSCH GmbH; technical bulletin.
[33] NETZSCH-Feinmahltechnik GmbH; technical bulletin.
[34] FRYMA-Maschinen AG; technical bulletin.
[35] DRAISWERKE GmbH; technical bulletin.

Filamentous Fungi of Biotechnological Importance: Facts and Perspectives on Strain Improvement by Genetic Engineering

by G. MOHR

Contents

Dr. G. MOHR,
Lehrstuhl für Allgemeine Botanik,
Ruhr-Universität Bochum, Postfach 10 21 48,
D-4630 Bochum 1, Fed. Rep. Germany
Present address: Department of Molecular Genetics and Biochemistry,
The Ohio State University, 484 West 12th Avenue,
Columbus, OH 43210, U.S.A

1 Introduction

Transformation is defined as the insertion of isolated, naked DNA into an organism. Griffith made the initial observation of transformation in 1928 [49] with *Pneumococcus*, but 40 years passed before efficient transformation systems were designed for *Escherichia coli* (*E. coli*) [22, 70]. Appropriate systems have now been developed for numerous prokaryotes and eukaryotes [119]; even cell organelles can be reproducibly transformed [15, 59]. The essential prerequisites for genetic engineering are the availability of transformation systems and the capability of modifying DNA *in vitro*.

There is considerable commercial interest in the applications of genetic engineering. As a single instance, the American company *Genentech* had $86 million in sales of genetically engineered human growth hormone in 1987 (*Bio/Technology*, Feb. 1988, p 119).

Filamentous fungi have a special place in biotechnology because of they have so many diverse industrial uses. Some excrete large amounts of low molecular weight metabolites, such as citric acid [11], or extracellular enzymes, such as glucoamylase [6]; some are efficient at performing complex chemical reactions, such as steroid biotransformation, and producing a multitude of medically important substances [25, 86, 89].

Although filamentous fungi are industrially so important, the first transformation systems were developed for *E. coli* [22] and *Saccharomyces cerevisiae* [7, 54]. Now, after an initial delay, filamentous fungi have been the objects of a corresponding surge of research effort. My purpose in this chapter is to present the possibilities that arise from these developments. I focus mainly on an integrative transformation system for filamentous fungi and on the use of fungi in basic research and applied biotechnology. I will also briefly compare genetic engineering with classical methods for strain improvement. To keep the topic within manageable limits, I restrict my remarks to filamentous ascomycetes and related deuteromycetes.

2 Transformation of filamentous fungi

2.1 Aspergillus nidulans as a model system

Transformation systems for *Aspergillus nidulans* have been quite well characterized and what we have learned can be applied to a large number of filamentous fungi. Many of the different transformation systems that were developed for *Aspergillus nidulans* during the past five years have also been used in the transformation of other fungi [41, 55, 60, 88, 94, 95, 120].

Several transformation systems have been described for *Neurospora crassa*, but the experience gained was not applied to fungi with relevance to biotechnology. I therefore shall not discuss *Neurospora crassa* further here. Reviews by Mishra [74] and by Rambosek and Leach treat that subject more fully [88].

2.1.1 Selectable markers and recipient strains

Genes that enable us to distinguish between transformed and nontransformed colonies are essential for transformation. These dominant, selectable indicator genes, called *selectable markers*, are of two types. The first, the genes of wild-type strains, such as the *amd*S$^+$ gene that encodes acetamidase [24], complement a mutation of the auxotrophic recipient; selection is done on minimal medium that supports the growth of prototrophic transformants only. (A nutritional mutant that has a requirement for a growth factor is an *auxotroph*; a strain with the same nutritional requirements as the wild-type parent from which the auxotroph was derived is a *prototroph*.) Genes of the second type, such as the *hph*^{R} gene that encodes hygromycin B phosphotransferase [50, 62], allow transformants to grow on substrates that contain toxins (hence they are called *resistance genes*); the selection is done on nutrient medium containing antibiotic. ESSER and DOHMEN have written a review of various genes and the mechanisms imparting resistance [37].

The source of the selectable marker is not important; genes from *Aspergillus nidulans* (*homologous*) or from other organisms (*heterologous*) can be used (Table 1). Selectable genes from ascomycetes or related deuteromycetes are expressed in *Aspergillus nidulans* without great difficulty. That is not true for the expression of corresponding genes from bacteria; in that case, homologous *expression signals* are required [4, 87].

Table 1 Homologous and heterologous marker genes for selecting transformants of *Aspergillus nidulans*

Selectable marker	Source	Recipient	Reference
*amd*S$^+$	*A. nidulans*	M	[102] [118]
*are*A$^+$	*A. nidulans*	M	[19]
*arg*B$^+$	*A. nidulans*	M	[58] [61] [106]
*oli*C^R	*A. nidulans*	wt	[112]
*pki*A$^+$	*A. nidulans*	M	[32]
prn$^+$	*A. nidulans*	M	[35]
*qut*A$^+$	*A. nidulans*	M	[10]
*qut*E$^+$	*A. nidulans*	M	[29]
*ribo*B$^+$	*A. nidulans*	M	[78]
*trp*C$^+$	*A. nidulans*	M	[121] [122]
hph^R/P$_{trpC}$	*E. coli*/*A. nidulans*	wt	[27]
hph^R/P$_{gpd}$	*E. coli*/*A. nidulans*	wt	[87]
nit2$^+$	*N. crassa*	M	[30]
pyr4$^+$	*N. crassa*	M	[4]
*trp*C$^+$	*A. niger*	M	[67]
*trp*C$^+$	*C. heterostrophus*	M	[105]

The gene symbol and source are indicated for each gene. If a heterologous promoter (P) is inserted upstream from the gene for expression in *Aspergillus*, its designation and source are indicated as well. The recipient column lists whether auxotrophic mutants (M) or wild types (wt) were used for transformation. Genera: *A.* = *Aspergillus*; *C.* = *Cochliobolus*; *E.* = *Escherichia*; *N.* = *Neurospora*.
Gene symbols are in the Appendix (Table 9)

Table 1 is a list of all selectable markers that have currently been reported for *Aspergillus nidulans*. The selection of transformants is determined predominantly by the complementing of the auxotrophic character of the recipient strain. The need for auxotrophic recipients limits the use of selection methods for the improvement of production strains, because industrial production strains are generally prototrophic and would have to be mutated to auxotrophic strains. This would involve the danger of introducing undesirable characteristics and consequently reducing the suitability for production.

If transformants are selected on the basis of antibiotic resistance, the use of specific recipient strains becomes unnecessary. Consequently, wild-type strains that have adequate sensitivity towards the antibiotic can be transformed.

2.1.2 Vector integration

It is clear from genetic and molecular analyses of numerous transformants that vectors integrate into nuclear DNA [41, 55, 60]. The following methods are used to demonstrate vector integration.

1. The total DNA content of transformants is digested with restriction enzymes. The digest is separated by agarose gel electrophoresis and immobilized on nitrocellulose filters. The DNA is hybridized on the filters with radioactively labeled probes. Changes in the restriction pattern are an indication of homologous integration [121].
2. The total DNA content of the transformants is subjected to partial digestion, then self-ligated and transformed in *E. coli*. Some of the resulting vectors are coupled to nuclear DNA [4].
3. If a transformant is crossed with a strain that has chromosomes marked by various mutations, linkage of a selectable marker with a chromosomal marker may be detected [116].

If a vector contains DNA that originates in a recipient, then integration takes place preferentially (in 70%–90% of cases studied) at the site of the homologous gene (*homologous integration*). This can happen in different ways: the entire vector may be integrated by a single crossing over, or only the homologous DNA region may be integrated in a double crossing over, or there may be a gene conversion [41, 60]. In all other cases, the integration is *heterologous*.

Moreover, which type of vector integration occurs depends on the recipient strain used [116, 118]. For example, the vector p3SR2 integrates in strain WG290 primarily in the chromosomal copy of the *amd*S^+ gene, but it integrates randomly at heterologous sites in the closely related strain MH1277. A cryptic mutation (probably located on chromosome III) may be responsible for this integration behavior [116, 118].

The number of integrated vector copies ranges from one to many, the latter case is characterized by a head-to-tail arrangement of individual monomers. The mechanism of integration of vector multimers is not understood at all.

The integration of a plasmid is independent of the presence of a selectable marker. Transformation with mixtures of two plasmids of which only one carries the selectable marker produces transformants in which almost all carry both vectors [117].

As far as we know, the transformation of *Aspergillus nidulans* is achieved exclusively by the integration of vectors. So far, autonomously replicating vectors could not be demonstrated without doubt.

2.1.3 Transformation rates and enhancers

Table 2 is a summary of the transformation rates obtained with different selectable markers. The first reported transformations (in 1983) had low rates, with the number of stable transformants growing on selection plates being fewer than ten per microgram of vector DNA [3, 102]. (Besides stable transformants, we also find a large number of abortive transformants on the selection medium. These are small colonies that, upon transfer to fresh selection medium, will not grow [102].) An increase in the transformation rate to 500 per microgram of vector DNA was obtained by optimizing the transformation procedure, or by reducing the size of the vectors [61, 118]. The discovery of

Table 2 Transformation rates with various selectable markers and the effect of transformation enhancers

Selectable marker	Source	Transformation enhancer	Transformation rate/μg DNA	Reference
*amd*S$^+$	*A. nidulans*	–	4	[102]
*amd*S$^+$	*A. nidulans*	–	400	[118]
*amd*S$^+$	*A. nidulans*	ans 1	6000	[115]
*arg*B$^+$	*A. nidulans*	–	10	[58]
*arg*B$^+$	*A. nidulans*	–	500	[61]
*arg*B$^+$	*A. nidulans*	unt	3000	[60]
*arg*B$^+$	*A. nidulans*	–	100	[28]
*arg*B$^+$	*A. nidulans*	ans 1	484	[28]
*oli*C^R	*A. nidulans*	–	10–70	[112]
*pki*A$^+$	*A. nidulans*	–	10–20	[32]
*qut*A$^+$	*A. nidulans*	–	1	[10]
*qut*E$^+$	*A. nidulans*	–	7	[29]
*ribo*B$^+$	*A. nidulans*	–	50	[78]
*trp*C$^+$	*A. nidulans*	–	10–20	[121]
*trp*C$^+$	*A. nidulans*	ans 1	2500	[115]
hph^R/P_{trpC}	*E. coli*/ *A. nidulans*	–	5	[27]
hph^R/P_{gpd}	*E. coli*/ *A. nidulans*	–	100	[87]
*pyr*4$^+$	*N. crassa*	–	4	[3]
*pyr*4$^+$	*N. crassa*	ans 1	5000	[4]
*pyr*4$^+$	*N. crassa*	EB 1	1000	[9]
*trp*C$^+$	*C. heterostrophus*	–	0.1	[105]

For abbreviations see Tables 1 and 9

transformation enhancers led to stable transformant frequencies of up to 6000 per microgram of vector DNA. Enhancers are DNA fragments that multiply the number of transformants when present in *cis* or *trans*. [4].

Three enhancers have been isolated to date from *Aspergillus nidulans*: *ans*1 and *unt* from nuclear DNA, and EB1 from mitochondrial DNA [4, 9, 28, 60]. The enhancer *ans*1 has a DNA sequence extraordinarily rich in adenine and thymine bases ($A + T$ is 81%); it is present in multiple copies in the *Aspergillus nidulans* genome and is located proximal to the centromere of coupling group I [28].

We have not yet succeeded in elucidating the mechanism involved in the enhancer-mediated increase in transformation rate. Perhaps the enhancer permits a limited autonomous replication of the vector prior to integration into the genome, or it may assist vector integration in some yet unknown manner [4].

If transformation of *Aspergillus nidulans* is efficient, gene cloning by complementing mutations is possible. However, the mechanism of vector integration is not understood, in spite of reproducibility and methodological competence in conducting transformations in *Aspergillus nidulans*.

2.1.4 Stability of integrated vectors

Stability of transformants is essential, because loss of a vector makes genetic analysis difficult and reduces the applicability of the technique to biotechnological problems.

Investigators have confirmed that integrated vectors are, in most cases, stable mitotically as well as meiotically in the absence of selection pressure [4, 102, 106, 118, 121], but there are exceptions. BALLANCE *et al.* [3] and UPSHALL [106] report the loss of an integrated vector in half of all colonies studied during vegetative replication without selection pressure; furthermore, 36% of all mycelia produced by ascospores (by selfing the transformants) no longer contained identifiable vector sequences [102, 106, 121].

We do not understand the reason for these conflicting results. Presumably strain-specific differences are involved, since different strains of *Aspergillus nidulans* were used in the studies. Clearly, if integrated vectors are to be used industrially, their stability needs to be investigated.

2.2 Transformation in basic research

At present, the potential of transformation systems is exploited primarily to answer questions in basic research.

2.2.1 Gene isolation, analysis, and regulation

All known mutations [21] can be inserted in recipient strains of *Aspergillus nidulans* by crossing; the appropriate genes can therefore be cloned in *Aspergillus nidulans* through complementation. For this, a suitable strain is transformed with the cloned DNA of a wild-type strain (which therefore acts as a genomic library). The DNA of the transformant that expresses the phenotype of the wild-type strain is transferred to *E. coli*, and the plasmids obtained are characterized. Table 3 is a list of the genes that have been cloned by transformation in *Aspergillus nidulans*.

Table 3 Genes of homologous and heterologous origin cloned in *Aspergillus nidulans*

Gene	Source	Phenotype or function	Reference
*acu*D	*A. nidulans*	isocitrate lyase	[5]
*brl*A	*A. nidulans*	"bristle", no conidia	[61]
*fw*A	*A. nidulans*	yellow-brown color of conidia	[4]
*nim*A	*A. nidulans*	induction of mitosis	[81]
*paba*A	*A. nidulans*	*p*-aminobenzoic acid biosynthesis	[122]
*ribo*B	*A. nidulans*	riboflavin biosynthesis	[78]
*y*A	*A. nidulans*	*p*-diphenol oxidase	[122]
*arg*B	*A. niger*	ornithine carbamoyltransferase	[18]
pda	*N. haematococca*	pisatin demethylase	[114]
*trp*C	*A. niger*	tryptophan biosynthesis	[67]
*trp*C	*C. heterostrophus*	tryptophan biosynthesis	[105]

Genera: *A.* = *Aspergillus*; *C.* = *Cochliobolus*; *N.* = *Nectria*

YELTON *et al.* isolated the *y*A gene in a particularly elegant way [122]. They cloned the nuclear DNA of the wild-type strain into a cosmid that contained the *trp*C$^+$ gene of *Aspergillus nidulans* as a selectable marker. With this library they transformed a yellow-colored spore mutant of *Aspergillus nidulans*. Among the approximately 1000 transformants, three contained green-colored spores (the wild phenotype). The DNA of these transformants was isolated and packaged *in vitro* into phage particles that were used to infect *E. coli* cells. When the cosmids from various *Aspergillus* transformants were analyzed, the *y*A gene could be localized [122].

Alternative techniques for isolating genes from filamentous fungi, besides cloning genes directly in *Aspergillus nidulans*, are feasible. One possibility is to complement mutations in *E. coli* with cloned DNA from *Aspergillus nidulans*. This option has been used for only a few genes so far, because in general it is not easy to express eukaryotic genes in prokaryotes [88]. A second possibility is to identify genes *in vitro* by using heterologous gene probes [94]. Which technique is used for a given gene must be decided based upon the conditions at hand.

Transformation systems are useful not only for isolating but also for analyzing individual genes or entire gene regions. Thus, the position of individual genes within complex transcriptional units (*gene clusters*) can be determined by transforming various genetically defined mutants with defined parts of the entire region [10, 29, 35].

Individual genes of the *qut* region in *Aspergillus nidulans*, which encodes regulatory proteins and the enzymes for quinate catabolism, have been localized by transforming genetically characterized *qut*$^-$ mutants with cloned partial sequences of the gene region. Subsequently, the individual mutants were complemented by certain gene regions. Finally, using this procedure, the physical and genetic maps of this region could be matched [10, 29, 48]. The *prn* region (proline metabolism) of *Aspergillus nidulans* was mapped similarly [35].

The analysis of expression signals and the regulation of gene activity are of extreme interest and receive new impetus from transformation systems. For example, *gene-fusions* are used to facilitate the observation of expression signals; the fusions contain a promoter (from *Aspergillus nidulans*) and an indicator gene that can be detected. The *lac*Z gene from *E. coli* that codes for β-galactosidase is often used. β-galactosidase activity can be either visually estimated or quantitatively determined [119]. Both gene regulation and gene expression in many organisms have been studied with *lac*Z fusions [97].

Some vectors have been constructed for the purpose of analyzing gene expression in *Aspergillus nidulans*. They have various promoters, such as P_{trpC}, P_{gpd}, or P_{amdS} (Table 9), upstream from the *lac*Z gene. These vectors have the essential property that only one copy will integrate at a defined locus of the genome upon recombination of homologous sequences [31, 108, 109].

VAN GORCOM *et al.* [109] studied various *Aspergillus nidulans* promoters using *lac*Z fusions. They observed that mycelium harbors ten times more β-galactosidase if the *gpd* promoter, rather than the *trp*C promoter, is inserted upstream from the *lac*Z gene. A detailed analysis of the promoter was obtained by fusing *trp*C promoters, containing deletions of various magnitudes, with the *lac*Z gene. The locations of a negatively acting controlling region and of a region essential for the correct initiation of transcription were determined [53]. DAVIS *et al.* [31] replaced the chromosomal copy of the *amd*S$^+$ gene with a fusion of *amd*S-*lac*Z. When the transformants were crossed with suitable mutants, strains were obtained in which the *amd*S-*lac*Z fusion was associated with different regulatory mutations. The experimenters could thus track the regulation of the *amd*S gene visually.

These examples are evidence of how transformation systems can significantly simplify the isolation and analysis of genes in *Aspergillus nidulans*. Because each gene that has a selectable phenotype can, in principle, be cloned directly in *Aspergillus nidulans*, it is only a matter of time until we have genes that catalyze industrially important reactions available in a cloned state. Elucidating the regulation and expression of fungal genes in *Aspergillus nidulans* with *lac*Z fusions brings us closer to the day when we can more easily monitor industrial processes.

2.2.2 Gene replacement

Almost every gene can be selectively replaced by an allele. The basics of gene replacement have been worked out for *Saccharomyces cerevisiae* [14, 100], and have been modified for *Aspergillus nidulans* [41, 73].

The prerequisites for replacing a gene are, first, that the gene in question exist in a cloned form, and, second, that an efficient transformation system, in which vectors integrate predominantly by homologous recombination, be at hand. The first step in every gene replacement is the transformation of a suitable recipient with the target gene that has been modified *in vitro*. There are two ways to go about this process (Figure 1).

1. The transformation can be done with a linear DNA fragment that is integrated into the genome by homologous recombination (double crossing over), thus eliminating the chromosomal copy (Figure 1, left). The frequency of this event is, of course, very low. Therefore, this one-step process can occur only in the selectable markers *per se* or in genes that have an integrated selectable marker.
2. In the two-step method on the right in Figure 1, gene replacement can occur. A recipient strain is transformed with a circular vector that carries both the target gene,

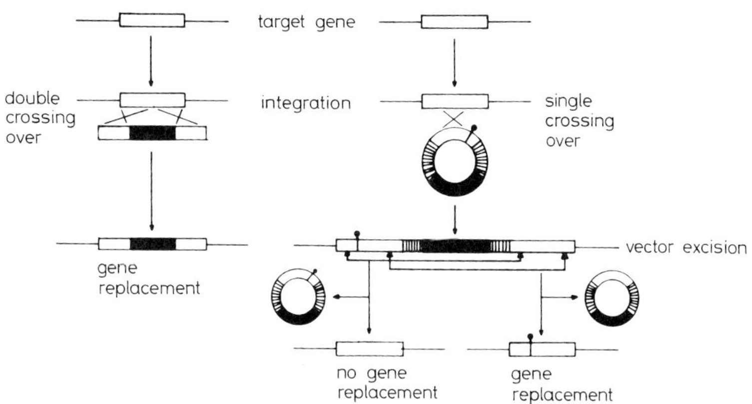

Fig. 1 Schematic representation of gene replacement in *Aspergillus nidulans* (modified from MILLER *et al.*, [73]). Left, one-step gene replacement; right, two-step gene replacement. Key to structures: black oblong region, selectable marker; white oblong region, gene to be replaced; hatched region, vector sequence; line, flanking chromosomal DNA; ●, mutation; intersecting lines, crossing over

mutated *in vitro* (● in the figure), and a selectable marker. The vector is integrated into the genome by a single crossing over. Transformants that have a selectable marker flanked on one side by the target gene and on the other side by the mutated copy of the target gene, are selfed or crossed with a suitable parent. During meiosis, the homologous sequences of the target gene recombine, liberating the vector, which disappears since it is unable to replicate by itself. Gene replacement may or may nor occur, depending on the recombination site. Colonies that have lost the vector can be easily recognized because the selectable marker is gone.

These techniques have been used in several gene replacement experiments to date. Several examples are listed in Table 4.

The methods described above permit the genetic modification of fungi to be targeted. The principal application for the one-step gene replacement is clearly the elimination of an undesirable trait, for example, the removal of the genes for enzymes involved in

Table 4 Examples of replacing a gene in *Aspergillus nidulans* by a gene modified *in vitro*

Gene	Replaced by	Method	Reference
*amd*S^{+}	*amd*S-*lac*Z	II	[31]
*arg*B^{+}	*arg*B^{-}	I	[73]
*ben*A^{+}	*ben*A^{R}	II	[34]
*Spo*C1 C1-C^{+}	C1-C^{-}	II	[73]
*trp*C^{-}	*trp*C^{+}	I	[73]
*trp*C^{+}	*trp*C-*lac*Z	I	[117]

*lac*Z fusions are identified by the added "*lac*Z"
I: one-step gene replacement; II: two-step gene replacement

product catabolism. In contrast, the two-step gene replacement method has wider potential, but in many of the fungi used industrially it may be difficult to carry out because nothing is known of the sexual cycle of the fungi and we know little about alternative methods for excising integrated vectors (this is discussed more fully in Section 2.3.2).

2.3 Transformation in biotechnology

2.3.1 Secretion of heterologous proteins

Herologous proteins are produced in a biologically active form by *Aspergillus nidulans* and are secreted into the culture medium. As I have already shown, *Aspergillus nidulans* is well studied genetically and transformation systems have been developed. Therefore, it is convenient to use current knowledge to produce heterologous proteins of commercial interest.

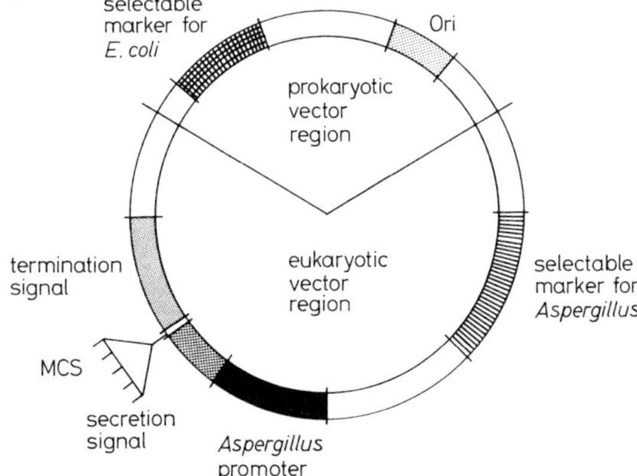

Fig. 2 Schematic representation of a vector that permits the expression and secretion of proteins (modified from ESSER and KÄMPER, [38]). See text for further explanation. MCS, multiple cloning site; Ori, replication origin

An *expression vector*, shown schematically in Figure 2, is necessary for the production of proteins. The plasmid consists of a eukaryotic and a prokaryotic region; the latter is needed only for replication in *E. coli* and is therefore of lesser interest. The eukaryotic region contains the selectable marker for *Aspergillus nidulans*, allowing recognition of the transformants which express the gene of interest. Site-specific integration of the vector can be achieved by using a homologous selectable marker.

If transcription is to be efficient, the *promoter* must be powerful, and highly inducible or repressible, since overexpression of proteins in *E. coli* often results in serious interference with the cell metabolism or in the formation of insoluble protein aggregates within the cells [71]. The promoter is followed by the *secretion signal*, which is a short sequence of DNA at the 5′ end of the gene that is translated (*signal peptide*) and causes the secretion of proteins. The signal peptide is split off during protein secretion [68, 111]. Some genes containing secretion signal sequences and partially regulable

Table 5 Cloned genes that contain regulatable promoters, a secretion signal, or both, from fungi of the genera *Aspergillus* and *Trichoderma*

Gene	Encoded protein	Source	Induced by	Repressed by	Reference
*alc*A	alcohol dehydrogenase I	*A. nidulans*	threonine/ fructose	glucose	[51]
*cbh*2	cellobiose hydrolase II	*T. reesei*	–	–	[83]
*egl*1	endoglucanase I	*T. reesei*	–	–	[101]
*egl*3	endoglucanase III	*T. reesei*	–	–	[93]
*gla*A	glucoamylase	*A. niger*	starch	xylose/glycerol	[12] [13]
*gla*A	glucoamylase	*A. awamori*	starch	xylose/glycerol	[77]

promoters have been cloned and sequenced. Table 5 contains a summary of these results. One or several single restriction enzyme scission sites (MCS in Figure 2) are usually located downstream of the secretion signal sequences; they are used to insert a gene that can be expressed in different reading frames. At the end of the expression sequence is a termination signal that shuts down transcription and contains a signal sequence for coupling a poly-A tail to the mRNA.

Three reports have appeared recently describing the expression and secretion of heterologous proteins in *Aspergillus nidulans* with specially engineered vector constructs [26, 52, 107]. The results described in those three papers are summarized in Table 6. All three used the same approach: an expression vector resembling the one in Figure 2 was constructed, and a gene was inserted at the same time. Suitable auxotrophic recipient

Table 6 Production and secretion of heterologous proteins by *Aspergillus nidulans* using expression vectors

Recipient	*A. nidulans pyr*G$^-$	*A. nidulans arg*B$^-$		*A. nidulans arg*B$^-$
Selectable marker	*pyr*4$^+$	*arg*B$^+$		*arg*B$^+$
Transformation	integrative	integrative		integrative
Protein	chymosin	IFN $\alpha 2$	egl	tPA
Source	bovine	human	*C. fimi*	human
Promoter	*gla*A	*alc*A/*gla*A		*tpi*A
Terminator	*gla*A	*alc*A/*gla*A		*tpi*A
Fermentation	shake flask	shake flask		shake flask
Yield	1 mg/l	1 mg/l	20 mg/l	1 mg/l
Comments	biologically active	biol. active	biol. active	not hyperglyco-sylated
Reference	[26]	[52]		[107]

Key to abbreviations: IFN $\alpha 2$ = interferon $\alpha 2$; egl = endoglucanase; tPA = tissue plasminogen activator; *C. fimi* = *Cellulomonas fimi*; for additional details see text strains (*arg*B$^-$ or *pyr*G$^-$) were transformed, and several transformants were isolated and analyzed for the expression of heterologous protein. In the simplest case (bacterial endoglucanase), gene expression was evaluated by a plating test [52]. Where this was infeasible (for example, for interferon $\alpha 2$), more involved tests such as antibody tests were used [52]. In all cases, the vectors integrated into the genome were stable and their presence could be detected with gel electrophoresis and Southern blot hybridization.

Only small amounts of protein (1 mg/l bovine chymosin, interferon $\alpha 2$, or tissue plasminogen activator) were secreted (Table 6). There are several reasons for this. First, laboratory strains were used for these experiments, and they are unlikely to be suitable for optimal production of protein. Second, all the reported results are from shake-flask experiments using minimal medium; a significant improvement in yield could probably be achieved under different culture conditions. Third, the DNA sequence in the environment of the ATG initiator codon of the secretory gene was changed during cloning in all the vector constructs used. (GWYNNE *et al.* [52] have observed that changes in the DNA sequence immediately in front of the ATG codon reduce the yield of interferon by half, and the addition of 12 triplets immediately after the initiator codon of the interferon gene reduces yields by 95%.)

These examples serve to encourage us that, in spite of the difficulties discussed above, *Aspergillus nidulans* is quite suitable for the commercial production of valuable proteins. This statement is based on two observations.

1. Heterologous proteins (chymosin, interferon $\alpha 2$, and bacterial endoglucanase) are present in a biologically active form in the culture supernatant. The yield of interferon $\alpha 2$ is already, without any optimization, at the level described for the best bacterial production system reported to date [52].
2. Tissue plasminogen activator (tPA) is not hyperglycosylated in *Aspergillus nidulans* (in *Saccharomyces cerevisiae* it is). Glycosylation may operate similarly in mammals and filamentous fungi [107].

The possibilities just discussed for expressing heterologous proteins are not limited to *Aspergillus nidulans*. FINKELSTEIN *et al.* describe the expression of active rat pancreatic ribonuclease in *Aspergillus niger* [44]. The expression vector they used also corresponds to the schematic representation in Figure 2, but the yield of pancreatic ribonuclease was not reported.

In this section I have shown that filamentous fungi can be used to produce interesting heterologous proteins. To what extent these findings can be incorporated into production processes remains to be seen, however.

2.3.2 Strain improvement

Using transformation systems for strain improvement has great potential. Some relatively simple measures include increasing the production of an important enzyme by gene dosage, or making complex modifications by altering regulatory or structural properties of enzymes or genes.

The gene dosage of a given gene can be increased by transforming a production strain with a vector that contains this gene in addition to a dominant selectable marker, and isolating transformants that carry multiple copies of the vector. Furthermore, new traits such as utilization of additional substrates can be inserted into strains by integrating heterologous genes. If we wish to destroy genes for enzymes that catabolize or modify the desired product, or to increase the expression of individual genes (by altering promoters), or to eliminate an undesired negative regulation (such as product inhibition of enzymes) by modifying the effector binding sites of the enzymes, we will need to use the technique of gene replacement.

BULL *et al.* [16] were the first to describe the increase in gene dosage for isopenicillin-*N* synthase (IPNS), an enzyme that participates in penicillin biosynthesis by catalyzing the formation of isopenicillin *N*, the central precursor in the biosynthesis of penicillins and cephalosporins [76]. They isolated transformants of *Penicillium chrysogenum* that contained numerous copies of a vector carrying the IPNS gene which was cloned by hybridizing it with a heterologous gene probe [16]. The homologous *oli*C^R gene (which permits transformation of prototrophic strains) was used as a selectable marker [112, 113]. The experimenters found upon molecular analysis that the vector inserted primarily by homologous integration, with some transformants receiving more than one copy of the vector, but they did not report whether the altered number of IPNS genes had any effect on penicillin yields.

In contrast, FINKELSTEIN *et al.* demonstrated that transformants of *Aspergillus niger* that contain numerous copies of the *gla*A gene secrete significantly more glucoamylase than does the nontransformed recipient. The authors developed a vector construct that contained, in addition to the selectable marker (*neo*^{R}), the cloned *gla*A gene from *Aspergillus niger*. Numerous transformants were isolated and the copy number of the integrated vectors determined. Aside from a few transformants that secreted less enzyme even though they had additional copies of the glucoamylase gene, most of the transformants had increased glucoamylase production. The highest values were attained by two transformants that each contained more than thirty additional copies of the gene and produced about 800% more glucoamylase [44].

These instances are the first description of an increase in gene dosage through transformation. The increase in yield associated with the higher gene dosage is a quite specific demonstration that strain improvement can be accomplished with genetic engineering methods.

2.3.3 Transformation of industrially important filamentous fungi

Integrative transformation systems have been developed for some common industrial filamentous fungi. It was an obvious step to transfer the knowledge obtained with *Aspergillus nidulans* to filamentous fungi of industrial interest. It is not surprising, therefore, that there are many reports of transformation of these fungi. Table 7 contains a list of published data. It is apparent that most of the transformed strains are wild-type; this is especially so for *Aspergillus ficuum*, *Cephalosporium acremonium*, *Curvularia lunata*, *Penicillium chrysogenum*, and *Trichoderma reesei* and is because resistance genes (such as *hph*^{R} or *oli*C^R) or the *amd*S$^+$ gene have usually been used to select the transformants. (The *amd*S$^+$ gene can be used to select transformants without isolating auxotrophic mutants because wild-type strains of most fungi are unable to utilize acetamide as a carbon or nitrogen source [24, 56, 63].)

In only a few cases is it possible to use homologous selectable markers (Table 7), namely, *pyr*A$^+$ in *Aspergillus niger*, *oli*C^R and *trp*C$^+$ in *Penicillium chrysogenum*, because the molecular analysis of most of these organisms is not sufficiently advanced [16, 47, 85, 110]. The selectable markers suitable for *Aspergillus nidulans* (such as *amd*S$^+$ and *arg*B$^+$) can be expressed without difficulty in many fungi, but the rates of

Table 7 Integrative transformation of filamentous fungi important in biotechnology

Organism	Selectable marker	Source	Recipient	Transformation rate/µg DNA	Ref.
Aspergillus ficuum	*amd*S$^+$	*A. nidulans*	wt	5	[75]
	hph^R/P$_{gpd}$	*E. coli*/ *A. nidulans*	wt	15	[75]
Aspergillus niger	*arg*B$^+$	*A. nidulans*	M	4	[17]
	*pyr*4$^+$	*N. crassa*	M	2	[113]
	*pyr*A$^+$	*A. niger*	M	50	[47]
	*pyr*A$^+$	*A. niger*	M	100	[110]
	*amd*S$^+$	*A. nidulans*	wt	4	[63]
	neo^R/P$_{am}$	*E. coli*/*N. crassa*	wt	10	[44]
	*oli*C^R	*A. niger*	wt	0.3	[113]
	hph^R/P$_{gpd}$	*E. coli*/ *A. nidulans*	wt	5–20	[87]
Aspergillus oryzae	*pyr*A$^+$	*A. niger*	M	16	[72]
	*arg*B$^+$	*A. nidulans*	M	0.7	[46]
Cephalosporium acremonium	*hph*^{R}/P$_{pcbC}$	*E. coli*/ *C. acremonium*	wt	500	[98]
	hph^R/P$_{pgk}$	*E. coli*/ *S. cerevisiae*	wt	4–5	[98]
Curvularia lunata	*hph*^{R}/P$_{gpd}$	*E. coli*/ *A. nidulans*	wt	0.1	[80]
Penicillium chrysogenum	*pyr*4$^+$	*N. crassa*	M	1000–4000	[20]
	*trp*C$^+$	*P. chrysog.*	M	300–1800	[85]
	*amd*S$^+$	*A. nidulans*	wt	2–20	[8]
	*amd*S$^+$	*A. nidulans*	wt	120	[66]
	*oli*C^R	*P. chrysog.*	wt	1–5	[16]
	ble^R/P$_{gpd}$	*E. coli*/ *A. nidulans*	wt	20	[66]
Trichoderma reesei	*arg*B$^+$	*A. nidulans*	wt	600	[84]
	*amd*S$^+$	*A. nidulans*	wt	50–100	[84]

The selectable markers used in the transformation of the fungi and their sources are listed. If a heterologous promoter (P) is responsible for the transcription of the selectable marker, then its origin is also indicated. Key to abbreviations: wt = wild type; M = auxotrophic mutant; for other abbreviations see Tables 1 and 9

transformation obtained are quite variable (see Table 7). The same vector inserted into *Aspergillus niger* and *Curvularia lunata* results in more than 200 times as many transformants of *A. niger* as of *C. lunata* (Table 7 [80, 87]). Similar things happen when different strains of a species are used (such as *Penicillium chrysogenum* [8, 66]).

When transformants of *Aspergillus niger*, *Penicillium chrysogenum*, and *Trichoderma reesei* were subjected to molecular analysis, no substantial differences between them and *Aspergillus nidulans*, which can be considered a model system, were found; the vectors are integrated in tandem in the nuclear DNA [63, 66, 84]. Furthermore, transformants of *Trichoderma reesei* have different stabilities, depending on whether they contain different vectors [84]. Transformants containing the *arg*B$^+$ gene (from *Aspergillus nidulans*) were quite stable. By contrast, 40% of the transformants that contained the *amd*S$^+$ gene (also from *Aspergillus nidulans*) were unstable. A related observation, by CANTORAL *et al.* [20], is that when the transformation enhancer *ans*1 was present, not only was the transformation rate of *Penicillium chrysogenum* tripled (Table 7) but also in 70% of the cases studied the vector remained stable upon mitosis. If the same vector but lacking the *ans*1 sequence was used in transformation, only 15% of the transformants were stable in mitosis.

Integrated vectors tend to be stable; no unstable transformants of *Aspergillus niger* are reported [17, 47, 63, 110]. A detailed analysis of vector integration into homologous or heterologous sequences of *Aspergillus niger* yielded surprising results [113]: the integration of a single vector into homologous DNA was as stable as the integration of many copies, arranged in tandem, into heterologous sequences, but several vector copies integrated into homologous sequences were not stable. The integration of several copies into homologous sequences is done by replacing the chromosomal copy of the *oli*C gene with the *oli*C^R allele of the vector.

WARD *et al.* [113] studied three transformants and demonstrated the homologous integration of the vector that carries the *oli*C^R gene (the oligomycin-resistant subunit 9 of ATP synthase) into the chromosomal copy (*oli*C^S). Two transformants contained several vector copies in tandem arrangements and were initially semiresistant to oligomycin because both the sensitive and the resistant subunits of ATP synthase were formed; however, both subsequently developed some completely resistant sectors. All vector copies in these sectors were found to have deletions, but the *oli*C^R rather than the *oli*C^S allele remained in the chromosome.

The third of WARD's transformants contained only one vector copy but was also sectored. In this case the vector was not deleted. Rather, both copies of the *oli*C gene were identical: they each contained the *oli*C^R allele sequence (*gene conversion* had taken place). This is the first description of a gene replacement (conversion) in *Aspergillus niger* and is evidence that the event of gene replacement (conversion) can occur in the absence of a sexual cycle (Section 2.2.1). The possibilities associated with this finding are immense because practically any gene in any fungus can be converted.

So far we have nothing to gainsay that the concepts developed from findings with *Aspergillus nidulans* as a model system can be transferred to many fungi. Using existing transformation systems to answer questions in basic research or for strain improvement appears promising. Nevertheless, it will doubtless be necessary to test current knowledge empirically with each new fungus.

2.3.4 Transformation of phytopathogenic filamentous fungi

Transformation systems for phytopathogenic fungi are useful for elucidating the phenomenon of pathogenicity. The interrelationships between plants and their fungal pathogens are often not understood, but this is exactly the area in which a strategic attack on plant pathogens will have great impact.

The first step is to isolate and analyze the genes that determine pathogenicity; these are the genes that permit the parasite to overcome the plant's defense mechanisms. Transformation can be used to detect the genes, since nonpathogenic strains can be transformed with DNA sequences from pathogenic lines and pathogenic transformants can then be isolated. There appears to be little hope of identifying genes for pathogenicity by inactivation with an integrated vector: firstly, little is known about the mechanism of vector integration, and, secondly, we know of several fungal virulence genes that are present as multiple copies within a genome (for example, the *pda* gene of *Nectria haematococca* [123]).

The transformation systems that might be useful in work with fungal virulence genes in plants are summarized in Table 8. As in the selection of transformants for biotechnologically important fungi (Section 2.3.3), transformants are generated exclusively with heterologous genes that permit transformation of wild-type strains (hph^R or $amdS^+$).

Table 8 Transformation systems for phytopathogenic filamentous fungi

Organism	Host	Selectable marker	Source	Transformation rate/µg DNA	Ref.
Cochliobolus heterostrophus	maize	$amdS^+$ hph^R/P_{Ch}	*A. nidulans* *E. coli/C. heterostrophus*	0.06 0.3–10	[103] [104]
Colletotrichium trifolii	alfalfa	hph^R/P_{Ch} $benA^R$	*E. coli/C. heterostrophus* *N. crassa*	5–10 0.4–1	[33] [33]
Fulvia fulva	tomato	hph^R/P_{gpd}	*E. coli/A. nidulans*	1–18	[79]
Fusarium oxysporum	various	hph^R/P_{trpC}	*E. coli/A. nidulans*	2	[65]
Glomerella cingulata	bean	$amdS^+$ hph^R/P_{Ch}	*A. nidulans* *E. coli/C. heterostrophus*	0.2 0.2	[90] [90]
Leptosphaeria maculans	Brassicaceae	hph^R/P_{gpd}	*E. coli/A. nidulans*	60	[43]
Magnaporthe grisea	rice	$argB^+$	*A. nidulans*	35	[82]
Septoria nodorum	wheat	hph^R/P_{gpd}	*E. coli/A. nidulans*	2–25	[23]

Key to abbreviations: P_{Ch} = undefined promoter from *C. heterostrophus*; for other abbreviations see Table 9

The rates of transformation given in Table 8 are quite low, so the efficient cloning of genes in fungal plant pathogens seems unlikely; for example, a yield of 0.2 transformants per microgram of DNA, reported by RODRIGUEZ and YODER [90] for *Glomerella cingulata*, seems quite inadequate for any useful application. The transformation rate is apparently dependent on the selection method: in both *Cochliobolus heterostrophus* [103, 104] and *Colletotrichium trifolii* [33], more transformants grow when hygromycin B resistance is being selected than with another system. We do not know why this should be so, but evidently the transformation rate can be increased significantly if the method of selection is chosen carefully.

Transformants that are analyzed are usually found to have many vector copies integrated in tandem, and, as a rule, the integration is heterologous. *Magnaporthe grisea* is capable of integrating vectors into homologous sequences, but the vector must contain homologous sequences and must be linearized with a restriction enzyme in this DNA region [82].

Integrated vectors are very stable. Transformants of *Septoria nodorum* still contain the vector even after a passage through the plant [23].

Vector integration in transformants has been found not to affect pathogenicity. Thirty transformants of *Glomerella cingulata* were as virulent in various races of bean as was the recipient strain [90].

In summary, the experience gained with *Aspergillus nidulans* is widely applicable to phytopathogenic fungi; moreover, we can expect that the use of transformation systems will significantly deepen our understanding of molecular mechanisms involved in the fungal infection of plants.

3 Comparative evaluation of strain improvement by genetic engineering and classical methods

In the preceding sections of this chapter I have described the potential merits of transformation systems for strain improvement in biotechnologically important filamentous fungi. However, I have not yet addressed the question of what advantages genetic engineering may offer over the successful classical methods (selection, alternating selection and mutation, and recombination by sexual or parasexual processes). I will answer this question by briefly presenting the mechanisms upon which the classical methods of strain improvement rest and giving examples of their use.

Numerous review articles describe at length the classical methods of strain improvement [25, 39, 40, 42, 57, 89]; consequently, I will not discuss these methods here.

Mutagenesis is the technique used to alter certain traits in biotechnologically important organisms. Single cells (spores or protoplasts) are treated either with chemical mutagens (such as nitrite) or with physical mutagens (such as UV light). (Detailed summaries of mutagens and their mode of action are available in the literature [1, 25, 39, 57, 89]). Mutants with improved traits are subsequently identified using appropriate selection methods that permit prompt detection of a trait in many individual clones. The success of each strain improvement depends upon such efficient screening [36]. By sequentially cycling mutagenesis and selection of the best strains, an increase in yield of 10–15% per cycle can be attained, but the probability of finding better variant lines decreases steadily [36].

A classic example of yield improvement by mutation and selection is *Penicillium chrysogenum*. Initial yields of penicillin from production strains were in the range of 1 mg/l; today, yields of up to 50 g/l are obtained and a further increase appears feasible [25].

Because mutations occur randomly, many lines must be studied, and consequently the improvement of traits for which there are no selection methods is involved and expensive. An additional difficulty is that undesired mutants that complicate the use of these strains in production processes may be generated. The production strains of *Penicillium chrysogenum*, for example, are aptly termed genetic misfits because they contain a multitude of genetic defects.

Ball *et al.* [2] reported a significant increase in glucoamylase yield for an *Aspergillus niger* production strain after several cycles of mutagenesis and selection, but the culture became so loaded with single cells and debris from their lysis that it was difficult to filter. Additional mutations did not help to improve the filtration properties of the strain.

The recombination of different mutations in a cell also leads to improvements in yield. Genetically different genes can recombine following sexual or parasexual events, provided they all occur in one cell. By definition, only individuals of a given species can be crossed sexually; parasexual events occur when species boundaries are crossed, as in protoplast fusion [39].

There are two avenues to follow for strain improvement. In the first, diploid strains (from crossing compatible parent strains) are selected that carry both parental genomes in a cell, provided they are stable. Because of the increased gene dosage, the new cell line can be expected to have a higher synthesizing potential. Kirimura *et al.* [64] isolated 50 diploid mycelia of *Aspergillus niger*, starting with auxotrophic mutants of citric acid producers (on solid medium or in shake flasks) that were forced to form heterokaryons by anastomosis. None of these diploid mycelia produced yields equaling that of the parents in shake culture, but on solid media 20 strains produced about 20% more citric acid.

The second alternative comes about because during meiosis or during the parasexual cycle there may be a single crossing over (during meiosis or mitosis) or a new combination of complete chromosomes [92], with progeny that combines the parental traits in one cell. Ball *et al.* [2] report that they obtained a variant of *Aspergillus niger* with improved glucoamylase production by taking advantage of the parasexual cycle. They crossed a productive strain that could not be filtered easily and a poorly producing strain that was easy to filter. Among the progeny were variants that had the good filtration properties of one parent and the high glucoamylase productivity of the other parent.

In contrast to the methods just described, genetic engineering methods, if they are to be used intelligently, must include precise knowledge of a gene, the form it is in, and how it must be altered to improve yields. With *in vitro* mutagenesis, we have the tool for introducing a desired mutation at *any* locus within any gene, and to multiply the mutant molecule as desired. (Reviews describing the methods of *in vitro* mutagenesis are in the literature [45, 96, 99]).

Genetic engineering in this case includes roughly the following steps: first, a specific gene is isolated and cloned into *E. coli* so that its structure (DNA sequence) may be analyzed; site-specific changes are introduced into the gene by *in vitro* mutagenesis, and the gene is then inserted into the recipient organism by transformation; the transformants are characterized and the effects of the *in vitro* mutagenesis are explored. We now have the tools so that every gene, regardless of its origin, can be expressed in a filamentous fungus when homologous expression signals are inserted upstream (Section 2.3.1).

Because our knowledge of the regulation of metabolic processes is lacking or incomplete, only simple gene technological changes are worthwhile. The integration of a heterologous gene into the fungal genome is an example of a simple change. This procedure has already been accomplished (Section 2.3.2) and demonstrates the power of genetic engineering: the genome of an organism is expanded beyond its natural composition. Classical methods are quite unsuitable for this, because the whole natural gene complement will only be rearranged or varied.

Changing complex biosynthetic pathways using genetic engineering strategies is not currently feasible because the basic principles are still unknown. A further complication is that the cell incorporates what appears to be a network of independent processes with effects that can rarely be predicted. Even after 65 years of industrial production of citric acid with *Aspergillus niger*, we do not understand precisely why the cells perform this metabolically useless feat [69, 91].

In summary, genetic engineering can certainly be used for strain improvement because the basic principles are known. Gene technology, however, is not the only answer because in some areas the classical methods of strain improvement are just as good. It is therefore more reasonable to consider genetic engineering as a new technique for strain improvement to supplement those already available.

4 Appendix

Table 9 Index of gene and gene product abbreviations

Gene symbol	Phenotype or function	Organism
*acu*D	isocitrate lyase	*Aspergillus nidulans*
*alc*A	alcohol dehydrogenase I	*Aspergillus nidulans*
am	NADP-dependent glutamate dehydrogenase	*Neurospora crassa*
*amd*S	acetamidase	*Aspergillus nidulans*
*are*A	repression by nitrogen metabolites	*Aspergillus nidulans*
*arg*B	ornithine carbamoyltransferase	*Aspergillus nidulans*, *Aspergillus niger*
*ben*A^R	β-tubulin; causes benomyl resistance	*Aspergillus nidulans*
*ble*R	phleomycin[1] resistance	*Escherichia coli*

[1] Phleomycin is an antibiotic in the same family as bleomycin (continued on next page)

Table 9 (continued)

Gene symbol	Phenotype or function	Organism
*brl*A	"bristle", no conidia	*Aspergillus nidulans*
cbh2	cellobiohydrolase II	*Trichoderma reesei*
egl	endoglucanase	*Cellulomonas fimi*
*egl*1	endoglucanase I	*Trichoderma reesei*
*egl*3	endoglucanase III	*Trichoderma reesei*
*fw*A	yellow-brown conidia	*Aspergillus nidulans*
glaA	glucoamylase	*Aspergillus niger*
gpd	glyceraldehyde-3-phosphate dehydrogenase	*Aspergillus nidulans*
hph^R	hygromycin B phosphotransferase; causes hygromycin B resistance	*Escherichia coli*
ifn α2	interferon α2	*Homo sapiens*
*lac*Z	β-galactosidase	*Escherichia coli*
neo^R	neomycin phosphotransferase; causes neomycin resistance	*Escherichia coli*
*nim*A	induction of mitosis	*Aspergillus nidulans*
*nit*2	repression of nitrogen metabolites	*Neurospora crassa*
oliC^R	ATP synthase, subunit 9; causes oligomycin resistance	*Aspergillus nidulans*/*Aspergillus niger*
*paba*A	*p*-aminobenzoic acid biosynthesis	*Aspergillus nidulans*
*pcb*C	isopenicillin-*N* synthase	*Cephalosporium acremonium*
pda	pisatin demethylase	*Nectria haematococca*
pgk	phosphoglycerate kinase	*Saccharomyces cerevisiae*
*pki*A	pyruvate kinase	*Aspergillus nidulans*
prn	proline metabolism	*Aspergillus nidulans*
*pyr*A	orotidine-5′-phosphate decarboxylase	*Aspergillus niger*
*pyr*4	orotidine-5′-phosphate decarboxylase	*Neurospora crassa*
*qut*A	quinate metabolism	*Aspergillus nidulans*
*qut*E	3-dehydroquinase	*Aspergillus nidulans*
*ribo*B	riboflavin biosynthesis	*Aspergillus nidulans*
*Spo*C	sporogenesis	*Aspergillus nidulans*
tPA	tissue plasminogen activator (tPA)	*Homo sapiens*
*tpi*A	triose-phosphate isomerase	*Aspergillus nidulans*
*trp*C	tryptophan biosynthesis	*Aspergillus nidulans, Aspergillus niger, Cochliobolus heterostrophus*
*y*A	*p*-diphenol oxidase	*Aspergillus nidulans*

5 Acknowledgements

I thank Prof. Dr. K. Esser for numerous discussions, for his constant interest, and for review of the manuscript. I gratefully acknowledge the financial support of the Max Bucher Research Foundation, Frankfurt/Main, Fed. Rep. Germany.

6 References

[1] Auerbach, C.: *Mutation Research*, Chapman and Hall, London (1976).
[2] Ball, C.; Lawrence, A. J.; Butler, J. M.; Morrison, K. B.; Improvement in Amyloglucosidase Production Following Genetic Recombination of *Aspergillus niger* Strains; *European J. Appl. Microbiol.* **5** (1978) 95–102.
[3] Ballance, D. J.; Buxton, F. P.; Turner, G.: Transformation of *Aspergillus nidulans* by the Orotidine-5'-phosphate Decarboxylase Gene of *Neurospora crassa*; *Biochem. Biophys. Res. Comm.* **112** (1983) 284–289.
[4] Ballance, D. J.; Turner, G.: Development of a High-Frequency Transforming Vector for *Aspergillus nidulans*; *Gene* **36** (1985) 321–331.
[5] Ballance, D. J.; Turner, G.: Genetic Cloning in *Aspergillus nidulans*: Isolation of the Isocitrate Lyase Gene (*acu*D); *Mol. Gen. Genet.* **202** (1986) 271–275.
[6] Barbesgaard, P.: Industrial Enzymes Produced by Members of the Genus *Aspergillus*. In: Smith, J. E.; Pateman, J. A., Eds.; *Genetics and Physiology of Aspergillus*, Academic Press, London, New York, San Francisco (1977).
[7] Beggs, J. D.: Transformation of Yeast by a Replicating Hybrid Plasmid; *Nature (London)* **275** (1978) 104–109.
[8] Beri, R. K.; Turner, G.: Transformation of *Penicillium chrysogenum* Using the *Aspergillus nidulans amd*S Gene as a Dominant Selective Marker; *Curr. Genet.* **11** (1987) 639641.
[9] Beri, R. K.; Lewis, E. L.; Turner, G.: Behaviour of a Replicating Mitochondrial DNA Sequence from *Aspergillus amstelodami* in *Saccharomyces cerevisiae* and *Aspergillus nidulans*; *Curr. Genet.* **13** (1988) 479–486.
[10] Beri, R. K.; Whittington, H.; Roberts, C. F.; Hawkings, A. R.: Isolation and Characterization of the Positively Acting Regulatory Gene *qut*A from *Aspergillus nidulans*; *Nucl. Acids Res.* **15** (1987) 7991–8001.
[11] Berry, D. R.; Chmiel, A.; Al obaidi, Z.: Citric Acid Production by *Aspergillus niger*. In: Smith, J. E.; Pateman, J. A., Eds.; *Genetics and Physiology of Aspergillus*, Academic Press, London, New York, San Francisco (1977).
[12] Boel, E.; Hansen, M. T.; Hjort, I.; Hegh, I.; Fiil, N. P.: Two Different Types of Intervening Sequences in the Glucoamylase Gene from *Aspergillus niger*; *EMBO J.* **3** (1984) 1581–1585.
[13] Boel, E.; Hjort, I.; Svensson, B.; Norris, F.; Norris, K. E.; Fiil, N. P.: Glucoamylases G1 and G2 from *Aspergillus niger* are Synthesized from Two Different but Closely Related mRNAs; *EMBO J.* **3** (1984) 1097–1102.
[14] Botstein, D.; Davis, R. W.: Recombinant DNA Research with Yeast. In: Strathern, J. N.; Jones, E. W.; Broach, J. R., Eds.; *The Molecular Biology of the Yeast Saccharomyces cerevisiae: Metabolism and Gene Expression*, Cold Spring Harbor Laboratory, Cold Spring Harbor, NY (1982).

[15] Boynton, J. E.; Gillham, N. W.; Harris, E. H.; Hosler, J. P.; Johnson, A. M.; Jones, A. R.; Randolph-Anderson, B. L.; Robertson, D.; Klein, T. M.; Sharke, K. B.; Sanford, J. C.: Chloroplast Transformation in *Chlamydomonas* with High Velocity Microprojectiles; *Science* **240** (1988) 1534–1538.

[16] Bull, J. H.; Smith, D. J.; Turner, G.: Transformation of *Penicillium chrysogenum* with a Dominant Selectable Marker; *Curr. Genet.* **13** (1988) 377–382.

[17] Buxton, F. P.; Gwynne, D. I.; Davies, R. W.: Transformation of *Aspergillus niger* Using the *arg*B Gene of *Aspergillus nidulans*; *Gene* **37** (1985) 207–214.

[18] Buxton, F. P.; Gwynne, D. I.; Garven, S.; Sibley, S.; Davies, R. W.: Cloning and Molecular Analysis of the Ornithine Carbamoyl Transferase Gene of *Aspergillus niger*; *Gene* **60** (1987) 255–266.

[19] Caddick, M. X.; Arst, Jr., H. N.; Taylor, L. H.; Johnson, R. I.; Brownlee, A. G.: Cloning of the Regulatory Gene *are*A Mediating Nitrogen Metabolite Repression in *Aspergillus nidulans*; *EMBO J.* **5** (1986) 1087–1090.

[20] Cantoral, J. M.; Diez, B.; Barredo, J. L.; Alvarez, E.; Martin, J. F.: High Frequency Transformation of *Penicillium chrysogenum*; *Bio/Technology* **5** (1987) 494–497.

[21] Clutterbuck, A. J.: Loci and Linkage Map of the Filamentous Fungus *Aspergillus nidulans* (Eidam) Winter (n = 8). In: O'Brien, S. J., Ed., *Genetic Maps*, Cold Spring Harbor Laboratory, Cold Spring Harbor, NY (1984); pp 265–273.

[22] Cohen, S. N.; Chang, A. C. Y.; Hsu, L.: Nonchromosomal Antibiotic Resistance in Bacteria: Genetic Transformation of *E. coli* by R-Factor DNA; *PNAS USA* **69** (1972) 2110–2114.

[23] Cooley, R. N.; Shaw, R. K.; Franklin, F. C. H.; Caten, C. E.: Transformation of the Phytopathogenic Fungus *Septoria nodorum* to Hygromycin B Resistance; *Curr. Genet.* **13** (1988) 383–389.

[24] Corrick, C. M.; Twomey, A. P.; Hynes, M. J.: The Nucleotide Sequence of the *amd*S Gene of *Aspergillus nidulans* and the Molecular Characterization of 5′ Mutations; *Curr. Genet.* **12** (1987) 231–233.

[25] Crueger, W.; Crueger, A.: *Lehrbuch der angewandten Mikrobiologie*, Akademische Verlagsanstalt, Wiesbaden, FRG (1982).

[26] Cullen, D.; Gray, G. L.; Wilson, L. J.; Hayenga, K. J.; Lamsa, M. H.; Rey, M. W.; Norton, S.; Berka, R. M.: Controlled Expression and Secretion of Bovine Chymosin in *Aspergillus nidulans*; *Bio/Technology* **5** (1987) 369–376.

[27] Cullen, D.; Leong, S. A.; Wilson, L. J.; Henner, D. J.: Transformation of *Aspergillus nidulans* with the Hygromycin Resistance Gene, *hph*; *Gene* **57** (1987) 21–26.

[28] Cullen, D.; Wilson, L. J.; Grey, G. L.; Henner, D. J.; Turner, G.; Ballance, D. J.: Sequence and Centromere Proximal Location of a Transformation Enhancing Fragment *ans*1 from *Aspergillus nidulans*; *Nucl. Acids Res.* **15** (1987) 9164–9175.

[29] Da Silva, A. J. F.; Whittington, H.; Clements, J.; Roberts, C.; Hawkings, A. R.: Sequence Analysis and Transformation by the Catabolic 3-Dehydroquinase (QUTE) Gene from *Aspergillus nidulans*; *Biochem. J.* **240** (1986) 481–488.

[30] Davis, M. A.; Hynes, M. J.: Complementation of *are*A-Regulatory Gene Mutations of *Aspergillus nidulans* by the Heterologous Regulatory Gene *nit*-2 of *Neurospora crassa*; *PNAS USA* **84** (1987) 3753–3757.

[31] Davis, M. A.; Corbet, C. S.; Hynes, M. J.: An *amd*S-*lac*Z Fusion for Studying Gene Regulation in *Aspergillus*; *Gene* **63** (1988) 199–212.

[32] De Graaff, L.; Van den Broek, H.; Visser, J.: Isolation and Transformation of the Pyruvate Kinase Gene of *Aspergillus nidulans*; *Curr. Genet.* **13** (1988) 315–321.

[33] Dickmann, M. B.; Whole Cell Transformation of the Alfalfa Fungal Pathogen *Colletotrichum trifolii*; *Curr. Genet.* **14** (1988) 241–246.

[34] Dunne, P. W.; Oakley, B. R.: Mitotic Gene Conversion, Reciprocal Recombination

and Gene Replacement at the *ben*A, Beta-Tubulin, Locus of *Aspergillus nidulans*; *Mol. Gen. Genet.* **213** (1988) 339–345.

[35] DURRENS, P.; GREEN, P. M.; ARST, JR., H. N.; SCAZZOCCHIO, C.: Heterologous Insertion of Transforming DNA and Generation of New Deletions Associated with Transformation in *Aspergillus nidulans*; *Mol. Gen. Genet.* **203** (1986) 544–549.

[36] ELANDER, R. P.; CHANG, L. T.: Microbial Culture Selection. In: PEPPLER, H. J.; PERLMAN, D., Eds.; *Microbial Technology*, **Vol 2**, 2nd ed., Academic Press, New York, San Francisco, London (1979).

[37] ESSER, K.; DOHMEN, G.: Drug Resistance Genes and Their Use in Molecular Cloning; *Process Biochem.* **22** (1987) 144–148.

[38] ESSER, K.; KÄMPER, J.: Transformation Systems in Yeast: Fundamentals and Applications in Biotechnology. *Process Biochem.* **23** (1988) 36–42.

[39] ESSER, K.; MEINHARDT, F.: Mikroorganismen. In: PRÄVE, P.; FAUST, U.; SITTIG, W.; SUKATSCH, D., Eds.; *Handbuch der Biotechnologie*, Akademische Verlagsanstalt, Wiesbaden, FRG (1982).

[40] ESSER, K.; MEINHARDT, F.: Genetics of Strain Improvement in Filamentous Fungi with Respect to Biotechnology. In: VANK, Z.; HOLEK, Z., Eds.; *Overproduction of Microbial Metabolites*, Butterworth Publ., Boston, MA (1986).

[41] ESSER, K.; MOHR, G.: Integrative Transformation of Filamentous Fungi with Respect to Biotechnological Application; *Process Biochem.* **21** (1986) 153–159.

[42] ESSER, K.; STAHL, U.: Hybridization. In: REHM, H. J.; REED, G., Eds.; *Biotechnology*, **Vol. 1**, VCH, Weinheim, Deerfield Beach, Basel (1981).

[43] FARMAN, M. L.; OLIVER, R. P.: The Transformation of Protoplasts of *Leptosphaeria maculans* to Hygromycin B Resistance; *Curr. Genet.* **13** (1988) 327–330.

[44] FINKELSTEIN, D. B.; RAMBOSEK, J.; CRAWFORD, M. S.; SOLIDAY, C. L.; MCADA, P. C.; LEACH, J.: Protein Secretion in *Aspergillus niger*. In: HERSCHBERGER, C. L.; QUEENER, S. M.; HEGEMAN, G.; *Genetics and Molecular Biology of Industrial Microorganisms*, American Society for Microbiology, Washington, DC (1989); pp 295–300.

[45] GATZ, C.; HILLEN, W.: Mutagenesis; *Progress in Botany* **46** (1984) 208–225.

[46] GOMI, K.; IIMURA, Y.; HARA, S.: Integrative Transformation of *Aspergillus oryzae* with a Plasmid Containing the *Aspergillus nidulans argB* Gene; *Agric. Biol. Chem.* **51** (1987) 2549–2555.

[47] GOOSEN, T.; BLOEMHEUVEL, G.; GYSLER, C.; DE BIE, D. A.; VAN DEN BROEK, H. W. J.; SWART, K.: Transformation of *Aspergillus niger* Using the Homologous Orotidine-5′-phosphate-decarboxylase Gene; *Curr. Genet.* **11** (1987) 499–503.

[48] GRANT, S.; ROBERTS, C. F.; LAMB, H.; STOUT, M.; HAWKINGS, A. R.: Genetic Regulation of the Quinic Acid Utilization (*qut*) Gene Cluster in *Aspergillus nidulans*; *J. Gen. Microbiol.* **134** (1988) 347–358.

[49] GRIFFITH, F.: The Significance of Pneumococcal Types; *J. Hyg.* **27** (1928) 113–159.

[50] GRITZ, L.; DAVIS, J.: Plasmid-Encoded Hygromycin B Resistance: The Sequence of Hygromycin B Phosphotransferase Gene and Its Expression in *Escherichia coli* and *Saccharomyces cerevisiae*; *Gene* **25** (1983) 179–188.

[51] GWYNNE, D. I.; BUXTON, F. P.; SIBLEY, S.; DAVIS, R. W.; LOCKINGTON, R. A.; SCAZZOCCHIO, C.; SEALY-LEWIS, H. M.: Comparison of the *cis*-Acting Control Regions of Two Coordinately Controlled Genes Involved in Ethanol Utilization in *Aspergillus nidulans*; *Gene* **51** (1987) 205–216.

[52] GWYNNE, D. I.; BUXTON, F. P.; WILLIAMS, S. A.; GARVEN, S.; DAVIES, R. W.: Genetically Engineered Secretion of Active Human Interferon and a Bacterial Endoglucanase from *Aspergillus nidulans*; *Bio/Technology* **5** (1987) 713–719.

[53] HAMER, J. E.; TIMBERLAKE, W. E.: Functional Organization of the *Aspergillus nidulans trpC* Promotor; *Mol. Cell. Biol.* **7** (1987) 2352–2359.

[54] HINNEN, A.; HICKS, J. B.; FINK, G. R.: Transformation of Yeast; *PNAS USA* **75** (1978) 1929–1933.
[55] HYNES, M. J.: Transformation of Filamentous Fungi; *Exp. Mycol.* **10** (1986) 1–8.
[56] HYNES, M. J.; CORRICK, C. M.; KING, J. A.: Isolation of Genomic Clones Containing the *amd*S Gene of *Aspergillus nidulans* and Their Use in the Analysis of Structural and Regulatory Mutations; *Molec. Cell. Biol.* **3** (1983) 1430–1439.
[57] JACOBSON, G. K.: Mutations. In: REHM, H. J.; REED, G., Eds.; *Biotechnology*, **Vol. 1**, VCH, Weinheim, Deerfield Beach, Basel (1981).
[58] JOHN, M. A.; PEBERDY, J. F.: Transformation of *Aspergillus nidulans* Using the *arg*B Gene; *Enzyme Microb. Technol.* **6** (1984) 386–389.
[59] JOHNSTON, S. A.; ANZIANO, P. Q.; SHARK, K.; SANFORD, J. C.; BUTOW, R. A.: Mitochondrial Transformation in Yeast by Bombardment with Microprojectiles; *Science* **240** (1988) 1538–1541.
[60] JOHNSTONE, I. L.: Transformation of *Aspergillus nidulans*; *Microbiol. Sci.* **2** (1985) 307–311.
[61] JOHNSTONE, I. L.; HUGHES, S. G.; CLUTTERBUCK, A. J.: Cloning an *Aspergillus nidulans* Gene by Transformation; *EMBO J.* **4** (1985) 2307–2311.
[62] KASTER, K. R.; BURGETT, S. G.; RAO, R. N.; INGOLIA, T. D.: Analysis of a Bacterial Hygromycin B Resistance Gene by Transcriptional and Translational Fusions and by DNA Sequencing; *Nucl. Acids Res.* **11** (1983) 6895–6911.
[63] KELLY, J. M.; HYNES, M. J.: Transformation of *Aspergillus niger* by the *amd*S Gene of *Aspergillus nidulans*; *EMBO J.* **4** (1985) 475–479.
[64] KIRIMURA, K.; NAKAJIMA, I.; LEE, S. P.; KAWABE, S.; USAMI, S.: Citric Acid Production by the Diploid Strains of *Aspergillus niger* Obtained by Protoplast Fusion; *Appl. Microbiol. Biotechnol.* **27** (1988) 504–506.
[65] KISTLER, H. C.; BENNY, U. K.: Genetic Transformation of the Fungal Plant Wilt Pathogen, *Fusarium oxysporum*; *Curr. Genet.* **13** (1988) 145–149.
[66] KOLAR, M.; PUNT, P. J.; VAN DEN HONDEL, C. A. M. J. J.; SCHWAB, H.: Transformation of *Penicillium chrysogenum* Using Dominant Selection Markers and Expression of an *Escherichia coli lac*Z Fusion Gene; *Gene* **62** (1988) 127–134.
[67] KOS, A.; KUIJVENHOVEN, J.; WERNARS, K.; BOS, C. J.; VAN DEN BROEK, H. W. J.; POUWELS, P. H.; VAN DEN HONDEL, C. A. M. J. J.: Isolation and Characterization of the *Aspergillus niger trp*C Gene; *Gene* **39** (1985) 231–238.
[68] KREIL, G.: Transfer of Proteins across Membranes; *Ann. Rev. Biochem.* **50** (1981) 317–348.
[69] KUBICEK, C. P.: Regulatory Aspects of the Tricarboxylic Acid Cycle in Filamentous Fungi – A Review; *Trans. Br. Mycol. Soc.* **90** (1988) 339–349.
[70] MANDEL, M.; HIGA, A.: Calcium Dependent Bacteriophage DNA Infection; *J. Mol. Biol.* **53** (1970) 154–162.
[71] MARSTON, F. A. O.: The Purification of Eukaryotic Polypeptides Expressed in *Escherichia coli*. In: GLOVER, D. M., Ed.; *DNA Cloning*, **Vol. III**, IRL Press, Oxford, Washington, DC (1987).
[72] MATTERN, I. E; UNKLES, S.; KINGHORN, J. R.; POUWELS, P. H.; VAN DEN HONDEL, C. A. M. J. J.: Transformation of *Aspergillus oryzae* Using the *A. niger pyr*G Gene; *Mol. Gen. Genet.* **210** (1987) 460–461.
[73] MILLER, B. L.; MILLER, K. Y.; TIMBERLAKE, W. E.: Direct and Indirect Gene Replacement in *Aspergillus nidulans*; *Mol. Cell. Biol.* **5** (1985) 1714–1721.
[74] MISHRA, N. C.: Gene Transfer in Fungi. In: CASPARI, E. W.; SCANDALIO, J. G., Eds.; *Advances in Genetics*, **Vol. 23**, Academic Press, New York, London, Tokyo (1985).
[75] MULLANEY, E. J.; PUNT, P. J.; VAN DEN HONDEL, C. A. M. J. J.: DNA Mediated Transformation of *Aspergillus ficuum*; *Appl. Microbiol. Biotechnol.* **28** (1988) 451–454.

[76] NÜESCH, J.; HEIM, H.; TREICHLER, H. J.: The Biosynthesis of Sulfur-Containing β-Lactam Antibiotics; *Ann. Rev. Microbiol.* **41** (1987) 51–75.

[77] NUNBERG, J. H.; MEADE, J. H.; COLE, G.; LAWYER, F. C.; MCCABE, P.; SCHWEICKART, V.; TAL, R.; WITTMAN, V. P.; FLATGAARG, J. E.; INNIS, M. A.: Molecular Cloning and Characterization of the Glucoamylase Gene of *Aspergillus awamori*; *Molec. Cell. Biol.* **4** (1984) 2306–2315.

[78] OAKLEY, C. E.; WEIL, C. F.; KRETZ, P. L.; OAKLEY, B. R.: Cloning of the *ribo*B Locus of *Aspergillus nidulans*; *Gene* **53** (1987) 293–298.

[79] OLIVER, R. P.; ROBERTS, I. N.; HARLING, R.; KENYON, L.; PUNT, P. J.; DINGEMANSE, M. A.; VAN DEN HONDEL, C. A. M. J. J.: Transformation of *Fulvia fulva*, a Fungal Pathogen of Tomato, to Hygromycin B Resistance; *Curr. Genet.* **12** (1987) 231–233.

[80] OSIEWACZ, H. D.; WEBER, A.: DNA Mediated Transformation of the Filamentous Fungus *Curvularia lunata* Using a Dominant Selectable Marker; *Appl. Microbiol. Technol.* **30** (1989) 375–380.

[81] OSMANI, S. A.; MAY, G. S.; MORRIS, N. R.: Regulation of the mRNA Levels of *nim*A, a Gene Required for the G2-M Transition in *Aspergillus nidulans*; *J. Cell. Biol.* **104** (1987) 1495–1504.

[82] PARSONS, K. A.; CHUMLEY, F. G.; VALENT, B.: Genetic Transformation of the Fungal Pathogen Responsible for Rice Blast Disease; *PNAS USA* **84** (1987) 4161–4165.

[83] PENTTILÄ, M.; LEHTOVAARA, P.; NEVALAINEN, H.; BHIKHABHAI, R.; KNOWLES, J.: Homology between Cellulase Genes of *Trichoderma reesei*: Complete Nucleotide Sequence of the Endoglucanase I Gene; *Gene* **45** (1986) 253–263.

[84] PENTTILÄ, M.; NEVALAINEN, H.; RÄTTÖ, M.; SALMINEN, E.; KNOWLES, J.: A Versatile Transformation System for the Cellulolytic Filamentous Fungus *Trichoderma reesei*; *Gene* **61** (1987) 155–164.

[85] PICKNETT, T. M.; SAUNDERS, G.; FORD, P.; HOLT, G.: Development of a Gene Transfer System for *Penicillium chrysogenum*; *Curr. Genet.* **12** (1987) 449–455.

[86] PRÄVE, P.; FAUST, U.; SITTIG, W.; SUKATSCH, D. A.: *Handbuch der Biotechnologie*; Akademische Verlagsanstalt, Wiesbaden, FRG (1982).

[87] PUNT, P. J.; OLIVER, R. P.; DINGEMANSE, M. A.; POUWELS, P. H.; VAN DEN HONDEL, C. A. M. J. J.: Transformation of *Aspergillus* Based on the Hygromycin B Resistance Marker from *Escherichia coli*; *Gene* **56** (1987) 117–124.

[88] RAMBOSEK, J.; LEACH, J.: Recombinant DNA in Filamentous Fungi: Progress and Prospects. In: *CRC Critical Reviews in Biotechnology*, **Vol. 6,** CRC Press Inc., Boca Raton, FL (1987).

[89] REHM, J. J.: *Industrielle Mikrobiologie*, 2nd ed., Springer Verlag, Berlin, Heidelberg, New York (1980).

[90] RODRIGUEZ, R. J.; YODER, O. C.: Selectable Genes for Transformation of the Fungal Plant Pathogen *Glomerella cingulata* f. spec. *phaseoli* (*Colletotrichum lindemuthianum*); *Gene* **54** (1987) 73–81.

[91] RÖHR, M.; KUBICEK, C. P.; KOMINEK, J.: Citric Acid. In: REHM, H. J.; REED, G., Eds.; *Biotechnology*, **Vol. 3**, VCH, Weinheim, Deerfield Beach, Basel (1983).

[92] ROPER, J. A.: Mechanisms of Inheritance. 3. The Parasexual Cycle. In: AINSWORTH, G. C.; SUSSMAN, A. S., Eds.; *The Fungi*, **Vol. II**, Academic Press, New York, London (1966).

[93] SALOHEIMO, M.; LEHTOVAARA, P.; PENTTILÄ, M.; TEERI, T. T.; STÅHLBERG, J.; JOHANSON, G.; PETTERSON, G.; CLAEYSSENS, M.; TOMME, P.; KNOWLES, J. K. C.: EGIII, a New Endoglucanase from *Trichoderma reesei*: The Characterization of Both Gene and Enzyme; *Gene* **63** (1988) 11–21.

[94] SAUNDERS, G.; TUITE, M. F.; HOLT, G.: Fungal Cloning Vectors; *Trends Biotechnol.* **4** (1986) 93–98.

[95] SCHIEMANN, J.: Gentransfer bei filamentösen Pilzen; *Biol. Zentbl.* **106** (1987) 533–546.

[96] SHORTLE, D.; DIMAIO, D.; NATHANS, D.: Directed Mutagenesis; *Ann. Rev. Genet.* **15** (1981) 265–294.

[97] SILHAVY, T. J.; BECKWITH, J. R.: Uses of *lac* Fusions for the Study of Biological Problems; *Microbiol. Rev.* **49** (1985) 398–418.

[98] SKATRUD, P. L.; QUEENER, S. W.; CARR, L. G.; FISHER, D. L.: Efficient Integrative Transformation of *Cephalosporium acremonium*; *Curr. Genet.* **12** (1987) 337–348.

[99] SMITH, M.: Site-Directed Mutagenesis; *Trends Biochem. Sci.* **7** (1982) 440–442.

[100] STRUHL, K.: The New Yeast Genetics; *Nature (London)* **305** (1983) 391–397.

[101] TEERI, T. T.; LEHTOVAARA, P.; KAUPPINEN, S.; SALUVUORI, I.; KNOWLES, J.: Homologous Domains in *Trichoderma reesei* Cellulolytic Enzymes: Gene Sequence and Expression of Cellobiohydrolase II; *Gene* **51** (1987) 43–52.

[102] TILBURN, J.; SCAZZOCCHIO, C.; TAYLOR, G. G.; ZABICKY-ZISSMAN, J. H.; LOCKINGTON, R. A.; DAVIES, R. W.: Transformation by Integration in *Aspergillus nidulans*; *Gene* **26** (1983) 205–221.

[103] TURGEON, B. G.; GARBER, R. C.; YODER, O. C.: Transformation of the Fungal Maize Pathogen *Cochliobolus heterostrophus* Using the *Aspergillus nidulans amd*S Gene; *Mol. Gen. Genet.* **201** (1985) 450–453.

[104] TURGEON, B. G.; GARBER, R. C.; YODER, O. C.: Development of a Fungal Transformation System Based on Selection of Sequences with Promotor Activity; *Mol. Cell. Biol.* **7** (1987) 3297–3305.

[105] TURGEON, B. G.; MACRAE, W. D.; GARBER, R. C.; FINK, G. R.; YODER, O. C.: A Cloned Tryptophane-Synthesis Gene from the Ascomycete *Cochliobolus heterostrophus.* Functions in *Escherichia coli*, Yeast and *Aspergillus nidulans*; *Gene* **42** (1986) 79–88.

[106] UPSHALL, A.: Genetic and Molecular Characterization of *arg*B$^+$ Transformants of *Aspergillus nidulans*; *Curr. Genet.* **10** (1986) 593–599.

[107] UPSHALL, A.; KUMAR, A. A.; BAILEY, M. C.; PARKER, M. D.; FAVREAU, M. A.; LEWISON, K. P.; JOSEPH, M. L.; MARAGANORE, J. M.; MCKNIGHT, G. L.: Secretion of Active Human Tissue Plasminogen Activator from the Filamentous Fungus *Aspergillus nidulans*; *Bio/Technology* **5** (1987) 1301–1304.

[108] VAN GORCOM, R. F. M.; POUWELS, P. H.; GOOSEN, T.; VISSER, J.; VAN DEN BROEK, H. W. J.; HAMER, J. E.; TIMBERLAKE, W. E.; VAN DEN HONDEL, C. A. M. J. J.: Expression of an *Escherichia coli β*-Galactosidase Fusion Gene in *Aspergillus nidulans*; *Gene* **40** (1985) 99–106.

[109] VAN GORCOM, R. F. M.; PUNT, P. J.; POUWELS, P. H.; VAN DEN HONDEL, C. A. M. J. J.: A system for the Analysis of Expression Signals in *Aspergillus*; *Gene* **48** (1986) 211–217.

[110] VAN HARTINGSVELDT, W.; MATTERN, E. E.; VAN ZEIJL, C. J. J.; POUWELS, P. H.; VAN DEN HONDEL, C. A. M. J. J.: Development of a Homologous Transformation System for *Aspergillus niger* Based on the *pyr*G Gene; *Mol. Gen. Genet.* **206** (1987) 71–75.

[111] WALTER, P.; GILMORE, R.; BLOBEL, G.: Protein Translocation across the Endoplasmic Reticulum; *Cell* **38** (1984) 5–8.

[112] WARD, J.; WILKINSON, B.; TURNER, G.: Transformation of *Aspergillus nidulans* with a Cloned Oligomycin-Resistant ATP Synthase Subunit 9 Gene; *Mol. Gen. Genet.* **202** (1986) 265–270.

[113] WARD, M.; WILSON, L. J.; CARMONA, C. L.; TURNER, G.: The *oli*C3 Gene of *Aspergillus niger*: Isolation, Sequence and Use as a Selectable Marker for Transformation; *Curr. Genet.* **14** (1988) 37–42.

[114] WELTRING, K. M.; TURGEON, B. G.; YODER, O. C.; VAN ETTEN, H. D.: Isolation of a Phytoalexin-Detoxification Gene from the Plant Pathogenic Fungus *Nectria haematococca* by Detecting Its Expression in *Aspergillus nidulans*; *Gene* **68** (1988) 335–344.

[115] WERNARS, K.: *DNA-Mediated Transformation of the Filamentous Fungus Aspergillus nidulans*; Proefschrift (Thesis) Landbouwhogeschool Wageningen, Niederlande (1986).
[116] WERNARS, K.; GOOSEN, T.; SWART, K.; VAN DEN BROEK, H. W. J.: Genetic Analysis of *Aspergillus nidulans amd*S$^+$ Transformants; *Mol. Gen. Genet.* **205** (1986) 312–317.
[117] WERNARS, K.; GOOSEN, T.; WENNEKES, L. M. J.; SWART, K.; VAN DEN HONDEL, C. A. M. J. J.; VAN DEN BROEK, H. W. J.: Cotransformation of *Aspergillus nidulans*: A Tool for Replacing Fungal Genes; *Mol. Gen. Genet.* **209** (1987) 71–77.
[118] WERNARS, K.; GOOSEN, T.; WENNEKES, L. M. J.; VISSER, J.; BOS, C. J.; VAN DEN BROEK, H. W. J.; VAN GORCOM, R. F. M.; VAN DEN HONDEL, C. A. M. J. J.; POUWELS, P. H.: Gene Amplification in *Aspergillus nidulans* by Transformation with Vectors Containing the *amd*S Gene; *Curr. Genet.* **9** (1985) 361–368.
[119] WINNACKER, E. L.: *Gene und Klone*, VCH, Weinheim, Deerfield Beach, Basel (1985).
[120] WÖSTEMEYER, J.; BURMESTER, A.: Genetische Manipulation von Pilzen; *BIUZ* **17** (1987) 114–121.
[121] YELTON, M. M.; HAMER, J. E.; TIMBERLAKE, W. E.: Transformation of *Aspergillus nidulans* by Using a *trp*C Plasmid; *PNAS USA* **81** (1984) 1470–1474.
[122] YELTON, M. M.; TIMBERLAKE, W. E.; VAN DEN HONDEL, C. A. M. J. J.: A Cosmid for Selecting Genes by Complementation in *Aspergillus nidulans*. Selection of the Developmentally Regulated *y*A Locus; *PNAS USA* **82** (1985) 834–838.
[123] YODER, O. C.; WELTRING, K.; TURGEON, B. G.; GARBER, R. C.; VAN ETTEN, H. D.: Technology for Molecular Cloning of Fungal Virulence Genes. In: BAILEY, J. A., Ed.; *Biology and Molecular Biology of Plant-Pathogen Interactions*, NATO ASI Serie H, **Vol. 1** (1986).

Applied Microbiology and Biochemistry

Immobilized Systems: Three Applications

by W. TISCHER

Contents

Dr. W. TISCHER,
Boehringer Mannheim GmbH,
Nonnenwald 2,
D-8122 Penzberg, Fed. Rep. Germany

1 Introduction

The term *immobilized systems* is presumably derived from the term *immobilized enzymes*. Enzymes are described as *immobilized* when they are bound chemically or by adsorption to carrier materials, or are entrapped in soluble form in devices, such as microcapsules or membranes, that are impermeable to the enzyme, ensuring a continuous exchange of substrate and product [1].

The thread that ties all parts of this definition together, and is the result of the technological application of these products, is the possibility of reusing the enzyme. This is taken to mean that the enzyme can be separated from the reaction participants by simple processes and is thus available for reuse [2]. Immobilized enzymes are used in many analytical measuring systems of widely different natures, as well as in bioengineering applications [3]. The features required of immobilizations can be very different, and both the strength of the immobilization and the extent of reusability may vary widely.

Since immobilized enzymes, as well as immobilized cells, organ cells, or proteins, are subjected to such a variety of requirements, it appears sensible to consider not just the immobilized biological components, but the entire *immobilized system*. This encompasses all aspects of the mechanism of action of biocatalysts in the reactors or analytical measuring instruments in which they are immobilized. The use of these systems in bioengineering and analysis will be described with three examples. The first example is the industrial scission of penicillin by covalently immobilized penicillin G amidase; the

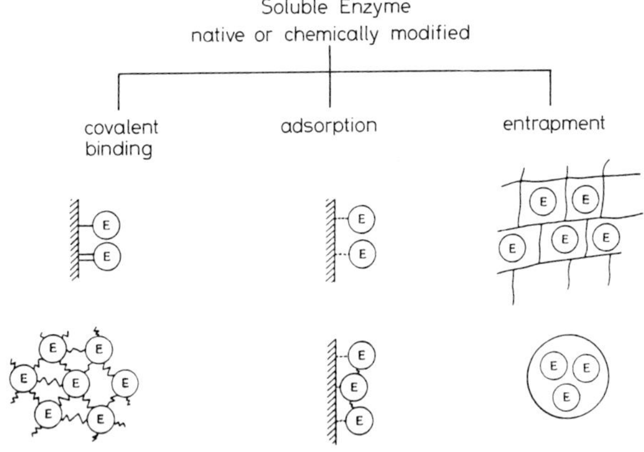

Fig. 1 Classification of immobilized enzymes and their schematic representation
In covalent binding, the enzyme may be bound to a specified water-insoluble polymer, or it may be incorporated into a growing polymer matrix (upper diagram). Alternatively, the enzyme may be cross-linked by bifunctional reagents to form an insoluble structure (lower diagram).
As an example of *adsorption*, one of the possible combinations of various types of binding is shown, namely adsorption plus cross-linking. *Entrapment* can be in a water-insoluble gel matrix (upper diagram) or in microcapsules (lower diagram). The use of a special reaction vessel that enables the exchange of materials across a semipermeable membrane also is an example of entrapment

second is the detection of low concentrations of analytes using adsorbed antibodies; and the third, from clinical chemistry, involves the analysis of certain parameters with test strips onto which the necessary enzymes are bound by entrapment.

The selected examples thus come from both bioengineering and analytical applications and are categorized according to the method used to immobilize the enzymes or antibodies. The classification is based on the type of chemical or physical treatment of the enzyme (Figure 1). If the enzymes are *bound*, functional groups on the enzyme surface are used for covalent or adsorptive bonding; if they are *included*, they can be used in their native form without any surface treatment. In principle, however, chemically modified soluble enzymes such as those described by SCHMID [4] may also be immobilized. The enzymes can be derivatized or cross-linked by chemical reagents and can be endowed with improved properties for the particular use while still in soluble form.

The classification of immobilization methods used here is derived from that of ZABORSKY [1] and is shown schematically in Figure 1. An example of each type is described below.

2 Covalently bound penicillin G amidase

Immobilization by covalent binding is used in analyses and determinations, particularly with biosensors [3, 5] and with immunological measuring methods (Section 3); in this latter area, adsorption and entrapment techniques compete strongly with covalent attachment.

Bioengineering processes frequently use immobilized biocatalysts [6], the most important examples being glucose isomerase and penicillin amidase [7]. Glucose isomerase is used either in the form of carrier-fixed microorganisms, generally cross-linked to an added inert material; alternatively, the isolated enzyme is adsorbed onto a support material and also cross-linked [8]. Penicillin amidase will be described in detail as an example of a biocatalyst, in the form of a covalently bound enzyme, that is used in engineering. Particular attention will be given to the course of the reaction and the special features of this enzymatic hydrolysis reaction.

2.1 Special features of the enzymatic hydrolysis by penicillin G amidase [9]

As the reaction proceeds in the hydrolysis direction (shown in Figure 2), phenylacetic acid is produced and the reaction solution is thus constantly acidified.

Even an extremely low fraction of product formed causes the acidification, since penicillin solutions in bioengineering applications are often about 0.2 M: the formation of only 10^{-6} M phenylacetic acid, corresponding to a turnover of $5 \times 10^{-4}\%$, leads to a proton concentration of 10^{-6} M and thus a pH value of 6. Further acidification leads to an equilibrium between forward and back reactions at a pH of between 4 and 5, and the hydrolysis practically comes to a standstill. Clearly, the reaction must proceed under

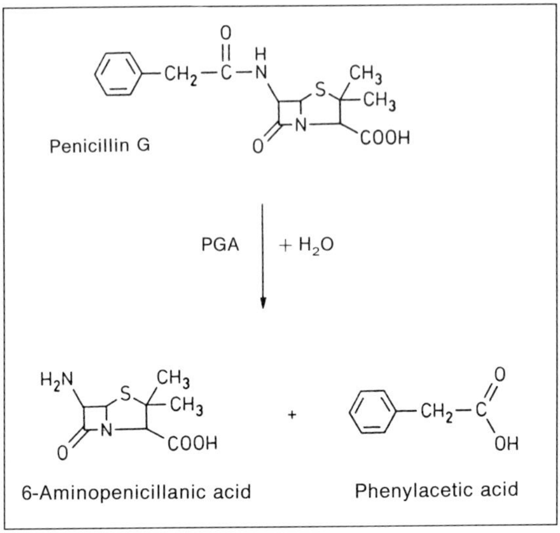

Penicillin G

PGA $+ H_2O$

6-Aminopenicillanic acid + Phenylacetic acid

Fig. 2
Hydrolysis of penicillin G
Penicillin is used as the soluble potassium or sodium salt. At the reaction conditions, the phenylacetic acid produced is in the fully dissociated form

constant, and the most efficient possible, neutralization. A pH of 8 is usually chosen as the most suitable; at that pH value the thermodynamic equilibrium of the hydrolysis lies at 97–98% – depending on the other reaction conditions – and the desired reaction product, namely 6-aminopenicillanic acid (6-APA), is stable in the reaction solution for several hours without significant losses.

The two products (6-APA and phenylacetic acid) cause substantial product inhibition, in addition to the inhibition arising as a result of the acidification during the reaction. Both products have inhibition constants in the millimolar range. Their steady accumulation leads to a drastic decrease in the reaction rate even at the start of the hydrolysis (Figure 3a, soluble enzyme). The reaction is thus controlled by product inhibition.

If a carrier-fixed enzyme is used, the effect of mass transfer must also be considered. The most significant factor is the influence of diffusion through the pores; diffusion depends on the particle size, the pore diameter, and the activity of bound enzyme. However, with particles of normal size (< 0.25 mm), the effect of substrate concentration should be insignificant, since the substrate concentration is about a factor of 200 larger than the Michaelis constant, K_M. This is the same factor by which the efficiency of the reaction [10] should compare with that of the soluble enzyme.

If enzyme activity is determined with an analytical test system in which only the initial activity at low substrate consumption and equally low product formation is measured, the obtained activities depend on the particle size and enzyme loading (Table 1). This is a clear indication that mass transfer is important in this system. If buffer is added, there is a marked increase in enzyme activity, indicating that catalyst particles are acidified. Buffer can increase proton transfer 10-fold [11] and can considerably reduce the time taken to reach steady state, which otherwise may be as long as several minutes, depending on the particle size [12]. If, however, a comparison is made

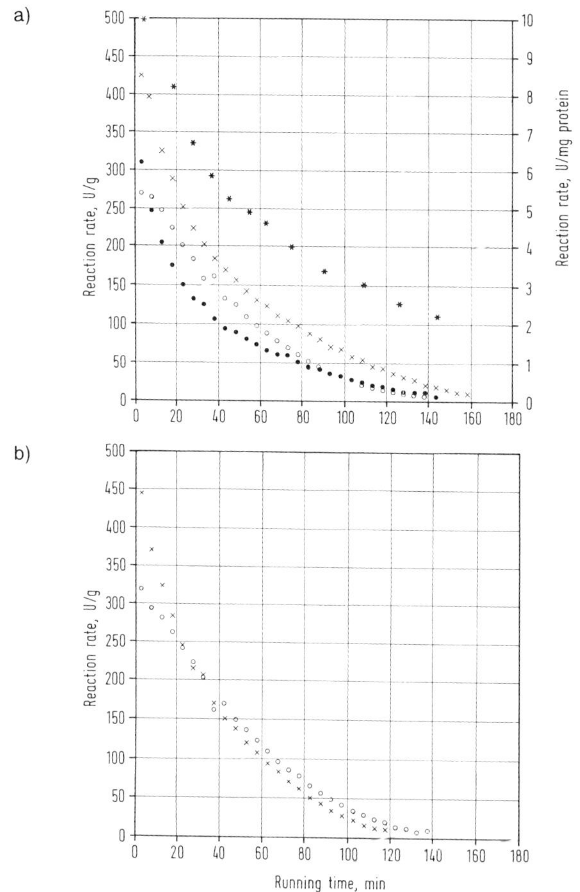

Fig. 3 Rate of reaction of hydrolysis of penicillin G in a laboratory reactor
Penicillin solution, approximately 0.2 M, is enzymatically hydrolyzed under conditions corresponding to those used in an industrial controlled reaction. The enzyme activity is continuously measured during the course of the reaction, and is expressed as activity of the dried carrier enzyme in U/g or of the soluble enzyme (U/mg protein). The reaction is stopped when the added penicillin has been 97% consumed, except for the run with soluble enzyme at 5000 U/l, which was stopped before that end point had been reached.

The individual curves correspond to the following starting materials:

⋆–⋆–⋆: soluble penicillin G amidase, 5000 U/l;
●–●–●: soluble penicillin G amidase, 12000 U/l;
×–×–×: carrier-fixed penicillin G amidase with a particle size $\leq$ 0.25 mm; in Figure 3a, 5000 U/l; in Figure 3b, initial weight the same as for the particle size $\geq$ 0.4 mm;
o–o–o: carrier-fixed penicillin G amidase with a particle size of $\geq$ 0.4 mm, 5000 U/l.

The figures for enzyme activity are analytically determined values

Table 1 The influence of particle size and addition of phosphate buffer on activity of penicillin G amidase. The enzyme activity was determined with an analytical test system using the pH-Stat method [9]

Particle size, mm	Phosphate concentration, mM	Enzyme activity, U/g, with	
		Standard quality enzyme	Enhanced quality enzyme
0.1–0.25	0.5	218	322
	200	380	671
0.4–0.63	0.5	145	241
	200	305	492

between the courses of the reaction with and without addition of buffer, under conditions comparable to those of an industrial controlled reaction (see Section 2.2), virtually no differences in the progress of the reaction are observed. Similarly, the use of different particle sizes leads to only very limited differences if the activity of added enzyme is normalized to the actual activity present (intrinsic activity) and not to the activity calculated from analytical results. Only the activity in the initial phase, which is markedly higher with smaller particles than with larger ones, differs (Figure 3b). Thus under industrial conditions the mass transfer effect is observed only until a steady state is reached, or until the concentration of products is high enough to inhibit the reaction rate.

2.2 Simulated industrial control of the hydrolysis

Because of the need for continuous efficient neutralization and the limited stability of 6-APA, satisfactory operation is most readily achieved in a technically simple manner by a batch process in a well-stirred reactor vessel equipped with a bottom filter screen that retains the spherical, porous catalyst particles when the vessel is emptied. The catalyst must satisfy the following requirements:

- high productivity, lasting for several hundred runs of the batch process;
- high mechanical strength, with good elasticity and low abrasion;
- a support stable in the presence of the added base and of microbial impurities, and, if necessary, able to withstand decontamination procedures;
- a carrier material that is easy to transport and store, under the user's conditions, even with interruptions in the production process.

2.2.1 Reactor vessel and process control

In the following paragraphs I describe the requirements for the reactor vessel and the control arrangements for round-the-clock operation on a laboratory scale, including the specially developed and hence most suitable catalyst.

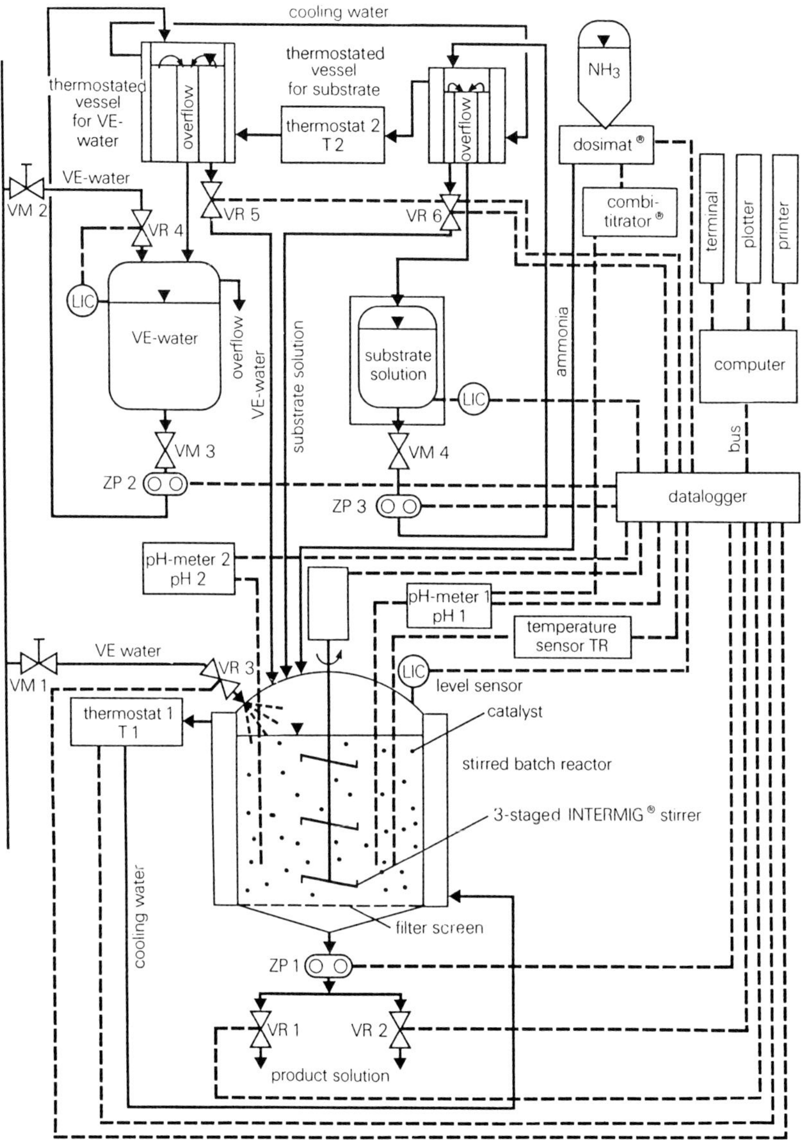

Fig. 4 Laboratory system for the enzymatic hydrolysis of penicillin
VM, VR = mechanically or electrically actuated control valves;
ZP = metering pumps;
dosimat®, combi-titrator® = titration apparatus (Metrohm);
LIC = level indicator and controller

The reactor vessel has a working volume of 1 or 3 l of penicillin solution (Figure 4). For optimal mixing, baffles and a 2- or 3-stage Intermig stirring system (Ekato, Schopfheim) are incorporated. The bottom of the vessel is fitted with a nylon or stainless steel filter screen with mesh openings in the order of 40 to 80 μm. To monitor pH, electrodes are incorporated into the upper and lower third of the reactor vessel, with the upper one as a back-up only. The lower pH electrode is coupled to an NH_3 titration apparatus (Metrohm, Filderstadt) that controls the addition of base and shows the volume added.

The required reagents are fed in from temperature-controlled (thermostated) tanks, arranged as overflow vessels, so that the volumes of deionized water and cooled, highly concentrated penicillin solution required can be maintained in constant volume, independent of the performance of feed pumps.

The required pumps and control valves are regulated by a process controller (Texas Instruments 9900 or HP 9000 & HP 38525) as shown in Figure 4.

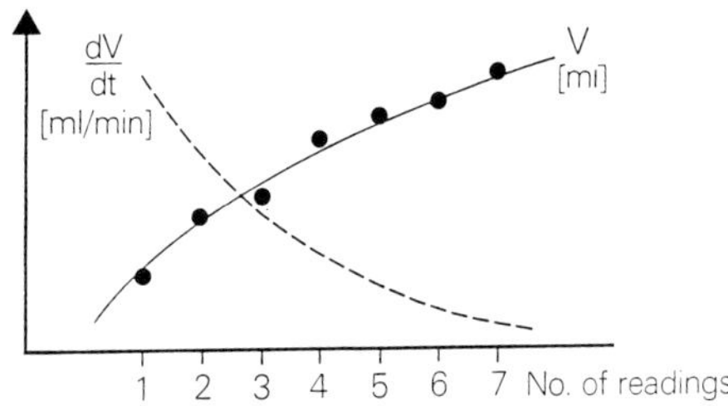

Fig. 5 Calculation of the reaction rate
The volume of ammonia V that is added over the reaction time t is periodically read. A set number of readings is totalled and used for the calculation of a compensating parabola ($V = a_0 + a_1 t + 2a_2 t$, ——). If this function is mathematically differentiated, the instantaneous reaction rate is obtained ($dV/dt = a_1 + 2a_2 t$, ---)

The most important parameter to monitor is the ammonia consumption. The amount used is periodically read from the titration apparatus and smoothed by a compensating parabola (Figure 5). Since the parabola gives the volume used as a function of reaction time, the reaction rate is obtained by mathematical differentiation. Thus a record of the reaction rate is continuously available and can be used to determine the initial rate, furthermore to monitor the course of the reaction, and, through connections to the various valves and pumps, to make any necessary alterations and to terminate the reaction. Control of the reaction is therefore achieved by monitoring not the amounts, but the rates. This seems logical, since it avoids unnecessary variations in the course of the reaction resulting from premature or delayed slowing of the reaction because of variations in the concentrations of ammonia and penicillin. It is worth noting that at the end of the reaction there is always a small residual consumption of base as a result of CO_2 entering from the air and of nonenzymatic hydrolysis and breakdown reactions of the reacting mixture.

Figure 6 is a simple flow chart of the course of a batch process. Emptying the vessel takes about 3 min, and the refilling time is 1–3 min, giving a total dead time of about

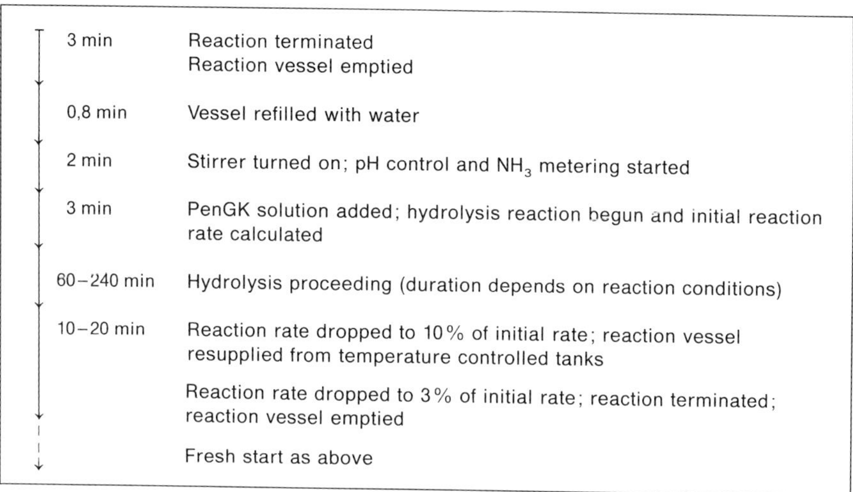

Fig. 6 Time course of penicillin hydrolysis in a batch reactor

5 min per batch. Three minutes after the start of the hydrolysis reaction (beginning with the inflow of penicillin solution) a reaction rate is measured and calculated; this is defined as the initial reaction rate and serves as a reference parameter for further control steps. The curves shown in Figure 3 were obtained from a pilot series of replicated experimental runs with starting materials at the stated conditions; congruent curves enabled us to be certain we had a stable system. HPLC [13] was used to determine both the residual penicillin content in the starting materials and the products obtained, as well as for establishing the shutdown conditions (the termination of the reaction depended on the prevailing reaction rate).

The enzyme activities are given in units of μmol NH_3 consumption per min at 28 °C and with 5% PenGK [13].

2.3 Catalysts

The experiments reported here were carried out with penicillin G amidase Enzygel (Boehringer Mannheim GmbH) [13]. The standard quality has an activity of > 150 U/g dried catalyst (measured at 28 °C with 5% penicillin solution). The particle size is between 100 and 600 μm, measured by sieve analysis in the dry state. 1 g of dry catalyst swells in the wet state to about 6–7 cm^3, which is to say, the particle radius roughly doubles. The particles show only a small proportion of fines: in the wet state, only about 1% of the particles are less than 250 μm. Generally 5000 U of this carrier-bound enzyme is added to a volume of 1 l of penicillin solution.

3 Adsorbed antibodies

Adsorption to immobilize microorganisms has been applied on an industrial scale only in sewage treatment plants, where activated charcoal, microporous glass, and particles of plastic foam are used [14]. Adsorption to fix enzymes for bioengineering purposes is worthwhile if a cheap carrier material and simple binding methods lead to a stable product with high activity, and also if the materials and techniques are nonhazardous and can be used for foodstuffs too. Because desorption often occurs more readily than with covalent binding, adsorption may also be combined with cross-linking of the enzyme. The binding of lipase to Celite® [15] and of aminoacylase to tannin-modified cellulose [16] are two examples of the adsorption of enzymes.

The binding of antibodies to plastic surfaces (tubes, rods, beads, latex particles, and microtiter plates are just a few examples) by adsorption is widely used in the analytical field. The antibody molecules thus bound are capable of forming a complex with the associated antigens, which are generally the analytes of interest. By this means, the analyte is selectively bound to the plastic surface from a mixture of many substances, such as blood serum, and so is removed from the reaction solution. In a further immune reaction, the analyte is then bound to an enzyme that induces a photometrically evaluable color signal in the biocatalyst of interest. Adsorption plays a key role in the so-called ELISA (*enzyme-linked immunosorbent assay*), which is described in detail below.

3.1 Special requirements of adsorption in analytical measuring systems

Purely adsorptive binding has various advantages and disadvantages [17, 18]; as is the case with most applications of immobilized enzymes and proteins, these must be viewed in conjunction with the requirements of the system as a whole. In terms of the immobilization itself, the simplicity of the process for making adsorbed proteins is a great advantage. In contrast to the expensive and time-consuming multistage treatment and activation of the plastic surface that may be necessary with covalent binding, adsorption involves merely soaking or filling and then incubating an antibody solution.

The disadvantage of purely adsorptive binding is the ready desorption (compared to covalent binding), which may lead to problems in the sensitivity and reproducibility of the measuring system [19].

If an antibody or an enzyme is to be adsorptively bound to a plastic surface, then the bound molecule must, in principle, be able to do two things. Firstly, the functional groups on the surface of the protein must bind firmly to the plastic surface; and secondly, the bound molecule must be capable of exercising its specific function, namely to bind antigens or convert substrates. Since the structure of protein molecules varies so widely, these two properties are not always optimally present, and it is not unusual to find tightly bound proteins that are not active, or poorly bound proteins that maintain their full functional capability. There are instances in which neither one is achieved, and these occurrences are becoming increasingly common in immunological analysis, largely because monoclonal antibodies are being used so extensively [20]. In contrast to the previously employed polyclonal antibodies, there is no longer a mixture of different molecular species; instead, the molecules are uniform, with individually different properties. This situation is similar to that in enzymology.

In spite of these disadvantages of adsorptive binding, it is easy to see why we continue to search for ways of improving the binding and functional capacity of proteins. In addition to the possibilities of using the other types of binding described elsewhere in this chapter (Figure 1) and of combining them, it is also possible to pretreat the protein so that its binding capacity is improved and its function is maintained. This technique should have the advantage that the necessary chemical pretreatment of the protein occurs in the easily controlled aqueous phase, in a small volume, and the resulting treated protein should still be easy to characterize and to bind to the carrier material. In Section 3.3 I describe the manufacture of such a product, after giving a little basic information about adsorption.

3.2 Fundamentals of adsorption of proteins

Adsorptive binding is often regarded with disfavor because it brings about the unwanted nonspecific binding of proteins in chromatography or in immunological test systems, and like effects. Nevertheless, it is useful in protein purification (as hydrophobic chromatography) and in the deliberate adsorption of proteins and enzymes in various applications such as biosensors. For analytical purposes, the binding should be as irreversible as possible (in contrast to what is wanted in chromatographic systems). Strong binding is the result of the following forces [21]:

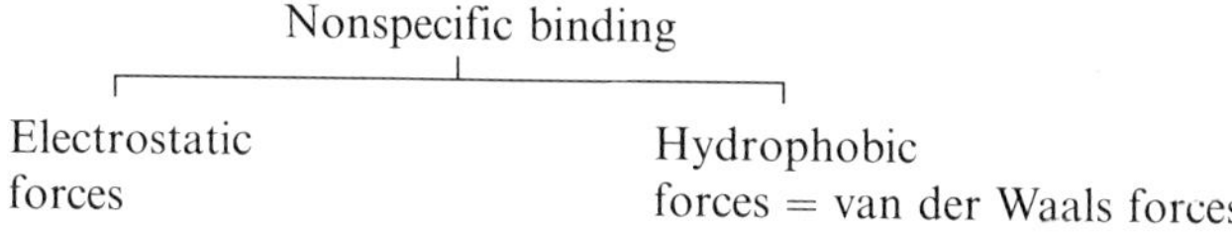

Affected by

Changes in pH;	Shifts in the
Ionic strength;	van der Waals constant
Multivalent ions;	(= Hamaker coefficient),
Formation of complexes.	by alteration of the surface tension.

3.2.1 Principles of protein adsorption [22]

There is no complete theory. The Langmuir adsorption isotherm applies only to substances of low molecular weight, since it deals only with dispersion forces (dipole interactions). The adsorption of polymers involves many other factors that cannot be discussed in detail here.

A unified presentation of protein adsorption is therefore not possible, but a highly simplified picture, which considers the internal coherence (stability of the conformation) of the protein and allows a classification of practical relevance, is useful. The following dichotomous chart is the result.

Loose structure	*Compact structure*
e.g., albumin	*e.g.*, ribonuclease
Partially unfolded, adsorption shows a measurable pH maximum.	Adsorbed as a globule, *i.e.*, in native conformation.
Surface tension plays a minor role, *i.e.*, the nature of the surface has a only a small effect.	The surface tension is important.

3.2.2 *The influence of surface tension* [23, 24]

Surface tension is created not only by liquids (at a gas/liquid interface) but also by solids (in this case, the surface tension at polymer surfaces is of interest) and dissolved substances such as proteins and enzymes.

Usually, plastics such as polystyrene, teflon, nylon, etc., have relatively low surface tensions (40 erg/cm^2), while proteins and aqueous solutions both have similar high values, in the range of 70 to 80 erg/cm^2. I will not attempt a full definition and description of surface tension here; suffice it to say that surface tension and hydrophobicity are correlated [21], with a low surface tension corresponding to a high hydrophobicity.

3.3 Use of an antibody-protein conjugate in immunological analysis

The example is intended to illustrate how a product with improved adsorption properties can be produced and used.

The first task was to seek a type of *attachment protein*, with more hydrophobic properties than required in its antibody function, and which could therefore be expected to show strengthened adsorptive binding. The increased hydrophobicity might be achieved by the selection of natural proteins from a range of hydrophobicities, or by chemical or thermal pretreatment of hydrophilic proteins. If serum proteins are ranked in order of hydrophobicity [25], the albumins are less hydrophobic than the immunoglobulins; they also have a lower molecular weight.

Albumins are not naturally suitable for use as attachment factors for hydrophobic antigens, but when bovine serum albumin is agglomerated by heat treatment, the result has a higher molecular weight and is more hydrophobic than are the immunoglobulins, while remaining adequately water-soluble. An increase in molecular weight may also be achieved through chemical cross-linking, and greater hydrophobicity by derivatization with chemical reagents or partial denaturation with acids, bases or chaotropic reagents [25].

The success of these measures can be monitored using simple chromatographic separation techniques: an increased retention time reflects increased hydrophobicity (Table 2) and hence improved wall adhesion. An attachment protein obtained in this way is then provided with functional groups, such as -SH groups, through derivatization with a reagent containing an -SH group (Figure 7).

Finally, in a manner analogous to the manufacture of enzyme-antibody conjugates, the attachment protein is coupled to the antibody using a heterobifunctional reagent, which reacts both with -SH groups and with amino groups. If the resulting antibody-protein conjugate is adsorptively bound to the walls of polystyrene tubes, the desired effect can be demonstrated by reduced desorption on addition of detergents (not shown here).

Table 2 Retention times of various proteins in hydrophobic interaction chromatography (HIC) The hydrophobicity of individual proteins, whether native or treated chemically or thermally, was determined by HIC as described in [25]

Protein	Retention time t_R, min
F_{ab} ⟨TSH⟩	41.5
BSA	23.2
γ-Globulin	32.0
β-Galactosidase	39.6
Cross-linked β-galactosidase	40.2
Streptavidin	38.3
Heat-treated BSA	cannot be eluted
Cross-linked γ-globulin	cannot be eluted

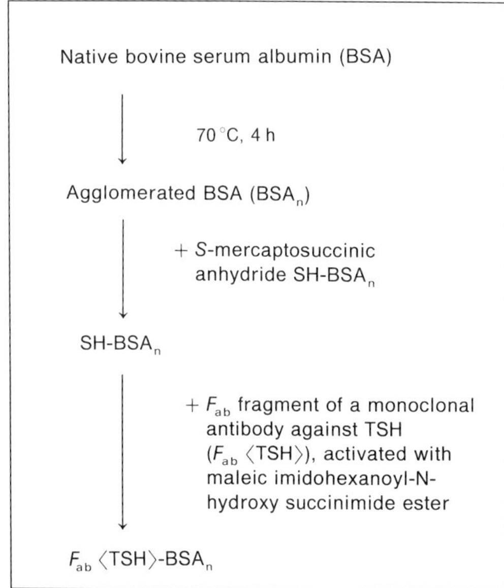

Fig. 7 Production of an antibody-protein conjugate

Furthermore, as can be illustrated by the example of an antibody against TSH (thyroid stimulating hormone) (Figure 8), the standard curve of a conjugate produced by this method has a greater slope than do the curves of antibodies not so treated. We conclude that either more antibody is bound for the same test reading or that the bound antibody has a greater functional capacity.

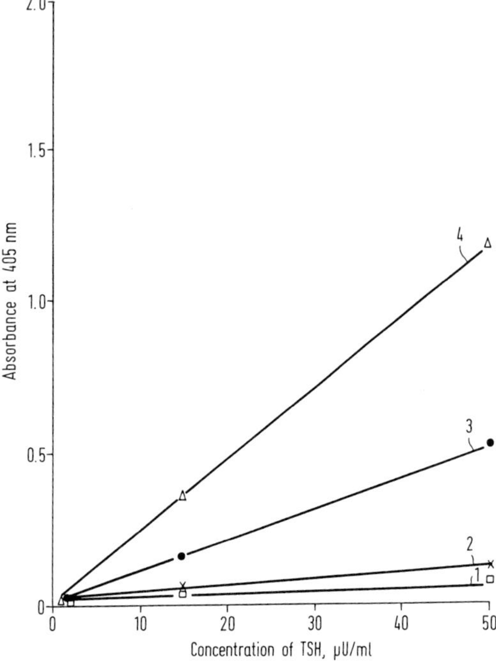

Fig. 8 Calibration curves for the determination of TSH in blood serum
The tubes, coated as described in [27], are used according to the conditions of the TSH-Enzymun® Test (Boehringer Mannheim GmbH). Calibration curves are constructed according to the test protocol, with the following results:
Curve 1: Untreated F_{ab} fragment of a monoclonal antibody to TSH;
Curve 2: Chemically cross-linked F_{ab} fragment;
Curve 3: F_{ab} fragment bound to an untreated globulin
Curve 4: F_{ab} fragment bound to previously cross-linked globulin or agglomerated bovine serum albumin

If instead of an antibody we use a universal binding protein such as avidin or streptavidin, which can enter into an extremely tight, almost irreversible bond with all the substances labeled with its complementary binding partner biotin [26], a plastic surface can be coated with virtually anything. Consequently, the number of possible reactions is very large. For example, medical diagnosis can now include the determination of numerous blood substances such as thyroid hormones, tumor markers, infectious pathogens, and so forth.

3.4 Method for the determination of TSH [27]

TSH (thyroid stimulating hormone) antigen is the analyte of interest in this system. The TSH is selectively immunologically bound in antibody-coated tubes, and selectively binds another antibody in a further immune reaction. An enzyme is chemically coupled to this second antibody. This enzyme, here horseradish peroxidase (POD), catalyzes a color reaction leading to a photometrically detectable color signal. This reaction of peroxidase on the tube surface is the actual biologically catalyzed event in this measuring system. In the second part of the system, the color signal produced is quantified and evaluated. The measurement may be performed in largely automatic photometers, if the number of samples is large.

The course of this reaction follows the so-called *sandwich principle*, and follows the sequence shown in Figure 9:

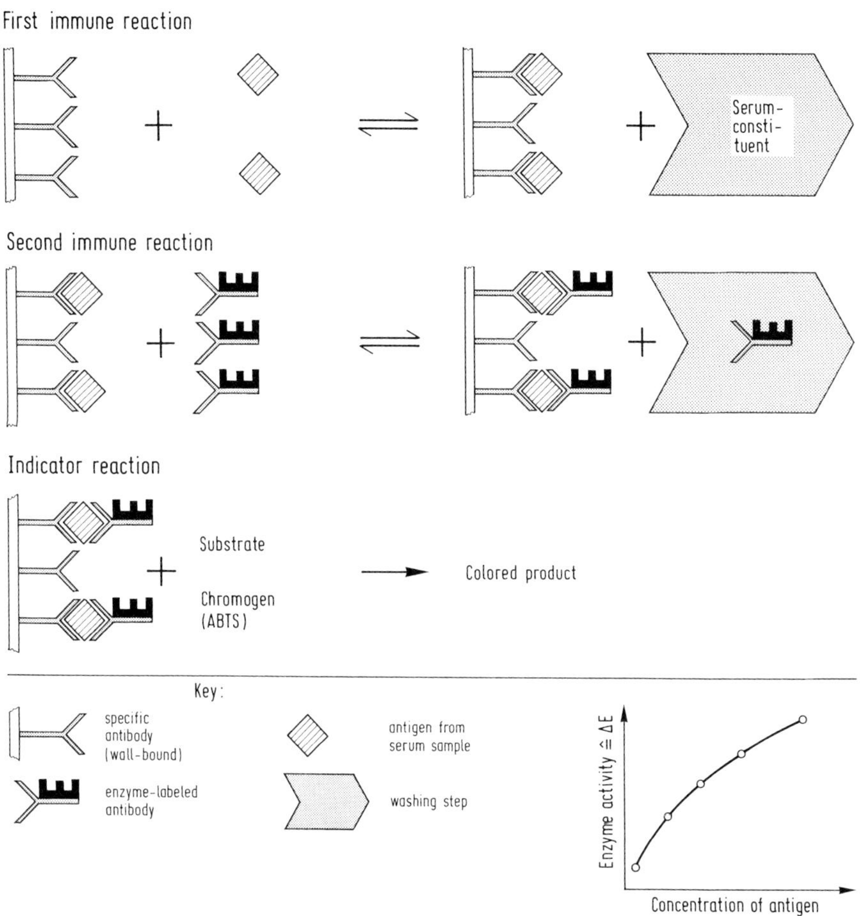

Fig. 9 Determination of TSH by the sandwich principle

In the *first immune reaction* the free serum antigen (TSH) first reacts with the excess antibody that is fixed to the wall of the tube. Serum constituents and any interfering factors are then washed out.

In the *second immune reaction* a sandwich complex is formed with excess specific enzyme-labeled (POD) antibodies: the antigen is bound to antibodies on both sides. The more antigen there is in the sample, the more enzyme-labeled antibodies are bound. The amount of sandwich-enzyme-antibody complex formed is thus an index of the antigen content of the sample. The POD conjugate that did not bind is removed in another separation step.

The *indicator reaction* takes place when substrate and chromogen are added. The activity of the wall-bound POD enzyme is determined photometrically. The intensity of color produced in the indicator reaction is proportional to the quantity of antigen in the serum sample. Therefore the measured enzyme activity increases with increasing antigen concentration. Results are quantified with calibration curves produced with standards of known concentration.

4 Entrapped enzymes on reagent carriers

Entrapment techniques are used both in bioengineering and in analyses. Typical of the former are methods for the immobilization of cells [28] and the incorporation of enzymes in membrane reactors [29]; entrapping of enzymes used as biosensors in dialysis membranes and other materials is an example of the latter [5].

Reagent carrier systems for the measurement of blood substances in clinical chemistry constitute a particular type of analytical application. These reagent carriers, generally known by the term *dry chemistry*, have their origin in a film technique used in photography, in which several reactive layers overlaid on each other produce a color signal that can be quantified by reflectance photometry. Another technique is the result of improvements in the test strips used in the measurement of blood glucose and in various urine tests [30]. Those fixing methods are actually entrapments in another setting, or rather they are compartmentalizations, designed for one-time use, the salient feature of which is that they guarantee a time-determined sequence of reactions. These latter systems are described in detail below.

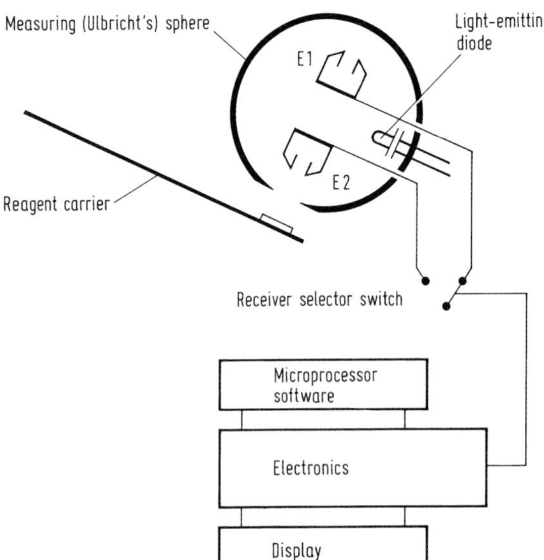

Fig. 10 Reflectance photometric measurement of the color developed on a reagent carrier. The reagent carrier is aligned with the opening of the Ulbricht's sphere (circular structure). A beam of white light from a light-emitting diode is flashed at and reflected from the reagent carrier, is absorbed by the receivers (E1 and E2), and is processed

4.1 Principle of the measurement with a "dry chemistry" reagent carrier [31]

The reagent carrier/test strip combinations mentioned above provide color signals at fixed points on their surface. Such color signals originate in a sequence of reactions of the analyte with color-forming components. The resulting color is quantified by reflectance photometry in a specially developed measuring device. The optics used are based on what is known as an *Ulbricht's sphere* (Figure 10) containing light-emitting diodes (LEDs). These LEDs emit light pulses, which are reflected from the test field of the reagent carrier into the cavity of the Ulbricht's sphere. The amount of reflected light is proportional to the concentration of analyte in the sample and is converted into a digitally displayed value by the electronics incorporated in the measuring device.

4.2 Construction and operation of a reagent carrier [31]

The reagent carriers are used to quantify various parameters of importance in the clinical chemistry of whole blood, serum, or plasma. The reagent carriers do this by performing the following four functions, the reading and transmission of the information being the fifth step:

1. Plasma procurement
2. Plasma transport
3. Preincubation
4. Reaction and color formation
5. Information transfer (magnetic tape)

The details of construction of the reagent carrier/test strip are shown in Figure 11 a, and the operation of the system in Figures 11 b–11 d. The first four functions enumerated above will be discussed with reference to the figure.

Procuring the plasma sample

Since plasma constituents are the participants in the later reaction, the first step is a complete separation of the erythrocytes from the plasma. The separating media are arranged so that no hemolysis occurs. Glass fiber mats of a certain fiber strength, capable of quantitatively retaining erythrocytes, were found to be the most suitable material for performing the separation.

Blood samples of a given volume may not have the same volume of plasma (since they may have different hematocrit readings). Therefore the test system must be designed so that it provides a constant volume of plasma: on the one hand, an adequate amount of plasma is available for the chemical reaction, and, on the other, excess plasma is disposed of automatically.

Plasma transport

This is achieved by the *transport pad.* Plasma separated by the barrier layer passes through this porous pad into the reaction region below, where it forms a plasma

a

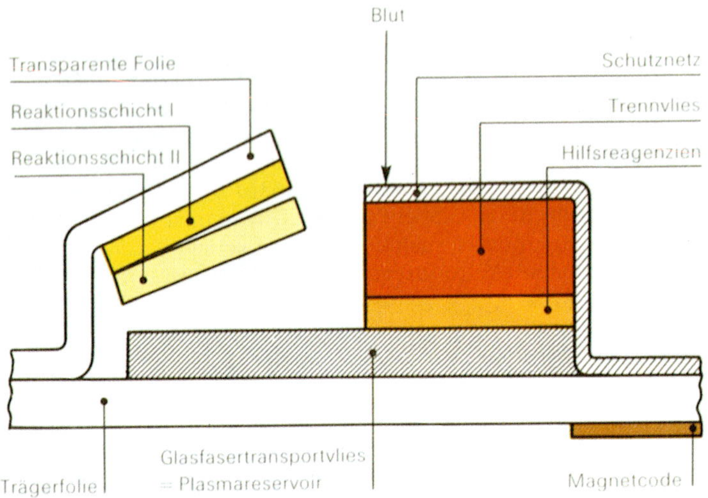

b

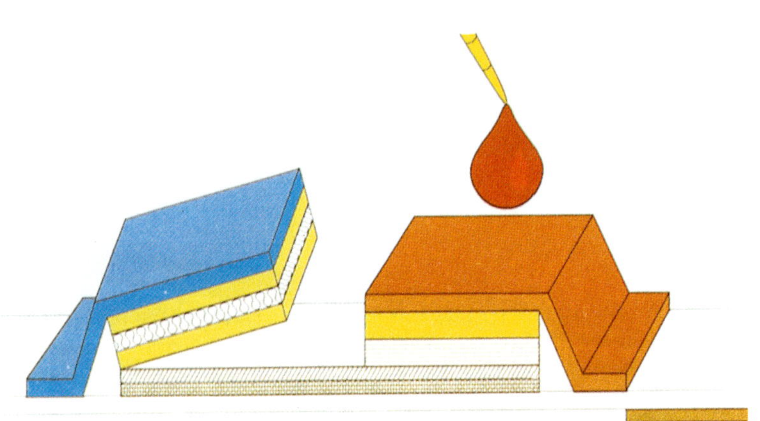

▲

Fig. 11 Construction and function of a reagent carrier ▶

(a) *Details of the construction of a reagent carrier/test strip system*
(Transparente Folie: Transparent film; Reaktionsschicht I: Reaction layer I; Reaktionsschicht II: Reaction layer II; Blut: Blood; Schutznetz: Protective mesh; Trennvlies: Barrier layer; Hilfsreagenzien: Auxiliary reagents; Trägerfolie: Backing layer; Glasfasertransportvlies = Plasmareservoir: Glass fiber transport pad = plasma reservoir; Magnetcode: Magnetically coded strip)

(b) *Application of the blood sample*

(c) *Separation of plasma*

(d) *Initiation of the color reaction* (400 g)

(e) *Measurement of the color intensity*

c

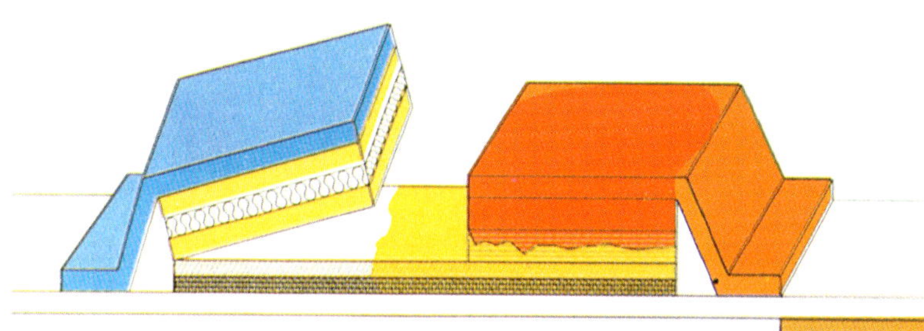

d

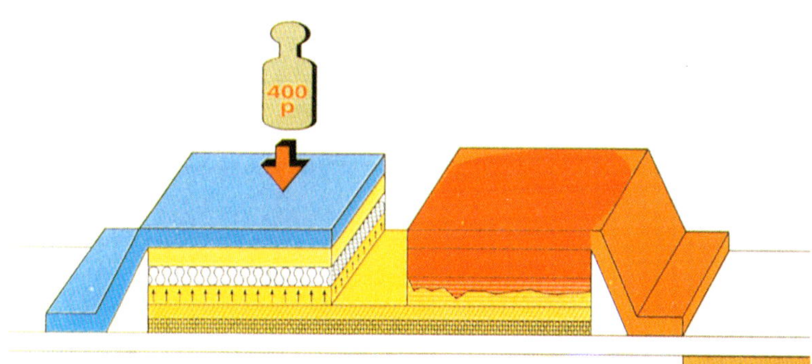

e

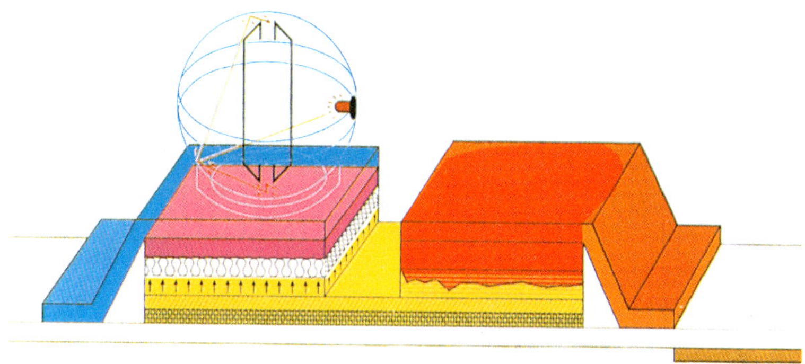

reservoir. The reaction is started by pressing the reaction region into the plasma reservoir and removing the requisite volume of sample. The excess plasma is pushed outside the reaction zone.

By this means it is possible to apply various sample materials such as blood, plasma, and serum onto the protective mesh, and still achieve an accurate plasma sample volume of 30 μl, within a combined tolerance of $\pm 5\%$ (28.5–31.5 μl).

Preincubation

The interim storage of the plasma has further advantages:

(a) The temperature of the sample is adequately stabilized before the main reaction is started;
(b) The reaction can be started at a specified time. This occurs when the reagent paper is dipped (pressed) into the plasma reservoir;
(c) Any interfering substances can be eliminated by reactions in the space below the separating system (examples include ascorbic acid, removed by ascorbic oxidase, and endogenous creatine, which would interfere with the reaction of the creatinine test)

The analysis begins, as shown in Figure 11, when 30 μl of whole blood is applied to the red-marked field with a suitable pipette (Figure 11 b). The erythrocytes are separated by the glass fiber barrier layer. Interfering substances can be separated, or reagents (such as EDTA) can be added to the plasma. The plasma then passes into the transport pad (Figure 11 c).

Course of the reactions

It generally takes about 1 minute to procure the plasma sample and complete any preliminary reactions, and then the preincubation phase ends. The reagent layers impregnated with chemicals (they appear dry) are pressed together at a specified time to start the parameter-specific reactions (Figure 11 d).

The intensity of the color can be measured (Figure 11 e) at timed intervals (kinetic measurement) or after the reaction has ended (end-point measurement). The color of the product is a function of the concentration.

4.3 Example of a determination using the reagent carrier/test strip method

The cholesterol content of blood is a characteristic and clinically important parameter, of great significance as a risk factor. The measurement of cholesterol proceeds by a chain of enzymatic reactions according to the following scheme:

$$\text{cholesterol ester} + H_2O \xrightarrow[\text{hydrolase}]{\text{cholesterol ester}} \text{cholesterol} + \text{RCOOH}$$

$$\text{cholesterol} + O_2 \xrightarrow[\text{oxidase}]{\text{cholesterol}} \Delta^4\text{-cholestenone} + H_2O_2$$

$$\text{3,3}',\text{5,5}'\text{-tetramethylbenzidine} + H_2O_2 \xrightarrow{\text{peroxidase}} \text{blue color}$$

The required enzymes are in the reaction layers (Figure 11) as a film on a paper-like carrier material.

The result of the measurement, combined with another test, namely the triglyceride determination, is used to make clinical recommendations (Table 3).

Table 3 Clinical recommendations at various serum cholesterol and triglyceride levels

	Treatment required No	 Depending on values of HDL- and/or LDL-cholesterol	 Yes
Cholesterol	< 200 mg/dl < 5.2 mmol/l	200–300 mg/dl 5.2–7.8 mmol/l	> 300 mg/dl > 7.8 mmol/l
Triglycerides	< 200 mg/dl < 2.3 mmol/l	> 200 mg/dl > 2.3 mmol/l	> 500 mg/dl * > 5.7 mmol/l *

* Because of the danger of pancreatitis

This system is useful for the measurement not only of clinically important parameters but also of other variables in a wide range of fluids. For example, a very simple glucose measurement system has been successfully demonstrated. Glucose in the medium can be measured during the course of a fermentation after a simple pipetting step within the normal reaction time of 3 minutes, without separating the cells. In this case it is necessary to draw a calibration curve that takes into account the composition and cell content of the medium (Figure 12).

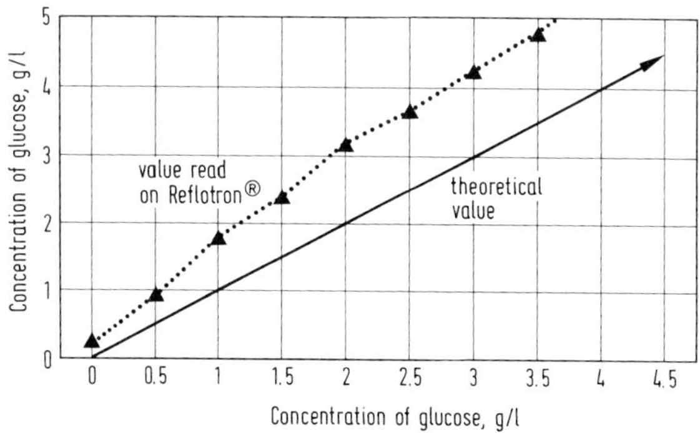

Fig. 12 Determination of the glucose content in a fermentation medium
The calibration curve is plotted from a reference measurement (——) and a measurement with a reagent carrier (▲ ... ▲). The samples were from a yeast fermentation, and were added to the test system without cell separation or other pretreatment

5 References

[1] Zaborsky, O. R.: *Immobilized Enzymes*, CRC Press, Cleveland,Ohio (1973); pp 1–3.

[2] Zlokarnik, M.: Verfahrenstechnische Aspekte der Immobilisierung – eine Einführung in die Problematik. In: *GVC VDI-Gesellschaft Verfahrenstechnik und Chemieingenieurwesen* (Ed.): Preprints "*Verfahrenstechnische Aspekte der Immobilisierung von Enzymen und ganzen Zellen*", GVC Symposium May 9–10, 1988, Heidelberg, FRG; pp 1–9.

[3] Mosbach, K., Ed.: *Immobilized Enzymes and Cells*, Part D; *Meth. Enzymol.* **137**, Academic Press, San Diego (1988).

[4] Schmid, R. D.: Stabilized Soluble Enzymes; *Adv. Biochem. Eng.* **12** (1979) 41–118.

[5] Barker, S. A.: Immobilization of the Biological Component of Biosensors. In: Turner, A. P. F.; Karube, I.; Wilson, G. S., Eds.; *Biosensors, Fundamentals and Applications*, Oxford University Press (1987); pp 95–99.

[6] Gestrelius, S.; Mosbach, K.: Overview – Enzyme Engineering (Enzyme Technology); *Meth. Enzymol.* **136** (1987) 353–356.

[7] Buchholz, S.; Klein, J.: Characterization of Immobilized Biocatalysts; *Meth. Enzymol.* **135** (1987) 3–30.

[8] Jensen, V. J.; Rugh, S.: Industrial-Scale Production and Application of Immobilized Glucose Isomerase; *Meth. Enzymol.* **136** (1987) 356–370.

[9] Tischer, W.; Botsch, J.; Gloger, M.; Kresse, G.-B.: Trägerfixierte Penicillin G-Amidase: Verhalten unter produktionsnahen Bedingungen. In: Ref. [2], pp 19–45.

[10] Gloger, M.; Tischer, W.: Determination of the Catalytic Activity of Immobilized Enzymes. In: Bergmeyer, H. U., *et al.*, Eds.; *Methods of Enzymatic Analysis*, Verlag Chemie, Weinheim (1983); Vol. I, 142–163.

[11] Engasser, J. M.; Horvath, C.: Buffer-Facilitated Proton Transport. pH Profile of Bound Enzymes; *Biochim. Biophys. Acta* **358** (1974) 178–192.

[12] Miyawaki, O.; Wingard, Jr., L. B.: Concentration Step Determination of Diffusion and Partition Coefficients of Ferricyanide for Membrane-Coated Rotating Disk Electrode; *Biotechnol. Bioeng.* **31** (1988) 179–182.

[13] Product description: *Penicillin G-Amidase*, Boehringer Mannheim GmbH, 6800 Mannheim 31.

[14] Zlokarnik, M.: Immobilisierung ganzer Zellen – eine Bestandsaufnahme aus bioverfahrenstechnischer Sicht; *BTF – Biotech Forum – Intern. Zeitschrift für Biotechnologie* **3** (Number 4) (1986).

[15] Yamanaka, S.; Tanaka, T.: Regiospecific Interesterification of Triglyceride with Celite®-Adsorbed Lipase; *Meth. Enzymol.* **136** (1987) 404–411.

[16] Watanabe, T.; Mori, T.; Tosa, T.; Chibata, T.: Immobilization of Aminoacylase by Adsorption to Tannin Immobilized on Aminohexyl Cellulose; *Biotechnol. Bioeng.* **21** (1979) 477–486.

[17] Parsons, G. H.: Antibody-Coated Plastic Tubes in Radio- immunoassay; *Meth. Enzymol.* **73** (1981) 224–239.

[18] Herrmann, J. E.: Quantitation of Antibodies Immobilized on Plastics; *Meth. Enzymol.* **73** (1981) 239–244.

[19] Lehtonen, O. P.; Viljanen, M. K.: Antigen Attachment in ELISA; *J. Immunol. Meth.* **34** (1980) 61–70.

[20] Seiler, F. R.; Gronski, P.; Kurrle, R.; Lüben, G.; Harthus, H.-P.; Ax, W.; Bosslet, T.; Schwick, H.-G.: Monoklonale Antikörper – Chemie, Funktion und Anwendungsmöglichkeiten. In: Präve, P.; Schlingmann, M.; Crueger, W.; Esser, K.; Thauer, R.; Wagner, F., Eds.; *Jahrbuch Biotechnologie Band 2*, Carl Hanser Verlag, München, Wien (1988/89); pp 209–262 (in German). Seiler, F. R.; *et al.*: Monoclonal

Antibodies – Their Chemistry, Function, and Possible Uses. In: FINN, R.K.; PRÄVE, P.; SCHLINGMANN, M.; CRUEGER, W.; ESSER, K.; THAUER, R.; WAGNER, F., Eds.; *Biotechnology Focus 2*, Hanser Publishers, Munich, Vienna, New York, (1988/89); pp 201–250 (in English).

[21] VAN OSS, C. J.; ABSOLOM, D. R.; NEUMANN, A. W.: Role of Attractive and Repulsive van der Waals Forces in Affinity and Hydrophobic Chromatography. In: GRIBNAU, T. C. J.; VISSER, J.; NIVARD, R. J. F., Eds.; *Affinity Chromatography and Related Techniques*, Elsevier Scientific Company, Amsterdam (1982) 29–37.

[22] NORDE, W.: Adsorption of Proteins from Solution at the Solid-Liquid Interface; *Adv. Colloid Interface Sci.* **25** (1986) 267–340.

[23] VAN OSS, C. J.; ABSOLOM, D. R.; NEUMANN, A. W.; ZINGG, W.: Determination of the Surface Tension of Proteins. I. Surface Tension of Native Serum Proteins in Aqueous Media; *Biochim. Biophys. Acta* **670** (1981) 64–73.

[24] JAMES, L. K.; AUGENSTEIN, L. G.: Adsorption of Enzymes at Interfaces: Film Formation and the Effect on Activity; *Adv. Enzymol.* **28** (1966) 1–40.

[25] TISCHER, W.; MAIER, J.; DEEG, R.: *Verfahren zur Bestimmung einer spezifisch bindefähigen Substanz*; Eur. Pat. Appl. 0,269,092 (1987).

[26] WILCHEK, M.; BAYER, E. A.: The Avidin-Biotin Complex in Immunology; *Immunology Today* **5** (1984) 39–43.

[27] *Enzymun-Test TSH, Klinik und Methode*; Publ. by the Scientific Department, Diagnostics, Boehringer Mannheim GmbH, 6800 Mannheim 31; *Enzymimmunoassays nach dem ELISA-Prinzip, Grundlagen und Anwendung*; Publ. by the Scientific Department, Diagnostics, Boehringer Mannheim GmbH, 6800 Mannheim 31.

[28] KLEIN, J.; VORLOP, K. D.: Immobilisierung von ganzen Zellen. In: CRUEGER, W., *et al.*, Eds.; *Jahrbuch Biotechnologie* **1986/87**, Carl Hanser Verlag, Munich/Vienna (1986) 369–380.

[29] KULA, M. R.; WANDREY, C.: Continuous Enzymic Transformation in an Enzyme-Membrane Reactor with Simultaneous NADH Regeneration; *Meth. Enzymol.* **136** (1987) 9–21.

[30] GREYSON, J.: Analytische Chemie auf trockenen Reagenzträgern – Probleme und Möglichkeiten; *Das Medizinische Laboratorium* **34** (1981) 209–215.

[31] *Reflotron® zur quantitativen Bestimmung von Parametern der klinischen Chemie – eine wissenschaftliche Produktinformation von Boehringer Mannheim*; FAHLBUSCH, R.; Publ. by Scientific Department, Diagnostics, Boehringer Mannheim GmbH, 6800 Mannheim 31. *Aufbau und Chemismus von Reagenzträgern in der Reflometrie*; WERNER, W.; Boehringer Mannheim GmbH, 6800 Mannheim 31; in English language available.

Microbial Corrosion *

by F. WIDDEL

Contents

Prof. Dr. F. WIDDEL,
Institut für Genetik und Mikrobiologie,
Universität München,
Maria-Ward-Straße 1a,
D-8000 München 19, Fed. Rep. Germany

* Dedicated to Professor Dr. Norbert PFENNIG on the occasion of his 65th birthday

1 Introduction

Living organisms are partly responsible for the processes of corrosion of rock and metal. Fungi and bacteria are particularly likely to dissolve carbonate rock because they often excrete organic acids. Rock can be dissolved by inorganic acids if ammonia is oxidized by nitrifying bacteria to nitric acid, or sulfur by sulfur bacteria to sulfuric acid; in both cases, oxygen from air is the oxidant. Of the types of metal corrosion that are biologically mediated, the one that is most important economically is that of iron at neutral pH in the absence of oxygen. This anaerobic corrosion is usually connected with the activities of sulfate-reducing bacteria, and they are probably the chief culprits. Hydrogen sulfide formed by sulfate-reducing bacteria is, because of its reactivity, an obviously important factor in anaerobic corrosion. Furthermore, the bacteria probably accelerate the cathodic (electron-consuming) reactions on the metal surface in contact with the aqueous surroundings.

Stone, metal, wood, and plant and animal fibers have been used as raw materials for thousands of years. In spite of the many synthetic polymers available, the classical materials remain indispensable, but they are more subject to weathering, aging, or biological damage than are the modern polymers. The deterioration of inorganic materials, namely, rock and metal, is termed *corrosion*; in the case of rock this is almost synonymous with *weathering*, but "corrosion" is a more technical word that applies to man-made objects.

Historical descriptions of protective measures against corrosion, indicating some knowledge of the causative factors, date back two or three thousand years [59, 60, 96, 115]. In LEVITICUS we find mention of patches of deterioration, probably due to lichen growth, on house walls; people dealt with the damage by removing and replacing the affected stones [60, 68]. The planking of Roman galleys was protected from shipworm by lead shielding, fastened on with copper nails. The Roman shipwrights also covered the nailheads with lead [115]; by preventing direct contact between copper and seawater, they precluded the dissolution of the less noble lead in a short-circuited galvanic cell. Not surprisingly, it was not until the 19th century that people began to understand corrosion processes on a physical/chemical level, and to develop and apply protective measures based on this new knowledge [59, 96, 115]. Until very recently, the corrosion of metals, especially of iron, continued to be of much greater interest than the corrosion of rock.

Introducing resistant alloys and plastics could reduce or prevent corrosive processes in many of the places they occur. Nevertheless, corrosion continues to give trouble in several areas. Iron and steel are used in vehicles, heating systems, and refineries, for example, and they cannot always be replaced by stainless materials. Nonferrous metals for alloys are expensive and less abundant than iron; plastics do not resist extreme thermal or mechanical stresses. A type of corrosion that we have begun to grasp fully only in the past decade or so is the deterioration of building stone by acidic or acid-forming atmospheric pollutants; a long-term solution is nowhere in sight.

The corrosion of rock or iron can be a purely chemical process, or it can be mediated by living organisms, especially fungi and bacteria. Those corrosion processes in which biological agents play a part are the subject of this chapter. I am referring to bacteria and fungi (and lichens as well) as microbes for these purposes (this is, of course, not a strict biological definition), and will therefore term their deteriorative effects on rock and metal *microbial corrosion*.

At first blush it seems strange that fungi and bacteria should attack rock and metal; in nature, these organisms consume substances of biological origin. The substrates of bacteria and fungi are thus natural oxidizable materials, mainly organic compounds and, in lesser amounts, ammonia and reduced sulfur compounds. Wood, paper, and other natural fibers can therefore be expected to rot and decay. Rock, however, is made up of oxides (CO_2, SiO_2, MgO, CaO, Al_2O_3, etc.) that are already in their highest oxidation state and cannot yield energy for metabolic processes. It is true that metallic iron readily oxidizes and liberates a great deal of energy if it reacts with oxygen, for example, but, as recently as a few thousand years ago, metallic iron (such as iron meteorites) was extremely rare in the biosphere, and hence is unlikely to be a natural substrate for microorganisms. Indeed, the corrosion of rock and metal by living organisms is much less direct than is the degradation of organic compounds with specific enzymes. Fungi and bacteria break down rock by excreting chemically aggressive metabolic products, mainly acids. These substances react with water-insoluble minerals to yield soluble ones, which do not necessarily serve as nutrients for the organisms living on the rock. Excreted acids can also dissolve metallic iron, but the most important biologically mediated corrosion of iron is a process at neutral pH in the absence of oxygen; in this anaerobic corrosion, bacteria reducing sulfate to sulfide are the key actors.

2 Corrosion of rock

The processes that lead to corrosion of rock are affected by the atmosphere, the weather, the composition of the rock material, and the nature of the organisms living on the rock. It is often unclear whether a rock is dissolving for purely chemical reasons, due to acidic compounds in the atmosphere, or whether the metabolic activity of organisms is involved. Nevertheless, relatively few chemical principles underlie the multifold phenomena of corrosion. When rock breaks down, it is usually being dissolved by acid. Other processes that can solubilize some components of rock include chelation, alkalization, and reduction of iron(III) and manganese(IV) compounds.

2.1 Fundamental chemical reactions and equilibria during corrosion of rock

2.1.1 Formation of inorganic acids from trace gases by chemical reaction

Corrosive inorganic compounds may have that property because they are acidic (they form H^+ ions). Inorganic acids may be formed as the end products of catabolism by certain microorganisms (see Section 2.2) or by purely chemical reactions of trace atmospheric gases. Acid-forming trace gases are CO_2, NO_X, and SO_2.

Dissolved CO_2 is a weak acid:

$$CO_2 + H_2O \rightleftharpoons H_2CO_3 \quad (1)$$

$$H_2CO_3 \rightleftharpoons H^+ + HCO_3^- \quad (2)$$

Only a very small proportion of the dissolved CO_2 (about 1 molecule in 650) is present as the molecular species of carbonic acid (H_2CO_3 or $(HO)_2CO$). Dissolved CO_2 (written as $CO_2(aq)$) and H_2CO_3 are normally treated together as the undissociated acid, and the H_2CO_3 can be neglected in most calculations. A good useful approximation to describe the acidic character is:

$$CO_2(aq) + H_2O \rightleftharpoons H^+ + HCO_3^- \quad pK_{a1} = 6.35\ (25\,^\circ C) \quad (3)$$

From the solubility constant and the dissociation constant, we can calculate [107] that the concentration of CO_2 in air (0.033%) lowers the pH of pure water to about 5.6 in an open system.

Of the nitrogen oxides ($NO_x = NO, NO_2$), only NO_2 disproportionates in water to yield acids [30] :

$$2\,NO_2 + H_2O \longrightarrow 2\,H^+ + NO_2^- + NO_3^- \quad (4)$$

Neutralization shifts the reaction completely to the right.

NO itself does not form an acid; but as concentrated gas NO reacts rapidly with oxygen to form NO_2, in an oxidation that follows third-order reaction kinetics [30]. In air (at constant concentration of O_2), the reaction velocity varies as the square of the NO concentration, which is probably why traces of NO are rather stable in air. Low concentrations are oxidized readily only by ozone [113]. Trace amounts of NO do not usually bind to rock in a humid atmosphere, in contrast to NO_2 (which reacts according to Equation (4)) [7].

SO_2 is oxidized slowly in air to SO_3 [113], which combines with water to make the strong acid H_2SO_4. SO_2 dissolved in water reacts as an acid of medium strength (presumably via free sulfurous acid, H_2SO_3, present at very low concentration) [32]:

$$SO_2(aq) + H_2O \rightleftharpoons H^+ + HSO_3^- \quad pK_{a1} = 1.8 \quad (5)$$

$$HSO_3^- \rightleftharpoons H^+ + SO_3^{2-} \quad pK_{a2} = 7.0 \quad (6)$$

The aqueous solution, especially if neutralized, reacts more rapidly than gaseous SO_2 with O_2; traces of heavy metals accelerate the oxidation significantly [19, 113]. The final result of these reactions is always the net formation of sulfuric acid.

2.1.2 Dissolution of rock minerals

Alkaline earth carbonates such as dolomite, $MgCa(CO_3)_2$, or calcite, $CaCO_3$, are readily dissolved by acids. The dissolution equilibria for $CaCO_3$ are depicted in Figure 1. In pure water without CO_2, the solubility of $CaCO_3$ is 0.12 mmol/l (12 mg/l). Some of the CO_3^{2-} ions consume protons from water as they dissolve, forming HCO_3^- ions and raising the pH to 9.9 [108].

In an open system in air, the final equilibrium is slightly different. CO_2 dissolves and reacts to HCO_3^- and H^+ ions; the latter scavenge CO_3^{2-} ions to yield further

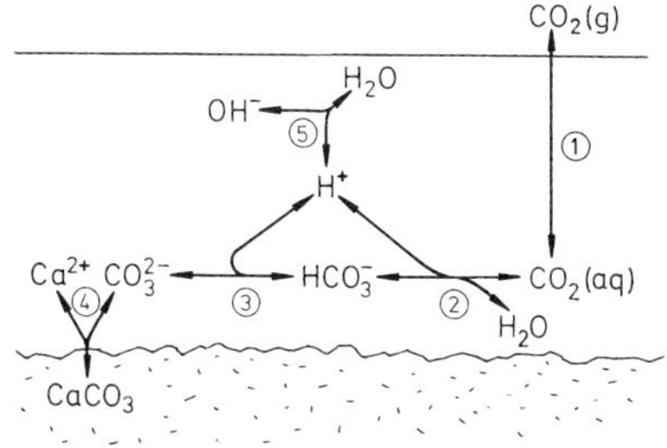

Fig. 1 Equilibrium reactions between limestone (lower, stippled region), aqueous phase (middle region), and air (upper region).
(1) Solubility constant of gaseous CO_2: $K_{CO_2} = 3.35 \times 10^{-4}$ M/kPa.
(2) $K_{a1} = 4.45 \times 10^{-7}$ M.
(3) $K_{a2} = 4.69 \times 10^{-11}$ M.
(4) Solubility constant (solubility product): $K_{CaCO_3} = 3.80 \times 10^{-9}$ M^2.
(5) Dissociation constant of water: $K_{H_2O} = 10^{-14}$ M^2.
All values at 25 °C. From [107, 108]

HCO_3^- ions. Because the CO_3^{2-} ions are removed, the solubility of $CaCO_3$ increases to 0.5 mmol/l (50 mg/l) and the pH at equilibrium is 8.4 [108].

If water in contact with minerals containing calcium carbonate has an equilibrium pH less than 8.4, acids other than atmospheric CO_2 must have acted upon the rock. These additional acids, which may be from fungi and bacteria, increase the Ca^{2+} concentration, according to the overall reaction of the neutralization:

$$CaCO_3 + 2H^+ \longrightarrow Ca^{2+} + CO_2 + H_2O \tag{7}$$

If there are fewer equivalents of acid than of calcite, some $CaCO_3$ remains undissolved. In this case, the pH is determined by the CO_2 content of the air and the Ca^{2+} concentration; these parameters determine, via the solubility product of $CaCO_3$ and the solubility constant for the solution of CO_2 gas in water, the concentration of CO_3^{2-} ions and of dissolved CO_2, respectively; hence, the concentrations of HCO_3^- and H^+ are also determined. For example, an aqueous phase (in air) that has 2 mmol/l (200 mg/l) of dissolved $CaCO_3$ (put into solution by an acid such as HNO_3) has a pH of 8.0, provided undissolved $CaCO_3$ is present. A pH of 7.0, in the presence of undissolved $CaCO_3$, can be achieved if as much as 220 mmol/l (22 g/l) $CaCO_3$ is dissolved [108]; this amount of $CaCO_3$ can neutralize 440 mmol/l (21 g/l) HNO_2 or (28 g/l) HNO_3 produced by denitrifying organisms. Carbonate-containing rocks thus present an ideal buffering environment for bacteria that produce strong acids, such as the nitrifyers or the sulfur bacteria.

Water, whether pure or acidic, does not dissolve the calculated amount of carbonate immediately upon contact with limestone. The speed with which the solid dissolves is limited by the rate at which ions detach themselves from the crystal surface. Dissolution slows as the reaction approaches equilibrium. Traces of organic substances or phosphate may have an additional retarding effect on the speed of dissolution [108].

Iron(III) and aluminum minerals that usually occur as hydrates of oxides, hydroxides, or silicates are far less soluble than alkaline earth carbonates. The concentration of Fe^{3+} ions in equilibrium with ferric hydroxide (which may be written in the simplified form $Fe(OH)_3$ for purposes of calculation) is only about 10^{-17} M at pH 7.0 (calculated from the solubility product [108]). Other soluble inorganic species that must be considered include ionized hydroxyl-containing complexes, but even the most abundant of these at a pH of 7.0, $Fe(OH)_2^+$, has an equilibrium concentration of only about $3 \cdot 10^{-9}$ M [108]. Compounds of aluminum are slightly more soluble than ferric iron compounds, but are still not nearly as soluble as calcium and magnesium minerals. Ferric and aluminum minerals therefore cannot be dissolved by acid alone in the presence of buffering carbonates. In the presence of chelating agents, however, aluminum and ferric minerals have a measurable solubility even at pH 7.0. One such chelating agent, often produced by fungi, is citric acid. At pH 7.0, a 10 mM citrate solution can keep about 1 mM Fe(III) in solution or can (theoretically) dissolve this amount [109]. The full amount cannot be kept in solution, however, in the presence of other ions that tend to form complexes and compete with the ligand. Complexes of magnesium and calcium are usually weaker than those of ferric iron, but, since the alkaline earth ions are far more abundant in rock as well as more soluble, they may reduce the solubility of iron(III) significantly [109].

If ferric minerals are reduced, they become considerably more soluble. Apart from sulfides, there are no Fe(II) compounds with extremely low solubility. At atmospheric CO_2 concentration and pH 7.0, Fe^{2+} ions are soluble up to 1.5 mM; above that concentration, they precipitate as $FeCaO_3$ (siderite) [108]. If oxygen is present, however, ferrous ions are easily (re)oxidized to ferric hydroxide. The insoluble manganese(IV) oxide, MnO_2, can also be reduced to the rather soluble Mn^{2+} ion.

Silicates are almost insoluble in acids because the anhydride of silicic acid, SiO_2, is poorly soluble and is not volatile like CO_2. Notwithstanding, alkali metal ions can be extracted from silicates by H^+ ions or chelation. The crystal lattice loses compactness and is gradually destroyed, which is how feldspars weather. Feldspars are among the most abundant minerals in the earth's crust (for example, there is potassium feldspar in granite). Potassium feldspar (orthoclase), and similarly sodium feldspar (albite), are converted (kaolinized) by protons to kaolinite and silicic acid [73, 110]:

$$2\,K[AlSi_3O_8] + 2\,H^+ + 9\,H_2O \longrightarrow 2\,K^+ + Al_2Si_2O_5(OH)_4 + 4\,Si(OH)_4 \qquad (8)$$

Below pH 9, the solubility of the undissociated silicic acid or amorphous SiO_2 is about 2 mM (200 mg/l). At higher pH values, with the formation of $SiO(OH)_3^-$ or $H_3SiO_4^-$ ions ($pK_a = 9.46$), the solubility increases [110]. The solubility of crystalline SiO_2 (quartz) is an order of magnitude lower. Biological processes may make a solution alkaline if, for example, neutralized organic acids are oxidized to CO_2 that diffuses into the atmosphere:

$$CH_3COO^- + 2\,O_2 \longrightarrow 2\,CO_2 + OH^- + H_2O \qquad (9)$$

Fungi and lichens in contact with silicate minerals can dissolve them, and, over long periods, can even dissolve glass and quartz [62]. The underlying physical and chemical reactions are not known. Silicic acid has little tendency to form complexes with organic compounds. One of the few organic complexes reported is formed with catechol under approximately neutral conditions [41]:

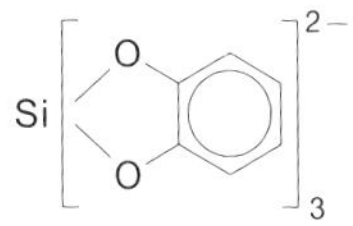

Soluble aromatic compounds with *o*-hydroxyl groups occur in higher plants [28, 39] and sometimes in fungi [9]; they may also be formed during microbial degradation of lignin if the methoxy groups are removed from the monomers by oxidation [21, 53, 100]. It is, however, not known whether they are involved in complexing silicic acid or dissolving silicate minerals.

2.2 Rock-colonizing organisms and their activities

Rocks may be colonized by algae, fungi, lichens (symbiotic associations of fungi with algae or cyanobacteria), cyanobacteria (blue-green algae), and numerous other types of bacteria. These organisms live not only on the surface but also in pores and subsurface cavities [27, 56, 74, 124]. Organisms living below the rock surface are termed *endolithic*. The endolithic rock dwellers exploit their habitat either by penetrating into existing crevices and pores or by actively creating their own cavities. In the latter case, the organisms have to dissolve the rock material, by mechanical or chemical action.

Rock-colonizing organisms may liberate minerals for growth by dissolving their substratum. Moreover, organisms in cavities, especially subsurface ones, may be protected to some extent from direct sunlight, rapid desiccation after a rain, or coming loose.

In some cases, however, the mineral may be dissolved simply as a result of metabolism rather than because of some special strategy for exploiting the rock as a habitat. This is true if the organisms' energy metabolism involves the net production of acids, as in nitrifying and sulfur-oxidizing bacteria.

By dissolving rock, the organisms in question accelerate the weathering process and thus play an important role in soil formation (pedogenesis). However, on the surfaces of buildings and other structures the activity of rock-dwelling organisms can be detrimental. The type of corrosion and the types of organisms involved depend on the available growth substrates, the composition of the rock, and the climate. For instance, the dominant organisms colonizing rocks in arid regions are lichens [56], whereas in livestock stables, nitrifying bacteria thrive in the humid, ammonia-laden atmosphere. In many instances of corrosion, however, the underlying biological and chemical processes are still not well understood. This is the case in the large-scale deterioration of old monuments and other buildings that has become widespread in recent decades, probably because of atmospheric pollution. Much of the corrosion is purely chemical, the result of acid formed from SO_2 (Section 2.1), but various colonizing organisms often accompany the corrosion, which suggests that the problem may be partly biological.

Rocks in which the quartz particles are interlinked only by a matrix of calcium carbonate display particularly severe and complex corrosion phenomena. The corroded outer layer of rock is often stratified [58]. The surface is like a crust, consisting of dust (soot, etc.) from the atmosphere, precipitated minerals leached from the rock, algae, fungal hyphae, bacteria, and polymers produced by living organisms or remaining after cells have died. Underneath the crust, there is a brittle, sandy layer from which the carbonate matrix has been dissolved. This layer grades into the original, uncorroded rock. Between the crust and the sandy layer there may be a brownish zone about 2 mm thick, rich in ferric hydroxide. The total thickness of the corroded layers outside the unaffected rock is 3 to 10 mm, or sometimes even more [58]. The crust scales off from time to time; it forms again on the newly exposed surface of the corroded rock. The crust is probably formed from minerals dissolved in deeper parts of the corroded layer, presumably transported by capillary water to the surface, where they precipitate and solidify upon evaporation of the water. During this process, minerals may also be dissolved by organic acids and precipitated when the carbon is biologically oxidized. The presence of iron and manganese oxides in the crust indicates that these must be dissolved by reduction and precipitated again by reoxidation.

We do not know precisely to what extent and by the agency of which metabolites the various organisms on rock contribute to corrosion under natural conditions. Most of our knowledge about potentially corrosive organisms stems from studies with pure cultures under defined conditions, which only distantly resemble those on weathering rocks. Nevertheless, the results of physiological studies are a prerequisite to our understanding the complex of activities on building stone. In the following sections, I discuss the rock-colonizing organisms in groups according to their major physiological characteristics.

2.2.1 Algae and cyanobacteria

Algae and cyanobacteria are *phototrophic* (photosynthetic) organisms; they use the energy of sunlight to synthesize their cell material from CO_2, H_2O, and mineral salts. These organisms either colonize existing depressions and crevices or create their own cavities. They can actively penetrate rock by exerting a constant mechanical stress on the rock with their growth and their gliding movements [124]. It may also be possible that they dissolve rock chemically [1 a].

2.2.2 Fungi, lichens, and chemoorganotrophic bacteria

Fungi, the fungal partners in lichens, and most bacteria are *chemoorganotrophic*, that is, they depend on organic substrates for their energy metabolism and cell synthesis. In lichens, the fungal partner obtains its organic nutrients from its phototrophic symbiont. Possible sources of organic substrates for fungi and chemoorganotrophic bacteria on rocks are:

(a) substances left by phototrophic rock-dwelling organisms: secreted low molecular weight compounds, polymers, or dead cells;
(b) organic inputs from the atmosphere: dust, or volatile compounds, for instance those evolved in biological decomposition processes;
(c) organic substances of biological origin, which have been included and chemically transformed during diagenesis of the rock [57].

There is little doubt that fungi and lichens dissolve rock mainly by chemical processes.

Many fungi excrete organic acids [8, 9, 26, 56, 57]. The best known of these is citric acid, which, in addition to its acidity, has considerable chelating capability. Acid production is somehow influenced by the availability of certain minerals: fungi increased their acid production when potassium was deficient [57].

The chemistry of mineral dissolution by lichens is not fully understood. The fungal partner can form oxalic acid and deposit large quantities (up to 60% of the dry mass) of calcium oxalate ($CaC_2O_4 \cdot H_2O$) either outside or inside the thallus [34, 37, 112]. Magnesium oxalate ($MgC_2O_4 \cdot 2\,H_2O$) is also rather commonly deposited. Since calcium oxalate is almost insoluble, oxalic acid probably destroys carbonate minerals by transforming their crystal structure rather than by dissolving and leaching calcium ions. Another possibility is that calcium oxalate, precipitated on or inside the thallus, is effectively a calcium sink, promoting a continuous dissolution of calcium ions from the carbonate matrix into the adherent water [112].

Whether *lichen substances* are significant in rock corrosion is controversial; they are aromatic metabolites with hydroxyl, aldehyde, and carboxyl groups, that may act as chelators [34, 37,89, 99, 112]. Some people think lichens destroy rock not only chemically but also by exerting mechanical effects: hyphae penetrate into the rock and gradually break it up because the tissues expand and contract as they take up or lose water [56, 103].

There are reports that lichens dissolve not only carbonates but also silicates [34, 126, 127], but the overall contribution of these organisms to weathering and pedogenesis remains difficult to assess. In humid regions, lichens preserve (rather than corrode) rocks, by protecting them from leaching by rain [37, 56, 57]. In contrast, in arid zones lichens obviously accelerate rock weathering; colonized rocks may be dotted with many small crater-like depressions [56, 57, 61]. Not only is $CaCO_3$ dissolved, but also iron and manganese minerals appear to be mobilized; these are redeposited on the surface as brown layers [56, 60]. Little is known about the speed at which lichens can dissolve rock. In a Mediterranean climate, lichen damage on buildings was obvious after 15 years of growth; blisters formed and material sloughed off [103]. In the Antarctic, lichens dissolve sandstone at an estimated maximal rate of 3 mm per century [34].

There is almost no information about the role of chemoorganotrophic bacteria in the weathering of rocks. In principle, bacteria may also dissolve rock by forming organic acids, which they produce for various reasons. If oxygen is absent, facultatively or obligately anaerobic bacteria ferment organic compounds to products such as short-chain fatty acids, lactic acid, succinic acid, and other acids. These acids are not particularly good chelators. The carbonate minerals are acidified and dissolved as a secondary effect as the organisms generate metabolic energy. The excreting of 2-ketogluconic acid by soil bacteria [122] appears to be a different situation, however. This acid is not an end product of fermentative metabolism, but rather is an intermediate in glucose metabolism in respiring bacteria. On the one hand, we may assume that accumulation of this acid outside the cell contributes to the solubilization of minerals for growth, but on the other hand, production of 2-ketogluconic acid by *Pseudomonas* species as a result of periplasmic glucose oxidation was observed only at unnaturally high concentrations of glucose (7 to 8 mM); if glucose was limiting, it was taken up directly, rather than being converted to the acid first [123 a]. Acid bacterial slimes and capsular materials may also dissolve rock [20]. Some of these materials contain carboxyl groups and are therefore acidic, cation-binding polyelectrolytes; if they are synthesized during growth on neutral substances, there is a net production of acid. Acidic components include uronic acids in alginate, teichuronic acids and some other polysaccharides, D-glutamate in polyglutamate, or pyruvate linked as a ketal to certain carbohydrate chains [44a, 72, 93, 93a, 106, 111a, 117].

2.2.3 Nitrifying bacteria and sulfur bacteria

Nitrifying bacteria and sulfur bacteria are termed *lithotrophic* because they obtain energy for growth by oxidizing reduced *inorganic* compounds; the former live on ammonia or nitrite, and the latter, on sulfur or sulfur compounds. The end products are nitric acid and sulfuric acid, respectively. The acid is produced, as in the fermentative bacteria, as a result of energy metabolism, not as a special strategy for exploiting rock as a habitat. Nitrifying and sulfur bacteria are also widespread in water and soil; they are key participants in the nitrogen and sulfur cycles [11, 15, 45, 46, 47, 65, 66]. Carbonate rocks are a particularly good substratum because they neutralize the acid formed, which would otherwise inhibit growth.

Nitrifying bacteria. The term *nitrifying bacteria* encompasses two bacterial groups, the ammonia oxidizers and the nitrite oxidizers [11, 118]. The former oxidize ammonia only as far as nitrite, never farther. Because these bacteria are active around neutral pH and not under acidic conditions, the net reaction should be written with the ions as they actually are present:

$$2\,NH_4^+ + 3\,O_2 \longrightarrow 4\,H^+ + 2\,NO_2^- + 2\,H_2O \tag{10}$$

The genera that perform this oxidation are *Nitrosococcus*, *Nitrosomonas*, *Nitrosolobus*, *Nitrosospira*, and *Nitrosovibrio* [11, 118].

Nitrite oxidizers catalyze the following net reaction:

$$2\,NO_2^- + O_2 \longrightarrow 2\,NO_3^- \tag{11}$$

Genera in this group are *Nitrobacter*, *Nitrococcus*, *Nitrospina*, and *Nitrospira*.

Although the overall equations make it look as if oxygen reacts directly with the nitrogen compounds, this is not what happens in most of the metabolic steps. Ammonia is the only compound that reacts directly with oxygen, yielding hydroxylamine, NH_2OH. This compound, as well as nitrite, is further oxidized not by a direct reaction with oxygen but by the withdrawal of electrons (reducing equivalents). These reactions involve water. The electrons are then transferred to oxygen in separate reactions and the energy is available for growth [38].

Almost all nitrifyers are obligately chemolithotrophic, that is, most are unable to grow by oxidizing organic substrates. These bacteria are obligately dependent on inorganic nitrogen compounds; some may oxidize urea instead of ammonia [11]. Nitrifyers synthesize cell material autotrophically from CO_2. The precise term for their metabolic type is thus *chemolithoautotrophic*. Organic compounds (formate, acetate, pyruvate, glycerol, and others) added to the medium may be incorporated to a certain extent and stimulate growth [63, 119]. *Nitrobacter* species constitute a physiological exception, since they can grow chemoorganotrophically with acetate, pyruvate, or glycerol as the sole carbon and energy source [11]. *Nitrobacter* can also grow by nitrate respiration in the absence of oxygen; in this case, nitrate, which otherwise is a product, is reduced by organic substrates to NO_2^-, NO or N_2O [11].

The concentrations of the substrates and products of nitrifyers, as well as the pH value, may be physiologically important. Some ammonia oxidizers may tolerate ammonium ions up to approximately 160 mM [74]. In a buffered medium, produced nitrite is usually tolerated up to about 7 mM [11], and some species may even produce up to 30 mM nitrite [74]. Growth in unbuffered media is inhibited fairly promptly, probably by the undissociated fraction of the nitrous acid, a chemically very reactive species. The optimum pH range is 7.5 to 8.0 [11, 118]. A species that grew even at pH 5.0 was isolated from corroded rock [10]. Nitrite oxidizers may be inhibited by high concentrations of ammonium ions [11] but tolerate their nitrite substrate up to 30 mM [74, 118]. In pure culture work, nitrate is still tolerated at 65 mM [11]. Best growth occurs at pH 7.0 to 8.0 [118].

Many nitrifyers use slime as a means for keeping cells together in aggregates [118]. Slime formed on the surface or in pores of rocks may help the cells adhere to the surface, so that they are not washed away during a rain.

The "true" nitrifyers I have discussed are not the only bacteria that can nitrify. A number of otherwise unremarkable chemoorganotrophic bacteria may oxidize nitrogen compounds to nitrite or nitrate in a side reaction [66].

Nitrous acid, with a pK_a of about 3.3 (25 °C), is a weaker acid than nitric acid ($pK_a = -1.4$). Nevertheless, at a pH of 5.0, the lowest that ammonia oxidizers can tolerate, 98% of the nitrous acid is in the dissociated state. We can therefore conclude that the ammonia-oxidizing bacteria are almost exclusively responsible for acidification and carbonate dissolution (Equation 10), but nitrite oxidizers are also important to acid production. They convert the reactive, generally antagonistic nitrite to the less reactive and more compatible nitrate. The result of this cooperation between two metabolic types of bacteria is that more acid can be produced than by ammonia oxidizers alone.

Nitrifying bacteria find favorable growth conditions on limestone, mortar, and concrete if these are exposed to gases from stables or sewers. Once nitrification has started and the pH is lowered, gaseous ammonia is readily scavenged from the air, so the process of nitrification should be self-stimulating. I know of no published measurements, but nitrate concentrations must be high, since "blooms" of calcium nitrate ($Ca(NO_3)_2 \cdot 4H_2O$) can appear on stable walls. Experiments in humid incubation chambers have simulated how nitrifying bacteria dissolve rock exposed to ammonium salts [10, 12]; small blocks of concrete were inoculated with a mixed culture of *Nitrosomonas* and *Nitrobacter*, and sprayed with an ammonium salt solution. The rate of formation of HNO_3 was 83 µmol (5.2 mg) per cm^2 per year, which theoretically can dissolve 4.2 mg $CaCO_3$. The sand grains on the surface of the blocks came loose because the cement matrix was dissolved, whereas uninoculated blocks were unaffected.

Experiments are under way to clarify how nitrifyers are involved in the usual type of building deterioration, which occurs even though the structures are not generally exposed to ammonia from nearby sources [10, 12, 74, 75]. This research is of practical importance since the corrosion of valuable monuments by atmospheric pollutants has accelerated significantly in recent decades. Samples of sandstone from the cathedral in

Cologne (Kölner Dom) (taken at a height of 40 m), the cathedral in Regensburg (Regensburger Dom) (30 m), and the Alte Pinakothek in Munich (15 m) [10, 11] were examined for nitrifyers. Stone from the Kölner Dom harbored ammonia oxidizers in high cell numbers (5×10^5/g of stone) and as a high proportion (about 6%) of the total bacterial cell polulation. In contrast, the other buildings had low counts of ammonia oxidizers. Members of the genus *Nitrosovibrio* were the dominant species of ammonia oxidizers in building stones [75]. Nitrite oxidizers in high numbers (up to 10^7/g) and as a high proportion (up to 60%) of the bacterial population were found at the Kölner Dom and the Alte Pinakothek, whereas the counts on the Regensburger Dom were low. Cell number as a function of depth was also measured in the Kölner Dom samples: cell numbers of nitrifyers were highest in the upper 5 mm of the stone; cell numbers decreased with depth, but the relative proportion of nitrifyers increased. The pH at the surface was around 5.0, and in the lightly colonized rock deeper than 5 cm it was above 8.0. Finding such high numbers of nitrifyers on the Kölner Dom indicates that they are indeed involved in corrosion. The substrate might be ammonium salts in atmospheric dust spread by wind. Ammonia that enters the atmosphere from biological decomposition processes and fertilizers can react with acids of nitrogen and sulfur to form salts [113]. Furthermore, the biological decomposition of proteins and nucleic acids from dead rock-colonizing organisms may be a source of ammonia. It is apparent that at the Alte Pinakothek, nitrite is being oxidized but ammonia is not. The high population of nitrite oxidizers on this building probably lives on atmospheric NO_2 (Equation 4). Buffering by the stone permits the NO_2 to disproportionate completely. Since the oxidation of nitrite by bacteria does not produce a significant amount of acid, the corrosion of this building should be purely chemical.

Corrosion of concrete in association with denitrifying bacteria has been observed on cooling towers of power plants [54].

Sulfur bacteria. Microorganisms of various genera that have in common the ability to grow aerobically with reduced sulfur species are lumped as a physiological group termed *sulfur bacteria* [64]. The most commonly used forms of sulfur are hydrogen sulfide (H_2S, often simply termed sulfide), elemental sulfur (S, S^0, or more exactly S_8) and thiosulfate ($S_2O_3^{2-}$) [50].

$$2\,H_2S + O_2 \longrightarrow 2\,S^0 + 2\,H_2O \tag{12}$$

$$H_2S + 2\,O_2 \longrightarrow 2\,H^+ + SO_4^{2-} \tag{13}$$

$$2\,S^0 + 3\,O_2 + 2\,H_2O \longrightarrow 4\,H^+ + 2\,SO_4^{2-} \tag{14}$$

$$S_2O_3^{2-} + 2\,O_2 + H_2O \longrightarrow 2\,H^+ + 2\,SO_4^{2-} \tag{15}$$

According to reactions (12) and (14), elemental sulfur can be formed from H_2S as an intermediate and may be oxidized further. The final result is always a net production of acid. The reactions are analogous to those we saw in the nitrifyers: most of the substrates of sulfur bacteria do not react directly with oxygen. Instead, the sulfur species donate electrons in reactions involving water; the electrons are transferred in separate reactions to oxygen, and the energy is conserved [49, 50]. The only exception is a reaction in which elemental sulfur can be directly oxygenated, with bisulfite (HSO_3^-) as an intermediate [49, 50].

The sulfur bacteria vary greatly with respect to their utilization of organic compounds. Some species are obligately chemolithotrophic, and are often also obligately autotrophic; organic compounds are assimilated only to a low extent or not at all (chemolithoautotrophic growth) [15, 50]. Other species can grow on organic substrates alternatively, or may even grow better on them [15, 50]. Furthermore, there are microorganisms that grow preferentially or exclusively chemoorganotrophically and catalyze the oxidation of a sulfur species more or less incidentally in a side reaction [49, 125]. It is a matter of opinion which types should be called sulfur bacteria and which should not.

The most abundant types of sulfur bacteria, seen in nature in the presence of reduced sulfur, are members of the genus *Thiobacillus* [15], the multicellular filamentous *Beggiatoa* species [45, 46, 47, 48], and the large, very motile cells of *Thiovulum* [48]. These bacteria are mesophilic, thriving below about 40 °C. Hot sulfur springs may harbor thermophilic sulfur bacteria that grow optimally at 70 to 90 °C. Among these thermophilic sulfur bacteria are *Thermothrix* [17] and the archaebacteria *Sulfolobus* and *Acidianus* [105]. *Beggiatoa* and *Thiovulum* are purely aquatic bacteria, and may thrive as dense populations in H_2S-O_2 gradients on or above anaerobic sediments. These bacteria have not been found on corroding rock. In contrast, thiobacilli live not only in aqueous habitats but also in soils and on moist rocks.

Thiobacilli may corrode rocks by producing acid, as in Equations (13) and (15). Some species, such as *Thiobacillus thiooxidans* and *Thiobacillus ferrooxidans*, are metabolically active down to a pH of about 1 and may make their habitat strongly acid [50, 51].

If sulfuric acid acts on rocks containing calcium carbonate, gypsum ($CaSO_4 \cdot 2\,H_2O$) may precipitate. The solubility of gypsum is 2.4 g/l [120], much greater than that of calcium carbonate, so gypsum erodes more readily (see Section 2.1).

Thiobacilli are not usually involved in the large-scale corrosion of buildings because reduced sulfur is absent there. It is true that vast quantities of hydrogen sulfide are given off by anaerobic sediments and volcanic sources, but the sulfide reacts chemically rather rapidly with oxygen, as well as being consumed biologically at the anaerobic/aerobic interface, so it does not accumulate in the atmosphere.

Thiobacilli corrode rock and concrete only if these materials are close to sites where H_2S is liberated from anaerobic processes [15]. For instance, in the sewer system in Hamburg the upper parts of the concrete walls, above the water level, were corroded to a depth of several cm [76]. The most heavily corroded places had a pH of 1 to 2; they contained gypsum and up to 2.8×10^8 *Thiobacillus* cells per gram of wall material. The thiobacilli predominated in the detectable bacterial population, with the rather acid-tolerant species *Thiobacillus thiooxidans* as the most abundant. Less heavily damaged places on the walls, where corrosion was probably still beginning, had populations in which other *Thiobacillus* species predominated; some of these latter species were facultatively chemoorganotrophic. The cell density of thiobacilli in the wastewater was several orders of magnitude less than that on the corroded concrete above the water level. The direct substrate of the thiobacilli was presumably not H_2S but elemental sulfur formed by chemical oxidation [76]. The chemical oxidation of sulfide is significantly stimulated by traces of heavy metals, such as nickel and iron, and can go to various oxidation states; the most common products are elemental sulfur and thiosulfate [18, 29a]. Both are substrates for thiobacilli, and the overall amount of acid produced is always the same, according to Equation (13).

In simulation experiments, pure cultures of thiobacilli were shown to corrode concrete [97].

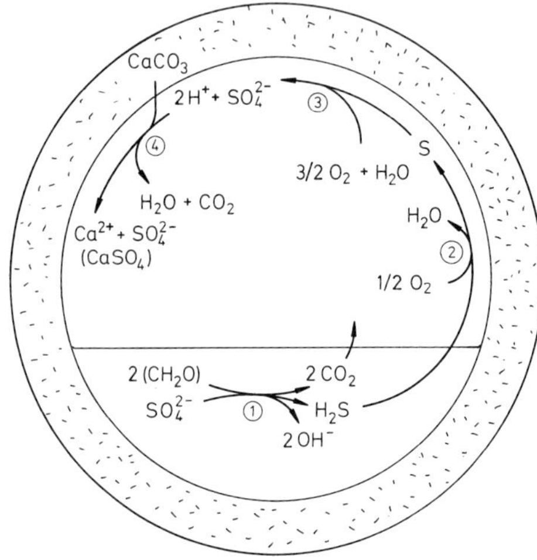

Fig. 2 Reactions by which unsubmerged concrete walls (above horizontal line) of sewer pipes are dissolved.
(1) Bacterial fermentation of organic substances, and reduction of sulfate to H_2S by the fermentation products.
(2) Chemical or biological oxidation of H_2S.
(3) Sulfur oxidation by thiobacilli.
(4) Dissolution of $CaCO_3$

The reactions leading to concrete corrosion in sewer systems are shown schematically in Figure 2. Since wastewater and sewage have a high biological oxygen demand, conditions in the sewers are anaerobic, and fermentative organisms thrive. The products of fermentation are oxidized by sulfate-reducing bacteria, which reduce dissolved sulfate to hydrogen sulfide (see Section 3.2). Corrosion occurs if the upper parts of the concrete walls are not under water, and are exposed to H_2S and to splashed or condensed water. Under those conditions, the reduction of sulfate, accompanied by increased alkalinity, and the oxidation of hydrogen sulfide, accompanied by increased acidity, are spatially separated from each other. In the natural sulfur cycle in waters, both these processes are in the same aqueous phase, and the acid is neutralized. Similarly, if the entire concrete surface were under water, local acidification would not occur. At a sampling site in the sewer system in Hamburg, treating the water with oxygen could inhibit the corrosion; this probably inhibited the anaerobic bacterial reduction of sulfate [76].

The sulfides of heavy metals and hence also sulfide-bearing ores constitute a special type of reduced sulfur compound utilized by some sulfur bacteria [40]. If pyrite is exposed to an aerobic, humid environment, acid is produced by a slow oxidation, either spontaneously or in a biological reaction catalyzed by sulfur bacteria:

$$4\,FeS_2 + 14\,O_2 + 4\,H_2O \longrightarrow 4\,Fe^{2+} + 8\,H^+ + 8\,SO_4^{2-} \qquad (16)$$

If conditions are neutral or slightly acid, the Fe^{2+} ions are oxidized mainly by chemical reaction; if acid continues to be produced and the pH drops further, the Fe^{2+} is oxidized almost exclusively by the appropriate sulfur bacteria:

$$4\,Fe^{2+} + O_2 + 4\,H^+ \longrightarrow 4\,Fe^{3+} + 2\,H_2O \qquad (17)$$

At pH values higher than about 2.5, ferric iron precipitates as the hydroxide. At lower pH values, the Fe^{3+} ions in solution oxidize pyrite, forming Fe^{2+} ions and sulfur, which are then oxidized by bacteria. The overall reaction is:

$$4\,FeS_2 + 15\,O_2 + 2\,H_2O \longrightarrow 4\,Fe^{3+} + 4\,H^+ + 8\,SO_4^{2-} \qquad (18)$$

The process is known as *leaching*. Sulfur bacteria able to carry out these reactions are *Thiobacillus ferrooxidans*, *Leptospirillum ferrooxidans*, and the thermophilic *Sulfolobus* species; all of them are tolerant of acid and of dissolved heavy metal ions. The total picture that must be considered includes not only the acidity of the H^+ ions formed stoichiometrically, but also the additional protons that are formed as a result of the Fe^{3+} ion tending to form hydroxo complexes ("hydrolyze"). Pyrite is not usually a constituent of stone buildings. Therefore, corrosion by bacterial leaching occurs only in tailings or when rock and concrete are exposed to acid mine drainage.

2.2.4 Bacteria that reduce iron and manganese

The reduction of iron(III) minerals and MnO_2 is yet another way in which constituents of rock are dissolved. However, on a global scale these reductive processes apparently corrode less rock than does the dissolution of the far more abundant alkaline earth carbonates.

Many bacteria and fungi reduce Fe(III) to Fe^{2+}, or MnO_2 to Mn^{2+} ions, respectively, in concentrations as high as those of catabolic products [29]. Often, the reductions are side reactions during aerobic or anaerobic growth on organic substrates, and do not confer any obvious benefit on the organisms in question [29]. A partial explanation of the reductions is that excreted metabolites react chemically with Fe(III) compounds and MnO_2. For instance, oxalate or pyruvate reduced these metals from their higher oxidation states to the divalent ions [29]. Other workers, however, have found pure cultures that grew by reducing Fe(III) compounds or MnO_2 with H_2 [4a, 5, 71] or acetate and other simple organic compounds [69, 70, 81]. These bacteria may be designated true iron or manganese reducers, or iron- or manganese-respiring organisms. The acetate-utilizing bacterium formed different types of end products: only soluble Fe(III) complexes and MnO_2 were reduced completely to the divalent state; Fe(III) hydroxide was reduced to the insoluble, very stable magnetite, $Fe_3O_4 (= FeO \cdot Fe_2O_3)$. The ability to reduce Fe(III) and MnO_2 is often but not always associated with the ability to reduce nitrate to N_2 or NH_4^+. Iron- and manganese-reducing bacteria could be regarded as a separate physiological group, active in mineral transformations. It is still not known to what extent these bacteria mobilize iron and manganese minerals during weathering of rock.

Almost all living organisms have mechanisms for solubilizing trace amounts of iron(III) from insoluble compounds, to supply their cells with this essential element.

The reoxidation of Fe^{2+} and Mn^{2+} with O_2 is either chemical or mediated by bacteria and fungi [83, 84, 128].

3 Corrosion of iron

The corrosion of iron is of great technological and economic importance, because iron is the metal we use most commonly and for the greatest number of different purposes. Iron is a base metal, easily oxidized, but iron oxides do not coat the metal in a cohesive protective layer, as do those of the even more base zinc and aluminum. The result is the well-known progressive rusting of iron. Stainless alloys are made by adding nonferrous metals such as chromium or nickel, but most iron is fabricated by cheap processes without these protective additives. Iron straight from the blast furnace contains 2.4% to 4% carbon, and is called *cast iron (or pig iron)*. If the carbon content is below about 1.7%, the iron is usually termed steel. Mild steel, with a carbon content of 0.4 to 1.7%, can be tempered; wrought iron, with less than 0.4% carbon, cannot. Alloys with a substantial proportion of other metals (such as chromium or nickel) are also called steels *(stainless steels)*.

Iron can be corroded in wet conditions, in hot gases, or in the presence of other molten materials. In water, iron is corroded not only by purely chemical reactions but also by the metabolic activity of microorganisms. I will start this discussion by presenting some electrochemical principles relevant to microbial corrosion.

3.1 Electrochemical principles of metal corrosion

3.1.1 Dissolution of metals

If a metal is immersed in water, some positive metal ions move into the solution, and electrons remain on the metal:

$$Me \longrightarrow Me^{z+} + z\,e^{-} \qquad (19)$$

The metal thus acquires a negative charge with respect to the solution [2]. This electron-producing process is called the *anodic reaction*. We can understand the tendency of the metal to liberate cations if we picture the positive ions already existing in the crystal lattice; the free electrons (the *electron gas*) are mobile within the lattice, and are the reason that metals are excellent conductors of electricity and heat. If the metal comes into contact with water, only the cations can be released; the electrons cannot exist freely in solution and are therefore confined to the metal. The metal/water interface thus constitutes a barrier permeable to ions but not to electrons. The high density of ions inside the metal, and the tendency of the ions to hydrate, can be regarded as the driving forces for the solution process. The separation of the charges causes a difference in the *electrical potential*, ϕ, (in what follows it is often called simply the *potential*) between the metal (subscript M) and the aqueous solution (subscript A), $\phi_M - \phi_A = \Delta\phi$. This difference in potential opposes the release of cations from the metal. Therefore, cations are redeposited on the metal, according to the electron-accepting *cathodic reaction*:

$$Me^{z+} + z\,e^{-} \longrightarrow Me \qquad (20)$$

The rate of this reaction increases with the concentration of dissolved metal ions. If the rate of escape is equal to the rate of return, an equilibrium is reached; even though the anodic and cathodic reactions continue, the net transfer of ions falls to zero. This state is characterized by the equilibrium potential difference, $\Delta\phi^*$. Increasing the concentration of dissolved ions by adding a salt of the metal would cause a net deposition of metal ions until a new equilibrium with another potential difference were reached. Therefore, a given concentration (or, to be exact, a given *activity*) of metal ions in solution always determines a certain $\Delta\phi^*$, which is a characteristic value for every metal. At the same cation activities, base metals (metals that tend to dissolve) have a higher $\Delta\phi^*$ than noble metals (metals that do not tend to dissolve). The potential difference between metal and solution, $\phi_M - \phi_A$, cannot be measured directly, but a physicochemically useful measure can be obtained if the electrical potential of the metal is determined against a reference electrode (subscript R), which has a defined, constant potential difference, $\phi_R - \phi_A$, against the same solution. Accordingly, the potential difference between metal and reference electrode, $\phi_M - \phi_R$, also has a defined, characteristic value. At equilibrium, the potential difference is designated as E (*electromotive force, EMF*). Potential differences are usually referred to the standard hydrogen electrode ($2\,H^+/H_2$, on platinum), even though that electrode is rarely used directly for measurement. The electrical potential of this electrode is zero by definition, and the *standard potential* of a metal, E^0, is the potential difference against the standard hydrogen electrode if the activity of metal ions is 1.0.

For Fe^{2+}/Fe, two different values of E^0 can be found in the literature, namely -0.440 V [4, 80, 109a] and -0.409 V [121]. The latter value is obtained by calculation from the free energy of solution of iron in acid with a proton activity of 1.0 (see [4, 111]), according to the following equation:

$$Fe + 2\,H^+ \longrightarrow Fe^{2+} + H_2 \qquad \Delta G^0 = -78.87 \text{ kJ} \tag{21}$$

The standard potential of Fe^{2+}/Fe is more of a theoretical value for calculations than one actually achieved. An activity of Fe^{2+} ions of 1.0 is not very likely in the water boilers, pipelines, ditches, etc., where corrosion occurs. In the first place, the activity of a polyvalent ion is markedly different from its concentration if this is above about 10 mM. The discrepancy is particularly significant if both the cations and the anions are polyvalent. For instance, the mean activity coefficient of a 0.5 M solution of $CuSO_4$ or $ZnSO_4$ is only about 0.063 [79]. In the second place, if the pH is near neutral or above, polyvalent cations precipitate readily as hydroxides or (in the case of divalent ions) as carbonates, so the solubility of metal ions in aqueous systems is often limited. The pH dependence of the value of E for metals that precipitate as hydroxides or other insoluble salts can be calculated from the Nernst equation and the solubility and dissociation constants. The potential of Fe^{2+}/Fe becomes especially negative if the presence of H_2S causes the ferrous iron to precipitate as the very poorly soluble FeS.

The following is the derivation of a formula for the pH dependence of the potential of a metal (Me), the cation of which reacts with a weak acid (H_2A) to yield a poorly soluble salt (MeA). Both the cation and anion are divalent, as is the case with sulfides and carbonates. From the definition of the solubility product, K_{MeA}, we can write:

$$[Me^{2+}] = \frac{K_{MeA}}{[A^{2-}]} \tag{22}$$

$[A^{2-}]$ is given by the dissociation constants of the acid, K_{a1} and K_{a2}:

$$K_{a1} = \frac{[H^+][HA^-]}{[H_2A]} \tag{23}$$

$$K_{a2} = \frac{[H^+][A^{2-}]}{[HA^-]} \tag{24}$$

If $[A_t]$ is the total concentration of the acid component A, that is, of the undissociated and dissociated acid, we can write the following equation for a closed system:

$$[H_2A] = [A_t] - [HA^-] - [A^{2-}] \tag{25}$$

Combining this with equation (23), we write:

$$[HA^-] = \frac{K_{a1}[A_t] - K_{a1}[A^{2-}]}{[H^+] + K_{a1}} \tag{26}$$

$[A^{2-}]$ is obtained from equation (24):

$$[A^{2-}] = \frac{K_{a1}K_{a2}[A_t]}{[H^+]^2 + K^{a1}[H^+] + K_{a1}K_{a2}} \tag{27}$$

The cation concentration is then:

$$[Me^{2+}] = \frac{K_{MeA}([H^+]^2 + K_{a1}[H^+] + K_{a1}K_{a2})}{K_{a1}K_{a2}[A_t]} \tag{28}$$

The value of the electrical potential is calculated from the Nernst equation:

$$E = E^0 + 0.0296 \cdot \log[Me^{2+}] \tag{29}$$

$$E = E^0 + 0.0296\{pK_{a1} + pK_{a2} + \log K_{MeA}\} - 0.0296\{\log[A_t] - \log([H^+]^2 + K_{a1}[H^+] + K_{a1}K_{a2})\} \tag{30}$$

For iron and hydrogen sulfide, with $E^0 = 0.44$ V, $\log K_{FeS} = -18.1$, $pK_{a1} = 7.02$, and $pK_{a2} = 13.9$ [108], equation (30) becomes:

$$E = -0.357 - 0.0296\{\log[A_t] - \log([H^+]^2 + 0.955 \cdot 10^{-7}[H^+] + 1.20 \cdot 10^{-21})\} \tag{31}$$

A new value for the pK_{a2} of H_2S of 18.58 was published in 1988 [101], but this value gives a high value for the solubility of metal sulfides that is not observed in reality. If the new pK_{a2} value is used, the solubility products of the sulfides also need to be revised.

The pH dependence of E in the presence of 0.001 M sulfide is shown in Figure 3 (line FeS/Fe); the line for the potential of Fe^{2+}/Fe is given for 0.001 M Fe^{2+}. The calculation was simplified by using concentrations instead of activities, which is a good approximation if the solutions are rather dilute. In the figure we can also see how E changes with increasing pH in the absence of sulfide (line $Fe(OH)_2/Fe$). That line is drawn using the maximum concentration of Fe^{2+} calculated from the solubility product of $Fe(OH)_2$ [108]. The oxidized forms are thermodynamically stable in the areas above the lines pertaining to iron, while the metal is stable below the lines. Therefore, if an EMF

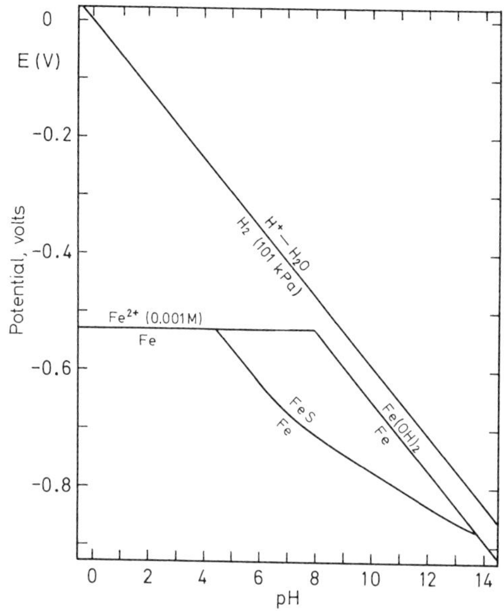

Fig. 3 Thermodynamic stability of iron, hydrogen, and their oxidized forms. Concentrations were used as approximations of activities for purposes of calculation. The FeS/Fe line was calculated for a concentration of *dissolved* sulfide of 0.001 M

corresponding to a value below the lines is artificially imposed on iron, the iron can no longer be corroded. This is the principle of *cathodic protection* [90, 102]. Corrosion can also be prevented by applying a rather positive potential, a measure named *anodic protection*, but this kind of protection is not of a thermodynamic nature because the metal is energetically unstable. Anodic protection relies on special forms of iron oxides or hydrated oxides that form on the surface and coat the metal like varnish [90, 102].

3.1.2 Reactions of the electrons

Even though electrons cannot be released as such into the medium, they can reduce compounds from the aqueous phase at the metal/water boundary. Electron acceptors might be oxygen, protons, or undissociated weak acids or water that may undergo the following cathodic reactions [91, 92]:

$$O_2 + 2\,H_2O + 4\,e^- \longrightarrow 4\,OH^- \quad (32)$$

$$O_2 + 4\,H^+ + 4\,e^- \longrightarrow 2\,H_2O \quad (33)$$

$$2\,H^+ + 2\,e^- \longrightarrow H_2 \quad (34)$$

$$2\,H_2S + 2\,e^- \longrightarrow 2\,HS^- + H_2 \quad (35)$$

$$2\,H_2O + 2\,e^- \longrightarrow 2\,OH^- + H_2 \quad (36)$$

The values of E allow us to predict which of these reactions can in principle be coupled to the anodic dissolution of a metal (Equation (19)). *Coupling* means that two different net reactions occur simultaneously, neither of which reaches its thermodynamic equilibrium; as a result, metal continues to dissolve. Each of the two net reactions in turn is composed of an anodic and a cathodic reaction (forward and back reaction), one of which dominates. The terms *anodic* and *cathodic* are thus used for the net reactions (iron dissolution is anodic; reduction of oxygen, protons, etc., is cathodic) as well as for the (partial) reactions that constitute the net reactions.

If some or all of the cathodic net reactions (32–36) are thermodynamically possible on a metal, the reaction kinetics rather than the free energies determine which process actually occurs or dominates. For instance, on iron surfaces at neutral pH, that is, at low proton concentrations, reaction (32) or (33) predominates. The ferrous ions formed then react with oxygen and water, producing brown hydrates of ferric oxide (rust). Alternatively, in strongly acidic solutions, hydrogen is the main cathodic product (reaction 34) even though a reaction with oxygen would be energetically more favorable. In oxygen-free water, the electrons can undergo only reactions (34), (35), and (36). Pure iron immersed in anoxic water at pH 7.0 can be expected, thermodynamically speaking, to liberate H_2 at a partial pressure of 101 kPa (1 atm). The solubility of H_2 in water is 0.76 mM (at 101 kPa and 25 °C). Since equimolar amounts of Fe^{2+} are formed (reaction 21), the E of Fe^{2+}/Fe at the concentration in question is -0.532 V, and the E of $2H^{+}/H_2$ is -0.414 V (see Figure 3).

The anodic and cathodic reactions, and hence the corrosion, may be distributed homogeneously over the metal surface. Often, however, the crystal lattice of the metal surface is inhomogeneous, so sites occur at which either the anodic or the cathodic reaction dominates [102]. This is how local galvanic cells are established, with a short circuit over the metal. Short-circuit currents may connect anodic and cathodic sites at different places on the same iron piece over distances of several meters, provided the sites are also in contact with the same aqueous phase.

3.1.3 Electrode processes and currents

If a metal is not electrically connected to other conductors, all currents within the metal of necessity cancel each other: there is no net current to or from the metal. To study the kinetics of corrosion reactions, iron immersed in solution is connected to an external current supply; the circuit is completed across an *auxiliary electrode* (for example, platinum) immersed in the same solution or connected to it by a salt bridge [90, 102]. The iron is termed the *working electrode*.

An artificially imposed electron flow to the iron boosts the cathodic and suppresses the anodic reaction; electron flow away from the iron causes the opposite shift in the electrode reactions. It is therefore possible to study the anodic or cathodic reactions with external electric instrumentation, and an apparatus for doing the measurements is depicted in Figure 4. The potential of the working electrode established by a given current is measured against a *reference electrode*. A high-impedance voltmeter must be used so that the current between working and reference electrode is negligible; a larger value of that current would reduce the value of the potential difference. An ohmic potential difference is generated between the working and auxiliary electrodes, from the resistance of the electrolyte. To reduce the influence of that on the measured potential difference, the distance between working and reference electrode has to be kept very short. This is done by constructing the opening of the reference electrode as a capillary

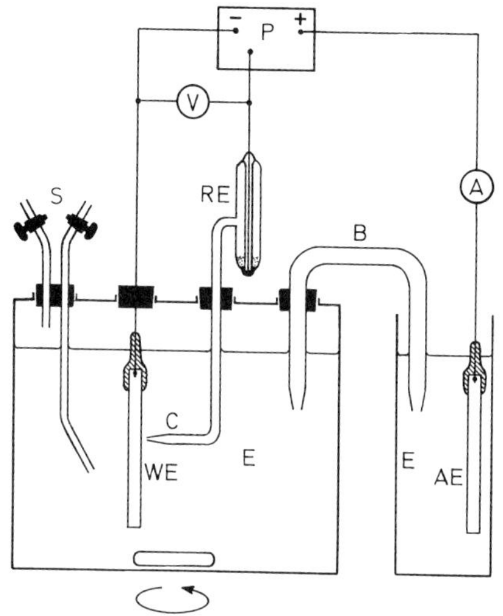

Fig. 4 Principle of an electrochemical device for measuring current density/potential characteristics of metal electrodes (modified from [86], [90], [102]). A, ammeter; AE, auxiliary electrode; B, salt bridge of concentrated KCl, thickened with agar or a similar substance; C, Haber-Luggin capillary; E, electrolyte; P, potentiostat; RE, reference electrode (*e.g.*, calomel electrode); S, ports with stopcocks, for additions, degassing, etc.; V, high-impedance voltmeter; WE, working electrode. The electrolyte can be stirred

(called a *Haber-Luggin* or *Luggin capillary*) that is brought close to the surface of the working electrode; too short a distance, or direct contact, has to be avoided since that would hinder the flow of ions. According to Ohm's law, the ohmic potential difference along the capillary is extremely small because the current is negligible; also, the resistance of the capillary is relatively low, because of its high concentration of salt.

When reporting measurements, it is convenient to convert the current, I (in units of A), to current density, j (in units of A/cm^2), by dividing by the surface area of the electrode; the values of j thus obtained are independent of the size of the particular electrode.

The arrangement shown in Figure 4 can be used to study various additives to the electrolyte. In one type of experiment, a constant electrode potential can be maintained and the resulting values of current be measured; the current supplied by the potentiostat is controlled by the measured potential difference. In a second type of experiment, a galvanostat may be used as the current supply; the current is kept constant, and the resulting electrode potential is measured. If j/E curves are to be determined, the potential or the current (current density) at each point must be kept constant for a while; it may take several minutes to establish the required steady state of a constant current density or of a constant potential difference, respectively [14], and only then should the measurement be made. Electrodes in an electrolyte thus do not behave the way solid ohmic resistances do, in which a given potential immediately causes the corresponding constant current, and vice versa. The response of an ohmic resistance to a voltage or current is *rigid*, in contrast to the *dynamic* behavior of electrodes in electrolytes [90].

The kinetics of the reactions at the electrode surface are complex. For the sake of simplicity, only one electrode reaction and its back reaction are considered in the following; an example might be metal dissolution and deposition (Equations 19 and 20). The rates of the reactions can be expressed as current densities, because of charge transfer. Reaction (19) causes the anodic current density, j_a, and reaction (20), the cathodic current density, j_c. At the equilibrium potential of the metal, both reactions occur at the same rate; j_a and j_c have the same value, called the *exchange current density*, j_0 [3, 80, 102].

$$|j_a^*| = |j_c^*| = j_0 \tag{37}$$

The asterisk is used for quantities at equilibrium conditions. The exchange current density cannot be measured directly, but may be calculated from the exchange of radioactive isotopes between a metal and a solution of its salt [80].

If an external current supply imposes an electron flow either to or from the metal, one of the reactions is accelerated and the corresponding back reaction is retarded. The measured *net current density* (j) is the difference between the cathodic and anodic current density, which now have different values:

$$j = j_a - j_c \tag{38}$$

If the transfer of charges were ideal, the imposed net electron flow would not change the electrode potential; the exchange of electrons between electrode and solution would occur unhindered. In reality, however, any net electron flow causes the electrode potential to deviate from the equilibrium potential. This deviation, which is a function of the current density, is called the *overpotential* η, and is a measure of the extent to which the anodic or cathodic reaction at an electrode is inhibited or limited.

Electrode reactions may be limited by various factors [3, 80, 90]. First, a metal immersed in water is surrounded by a so-called *double layer*, consisting of a layer of electrons in the metal and a layer of cations in the solution next to the metal; the layers resemble those in a capacitor. Ions reacting at the electrode, whether going inward or outward, have to pass through the double layer. This requires a certain amount of energy (the *activation energy*) which is observed as an overpotential called the *activation overpotential*. Second, electrode reactions may be limited by the finite rate of diffusion of the reactant to the metal surface, the result of which is the *diffusion overpotential*. Third, it is possible that the immediate reactant at the electrode is formed in a preceding chemical reaction, or that the product is removed in a subsequent reaction. This causes the *reaction overpotential*. The diffusion overpotential and the reaction overpotential taken together are the *concentration overpotential*, because the effects are closely related. If the electrode potential is made ever more positive or ever more negative than its equilibrium potential, the transfer of charges between metal and electrolyte is eventually limited by diffusion or by preceding or subsequent reactions. There are thus an anodic and a cathodic *limiting current density*, (j_{aL} and j_{cL}, respectively) that cannot be exceeded.

The following equation relates the current density to the overpotential, η [1] [90, 102]:

$$j = \frac{e^{\eta \alpha z F/RT} - e^{-\eta(1-\alpha) z F/RT}}{1/j_0 + (1/j_{aL})\, e^{\eta \alpha z F/RT} + (1/j_{cL})\, e^{-\eta(1-\alpha) z F/RT}} \tag{39}$$

[1] The dependence of the exchange and limiting current densities on the ion concentration has not been taken into account in the equations. In this discussion, the ion concentration is taken to be constant.

where z is the number of charges per ion, F is Faraday's constant (the magnitude of the charge of a mole of electrons, 96,485 °C/mol), R is the gas constant (8.3144 J/K · mol) and T is the thermodynamic (absolute) temperature. The parameter α is called the *symmetry factor* or *transfer coefficient*, and is explained as follows. If the electrode potential deviates to the same extent above and below the equilibrium potential, the deviations of j_a and j_c from the exchange current density are often not the same. This asymmetry between the anodic and cathodic reactions is α. If j_a and j_c respond symmetrically, α is equal to 0.5.

If the metal electrode is kept in a range that is neither close to the equilibrium potential nor close to the limiting current density, the exponential terms in the denominator and one of the exponential terms in the numerator are negligible or nearly so (because the limiting current densities, j_{aL} and j_{cL}, are much larger than the exchange current density, j_0). In this case, the current density can be described approximately by the simple exponential function

$$j = j_0 \cdot e^{\eta \alpha z F/RT} \tag{40}$$

or, if the polarity is inverse,

$$j = j_0 \cdot e^{-\eta(1-\alpha) z F/RT} \tag{41}$$

Therefore, a semilogarithmic plot of j and η in this range is almost linear. Because J. Tafel in 1905 was the first to do this empirically, the graph is called a *Tafel plot*. Extrapolating the Tafel plot to $\eta = 0$ is equivalent to mathematically extending the simple exponential form of the current density down into this range; therefore, the exchange current density j_0 is the intercept of this extrapolated line with the $\eta = 0$ line ($j = j_0$ because $e^0 = 1$) (Figure 5).

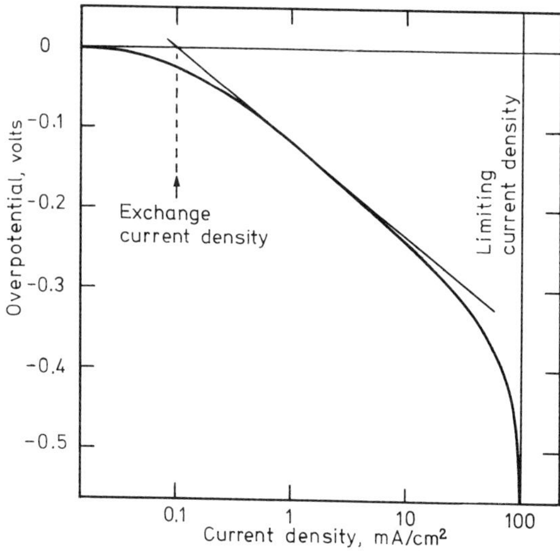

Fig. 5 Semilog (Tafel) plot of the current density and overpotential of a cathode, as calculated from Equation (39). The Tafel slope is indicated. Arbitrary values used in the calculation are as follows: symmetry factor, $\alpha = 0.5$; equilibrium (exchange) current density, $j_o = 0.1$ mA/cm^2; and limiting current density, $j_{aL} = j_{cL} = 100$ mA/cm^2

The electrode material has a strong influence on the value of the overpotential [80]. On many metals, forming H_2 by reducing H^+ ions (strictly speaking, H_3O^+ ions) causes a significant overpotential; the process is relatively complex. The newly produced hydrogen atoms must come together on the metal to form the H–H bond [92]. Platinum metals are known as catalysts for the activation of H_2; contrariwise, below a certain current density, the overpotential of H_2 formation on platinum metals is almost zero [80].

Often, more than one reaction (accompanied, of course, by its back reaction) takes place on a metal electrode. Iron is a good example: if an iron electrode is connected to a current supply, as in Figure 4, the net current density is composed of the currents connected with iron dissolution (subscript Fe) and hydrogen formation (subscript H), both reactions consisting again of their forward and back reactions:

$$j = j_{a,Fe} + j_{a,H} - j_{c,Fe} - j_{c,H} \tag{42}$$

One of the net reactions can be completely suppressed if the electrode potential is held at the equilibrium potential of that reaction by means of the potentiostat. In that case, j is composed solely of the contributions from the other net reaction. On freely corroding iron, without net current to or from the metal, both the net oxidation of the metal and the production of hydrogen occur at the same rate. The potential of free corrosion is thus somewhere between the equilibrium potentials (never attained) of the two net reactions. This potential is called the *rest potential* [102].

3.2 Microbial corrosion of iron

3.2.1 Microbial corrosion of iron by acid or oxygen

The corrosion of iron by acid according to Equation (21), in which acid-tolerant thiobacilli create a rather acid environment by oxidizing sulfur species, is a process that is easily understood (see Section 2.2). However, this type of microbial corrosion occurs only when iron structures are in contact with sulfurous waters, acid mine drainages, and the like.

Iron can be microbially corroded by O_2 at neutral pH in aqueous environments if anodic sites can be established, thus forming local galvanic cells. The oxygen tension may be very low under bacterial colonies or films adhering to the metal, especially if organic substrates are present. At these sites, only the anodic dissolution of Fe to Fe^{2+} takes place. The electrons flow to other sites on the iron surface, where they can reduce oxygen [77]. The iron thus dissolves in the anaerobic microniche under the bacterial layer, and corrosion is observed as pitting. If the anodic and cathodic net reactions are separated spatially, corrosion is obviously more severe than if the reactions occur evenly over the whole surface. The Fe^{2+} ions formed may be oxidized chemically or by iron-oxidizing bacteria [83] to hydrates of ferric oxide.

3.2.2 Anaerobic microbial corrosion

Iron also corrodes if neither oxygen nor acids are present. This anaerobic corrosion, of great economic significance, is largely mediated by microorganisms; the chief culprits seem to be sulfate-reducing bacteria.

Sulfate-reducing bacteria (sulfate reducers) are obligately anaerobic bacteria that gain energy for growth by oxidizing organic compounds or H_2 with SO_4^{2-} [67, 87, 88, 125], reducing the sulfate to H_2S. Most known species of sulfate-reducing bacteria are mesophilic and grow optimally between 20° and 40 °C. The mesophilic species described so far are classified in 10 genera [94, 125], the names of which all start with *Desulfo-*. One of these genera, the spore-forming *Desulfotomaculum*, also comprises a few thermophilic species that grow optimally between 55° and 65 °C [82, 125]. In addition, the thermophilic genus *Thermodesulfobacterium* [95, 129] and the extreme thermophile *Archaeoglobus fulgidus* [104] have been described. The latter, one of the archaebacteria, has a temperature optimum of 83 °C. Nutritional and biochemical characters can be used to distinguish two subgroups among the sulfate reducers. The incomplete oxidizers degrade organic substrates, such as lactate, to acetate as an end product:

$$2\,CH_3CHOH{\cdot}COO^- + SO_4^{2-} \longrightarrow 2\,CH_3COO^- + 2\,HCO_3^- + H_2S \tag{43}$$

In contrast, the complete oxidizers mineralize organic substrates, including acetate, to CO_2 or bicarbonate, as in these reactions:

$$2\,CH_3CHOH{\cdot}COO^- + 3\,SO_4^{2-} + 2\,H^+ \longrightarrow 6\,HCO_3^- + 3\,H_2S \tag{44}$$

$$CH_3COO^- + SO_4^{2-} \longrightarrow 2\,HCO_3^- + H_2S \tag{45}$$

The organisms commonly use low molecular weight compounds, including mono- and dicarboxylic acids, lactate (Equations (43) and (44)), alcohols, and aromatic compounds such as benzoate. They may also oxidize long-chain saturated hydrocarbons (alkanes) [1]. Many of the incomplete and complete oxidizers among the sulfate-reducing bacteria can also grow with H_2 as the sole electron donor for sulfate reduction. This capability plays a key role in the hypotheses and models of anaerobic corrosion, and its activating enzyme, hydrogenase, is also of great interest. Hydrogenases, both cytoplasmic ones and very active periplasmic ones, have been found in sulfate-reducing bacteria of the genus *Desulfovibrio* [67, 87]. The periplasmic hydrogenases are thought to contribute to the transmembrane proton gradient and thus to energy conservation during growth on H_2. This conclusion is founded on the belief that the electron-yielding H_2 oxidation by hydrogenase (equation (46)) and the electron-consuming sulfate reduction (Equation 47) are separated by the cytoplasmic membrane [67, 87].

$$4\,H_2 \longrightarrow 8\,H^+ + 8\,e^- \tag{46}$$

$$SO_4^{2-} + 10\,H^+ + 8\,e^- \longrightarrow 4\,H_2O + H_2S \tag{47}$$

The overall reaction is thus

$$4\,H_2 + SO_4^{2-} + 2\,H^+ \longrightarrow 4\,H_2O + H_2S. \tag{48}$$

The electrons generated can be transported across the cytoplasmic membrane by electron carriers (cytochromes, iron-sulfur proteins), and the protons by ATPase, driving the phosphorylation of ADP to ATP. Some species (*autotrophs*) can synthesize their cell carbon entirely from CO_2 when they grow on H_2. Other H_2-utilizing species depend on organic compounds such as acetate as their major source for organic cell constituents (*heterotrophs*) [16, 125].

The detailed mechanism by which sulfate-reducing bacteria corrode iron was first suggested by VON WOLZOGEN KÜHR and VAN DER VLUGT [116] when they explained the corrosion of pipes in wet soil by the activities of hydrogen-consuming sulfate reducers. Their proposed mechanism is known as the *depolarization theory*.

Anodic reaction $$4\,Fe \rightleftharpoons 4\,Fe^{2+} + 8\,e^- \tag{49}$$

Dissociation of water $$8\,H_2O \rightleftharpoons 8\,H^+ + 8\,OH^- \tag{50}$$

Cathodic reaction $$8\,H^+ + 8\,e^- \rightleftharpoons 8\,H \tag{51}$$

Hydrogen consumption $$8\,H + SO_4^{2-} + 2\,H^+ \longrightarrow 4\,H_2O + H_2S \tag{52}$$

Sulfide precipitation $$Fe^{2+} + H_2S \rightleftharpoons FeS + 2\,H^+ \tag{53}$$

Hydroxide precipitation $$3\,Fe^{2+} + 6\,OH^- \rightleftharpoons 3\,Fe(OH)_2 \tag{54}$$

Total reaction $$4\,Fe + SO_4^{2-} + 4\,H_2O \longrightarrow FeS + 3\,Fe(OH)_2 + 2\,OH^- \tag{55}$$

Equation (52) is written with atomic hydrogen, which was assumed to be a direct substrate for H_2-utilizing sulfate reducers. Much effort has been devoted to trying to find support for this theory. Many workers have regarded the consumption of hydrogen as the key reaction that enables or stimulates anaerobic corrosion. Nevertheless, there is to date no consensus on the proposed mechanism, and some alternative explanations of anaerobic corrosion have been suggested. In recent publications, it is not always clear whether depolarization is meant as the consumption of atomic or of molecular hydrogen. Often, the term *cathodic hydrogen* is used without further definition. A distinction between atomic and molecular hydrogen, however, is important for understanding the kinetics of the redox reactions on corroding iron.

There are good reviews in the extensive literature dealing with anaerobic microbial corrosion [e.g., 35, 77, 85]. I therefore do not intend to give another overview of the literature here. Rather, I attempt to give a brief presentation of some basic models of anaerobic corrosion and to use the principles outlined above (Section 3.1) to discuss some problems and contradictions that remain.

Depolarization experiments with sulfate reducers. The apparatus depicted in Figure 4 has been used to determine current density/potential curves, both without and with sulfate-reducing bacteria present, to discover whether the bacteria stimulate the cathodic reaction by the consumption of H or of H_2. The potential of the working electrode was kept more negative than its rest potential, so that electrons from the external supply flowed to the metal. The measured electrode potential was thus a function of the overpotential of the reduction of protons. The current supply was operated as a potentiostat. In most experiments the added sulfate reducers were species able to utilize H_2, of the genus *Desulfovibrio*.

In the experiments with sulfate as terminal electron acceptor, the sulfide precipitated as FeS, which could have caused an additional effect on electrode reactions. In other experiments, therefore, fumarate and benzyl viologen (BV^{2+}) were added as alternative electron acceptors. Many *Desulfovibrio* species can dismutate (ferment) fumarate in the absence of sulfate, producing succinate, acetate, and CO_2. If an electron donor such as H_2 is present, the chief product of fumarate reduction is succinate [6, 33, 78]. Benzyl viologen can withdraw electrons directly at the hydrogenase:

$$H_2 + 2\,BV^{2+} \xrightleftharpoons{\text{Hydrogenase}} 2\,H^+ + 2\,BV^+ \tag{56}$$

The experimenters found that at a given electrode potential, the current density in the presence of *Desulfovibrio* was always higher than in their absence [13, 14, 36, 86]; the more negative the electrode potential, the greater was the change in current density. Two examples are combined in Figure 6. It is striking how much current densities differ in independent experiments at the same potentials, so comparisons between j/E curves in the absence and presence of bacteria are meaningful only if they have been recorded in the same apparatus. One can state the results the other way: at a particular current density, the electrode potential became less negative when sulfate reducers were added. The bacteria thus lowered the overpotential. This depolarization was observed with fumarate [13] and with benzyl viologen [14], as well as with sulfate. In other experiments, labeled $[^{35}S]-SO_4^{2-}$ was reduced to sulfide [36], and the sulfate reducers did indeed grow at an electrode potential more negative than -0.74 (referred to the standard hydrogen electrode) [86]. However, the depolarization was transitory, the working electrode gradually became repolarized [36], and the sulfate reducers ceased growing after 20 days [86]. In both experiments, a layer of FeS formed, which was supposed to slow the electrode reaction, and cause both the repolarization and the cessation of growth.

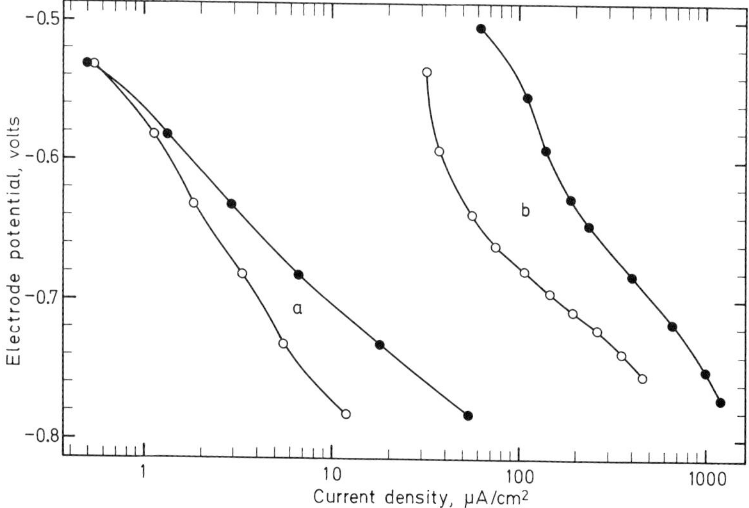

Fig. 6 Depolarization experiments with *Desulfovibrio*; the curves labeled a are redrawn from [36], b from [86]. For this figure, the potentials originally measured against a calomel electrode were converted into potentials against a standard hydrogen electrode. KEY: ○, without bacteria; ●, with *Desulfovibrio*. The rest potential was about -0.48 to -0.49 V

In experiments with *Desulfotomaculum orientis*, a spore-forming sulfate reducer, the electrode did not depolarize [14]. The explanation proffered was that the organism lacked hydrogenase. Later on, however, other workers found that *Desulfotomaculum orientis* does grow with H_2 [55], but the hydrogenase is in the cytoplasm, in contrast to that in *Desulfovibrio* [24]. So far, there are no reports on experiments to resolve whether the localization of the hydrogenase can explain the different results that the different species give on iron cathodes.

Depolarization experiments with ferrous sulfide. In a modified model of corrosion, FeS promotes rather than hinders the cathodic reaction [13, 52]. During current density/potential measurements using a horizontally mounted working electrode, depolarization occurred immediately after the FeS had sedimented on the electrode. The depolarization by FeS alone was more significant than that by *Desulfovibrio* cells with FeS [13]. Presumably, the H^+ ions were reduced at the FeS on the iron, rather than at the iron itself, allowing the H_2 to be formed with a lower overpotential. In this model, ferrous sulfide is an electric conductor and depolarizes the iron electrode, as does platinum metal in intimate contact. However, the depolarizing effect of FeS gradually disappeared. Depolarization was most effective if both FeS and *Desulfovibrio* were simultaneously present [13], and ferrous sulfide that had become ineffective regained its depolarizing effect if the sulfate reducer was added [52]. It was assumed that the sulfate reducer took atomic hydrogen from the ferrous sulfide and thus increased its depolarizing activity, the ferrous sulfide acting as a bridge for reducing equivalents between the metal surface and the bacteria. One might ask whether such a bridge could also permit electrons to flow directly into sulfate reduction inside the cells, without any involvement of H or H_2. One would have to postulate a component transporting electrons from the FeS attached on the outside (a microphotograph in [77] depicts this) to the hydrogenase on the other side of the outer membrane of *Desulfovibrio*.

Depolarization experiments with hydrogen sulfide. The results of depolarization experiments by Costello [23] have led people to wonder whether the consumption of hydrogen is indeed responsible for the depolarization by sulfate-reducing bacteria. When making current density/potential curves, Costello observed a depolarization upon adding H_2S or cell-free culture supernatant to the electrolyte. Washed cells of sulfate reducers or supernatants from cultures from which sulfide had been removed by sparging with N_2 did not depolarize the cathode. He suggested that the depolarizing mechanism might be the reduction of H_2S to H_2 with formation of HS^- ions, as in Equation (35). He explained the depolarization observed previously by Booth and Tiller [14] in the presence of sulfate reducers and benzyl viologen as a direct reaction of the dye at the iron surface. Even though Booth and Tiller had found that benzyl viologen was not reduced in the absence of sulfate reducers, Costello [23] demonstrated that depolarization increased with the amount of added benzyl viologen. Supposedly, the benzyl viologen was reoxidized at the site of the hydrogenase of the organism if a physiological electron acceptor (sulfate, fumarate) was present. According to this model, benzyl viologen cycles as a mediator between iron and the bacteria and thus maintains the depolarization. We have found that iron powder mixtures with benzyl viologen are reduced to the blue-violet state, but for only a moment, and quickly become colorless as the dye reacts further (unpublished results of VON BÜNAU and WIDDEL).

Experiments with freely corroding iron. Depolarization experiments in electrochemical cells (Figure 4) do not accurately reproduce the conditions of free corrosion for the following reason. The external current supply keeps the iron artificially more negative than its rest potential; the iron is prevented from going into solution, so that the situation is actually cathodic protection; and the formation of H_2 occurs at a more negative overpotential than during free corrosion.

If the experimenter set the electrode potential to a value close to that of the rest potential, the depolarizing effect decreased markedly; this was true whether sulfate-reducing bacteria [13, 36], FeS [13], or H_2S [23] was present. None of these added agents caused the rest potential to change markedly.

Experiments were done to test the effect of sulfate reducers or FeS on the free corrosion of iron. Specimens of steel were incubated anaerobically under various conditions in sulfate-free medium with fumarate [13]. The weight loss due to corrosion was followed over time. If FeS was allowed to deposit on the steel specimens, the rate of weight loss was strongly accelerated, especially if 1% NaCl was also present. The addition of sulfate reducers had no effect.

Steel wool should corrode at significant rates that are easily measured, since it has a large surface area relative to the volume of the incubation vessel. We used it as the iron source in free corrosion experiments with cultures of a H_2-utilizing *Desulfovibrio* species that incompletely oxidizes organic substrates [22]. At the end of the incubation period we determined the total amount of sulfide formed (FeS plus dissolved sulfide). If steel wool and the bacteria were incubated in the absence of other hydrogen donors, no measurable sulfide had formed at the end of two weeks. Obviously, hydrogen was not evolved from the steel wool in that time. However, if a defined amount of lactate was added, the amount of sulfide formed within two weeks in the presence of steel wool was higher by 30 to 50% than in controls with lactate alone. When the incubation period was 60 days, the amount of sulfide formed was more than 90% greater in the presence of steel wool than in its absence. The additional sulfide must have been formed by sulfate reduced with reducing equivalents from the steel wool. With *Desulfovibrio sapovorans*, which cannot use H_2, no additional sulfide was formed. Obviously, the steel wool could provide measurable amounts of reducing equivalents for sulfate reduction only if the bacteria had an additional electron donor, in this case, lactate. The explanation may lie with several different, probably cumulative, factors. It is possible that the bacteria significantly depolarized the steel, according to the overall reaction (Equation (55)), only when they grew on lactate. This is in agreement with a report that steel specimens were more heavily corroded when they were in continuously growing cultures of sulfate reducers than when they were in stationary cultures [123]. Another possibility is that H_2S formed during lactate oxidation first liberated H_2 according to

$$Fe + H_2S \longrightarrow FeS + H_2 \tag{57}$$

The H_2 would then be an additional electron donor for the *Desulfovibrio* incubated with the steel wool. Finally, FeS produced at the electrode and in contact with it may have favored the cathodic depolarization [13, 52].

In subsequent experiments we learned that steel wool from different batches corrodes at significantly different rates; it is probably inhomogeneous and contains variable proportions of rather corrosion-resistant alloying constituents (unpublished results of von Bünau and Widdel). We therefore used pure iron powder in further corrosion experiments. This pure iron evolved hydrogen chemically in sulfide-free mineral medium buffered with CO_2/bicarbonate; indeed, H_2 formation is what we expect thermodynamically (Section 3.1 and Figure 3). In comparison with the iron powder, the steel samples of earlier experiments [22] seemed to be more resistant to corrosion in the absence of sulfide. A neutralized sulfide solution significantly increased the rate at which H_2 was chemically evolved by iron powder (Figure 7). About the time the sulfide had been consumed, the rate decreased to approximately the rate measured in a sulfide-free control. The newly formed FeS attached to the iron particles seemed to have no significant effect on the rate of H_2 formation in this experiment. We conclude from the data that the rate of anaerobic corrosion of iron increases if free H_2S is present to react with the iron according to the overall reaction (57). Therefore, the reduction of sulfate with organic substrates, even by hydrogenase-negative sulfate reducers, should also promote anaerobic corrosion. As for the mechanism by which free H_2S stimulates the dissolution of iron, it is possible that H_2S reacts readily as an undissociated, neutral

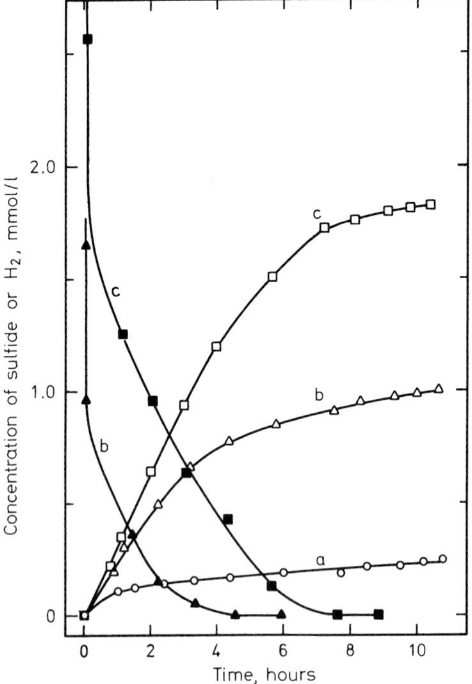

Fig. 7 Acceleration by sulfide of the chemical formation of H_2 from water and by anaerobically immersed, powdered Fe.
KEY: open symbols, H_2 in the gas phase; closed symbols, dissolved sulfide.
Curve a, control without sulfide; curves b, 2 mM sulfide added; curves c, 4mM sulfide added.
The measurements were carried out in stoppered vials with pure iron (300 mg) in O_2-free, CO_2/$NaHCO_3$-buffered (30 mM, pH 7.1) medium (30 ml), under a head space (30 ml) of N_2 and CO_2, at 22 °C. Sulfide was added as a neutral H_2S/NaHS solution. The vials were shaken. Part of the added sulfide was immediately bound by oxidized forms of iron initially present
(unpublished results of von Bünau and Widdel)

molecule on the iron surface according to reaction (35); another possibility is that free sulfide scavenges the released Fe^{2+} ions by precipitating them as FeS. The positive ions presumably form a layer surrounding the negative metal [3, 80]. If the formation of a precipitate destroyed this layer, it might stimulate the dissolution of the metal.

If iron is immersed in an anoxic, sulfide-free medium without organic electron donors, only one-fourth mole of sulfate can be reduced per mole of iron oxidized. Therefore, only one-fourth of the Fe^{2+} ions generated can be precipitated as FeS (Equation 55); if that is the only mechanism by which sulfide is being formed, corrosion should be stimulated only slightly.

Corrosion by polysulfide. Yet another corrosion reaction can occur if small amounts of O_2 occasionally get into anaerobic habitats without making them aerobic. This is rather common, since every anaerobic habitat comes into contact with oxygen or oxidizing surroundings somewhere. Under these circumstances, sulfide oxidizes as far as sulfur (as in Equation (13)) or polysulfide (HS_n^-) [18, 29 a]. For instance, disulfide can be formed according to the following reaction:

$$4\,H_2S + O_2 \longrightarrow 2\,H^+ + 2\,HS_2^- + 2\,H_2O \tag{58}$$

Polysulfides are fairly stable under alkaline conditions but decompose readily to sulfur and hydrogen sulfide if acidified. Contrariwise, sulfide and sulfur combine readily at high pH to form polysulfides [29 a, 88 a]. Sulfur and particularly polysulfides are highly corrosive agents [98], as exemplified in this reaction:

$$HS_2^- + Fe \longrightarrow HS^- + FeS \tag{59}$$

Polysulfides even convert several rather noble metals, such as silver, rapidly to sulfides.

Growth of methanogenic bacteria in the presence of iron. Methanogenic bacteria (methanogens) are also believed to corrode materials by consuming hydrogen [25]. They are, like sulfate reducers, obligate anaerobes. The most common energy-yielding reaction among the many species of these bacteria is the reduction of CO_2 to CH_4 with H_2 [114]:

$$CO_2 + 4\,H_2 \longrightarrow CH_4 + 2\,H_2O \tag{60}$$

These bacteria grew and formed methane in medium with iron powder without other hydrogen donors [25]. One hypothesis given was that iron is thermodynamically capable of dissolving because the methanogens use molecular hydrogen; however, the calculations were based on an activity for the Fe^{2+} of 1.0, which is far from real conditions. The calculated free energy corresponded to a value of E^0 of 0.409 V, which is the less negative of the two different values found in the literature (see Section 3.1). The amounts of methane and Fe^{2+} formed in the presence of steel pellets were in agreement with the overall reaction

$$8\,H^+ + 4\,Fe + CO_2 \longrightarrow CH_4 + 4\,Fe^{2+} + 2\,H_2O \tag{61}$$

If the methanogens were not present, no Fe^{2+} formed [25]. In the experiments with iron powder, chemically formed molecular hydrogen is probably used for methanogenesis; this seems plausible because the media incorporated sulfide as reductant, and sulfide stimulates the oxidation of iron (see Figure 7). However, the steel pellets did not release Fe^{2+} ions in the absence of bacteria, and must therefore resist corrosion better than does iron powder. Since the methanogens catalyzed the formation of Fe^{2+} from the steel pellets, the pellets may indeed have been corroded by bacterial depolarization on the metal surface.

Corrosion by phosphorus compounds. When experiments on the corrosion of iron were done in cultures of sulfate-reducing bacteria, iron phosphide (Fe_2P) was found, as well as FeS [42, 43, 44]. The Fe_2P phosphide is not a real salt, and is rather resistant to hydrolysis. It has been suggested that Fe_2P is precipitated by volatile phosphorus compounds formed by sulfate-reducing bacteria, such as one detected in the gas phase above cultures of sulfate-reducing bacteria. That compound was probably not phosphine (PH_3) [43]. However, small quantities of phosphine were reported to evolve when H_2S acted on sodium phosphate ($Na_3PO_4 \cdot 12\,H_2O$), sodium phosphite ($Na_2HPO_3 \cdot 5\,H_2O$), or sodium hypophosphite ($NaH_2PO_2 \cdot H_2O$) [43, 44]. More detailed investigations on the formation of volatile phosphorus compounds in cultures of sulfate reducers have not yet been reported.

WEIMER *et al.* found in corrosion experiments with sulfate reducers in lactate media that the rate of corrosion of steel specimens increased with increasing phosphate concentration [123]; vivianite ($Fe_3(PO_4)_2 \cdot 8\,H_2O$) was the product of corrosion. Vivianite should be formed only if there is more phosphate than free sulfide. The experimenters also found that small quantities of iron phosphide were formed from added phosphate.

The reduction of phosphate and other phosphorus species by physiological electron donors differs from the reduction of sulfate because it would require a very high input of energy and thus be a drain on the energy metabolism of the organism. I calculated redox potentials at pH 7.0 from the E^0 values of the redox reactions of phosphorus species and the dissociation constants of the respective acids [31]. The values of $E^{0'}$ are more negative than the $E^{0'}$ value of the $2\,H^+/H_2$ reaction (– 0.414 V):

$$(E^{0'})$$

$$H_2PO_4^- + 2H^+ + 2e^- \rightleftharpoons H_2PO_3^- + H_2O \qquad -0.64\ V \tag{62}$$

$$H_2PO_3^- + 2H^+ + 2e^- \rightleftharpoons H_2PO_2^- + H_2O \qquad -0.90\ V \tag{63}$$

$$H_2PO_2^- + 2H^+ + e^- \rightleftharpoons P + 2H_2O \qquad -1.27\ V \tag{64}$$

$$P + 3H^+ + 3e^- \rightleftharpoons PH_3 \qquad -0.48\ V \tag{65}$$

The dissimilatory reduction of protons to H_2 should take place more readily than the reduction of phosphate, the other acids, or phosphorus. Those doing corrosion experiments should also keep in mind that iron and steel often contain small amounts of phosphides.

Concluding remarks. In conclusion, we do not yet have a model of anaerobic corrosion that is in agreement with all the various experiments. Some of the findings and interpretations are contradictory.

Metallic iron immersed in water is thermodynamically unstable at neutral pH even in the absence of O_2. Hence, the role of anaerobic bacteria cannot be to render corrosion exergonic, as is sometimes assumed. Rather, the bacterial processes must be regarded as accelerants of corrosion, and their effects as problems in dynamic electrochemistry (the kinetics of electrode processes). The processes of anaerobic corrosion at neutral pH are complex, and are not quite as easily understood as is, for example, the corrosion of iron in acid. There are additional factors to be considered besides those outlined in the foregoing paragraphs; for example, cells may adhere to the metal surface with sticky polymers, or the CO_2/bicarbonate concentration may affect corrosion. It is my belief that two principal mechanisms are of prime importance in anaerobic corrosion. One of these is the acceleration of corrosion by H_2S, the end product of sulfate-reducing bacteria. The other is the bacterially catalyzed depolarization that is so often discussed, but contradictory published reports must be reconciled. One also has yet to demonstrate depolarization with freely corroding iron.

Mechanistically, it appears barely possible that a scavenging of molecular hydrogen, H_2, is responsible for the depolarization of iron. On freely corroding iron, and even more on iron cathodes (such as depicted in Figure 4), H_2 is formed at a significant overpotential, that is, rather far from equilibrium and thus with negligible back reaction. If bacteria utilize the H_2 there should be no major effect on the rate of H_2 formation. Hence, any effect that sulfate reducers had on the current density/potential curves was probably not due to their utilizing molecular hydrogen; the bacteria must have accelerated a preceding reaction. They may have done this by utilizing atomic hydrogen, as formulated in the original corrosion theory [116], or by favoring the condensation of H atoms to H_2. If the bacteria are indeed able to catalyze one of these reactions, they should live in very intimate contact with the metal surface. Furthermore, redox-active agents, either excreted by living cells or liberated from lysed cells, may also be involved in depolarization. Microbiologists should collaborate with physical chemists to shed more light on the fundamental questions.

4 References

[1] Aeckersberg, F.; Bak, F.; Widdel, F.: Anaerobic Oxidation of Saturated Hydrocarbons to CO_2 by a New Type of Sulfate-Reducing Bacterium; *Arch. Microbiol.* (1991), in press.

[1a] Alexandersson, E. T.: Marks of Unknown Carbonate-Decomposing Organelles in Cyanophyte Borings; *Nature (London)* **254** (1975) 237–238.

[2] Atkins, P. W.: *Physical Chemistry*, 3rd ed., Chapter 11 (Equilibrium Electrochemistry: Ions and Electrodes), Oxford University Press, Oxford (1988); pp 234–257.

[3] Atkins, P. W.: *Physical Chemistry*, 3rd ed., Chapter 32 (Dynamic Electrochemistry), Oxford University Press, Oxford (1988); pp 790–809.

[4] Atkins, P. W.: *Physical Chemistry*, 3rd ed., Tables of Data, Oxford University Press, Oxford (1988), 813–836.

[4a] Balashova, V. V.: The Use of Molecular Sulfur as an Agent Oxidizing Hydrogen by the Facultative Anaerobic *Pseudomonas* Strain; *Mikrobiologiya (Moscow)* **54** (1985) 324–326.

[5] BALASHOVA, V. V.; ZAVARZIN, G. A.: Anaerobic Reduction of Ferric Iron by Hydrogen Bacteria; *Mikrobiologiya (Moscow)* **48** (1979) 773–778.

[6] BARTON, L. L.; LE GALL, J.; PECK, H. D., JR.: Phosphorylation Coupled to Oxidation of Hydrogen with Fumarate in Extracts of the Sulfate Reducing Bacterium, *Desulfovibrio gigas*; *Biochem. Biophys. Res. Commun.* **41** (1970) 1036–1042.

[7] BAUMGÄRTNER, M.; REHMDE, A.; BOCK, E.; CONRAD, R.: Release of Nitric Oxide from Building Stones into the Atmosphere; *Atmos. Environ.* **24 B** (1990) 87–92.

[8] BERTHELIN, J.: Microbial Weathering Processes. In: KRUMBEIN, W. E., Ed.; *Microbial Geochemistry*, Blackwell, Oxford, London (1983); pp 223–262.

[9] BIRKINSHAW, J. H.: Chemical Constituents of the Fungal Cell. In: AINSWORTH, G. C.; SUSSMAN, A. S., Eds.; *The Fungi*, Academic Press, New York, London (1965); Vol. I, The Fungal Cell, pp 179–228.

[10] BOCK, E.: Biologisch induzierte Korrosion von Naturstein – starker Befall mit Nitrifikanten; *Bautenschutz Bautensanierung, Sonderausgabe* (**1987**) 42–45.

[11] BOCK, E.; KOOPS, H.-P.; HARMS, H.: Nitrifying Bacteria. In: SCHLEGEL, H. G.; BOWIEN, B., Eds.; *Autotrophic bacteria*, Science Tech Publishers, Madison; Springer-Verlag, Berlin, Heidelberg, New York (1989) pp 81–96.

[12] BOCK, E.; SAND, W.; MEINCKE, M.; WOLTERS, B.; AHLERS, B.; MEYER, C.; SAMELUCK, F.: Biologically Induced Corrosion of Natural Stones – Strong Contamination of Monuments with Nitrifying Organisms; *Biodeterioration* **7** (1987) 436–440.

[13] BOOTH, G. H.; ELFORD, L.; WAKERLEY, D. S.: Corrosion of Mild Steel by Sulphate-Reducing Bacteria: An Alternative Mechanism; *Br. Corros. J.* **3** (1968) 242–245.

[14] BOOTH, G. H.; TILLER, A. K.: Cathodic Characteristics of Mild Steel in Suspensions of Sulphate-Reducing Bacteria; *Corros. Sci.* **8** (1968) 583–600.

[15] BOS, P.; KUENEN, J. G.: Microbiology of Sulphur-Oxidizing Bacteria. In: *Microbial Corrosion*, The Metals Society, London (1983), pp 18–27.

[16] BRYSCH, K.; SCHNEIDER, C.; FUCHS, G.; WIDDEL, F.: Lithoautotrophic Growth of Sulfate-Reducing Bacteria, and Description of *Desulfobacterium autotrophicum* gen. nov., sp. nov.; *Arch. Microbiol.* **148** (1987) 264–274.

[17] CALDWELL, D. E.: Genus *Thermothrix*. In: STALEY, J. T.; BRYANT, M. P.; PFENNIG, N.; HOLT, J.G., Eds.; *Bergey's Manual of Systematic Bacteriology*, Williams & Wilkins, Baltimore, Hong Kong, London, Sydney (1989); Vol. 3, pp 1868–1871.

[18] CHEN, K. Y.; ASCE, A. M.; MORRIS, J. C.: Oxidation of Sulfide by O_2: Catalysis and Inhibition; *J. Sanit. Eng. Div., Proc. Am. Soc. Civ. Eng.* **98** (1972) 215–227.

[19] CHENG, R. T.; CORN, M.; FROHLIGER, J. O.: Contribution to the Reaction Kinetics of Water Soluble Aerosols and SO_2 in Air at ppm Concentrations. In: *The Chemistry of Air Pollution*, MSS Information Corporation, New York (1973); pp 15–36.

[20] CLAUS, D.; WITTMANN, H.; RIPPEL-BALDES, A.: Untersuchungen über die Zusammensetzung von Bakterienschleimen und deren Lösungsvermögen gegenüber schwerlöslichen anorganischen Verbindungen; *Arch. Mikrobiol.* **29** (1958) 169–187.

[21] COLBERG, P. J.: Anaerobic Microbial Degradation of Cellulose, Lignin, Oligolignols, and Monoaromatic Lignin Derivatives. In: ZEHNDER, A. J. B., Ed.; *Biology of Anaerobic Microorganisms*, John Wiley & Sons, New York (1988); pp 333–372.

[22] CORD-RUWISCH, R.; WIDDEL, F.: Corroding Iron as a Hydrogen Source for Sulphate Reduction in Growing Cultures of Sulphate-Reducing Bacteria; *Appl. Microbiol. Biotechnol.* **25** (1986) 169–174.

[23] COSTELLO, J. A.: Cathodic Depolarization by Sulphate-Reducing Bacteria; *S. Afr. J. Sci.* **70** (1974) 202–294.

[24] CYPIONKA, H.; DILLING, W.: Intracellular Localization of the Hydrogenase in *Desulfotomaculum orientis*; *FEMS Microbiol. Lett.* **36** (1986) 257–260.

[25] DANIELS, L.; BELAY, N.; RAJAGOPAL, B. S.; WEIMER, P. J.: Bacterial Methanogenesis and Growth from CO_2 with Elemental Iron as the Sole Source of Electrons; *Science* **237** (1987) 509–511.

[26] FOSTER, J. W.: *Chemical Activities of Fungi*, Academic Press, New York (1949).

[27] FRIEDMANN, E. I.; OCAMPO-FRIEDMANN, R.: Endolithic Microorganisms in Extreme Dry Environments: Analysis of a Lithobiontic Microbial Habitat. In: KLUG, M. J.; REDDY, C. A., Eds.; *Current Perspectives in Microbial Ecology*, American Society for Microbiology, Washington, D.C. (1984); pp 177–185.

[28] GEISSMAN, T. A.: The Biosynthesis of Phenolic Products. In: BERNFELD, P., Ed.; *Biogenesis of Natural Compounds*, Pergamon Press, Oxford, London, New York, Paris (1963); pp 563–616.

[29] GHIORSE, W. C.: Microbial Reduction of Manganese and Iron. In: ZEHNDER, A. J. B., Ed.; *Biology of Anaerobic Microorganisms*, Wiley, New York (1988); pp 305–331.

[29a] *GMELINS Handbuch der anorganischen Chemie*, 8th ed., Schwefel (Sulfur), Part B, (Wäßrige Lösung des Schwefelwasserstoffs und der Sulfid-Ionen), Verlag Chemie, Weinheim (1953); pp 81–106.

[30] GREENWOOD, N. N.; EARNSHAW, A.: *Chemistry of the Elements*, Chapter 11 (Nitrogen), Pergamon Press, Oxford, New York (1986); pp 466–545.

[31] GREENWOOD, N. N.; EARNSHAW, A.: *Chemistry of Elements*, Chapter 12 (Phosphorus), Pergamon Press, Oxford, New York (1986); pp 546–636.

[32] GREENWOOD, N. N.; EARNSHAW, A.: *Chemistry of the Elements*, Chapter 15 (Sulfur), Pergamon Press, Oxford, New York (1986); pp 757–881.

[33] GROSSMAN, J. P.; POSTGATE, J. R.: The Metabolism of Malate and Certain Other Compounds by *Desulphovibrio desulphuricans*; *J. Gen. Microbiol.* **12** (1955) 429–445.

[34] HALE, M. E., JR.: *The Biology of Lichens*, 3rd ed., Edward Arnold, London (1983).

[35] HAMILTON, W. A.: Sulphate-Reducing Bacteria and Anaerobic Corrosion; *Ann. Rev. Microbiol.* **39** (1985) 195–217.

[36] HARDY, J. A.: Utilization of Cathodic Hydrogen by Sulphate-Reducing Bacteria; *Br. Corros. J.* **18** (1983) 190193.

[37] HAWKSWORTH, D. L.; HILL, D. J.: *The Lichen-Forming Fungi*, Blackie, Glasgow, London (1984).

[38] HOOPER, A. B.: Biochemistry of the Nitrifying Lithoautotrophic Bacteria. In: SCHLEGEL, H. G.; BOWIEN, B., Eds.; *Autotrophic Bacteria*, Science Tech Publisher, Madison; Springer-Verlag, Berlin, Heidelberg, New York (1989); pp 239–265.

[39] HUMPHRIES, S. G.: The Biosynthesis of Tannins. In: BERNFELD, P., Ed.; *Biogenesis of Natural Compounds*, Pergamon Press, Oxford, London, New York, Paris (1963); pp 617–639.

[40] HUTCHINS, S. R.; DAVIDSON, M. S.; BRIERLEY, J. A.; BRIERLEY, C. L.: Microorganisms in Reclamation of Metals; *Ann. Rev. Microbiol.* **40** (1986) 311–336.

[41] ILER, R. K.: Hydrogen-Bonded Complexes of Silica with Organic Compounds. In: BENDZ, G.; LINDQVIST, I., Eds.; *Biochemistry of Silicon and Related Problems*, Plenum Press, New York, London (1977); pp 53–76.

[42] IVERSON, W. P.: Corrosion of Iron and Formation of Iron Phosphide by *Desulfovibrio desulfuricans*; *Nature (London)* **217** (1968) 1265–1267.

[43] IVERSON, W. P.; OLSON, G. J.: Anaerobic Corrosion by Sulfate-Reducing Bacteria Due to Highly Reactive Volatile Phosphorus Compounds. In: *Microbial Corrosion*, The Metals Society, London (1983); pp 46–53.

[44] IVERSON, W. P.; OLSON, G. J.: Anaerobic Corrosion of Iron and Steel: A Novel Mechanism. In: KLUG, M. J.; REDDY, C. J., Eds.; *Current Perspectives in Microbial Ecology*, American Society for Microbiology, Washington, D.C. (1984); pp 623–627.

[44a] Jarman, T.R.: Bacterial Alginate Synthesis. In: Berkeley, R. C. W.; Gooday, G. W.; Ellwood, D. C., Eds.; *Microbial Polysaccharides and Polysaccharases*, Academic Press, London, New York (1979); pp 35–50.

[45] Jørgensen, B. B.: The Microbial Sulphur Cycle. In: Krumbein, W. E., Ed.; *Microbial Geochemistry*, Blackwell, Oxford, London (1983); pp 91–124.

[46] Jørgensen, B. B.: Ecology of the Sulphur Cycle: Oxidative Pathways in Sediments. In: Cole, J. A.; Ferguson, S. J., Eds.; *The Nitrogen and Sulphur Cycles*, Cambridge University Press, Cambridge, New York (1988); pp 32–63.

[47] Jørgensen, B. B.: Biogeochemistry of Chemoautotrophic Bacteria. In: Schlegel, H. G.; Bowien, B., Eds.; *Autotrophic Bacteria*, Science Tech Publishers, Madison; Springer-Verlag, Berlin, Heidelberg, New York (1989); pp 117–146.

[48] Jørgensen, B. B.; Revsbech, N. P.: Colorless Sulphur Bacteria, *Beggiatoa* spp. and *Thiovulum* spp., in O_2 and H_2S Microgradients; *Appl. Environ. Microbiol.* **45** (1983) 1261–1270.

[49] Kelly, D. P.: Oxidation of Sulphur Compounds. In: Cole, J. A.; Ferguson, S. J., Eds.; *The Nitrogen and Sulphur Cycles*, Cambridge University Press, Cambridge, New York (1988); pp 65–98.

[50] Kelly, D. P.: Physiology and Biochemistry of Unicellular Sulfur Bacteria. In: Schlegel, H. G.; Bowien, B., Eds.; *Autotrophic Bacteria*, Science Tech Publisher, Madison; Springer-Verlag, Berlin, Heidelberg, New York (1989); pp 193–217.

[51] Kelly, D. P.; Harrison, A. P.: Genus *Thiobacillus*. In: Staley, J. T.; Bryant, M. P.; Pfennig, N.; Holt, J. G., Eds.; *Bergey's Manual of Systematic Bacteriology*, Williams & Wilkins, Baltimore, Hong Kong, London, Sydney (1989); Vol. 3, pp 1842–1858.

[52] King, R. A.; Miller, J. D. A.: Corrosion by the Sulphate-Reducing Bacteria; *Nature (London)* **233** (1971) 491–492.

[53] Kirk, T. K.; Farrell, R. L.: Enzymatic "Combustion": The Microbial Degradation of Lignin; *Ann. Rev. Microbiol.* **41** (1987) 465–505.

[54] Kirstein, K.-O.; Stiller, W.; Bock, E.: Mikrobiologische Einflüsse auf Betonkonstruktionen; *Beton- und Stahlbetonbau* **81** (8) (1986) 202–205.

[55] Klemps, R.; Cypionka, H.; Widdel, F.; Pfennig, N.: Growth with Hydrogen, and Further Physiological Characteristics of *Desulfotomaculum* species; *Arch. Microbiol.* **143** (1985) 203–208.

[56] Krumbein, W. E.: Über den Einfluß der Mikroflora auf die exogene Dynamik (Verwitterung und Krustenbildung); *Geol. Rundschau* **58** (1969) 333–365.

[57] Krumbein, W. E.: Über den Einfluß von Mikroorganismen auf die Bausteinverwitterung – Eine ökologische Studie; *Deutsche Kunst- und Denkmalpflege* **31** (1973) 54–71.

[58] Krumbein, W. E.: Biogene Krusten; *Bautenschutz Bausanierung, Sonderausgabe* (**1987**) 61–64.

[59] Krumbein, W. E.: Biotransformations in Monuments – a Sociobiological Study; *Durability of Building Materials* **5** (1988) 359–382.

[60] Krumbein, W. E.; Jens, K.: Biogenic Rock Varnishes of the Negev Desert (Israel), an Ecological Study of Iron and Manganese Transformation by Cyanobacteria and Fungi; *Oecologia* **50** (1981) 25–38.

[61] Krumbein, W. E.; Schönborn-Krumbein, C. E.: "Steinkrankheiten"; *Labor 2000* (**1987**) 79–92.

[62] Krumbein, W. E.; Werner, D.: The Microbial Silica Cycle. In: Krumbein, W. E., Ed.; *Microbial Geochemistry*, Blackwell, Oxford, London (1983); pp 125–127.

[63] Krümmel, A.; Harms, H.: Effect of Organic Matter on Growth and Cell Yield of Ammonia-Oxidizing Bacteria; *Arch. Microbiol.* **133** (1982) 50–54.

[64] Kuenen, J. G.: Colorless Sulfur Bacteria. In: Staley, J. T.; Bryant, M. P.; Pfennig,

N.; Holt, J. G., Eds.; *Bergey's Manual of Systematic Bacteriology*, Williams & Wilkins, Baltimore, Hong Kong, London, Sydney (1989); Vol. 3, pp 1834–1871.

[65] Kuenen, J. G.; Bos, P.: Habitats and Ecological Niches of Chemolitho(auto)trophic Bacteria. In: Schlegel, H. G.; Bowien, B., Eds.; *Autotrophic Bacteria*, Science Tech Publishers, Madison; Springer-Verlag, Berlin, Heidelberg, New York (1989); pp 53–80.

[66] Kuenen, J. G.; Robertson, L. A.: Ecology of Nitrification and Denitrification. In: Cole, J. A.; Ferguson, S. J., Eds.; *The Nitrogen and Sulphur Cycles*, Cambridge University Press, Cambridge, New York (1988); pp 161–218.

[67] Le Gall, J.; Fauque, G.: Dissimilatory Reduction of Sulfur Compounds. In: Zehnder, A. J. B., Ed.; *Biology of Anaerobic Microorganisms*, John Wiley & Sons, New York (1988); pp 587–639.

[68] *LEVITICUS*, Chapter 14, verses 37–45.

[69] Loveley, D. R.; Lonergan, D. J.: Anaerobic Oxidation of Toluene, Phenol, and *p*-Cresol by the Dissimilatory Iron-Reducing Organism, GS-15; *Appl. Environ. Microbiol.* **56** (1990) 1858–1864.

[70] Loveley, D. R.; Phillips, E. J. P.: Novel Mode of Microbial Energy Metabolism: Organic Carbon Oxidation Coupled to Dissimilatory Reduction of Iron or Manganese; *Appl. Environ. Microbiol.* **54** (1988) 1472–1480.

[71] Loveley, D. R.; Phillips, E. J. P.; Lonergan, D. J.: Hydrogen and Formate Oxidation Coupled to Dissimilatory Reduction of Iron or Manganese by *Alteromonas putrefaciens*; *Appl. Environ. Microbiol.* **55** (1989) 700–706.

[72] Marshall, K. C.: Mechanisms of Bacterial Adhesion at Solid-Water Interfaces. In: Savage, D. C.; Fletcher, M., Eds.; *Bacterial Adhesion*, Plenum Press, New York, London (1985); pp 133–161.

[73] Matthes, S.: *Mineralogie*, 2nd edition, Springer-Verlag, Berlin (1987); pp 261–276.

[74] Meincke, M.; Ahlers, B.; Krause-Kupsch, T.; Krieg, E.; Meyer, C.; Sameluck, F.; Sand, W.; Wolters, B.; Bock, E.: *Isolation and Characterization of Endolithic Nitrifiers*. In: VIth International Congress of Deterioration and Conservation of Stone, Proceeding Actes; Nicholas Copernicus University Press Department, Toruń (1988); pp 15–23.

[75] Meincke, M.; Krieg, E.; Bock, E.: *Nitrosovibrio* spp., the Dominant Ammonia-Oxidizing Bacteria in Building Sandstone; *Appl. Environ. Microbiol.* **55** (1989) 2108–2110.

[76] Milde, K.; Sand, W.; Wolff, W.; Bock, E.: Thiobacilli of the Corroded Walls of the Hamburg Sewer System; *J. Gen. Microbiol.* **129** (1983) 1327–1333.

[77] Miller, J. D. A.: Metals. In: Rose, A. H., Ed.; *Microbial Biodeterioration, Economic Microbiology*, Academic Press, London, New York (1981); Vol. 6, pp 149–202.

[78] Miller, J. D. A.; Wakerley, D. S.: Growth of Sulfate-Reducing Bacteria by Fumarate Dismutation; *J. Gen. Microbiol.* **43** (1966) 101–107.

[79] Moore, W. J.; Hummel, D. O.: *Physikalische Chemie*, 4th ed., Chapter 10 (Elektrochemie I: Ionen), Walter de Gruyter, Berlin, New York (1986); pp 506–571.

[80] Moore, W. J.; Hummel, D. O.: *Physikalische Chemie*, 4th ed., Chapter 12 (Elektrochemie II: Elektroden und Elektrodenreaktionen), Walter de Gruyter, Berlin, New York (1986); pp 620–676.

[81] Myers, C. R.; Nealson, K. H.: Bacterial Manganese Reduction and Growth with Manganese Oxide as the Sole Electron Acceptor; *Science* **240** (1988) 1319–1321.

[82] Nazina, T. N.; Ivanova, A. E.; Kanchaveli, L. P.; Rozanova, E. P.: A New Thermophilic, Methylotrophic Sulfate-Reducing Bacterium, *Desulfotomaculum kuznetsovii* sp. nov.; *Mikrobiologiya (Moscow)* **57** (1988) 823–827.

[83] Nealson, K. H., The Microbial Iron Cycle. In: Krumbein, W. E., Ed.; *Microbial Geochemistry*, Blackwell, Oxford, London (1983); pp 159–190.

[84] NEALSON, K. H.: The Microbial Manganese Cycle. In: KRUMBEIN, W. E., Ed.; *Microbial Geochemistry*, Blackwell, Oxford, London (1983); pp 191–221.

[85] PANKHANIA, I. P.; Hydrogen Metabolism in Sulphate-Reducing Bacteria and its Role in Anaerobic Corrosion; *Biofouling* **1** (1988) 27–47.

[86] PANKHANIA, I. P.: Utilization of Cathodic Hydrogen by *Desulfovibrio desulfuricans*; *J. Gen. Microbiol.* **132** (1986) 3357–3365.

[87] PECK, H. D., JR.; LISSOLO, T.: Assimilatory and Dissimilatory Sulphate Reduction: Enzymology and Bioenergetics. In: COLE, J. A.; FERGUSON, S. J., Eds.; *The Nitrogen and Sulphur Cycles*, Cambridge University Press, Cambridge, New York (1988); 99–132.

[88] POSTGATE, J. R.: *The sulphate-Reducing Bacteria*, 2nd ed., Cambridge University Press, Cambridge, London, New York (1984).

[88a] PRYOR, W. A.: *Mechanisms of Sulfur Reactions*, Chapter 2 (Elemental Sulfur), McGraw-Hill, New York (1962); pp 7–15.

[89] PURVIS, O. W.; ELIX, J. A.; BROOMHEAD, J. A.; JONES, G. C.: The Occurrence of Copper-Norstictic Acid in Lichens from Cupriferous Substrata; *Lichenologist* **19** (1987) 193–203.

[90] RAHMEL, A.; SCHWENK, W.: *Korrosion und Korrosionsschutz von Stählen*, Chapter 2 (Das System Werkstoff/Medium), Verlag Chemie, Weinheim (1977); pp 3–42.

[91] RAHMEL, A.; SCHWENK, W.: *Korrosion und Korrosionsschutz von Stählen*, Chapter 3 (Theorie und Untersuchungsverfahren), Verlag Chemie, Weinheim (1977); pp 43–85.

[92] RAHMEL, A.; SCHWENK, W.: *Korrosion und Korrosionsschutz von Stählen*, Chapter 4 (Korrosion in wäßrigen Medien), Verlag Chemie, Weinheim (1977); pp 87–192.

[93] ROBB, I. D.: Stereo-Biochemistry and Function of Polymers. In: MARSHALL, K. C., Ed.; *Microbial Adhesion and Aggregation*, Springer-Verlag, Berlin, Heidelberg, New York, Tokyo (1984); pp 39–49.

[93a] ROGERS, H. J., PERKINS, H. R., WARD, J. B.: *Microbial Cell Walls and Membranes*, Chapman and Hall, London, New York (1980); Chapter 7.1.2 (Teichuronic Acids) pp 222–223.

[94] ROZANOVA, E. P.; NAZINA, T. N.; GALUSHKO, A. S.: A New Genus of Sulfate-Reducing Bacteria and the Description of its New Species, *Desulfomicrobium apsheronum* gen. nov., sp. nov.; *Mikrobiologiya (Moscow)* **57** (1988) 634–641.

[95] ROZANOVA, E. P.; PIVOVAROVA, T. A.: Reclassification of *Desulfovibrio thermophilus*; *Mikrobiologiya (Moscow)* **57** (1988) 102–106.

[96] ROSE, A. H.: History and Scientific Basis of Microbial Biodeterioration of Metals. In: ROSE, A. H., Ed.; *Microbial Biodeterioration, Economic Microbiology*, Academic Press, London, New York (1981); Vol. 6, pp 1–18.

[97] SAND, W.: Importance of Hydrogen Sulfide, Thiosulfate, and Methylmercaptan for Growth of Thiobacilli During Simulation of Concrete Corrosion; *Appl. Environ. Microbiol.* **53** (1987) 1645–1648.

[98] SCHASCHL, E.: Elemental Sulphur as a Corrodent in Deaerated Neutral Aqueous Solutions; *Mat. Performance* **19** (1980) 9–12.

[99] SCHATZ, A.: Pedogenic (Soil-Forming) Activity of Lichen Acids; *Naturwiss.* **49** (1962) 518–519.

[100] SCHINK, B.: Principles and Limits of Anaerobic Degradation: Environmental and Technological Aspects. In: ZEHNDER, A. J. B., Ed.; *Biology of Anaerobic Microorganisms*, Wiley, New York (1988); pp 771–846.

[101] SCHOONEN, M. A. A.; BARNES, H. L.: An approximation of the Second Dissociation Constant for H_2S; *Geochim. Cosmochim. Acta* **52** (1988) 649–654.

[102] SCHWENK, W.: Grundlagen der Korrosion und des elektrochemischen Korrosions-

schutzes. In: VON BAECKMANN, W.; SCHWENK, W., Eds.; *Handbuch des kathodischen Korrosionsschutzes*, Verlag Chemie, Weinheim (1980); pp 25–61.

[103] SEAWARD, M. R. D.: Lichen Damage to Ancient Monuments: A Case Study; *Lichenologist* **20** (1988) 291–295.

[104] STETTER, K. O.: Group II. Archaeobacterial Sulfate Reducers. In: STALEY, J. T.; BRYANT, M. P.; PFENNIG, N.; HOLT, J. G., Eds.; *Bergey's Manual of Systematic Bacteriology*, Williams & Wilkins, Baltimore, Hong Kong, London, Sydney (1989); Vol. 3, pp 2216–2218.

[105] STETTER, K. O.: Order III. Sulfolobales ord. nov. In: STALEY, J. T.; BRYANT, M. P.; PFENNIG, N.; HOLT, J. G., Eds.; *Bergey's Manual of Systematic Bacteriology*, Williams & Wilkins, Baltimore, Hong Kong, London, Sydney (1989); Vol. 3, pp 2250–2253.

[106] STOTZKY, G.: Mechanisms of Adhesion to Clays, with Reference to Soil Systems. In: SAVAGE, D. C.; FLETCHER, M., Eds.; *Bacterial Adhesion*, Plenum Press, New York, London (1985); pp 195–253.

[107] STUMM, W.; MORGAN, J. J.: *Aquatic chemistry*, 2nd ed., Chapter 4 (Dissolved Carbon Dioxide), Wiley, New York (1981); pp 171–229.

[108] STUMM, W.,; MORGAN, J. J.: *Aquatic chemistry*, 2nd ed., Chapter 5 (Precipitation and Dissolution), Wiley, New York (1981); pp 230–322.

[109] STUMM, W.; MORGAN, J. J.: *Aquatic chemistry*, 2nd ed., Chapter 6 (Metal Ions in Aqueous Solutions: Aspects of Coordination Chemistry), Wiley, New York (1981); pp 323–417.

[109a] STUMM, W.; MORGAN, J. J.: *Aquatic chemistry*, 2nd ed., Chapter 7 (Oxidation and Reduction), Wiley, New York (1981); pp 418–503.

[110] STUMM, W.; MORGAN, J. J.: *Aquatic chemistry*, 2nd ed., Chapter 9 (The Regulation of the Chemical Composition of Natural Waters), Wiley, New York (1981); pp 523–598.

[111] STUMM, W.; MORGAN, J. J.: *Aquatic chemistry*, 2nd ed., Appendix, Wiley, New York (1981); pp 749–756.

[111a] SUTHERLAND, I. W.: Microbial Exopolysaccharides: Control of Synthesis and Acylation. In: BERKELEY, R. C. W.; GOODAY, G. W.; ELLWOOD, D. C., Eds.; *Microbial Polysaccharides and Polysaccharases*, Academic Press, London, New York (1979); pp 1–34.

[112] SYERS, J. K.; ISKANDAR, I. K.: Pedogenetic Significance of Lichens. In: AHMADJIAN, V.; HALE, M. E., Eds.; *The Lichens*, Academic Press, New York, London (1973); pp 225–248.

[113] URONE, P.: The Primary Air Pollutants – Gaseous, Their Occurrence, Sources and Effects. In: STERN, A. C., Ed.; *Air Pollution*, 3rd ed., Academic Press, New York, London (1976); Vol. I, Air Pollutants, Their Transformation and Transport, pp 23–75.

[114] VOGELS, G. D.; KELTJENS, J. T.; VAN DER DRIFT, C.: Biochemistry of Methane Production. In: ZEHNDER, A. J. B., Ed.; *Biology of Anaerobic Microorganisms*, John Wiley & Sons, New York (1988); pp 707–770.

[115] VON BAECKMANN, W.: Historische Entwicklung des Korrosionsschutzes. In: VON BAECKMANN, W.; SCHWENK, W., Eds.; *Handbuch des kathodischen Korrosionsschutzes*, Verlag Chemie, Weinheim (1980); pp 1–24.

[116] VON WOLZOGEN KÜHR, C. A. H.; VAN DER VLUGT, L. S.: De grafiteering van gietijzer als electrobiochemisch proces in anaerobe gronden; *Water (Den Haag)* **18** (1934) 147–165.

[117] WARD, J. B.; BERKELEY, R. C. W.: The Microbial Cell Surface and Adhesion. In: BERKELEY, R. C. W.; LYNCH, J. M.; MELLING, J.; RUTTER, P. R.; VINCENT, B., Eds.; *Microbial Adhesion to Surfaces*, Ellis Horwood, Chichester (1980); pp 47–66.

[118] WATSON, S. W.; BOCK, E.; HARMS, H.; KOOPS, H.-P.; HOOPER, A. B.: Nitrifying Bacteria. In: STALEY, J. T.; BRYANT, M. P.; PFENNIG, N.; HOLT, J. G., Eds.; *Bergey's*

Manual of Systematic Bacteriology, Williams & Wilkins, Baltimore, Hong Kong, London, Sydney (1989); Vol. 3, pp 1808–1834.

[119] WATSON, S.; BOCK, E.; VALOIS, F. W.; WATERBURY, J. B.; SCHLOSSER, U.: *Nitrospira marina* gen. nov., sp. nov.: A Chemolithotrophic Nitrite-Oxidizing Bacterium; *Arch. Microbiol.* **144** (1986) 1–7.

[120] WEAST, R. C.: *Handbook of Chemistry and Physics*, 58th ed., CRC Press, West Palm Beach, FL (1978); pp B85–B178.

[121] WEAST, R. C.: *Handbook of Chemistry and Physics*, 58th ed., CRC Press, West Palm Beach, FL (1978); pp D141–D146.

[122] WEBLEY, D. M.; DUFF, R. B.: The Incidence in Soils and Other Habitats of Microorganisms Producing 2-Ketogluconic Acid; *Pl. Soil* **22** (1966) 307–313.

[123] WEIMER, P. J.; VAN KAVELAAR, M. J.; MICHEL, C. B.; NG, T. K.: Effect of Phosphate on the Corrosion of Carbon Steel and on the Composition of Corrosion Products in Two-Stage Continuous Cultures of *Desulfovibrio desulfuricans*; *Appl. Environ. Microbiol.* **54** (1988) 386–396.

[123a] WHITING, P. H.; MIDGLEY, M.; DAWES, E. A.: The Role of Glucose Limitation in the Regulation of the Transport of Glucose, Gluconate and 2-Oxogluconate, and of Glucose Metabolism in *Pseudomonas aeruginosa*; *J. Gen. Microbiol.* **92** (1976) 304–310.

[124] WHITTON, B. A.; POTTS, M.: Marine Littoral. In: CARR, N. G.; WHITTON, B. A., Eds.; *The Biology of Cyanobacteria*, University of California Press, Berkeley, Los Angeles (1982); pp 515–542.

[125] WIDDEL, F.: Microbiology and Ecology of Sulfate- and Sulfur-Reducing Bacteria. In: ZEHNDER, A. J. B., Ed.; *Biology of Anaerobic Microorganisms*, John Wiley & Sons, New York (1988); pp 469–585.

[126] WILSON, M. J.; JONES, D.; MCHARDY, W. J.: The Weathering of Serpentinite by *Lecanora atra*; *Lichenologist* **13** (1981) 167–176.

[127] WIRTH, V.: *Die Silicatflechten-Gemeinschaften im außeralpinen Raum*, Dissertationes Botanicae, Vol. 17; Cramer, Lehre/Braunschweig (1972); pp 9–11.

[128] ZAVARZIN, G. A.: The Genus *Metallogenium*. In: STARR, M. P.; STOLP, H.; TRÜPER, H. G.; BALOWS, A.; SCHLEGEL, H. G., Eds.; *The Prokaryotes*, Springer, Berlin, Heidelberg, New York (1981); Vol. 1, pp 524–528.

[129] ZEIKUS, J. G.; DAWSON, M. A.; THOMPSON, T. C.; INGVORSEN, H.; HATCHIKIAN, E. C.: Microbial Ecology of Volcanic Sulphidogenesis: Isolation and Characterization of *Thermodesulfobacterium commune* gen. nov. and sp. nov.; *J. Gen. Microbiol.* **129** (1983) 1159–1169.

Modern Wastewater Technology

by J. R. LEMKE

Contents

Dr. J. R. LEMKE,
WV Umweltschutz[1],
Bayer AG,
D-5090 Leverkusen,
Fed. Rep. Germany

[1] Water Treatment, Environmental Protection

1 Introduction

The availability of good quality drinking water is vital, and even early peoples were aware of this. As urbanization spread and wells could not meet the demand, water conduits had to be built to supply the public and private baths and wells with water. In the ancient cities of Babylon and Nineveh, as well as in Egypt, water supply networks were excavated. The Greeks built large water conduits, such as the impressive water pipeline in Pergamon; there, by about 200 BC, they had constructed a pressurized water conduit to pipe drinking water through valleys and over mountains, negotiating differences in altitude of almost 200 m. The aqueducts built by the Romans are famous; the city of Rome, in its prime, was supplied with water from 11 aqueducts.

Where water is consumed, wastewater is generated. In Babylon, excavations have exposed the remains of wastewater sewers beneath the drinking water conduits. Similar sewers have also been discovered in ancient Greek cities. Part of a large sewer, the *cloaca maxima*, built about 600 BC, still exists today and empties into the Tiber River. When the flow or the velocity of the wastewater was inadequate, slaves had to help move, as well as remove, the sludge deposits in the *cloaca maxima*. In Cologne, remnants of the *cloaca maxima* that the Romans had installed during their occupation are still in existence. The Romans also invented wastewater taxation. Emperor Vespasian (69–79 AD) did not shy away from taxing latrines; regarding the revenues from his latrine tax, he is supposed to have said, "*Non olet*," meaning that it (money) does not stink.

During the Middle Ages, the intuitive know-how about water supply and wastewater treatment was lost. Used water and all the wastes from houses and farms were dumped onto the streets, where they mixed with street dirt and ran down open drains in the middle of the street. Any contagious disease organism introduced found optimal nutrients, consequently causing devastating epidemics.

In the first great plague epidemic of 1349–1351 (the Black Death), 25 million people fell victim in Europe – 25% of the total population of Europe at the time.

The techniques of wastewater treatment did not begin to develop, or even to be encouraged, until the middle of the 19th century; industrialization was spreading, and increasing wastewater production provided the impetus. The condition of rivers, particularly those in industrial centers, was alarming; something had to be done.

The first strategies to be developed for systematic wastewater treatment were natural ones, such as irrigation fields or irrigation meadows, in which the recycling of components of wastewater was one of the goals. Settling pits were another natural treatment system; in ignorance they were often installed in the immediate vicinity of wells.

Max von Pettenkofer set forth the founding principles of modern hygiene. He insisted on the separation of drinking water and wastewater and, accordingly, the abandonment of settling pits, but his scientific premise for this demand was incorrect: he postulated that people breathing the products of wastewater decomposition would fall ill. The proof that living organisms are the cause of infectious disease was furnished by Robert Koch only at the end of the 19th century. In 1876 he discovered the anthrax bacillus, in 1882 the tuberculosis bacterium, and in 1883 the cholera pathogen, which latter is transmitted primarily in drinking water [1].

2 Primary treatment of wastewater

The demands of PETTENKOFER and the findings of KOCH led to the construction of separate canal systems for wastewater. Rakes and screens removed the solid constituents; the remainder was discharged into the rivers, left to the self-cleaning power of nature. This separation alone led to a significant reduction in infectious disease [2].

Even today *rakes* and *screens*, as *mechanical clarifiers*, fulfill the important function of *primary treatment* or *pretreatment* in wastewater treatment. The operational safety of a purification plant depends heavily on these mechanical devices. Inadequately operating rakes or screens result in plugged pumps and pipelines obstructed by fibrous and coarse material. The large quantities of fibers, which can spin themselves into strands several meters long, are a particular problem in domestic wastewater today. Installations include rakes with adjustable gaps and screens of different design and pore size (0.5 mm to 6 mm diameter) [3].

After the coarse and fibrous material has been removed, sand and granular solid matter particulates are eliminated in subsequent steps. *Grit chambers* or *sand traps* are large flow basins with a fluid velocity of 30 cm/s, in which sand and solid matter settle out by gravity. Grit chambers are especially important for processing rain water and storm runoff in community treatment plants. They come in various designs, and there are three different types:

- long (or flat) sand traps,
- round sand traps, and
- aerated sand traps [4].

In the final primary treatment step, after the last grit chamber, the wastewater is funneled into the *sedimentation basin*. This step removes nearly all floating organic matter. A prerequisite for this is a further reduction in flow velocity; the settling time should be at least one hour.

In recent years the *flotation* method has been developed to remove suspended particulates by causing them to adhere to very fine gas bubbles and bringing them to the surface. Even suspended matter that is difficult to remove can be eliminated from wastewater by this method. Depending on the way the gas bubbles are generated and introduced into the wastewater, flotation processes fall into three categories:

- mechanical flotation,
- dissolved air flotation, and
- electroflotation.

In *mechanical flotation*, air bubbles are generated by an aerator that introduces air mechanically with a rapidly rotating impeller; a simple generator can be placed in a lagoon. In *dissolved air flotation*, the gas bubbles are generated in a pressure tank; the wastewater is initially saturated with air at 4 to 5 bars pressure and then quickly depressurized, allowing the gas to expand. In *electroflotation* the gas bubbles are generated by electrolyzing water; very small bubbles of hydrogen and oxygen are formed [5].

3 Methods for biological (secondary) treatment of wastewater

3.1 Fundamentals

The actual purification of wastewater, that is, the removal of all possible dissolved organic substances, is done biologically, and constitutes an acceleration of the natural processes that participate in the self-cleaning of the hydrosphere. Many different substances are simultaneously degraded. The purification is done by microorganisms and follows biological principles that apply to metabolism in general. All the different methods used in biological wastewater technology are variations of two basic systems,

- trickling-filter systems and
- aerated and activated-sludge systems.

In the *trickling-filter system*, the microorganisms grow on a solid substrate over which water flows constantly. The filter bed is biotechnologically a solid-bed fermentor with immobilized organisms. In *aerated systems* the microorganisms are suspended. The system is one of continuous fermentation with partial recycling of biomass. The objective of fermentation during the biological purification step is to utilize the substrates optimally to achieve the maximum degradation of organic substances present in the wastewater.

Designers and operators of these systems must know the nutrient requirements of microorganisms if they are to have maximum control of the substrate utilization process. As a prerequisite, they should know the basic regulatory mechanisms of the enzymes involved within the organisms. The cell is a self-regulating system; control is exerted through feedback. The cell's regulatory mechanisms coordinate the metabolic performances of individual enzyme systems for maximum economy, which is fundamental to the broad adaptability of microbial populations in a water-purification facility.

Regulation occurs at two levels: the rate of enzyme synthesis can be controlled by induction or repression, or enzyme activity can be lowered or increased. In induction, each enzyme of a given degradation sequence can be induced separately by the product of a preceding reaction, or, alternatively, all enzymes in a degradation sequence can be induced simultaneously by the substrate or by the first degradation product. In this manner, the various substrates induce their own metabolism; in a mixed population, this process can take place in several organisms.

Wastewater contains some substances that can be digested only by specific microorganisms, and often only when other more readily digestible material is no longer available. The organisms must initially adapt to perform the degradation. This fact has led to some erroneous concepts in wastewater technology. For example, many people arbitrarily refer to easy, difficult, and nondegradable substances.

When aerobic organisms metabolize a substrate with oxygen, the end products are carbon dioxide and water; the metabolic end products of facultative and obligate anaerobes, however, are methane and carbon dioxide. Only certain organisms are capable of oxidizing ammonia to nitrite, and nitrite to nitrate. Specialized strains can degrade hydrocarbons; the ability to degrade halogenated aromatic and aliphatic compounds is particularly rare. The degradation of these compounds has been described in a recent comprehensive compilation [6]. The composition of a wastewater and the fact that individual compounds are often at very low concentrations are the reasons that a biological system in a water treatment facility may not always successfully adapt, which is requisite for degradation. However, most compounds studied in soil percolation experiments do decompose biologically [7].

3.2 Trickling-filter systems

Soil filters have been used occasionally throughout history for wastewater purification, as have irrigation fields, since soil-mediated purification does not require the soil to be under cultivation with higher plants; rather, soil bacteria supplied with air are responsible for the purification. The next logical step in development, therefore, was the construction of trickling filters.

Trickling filters have been used since the turn of the twentieth century. They are effective as a purification step alone or in combination with aerated systems. A trickling filter is a cylindrical structure, open at the top, filled with crushed stone or synthetic material (*fixed-film system*), the surfaces of which are covered with microbial growth. The wastewater is evenly distributed onto the trickling filter, usually by a rotating sprinkler, and trickles through the bed. The *floc*, a biologically active film, develops on the surfaces of the filter bed, similar to the growth seen on rocks in a stream. The wastewater picks up the required oxygen across the large surface area of the bed and within the tank, from air introduced through appropriate openings at the bottom of the trickling filter and rising upward. The efficiency of a trickling filter depends strongly on temperature and drops significantly during cold weather. It is important to mechanically pretreat the wastewater scrupulously, to avoid clogging the trickling filter [8] (Figure 1).

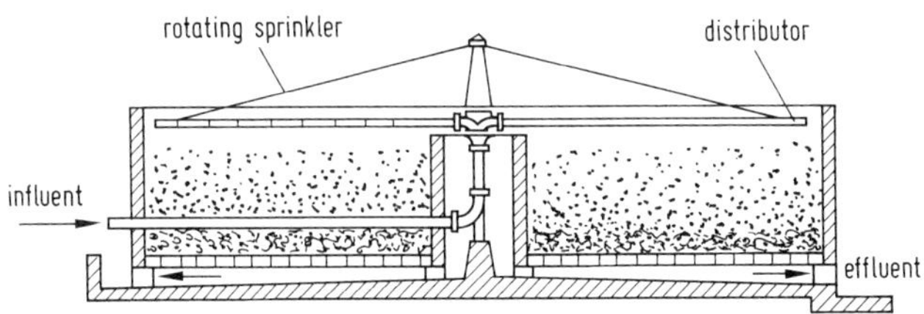

Fig. 1 Trickling filter

The *rotating biological contactor* (*RBC*) is a special type of trickling filter that has proved especially satisfactory for small facilities [9]. Plastic disks are arranged along a slowly rotating shaft; they are half submerged in a basin through which wastewater passes. The disks serve as growth surfaces for microorganisms that are exposed in alternating fashion to wastewater, from which they take organic matter, and to air, from which they take oxygen. A settling tank must be in line with either the trickling filter or the RBC to collect the flocs that slough off from the substrate.

3.3 Aerated and activated-sludge systems

The *aerated lagoon* is the classical method of biological wastewater treatment. The first such facility was developed in 1914 in England, and the method was first introduced into Germany at Essen in 1925 [10, 11]. After coarse and fine particulates are removed, the wastewater is channeled into lagoons or ditches, where it mixes with the microbial population, the *activated sludge*, and is aerated vigorously. The aerated lagoons are generally simple concrete basins (Figure 2).

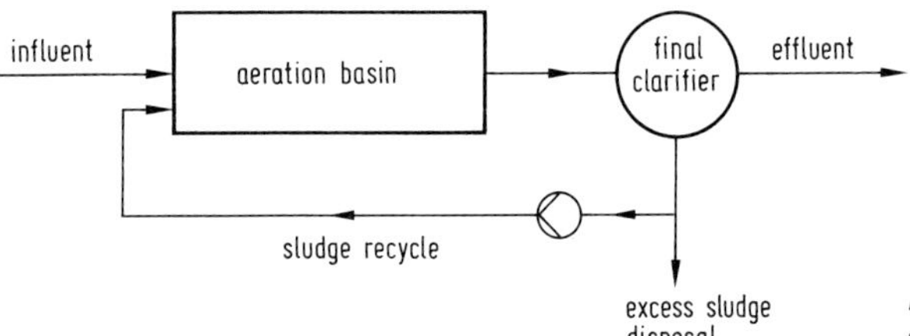

Fig. 2 Conventional aerated secondary treatment system

In principle, two different aeration systems are used: *mechanical surface aeration*, and *pressurization* or *air injection* [12, 13]. A rolling bar aerator or a rotor aerator is used for surface aeration. In a rolling bar aerator, steel rods mounted on a rotating horizontal shaft hit the water and splash it into the air. A rotor aerator turning at the water surface splashes the water into the air. In either case, the water picks up oxygen from the air. Surface aerators are suitable only for shallow basins no deeper than five meters. The efficacy of surface aerators is modest; oxygen incorporation is approximately proportional to the work expended, and at best the oxygen yield is 2 kg O_2/kWh.

A further disadvantage of the surface aerator is aerosol formation. This has led to the increased use of pressurized aerators in recent years. In a pressurized aerator, fine bubbles of compressed air are introduced at the bottom of the tank either through perforated pipes (coarse bubble injectors), cinder blocks, dome-shaped porous ceramic aerators (fine bubble injectors), or dual-jet nozzles. Coarse bubble injectors have values of oxygen yield of 2.5 to 3.5 kg O_2/kWh, and higher values can be achieved with dual-jet

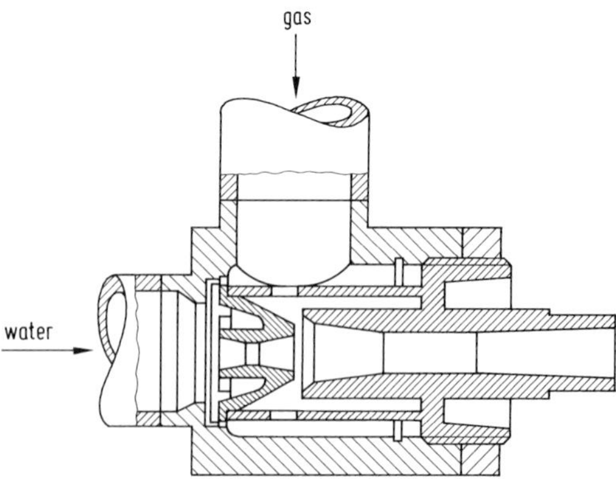

Fig. 3 Schematic view of Bayer injector [25]

nozzles. The Bayer injector (Figure 3) is an instance of using the kinetic energy of the fluid jet to disperse the gas into very fine bubbles. The injectors are arranged above the bottom of the aerated tank; the liquid level can be set at will and is therefore a parameter that can be optimized [14]. Detailed studies on the suitability and efficacy of surface aerators and injection aerators have been reported by ZLOKARNIK [15, 16].

There are various types of aerated systems. Designs include the rectangular *oxidation ditch* in which the wastewater is introduced only at one end; the concentration of suspended matter decreases towards the exit. The completely mixed tank, another type of design, has material homogeneously distributed. Still other types are recirculating tanks and *Carrousel* tanks. We will briefly discuss the *adsorption* aerated lagoon developed by BÖHNKE [17] for a multistep aeration facility as representative of these designs. It consists of two aerated lagoons connected in series with an intermediate clarifier. The incoming wastewater is channeled into the first lagoon or tank for the first treatment step, adsorption. This tank is designed for brief retention times and, consequently, is heavily loaded. After an intermediate clarification step, the wastewater flows into a second and larger lagoon with a longer retention time. The biological digestion of the wastewater constituents takes place primarily in the lightly loaded second lagoon. The effluent values for organic matter of lightly loaded facilities are usually very low.

3.4 Closed aerobic systems (bioreactors) with oxygen aeration

The first attempts to use oxygen-enriched air or pure oxygen to improve the efficacy with which a water treatment facility clarifies wastes were made on a laboratory scale nearly 50 years ago [18]. These early studies did not lead to any practical applications because the questions of economics could not be resolved satisfactorily. In spite of this, experiments with oxygen injection were continued in the USA. In 1972, the first wastewater treatment plants using the UNOX process, developed by the Union Carbide Company, went into operation in the USA. In the early seventies experiments were begun in Europe, because this process can be used to advantage with highly concentrated and odoriferous wastewaters [19,20].

Since the partial pressure of pure oxygen is 4.8 times greater than that of atmospheric oxygen, the oxygen content in aerated lagoons can be proportionately higher, and high biosludge concentrations can be obtained.

The wastewater is treated in closed tanks. Up to 90% of the (technically pure) oxygen added is consumed. The exhaust gas is still about 50% oxygen, so if it must be thermally treated for odor that can easily be done. When oxygen is used instead of air, only about 1% remains unconsumed, so odors of unspent nitrogen escaping from the stripping process are prevented.

The first small plants in Germany to use oxygen aeration went into operation in 1974. They were connected with dead animal disposal facilities and were able to eliminate the usual distressing smells. The plants operate even under extreme load fluctuations. The high degree of operational safety and the low space requirements of these facilities were also arguments in favor of installing them.

The first larger facility of this type in Germany was built by Bayer AG for the clarification of industrial wastewater and became operational in 1977 in Wuppertal-

Elberfeld [21]. Details of its operation are in Table 1. Since then other facilities using oxygen aeration for community, mixed community-commercial, and industrial plants have been installed [22].

Table 1 Performance data[1] for a treatment plant using oxygen aeration to treat industrial wastewater [21]

Wastewater volume	20,000 m^3/day
COD degradation capacity	61 t/day
BOD_5 degradation capacity	34 t/day
Oxygen introduction rate	
Primary aeration stage	51 t/day
Secondary aeration stage	7 t/day
Volumes	
Storage tanks, 2	4,000 m^3 each
Primary aeration stage	6,200 m^3
Secondary aeration stage	3,200 m^3
Intermediate clarifier	4,400 m^3
Final clarifier	4,800 m^3
Surface charge (surface loading)	
Intermediate clarifier	0.7 $m^3/m^2 \cdot h$
Final clarifier	0.6 $m^3/m^2 \cdot h$
Retention times	
Primary aeration stage	7.5 h
Secondary aeration stage	3.8 h
Intermediate clarifier	5.3 h
Final clarifier	5.8 h
Percent reduction	
Primary aeration stage	max. 80% COD, max. 93% BOD_5
Secondary aeration stage	max. 7% COD, max. 5% BOD_5
Overall, both stages	87% COD, 98% BOD_5
Sludge load	
Primary aeration stage	0.7 kg BOD_5/kg dried solids · d
Secondary aeration stage	0.2 kg BOD_5/kg dried solids · d
Volumetric load	
Primary aeration stage	5.6 kg $BOD_5/m^3 \cdot d$
Secondary aeration stage	0.8 kg $BOD_5/m^3 \cdot d$
Excess sludge	25% of BOD_5 load
Aeration unit	
Compressed air volume	12,000 m^3/h
Oxygen produced (pressure: 0.4 bar; 98% pure O_2)	2,000 m^3/h
Nitrogen produced (pressure: 5 bar; max. 20 ppm O_2 in N_2	2,100 m^3/h

[1] Abbreviations: COD = chemical oxygen demand; BOD_5 = 5-day biochemical oxygen demand; t = metric tons; h = hour; d = day

3.5 Closed aerobic systems (bioreactors) with forced-air aeration

As I mentioned in Section 3.3 for aerated systems of the open type, the liquid level is an important parameter in optimizing oxygen uptake. The deeper the liquid is, and the longer the retention time of the gas phase is, the greater is the oxygen uptake from air. These factors, along with the need to minimize space and the constraints of odorless operation, led to the development of closed aerated bioreactors, of either shaft or tower design.

The Deep Shaft process developed by the ICI company [23] has as the aerated bioreactor volume a shaft containing a pipe 20 to 200 meters long installed along the centerline of the shaft (Figure 4). The oxygen from the injected forced air is almost totally consumed in the fluid column. The first reactor of this design became operational in Germany in 1975 [24].

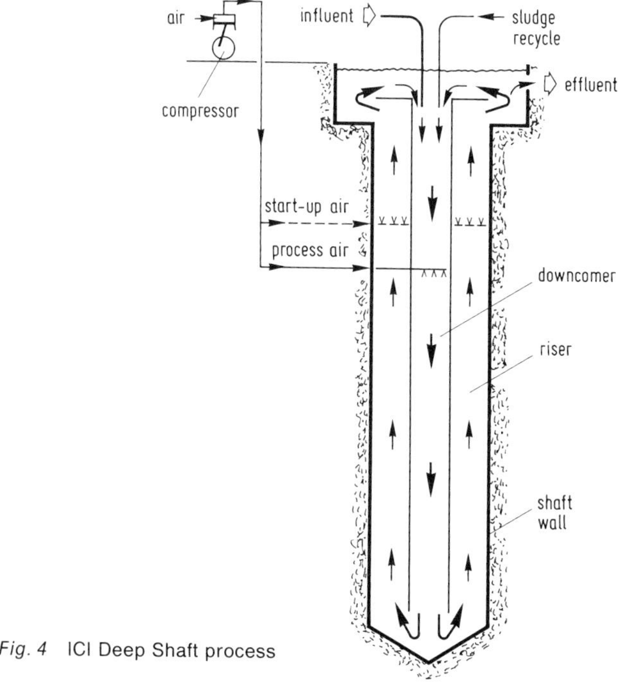

Fig. 4 ICI Deep Shaft process

Aerated bioreactors of the tower design are the *Turmbiologie* ("tower biology") of the Bayer AG and the *Biohoch-Reaktor* ("high bioreactor") of Hoechst AG. Bayer's Turmbiologie installation in Leverkusen consists of four cylindrical towers 26 meters in diameter and 30 meters high (Figure 5). Each tower has 72 dual-jet nozzles (wastewater/air injection nozzles). The aerated volume of each is about 13,000 m^3, with full use of the 26-m height. Under these conditions the exhaust air can contain as little as 4 vol% O_2 (81% of O_2 is consumed) [25] (Figure 6).

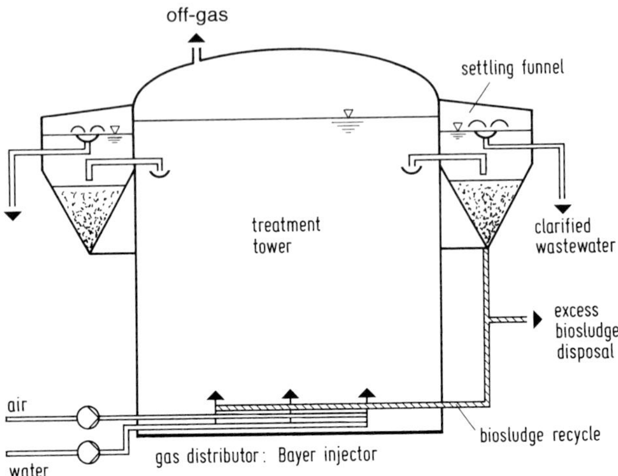

Fig. 5 *Turmbiologie* ("tower biology") wastewater treating tower of the Bayer AG

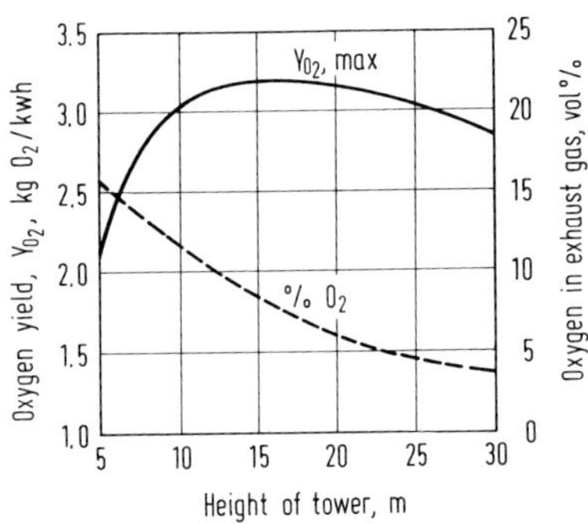

Fig. 6 Oxygen yield, Y_{O2}, and percent oxygen in exhaust gas as functions of bioreactor height [25]

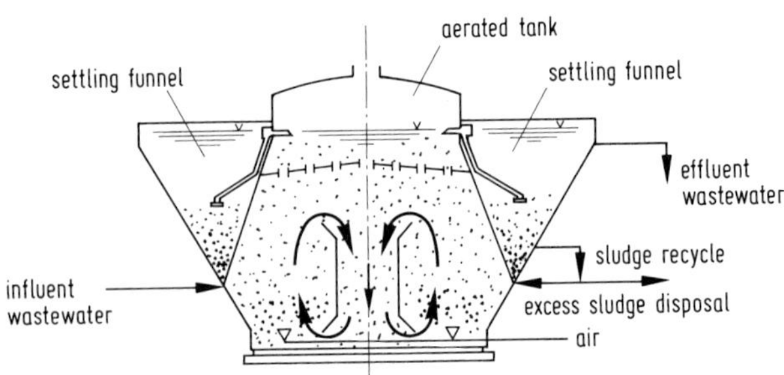

Fig. 7 The Hoechst *Biohoch-Reaktor*

The Biohoch-Reaktor also consists of an aeration column 22 meters high with a circular final clarifier on top (Figure 7). The difference between the Biohoch-Reaktor and the Bayer Turmbiologie is that the Biohoch-Reaktor aeration column has a central pipe or draft tube and an intermediate divider. The reactor can therefore be operated in different modes: exclusively as a bubble column, as a recirculating aerator, or in tandem with recirculation below and a bubble column above. The stationary mixers used in early models have been superseded by radial injection nozzles [26].

A more recent development is the lift-rod reactor. It is a vertical cylinder containing a rod along which perforated disks are mounted at intervals. The injected air is dispersed by the upward and downward movements of the rod and disks. Data on the performance of lift-rod reactors in large installations are not yet available [27].

3.6 Special bioreactor systems

The *sequencing batch reactor* (*SBR*) process can be seen as a special type of aeration system. This name is applied to a process during which a batch of activated sludge is dosed at regular intervals with wastewater [28]. Once the reactor is filled, oxygen (as air) is introduced. After the aeration has ceased, the activated sludge can settle out. The purified supernatant water is drawn off and the activated sludge is left behind. The reactor is again dosed with wastewater and the cycle repeated. The substrate concentration in the reactor can be controlled by the rate of liquid dosing. Switching aeration either on or off establishes the conditions for the aerobic or anaerobic phase. Aerobes, facultative anaerobes, and denitrifiers can be present in the same reactor. This is a relatively new process whose practical usefulness has yet to be demonstrated.

Another special method for biological wastewater clarification involves growth surfaces [29] or solid bodies [30] on which biomass accumulates in aerated lagoons. The purpose of this method is to limit the volume of aeration lagoons, to rehabilitate overloaded facilities, or to keep microorganisms with specific metabolic profiles confined to one clarification plant. Sand, charcoal, lignite [31], foam [32], or hard plastic sheets [33] can serve as growth surfaces.

3.7 Anaerobic systems

Higly concentrated, organically polluted wastewaters, which would be prohibitively costly in terms of energy to treat aerobically because of the large amount of oxygen required, can be pretreated anaerobically. The digestion of organic matter to methane and carbon dioxide by anaerobic bacteria has been studied in sufficient depth in recent years that an efficient anaerobic digestion process for wastewater clarification can be designed. We now know that three groups of bacteria (not just two as formerly believed) participate in the anaerobic conversion: the hydrolytic and acid-forming (fermenting) bacteria, the acetic acid and hydrogen-forming (acetogenic) bacteria, and the methane-forming (methanogenic) bacteria [34]. This understanding has been used in recent years to develop two-phase anaerobic digestion facilities where the acid phase is separate from the methane phase. The process was further improved by retaining the slow-growing anaerobes in the biomass using the growth surfaces mentioned in the preceding section. Examples of highly concentrated wastewaters that require anaerobic treatment are effluents from pulp mills, sugar refineries, yeast production plants, fermentation facilities, and coke plants [35, 36].

4 Final clarification

The purification efficacy of a wastewater treatment plant depends significantly on the final stage, the final clarification. The activated-sludge flocs must be completely eliminated from the water. A fraction of the biomass is recycled to the aerated lagoon, and the remaining fraction is removed. The number of free-floating organisms in the effluent should be minimal. Generally, final clarification basins are designed with either vertical or horizontal water flow. In basins where the water moves vertically, the wastewater flows along a sloping-plate clarifier, where fine flocs are retained [37]. An alternative to the clarification basins is the flotation method. The sludge flocs, suspended in water, are brought to the water surface by aeration. The flotation layer or flotate formed is removed mechanically [38, 39, 40].

5 Methods for additional (tertiary) treatment of wastewater

In the past, wastewater purification consisted of reducing oxygen-consuming organic matter. Current standards go beyond this; for example, nitrogen and phosphate compounds present in wastewater must also be removed.

5.1 Nitrification

Nitrogen exists in wastewater either in reduced form as the NH_4^+ ion, or in organic form as bound nitrogen. The bound nitrogen can, during degradation of organic matter by heterotrophic microorganisms, be converted to NH_4^+. Furthermore, NH_4^+ can be oxidized to nitrate by two groups of autotrophic microbes. Because autotrophic organisms grow at only one-tenth the rate of heterotrophs, nitrification usually proceeds only in a final clarifier basin with a long retention time and a low concentration of residual degradable organic carbon compounds. Data on nitrification have been obtained primarily from population centers [41]. A nitrification and denitrification facility processing wastewater from refineries has been operating successfully for several years [42].

5.2 Denitrification

Denitrification is the process by which nitrate is converted to elemental nitrogen in the absence of ambient oxygen. It occurs because many of the heterotrophic bacteria in the activated sludge are capable of switching their metabolism from oxygen respiration to nitrate respiration, if dissolved oxygen is absent. The greatest technical challenge, after the degradation of carbon compounds, is the development of tanks or operating-time regimes that permit aerobic nitrification and anaerobic denitrification. In Germany, nitrification and denitrification usually occur in two separate tanks. Two methods are possible: denitrification before or after wastewater treatment, or simultaneous denitrification. In denitrification before or after wastewater treatment, the water passes through an anaerobic denitrification tank. The organic compounds contained in primary-treated wastewater, or other added organic materials, can serve as carbon sources in the final clarifier effluent (Figure 8 and Table 2). In simultaneous denitrification, aerobic nitrifying and anaerobic denitrifying zones within the agitated tank are created by setting different environmental conditions [43].

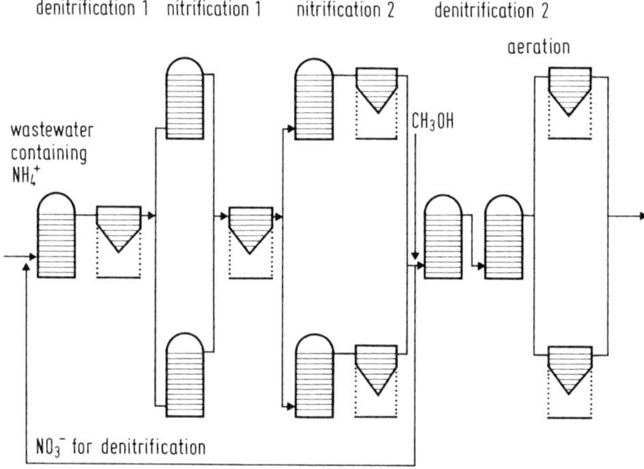

Fig. 8 Scheme for treating ammonia-containing wastewater with two-step nitrification and subsequent denitrification [42]

Table 2 Effluent values obtained with two-step nitrification and subsequent denitrification [42]

	COD, ppm	NH_4^+, ppm	NO_2^-, ppm	NO_3^-, ppm
Intake	900–1,600	700–1,100	–	–
After 1st nitrification	350–700	50–200	400–700	30–400
After 2nd nitrification	–	6–80	2–20	1,300–2,600
Effluent	60–170	3–15	0–2	0–12

5.3 Phosphate removal

The biological elimination of phosphate is not important industrially because flocculation and precipitation with iron and aluminum ions can remove residual phosphate concentrations. So far, studies of biological phosphate removal have been conducted primarily in the laboratory [44]. The individual processes involved in phosphate elimination are currently not clearly understood. Cells can be made to accumulate intracellular phosphate and store it in the form of long-chain polyphosphates only when the conditions of the aerobic and anaerobic milieu are cyclically changed. This process is expensive and is by itself not adequate to meet the specifications soon to be required for treatment plant effluents [45]. The phosphate concentrations currently mandated for effluents cannot be achieved by biological means, because a certain level of nitrogenous and phosphate compounds will always be needed for the degradation of organic matter.

6 Outlook

The biological treatment of wastewater is the method most frequently used to treat water containing organic pollutants from communities and industries. The various processing steps have been optimized. These systems are facing new demands that will require new procedural steps. Tightening the specifications of effluent quality required by law has as its primary targets the lowering of ammonia concentrations and the elimination of all nitrogen and phosphate. Given recent events in the North Sea and the Baltic Sea, we should be seeking to meet these demands as expeditiously as possible. The methods being used in current operations must be modified accordingly. We have numerous combinations of feasible processes available to do this. Even so, biological wastewater purification will be barely adequate to meet the specifications that will be demanded in future.

7 Acknowledgement

The translator and editor thank James F. Urek, P.E., of Waste Management of North America, Inc., Oak Brook, IL, USA, for his kind assistance with the translation and usage of technical terms.

8 References

[1] IMHOFF, K. R.: Die Entwicklung der Abwasserreinigung und des Gewässerschutzes seit 1868; *GWF Wasser/Abwasser* **120** (1979) 563–576.

[2] BÖHNKE, B.; HOFFMANN, J.; KÖHLHOFF, D.: Technische und gesetzliche Entwicklung der Abwassertechnik in Deutschland; *Wissenschaft und Umwelt* **1** (1980) 1–17.

[3] KLEFFNER, I.; MÖLLER, P.: Siebe als Möglichkeit zur Verbesserung und Vereinfachung von Kläranlagen; *Korrespondenz Abwasser* **23** (1976) 39–42.

[4] Imhoff, K.; Imhoff, K. R.: *Taschenbuch der Stadtentwässerung*, 25th ed., R. Oldenbourg-Verlag, München, Wien (1979).

[5] Gros, H.: Einsatz der Flotation mit Mikroblasen in Wasser- und Abwassertechnik; *Wasser-Luft-Betrieb* **22** (1978) 404–408.

[6] *Biologischer Abbau persistenter Substanzen. 5. Dechema-Fachgespräch Umweltschultz*, R. Oldenbourg-Verlag, München, Wien (1988).

[7] Charakterisierung von Abwassereinleitungen aus der Sicht der Trinkwasserversorgung; *Abschlußbericht zum Forschungsvorhaben* 02–WT 101, **Vol. 31**, Engler-Bunte-Institut der Universität Karlsruhe (1987).

[8] *Lehr- und Handbuch der Abwassertechnik*, **Vol. II**, 2nd ed., Wilhelm Ernst & Sohn, Berlin (1975).

[9] Hartmann, H.: *Untersuchungen über die biologische Reinigung von Abwasser mit Hilfe von Tauchtropfkörpern. 9. Stuttgarter Berichte zur Siedlungswasserwirtschaft*, Kommissionsverlag R. Oldenbourg, München (1960).

[10] Habeck-Tropfke, H. H.: *Abwasserbiologie; Werner-Ingenieur-Texte* **60**, Werner-Verlag Düsseldorf (1980).

[11] Uhlmann, D.: *Hydrobiologie*, Gustav Fischer Verlag, (1975).

[12] Zlokarnik, M.: Eignung und Leistungsfähigkeit von Oberflächenbelüftern für biologische Abwasserreinigungsanlagen; *Korrespondenz Abwasser* **27** (1980) 14–21.

[13] Zlokarnik, M.: Die Leistungsfähigkeit optimierter Injektoren für den Sauerstoffeintrag in biologischen Abwassereinigungsanlagen; *Verfahrenstechnik* **13** (1979) 601–604.

[14] Zlokarnik, M.: Verfahrenstechnik der aeroben Abwasserreinigung; *Chem. Ing. Tech.* **54** (1982) 939–952.

[15] Zlokarnik, M.: Eignung und Leistungsfähigkeit von Volumenbelüftern für biologische Abwasserreinigungsanlagen; *Korrespondenz Abwasser* **27** (1980) 194–209.

[16] Böhnke, B.: Das Adsorptions-Belebungsverfahren; *Korrespondenz Abwasser* **24** (1977) 33–42.

[17] Böhnke, B.: Vergleichende Betrachtungen von Versuchs- und Betriebsergebnissen der zweistufigen AB-Technik unter besonderer Berücksichtigung mikrobiologischer Reaktionsmechanismen; *Korrespondenz Abwasser* **30** (1983) 452–536.

[18] Hegemann, W.: Beitrag zur Anwendung von reinem Sauerstoff beim Belebungsverfahren; *Technisch-wissenschaftliche Schriftreihe der ATV*, Bonn (1974).

[19] Mack, K.: Biologische Reinigung von Abwasser mit Sauerstoff; *Wasser-Luft-Betrieb* **17** (1973) 1–4.

[20] Abwendung der Sauerstoffbegasung beim Belebungsverfahren; *Korrespondenz Abwasser* **22** (1975) 278–291.

[21] Lemke, J. R.; Mack, K.: Biologische Abwasserreinigung mit technisch reinem Sauerstoff in der Kläranlage Bayer AG, Werk Wuppertal-Elberfeld; *Chemie-Technik* **7** (1978) 193–196.

[22] Kläranlagen mit Sauerstoffbegasung in der Bundesrepubik Deutschland; *Gewässerschutz-Wasser-Abwasser, RWTH Aachen* **37** (1979) 1–36.

[23] *Tiefstrombelüftung*, Prospekt der Deutschen ICI GmbH, Frankfurt.

[24] Seyfried, C. F.: Biologische Abwasserreinigung nach dem ICI-Tiefstrombelüftungsverfahren – Erfahrungen mit der Kläranlage Emlichheim; *Gewässerschutz-Wasser-Abwasser* **25** (1978) 535–544.

[25] Fuhr, H.; Mann, T.: Abgedeckte Kläranlagen; *Chemische Industrie* **31** (6) (1979) 477–482.

[26] Müller, G.; Sell, G.; Bauer, A.; Leistner, G.: Abwasserreinigung in einem Bio-Hochreaktor; *Chemie-Technik* **7** (1978) 257–260.

[27] Brauer, H.; Sucker, D.: Abwasserreinigung in einem Hochleistungsreaktor; *Chemie-Ing.-Techn.* **50** (1978) 876.

[28] RUBIO, M. A.; GÖRG, S.; WILDERER, P. A.: Das "Sequencing-Batch-Reactor"-Verfahren; *Forum Mikrobiologie* **5** (1988) 169175.

[29] SCHLEGER, S.: Der Einsatz von getauchten Feststoffkörpern beim Belebungsverfahren; *GWF-Wasser/Abwasser* **127** (1986) 421–428.

[30] HEGEMANN, W.; ENGELMANN, E.: Belebungsverfahren mit Schaumstoffkörpern zur Aufkonzentrierung von Biomasse; *GWF-Wasser/Abwasser* **124** (1983) 233–238.

[31] OEHME, C.: Trägerbiologie in der Abwassertechnik; *Chem.-Ing.-Techn.* **56** (1984) 599–609.

[32] HEGEMANN, W.; WILDMOSER, A.: Sanierung einer Belebungsanlage durch den Einsatz von schwimmenden Aufwuchskörpern zur Biomassenanreicherung; *GWF-Wasser/Abwasser* **127** (1986) 415–421.

[33] HIROSE, M.: Ein neues System einer Kombination des Belebungsverfahrens mit sessilen Organismen auf Aufwuchsflächen; *GWF-Wasser/Abwasser* **124** (1983) 239–242.

[34] SAHM, H.: Biologie der Methan-Bildung; *Chem.-Ing.-Techn.* **53** (1981) 854–863.

[35] HASENBÖHLER, A.: Anaerobe biologische Abwasserreinigungsanlagen; *Zuckerind.* **107** (1982) 835–838.

[36] TSCHERSICH, J.: Biothane® – ein Verfahren zur anaeroben Abwasserreinigung und Biogas-Gewinnung; *Zuckerind.* **107** (1982) 838–842.

[37] RESCH, H.: Wirkungsweise der Nachklärung von Belebungsanlagen; *Österreichische Wasserwirtschaft* **39** (1987) 166–172.

[38] ZLOKARNIK, M.: Neue Flotationstechnik zur Abtrennung und Eindickung von Klärschlamm bei der biologischen Abwasserreinigung. Teil 1: Begasungsflotation; *Korrespondenz Abwasser* **32** (1985) 525–530.

[39] ZLOKARNIK, M.: Neue Flotationstechnik zur Abtrennung und Eindickung von Klärschlamm bei der biologischen Abwasserreinigung. Teil 2: Entgasungsflotation; *Korrespondenz Abwasser* **32** (1985) 598–603.

[40] SCHMITT, E.: Vollbiologische Betriebskläranlage. Bayer-Turmbiologie mit Flotationsanlage in der Königsbacher Brauerei AG, Köln; *Brauwelt* **123** (1983) 1830–1841.

[41] DAMIECKI, R.: Leistung konventioneller Kläranlagen bezüglich des Abbaus von Ammoniumstickstoff; *Gewässerschutz-Wasser-Abwasser, RWTH Aachen* **50** (1982) 353–374.

[42] PASCIK, I.: Biochemische Stickstoff-Elimination aus ammoniumreichem Raffinerieabwasser in der Bayer-Turmbiologie; *Chem.-Ing.-Techn.* **54** (1982) 1190–1192.

[43] PÖPEL, H. J.: Grundlagen und Bemessung der biologischen Stickstoffelimination; *GWF-Wasser/Abwasser* **128** (1987) 469–474.

[44] SEKOULOV, H.; FRUHEN, M.;VAN ASSCHE, J.: Laboruntersuchungen zur Bestimmung des biologischen Phosphoreliminationsvermögens von Belebtschlämmen; *Korrespondenz Abwasser* **34** (1987) 339–344.

[45] EKAMA, G. A.; MARAIS, G. I. R.: Zusätzliche biologische Phosphorelimination beim Belebungsverfahren – Erfahrungen in Südafrika; *GWF-Wasser/Abwasser* **126** (1985) 241–249.

Microbial Degradation of Aliphatic Hydrocarbons and Its Environmental Importance

by R. MÜLLER-HURTIG and F. WAGNER

Contents

Dr. R. MÜLLER-HURTIG and Prof. Dr. F. WAGNER,
Technische Universität Braunschweig,
Institut für Biochemie und Biotechnologie,
Konstantin-Uhde-Str. 5,
D-3300 Braunschweig, Fed. Rep. Germany

1 Introduction

The environmental problems we are now facing require that we come up with better biotechnological methods for degrading not only aromatic compounds [1] and cycloalkanes and chlorinated alkanes, but also mixtures of saturated, unsaturated, branched, and unbranched aliphatic alkanes. If we could improve our capabilities we could reclaim waste sites and eliminate hydrocarbon compounds from exhaust and stack gases.

Degradative pathways and their regulation need further research. As an example, when undecane is degraded, the product may be undecanoic acid, a toxic metabolite that inhibits the synthesis of long-chain fatty acids in various microorganisms [2]; since we currently have such inadequate knowledge about the degradation of toxic substances, it is possible that when we attempt to reclaim contaminated soil, the microbial oxidation of toxic substances could produce even more toxic metabolites. At present the toxic intermediates of degradation cannot be identified at the laboratory stage of a reclamation project.

Numerous review articles on the degradative pathways of alkanes have appeared since 1965 [3–8]. In this chapter, therefore, we will only briefly discuss known degradative pathways and instead will stress recent advances and problems in environmental biotechnology.

2 Degradative pathways in the microbial oxidation of alkanes

Currently known degraders of alkanes include many genera of bacteria and fungi, as well as some yeasts and green algae [9]. This relatively wide-ranging capability is not surprising considering how widespread the hydrocarbons are as biological substrates everywhere in the world and ever since plants began to biosynthesize them.

Aliphatic hydrocarbons, which are quite water insoluble, must first diffuse to the cytoplasmic membrane. How they get that far has been the subject of investigations since the end of the 1970s [10, 11].

Three different pathways for the oxidation of alkanes have been described to date, but we still lack a complete understanding of their mechanisms. The first enzymatic step, the oxidation of the alkane molecule at the cytoplasmic membrane, can occur at one or both of the terminal methyl groups or on a methylene group.

2.1 Monoterminal oxidation

Enzymatic oxidation at a terminal methyl group is a more frequent first step than is the subterminal oxidation at one of the methylene groups. Microorganisms form fatty acids from alkanes with more than ten carbon atoms, primarily monoterminally via the corresponding alcohol and aldehyde, and subsequently metabolize them by β-oxidation (Figure 1).

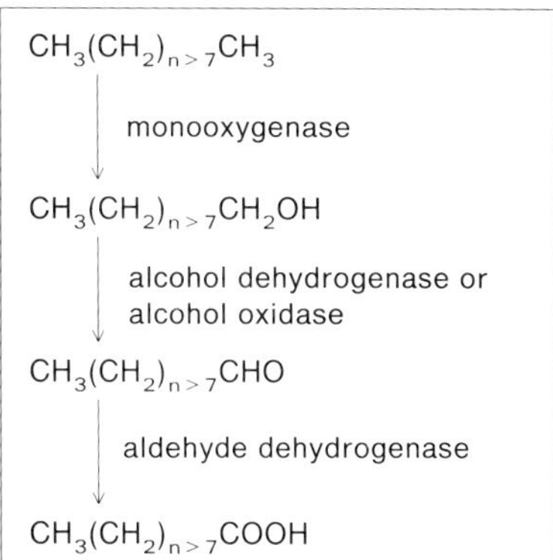

Fig. 1 Principal pathway for the monoterminal oxidation of alkanes

FINNERTY has described an alternative monoterminal oxidation pathway [12]. He was able to stabilize the intermediates by extracting the cells with concentrated sulfuric acid and identified the postulated labile fatty acid peroxides. Since he had direct proof of the existence of the peroxide and the enzymes involved, he proposed that the probable course for monoterminal oxidation in *Acinetobacter* H01-N is from an alkane peroxide, via a fatty acid peroxide, then (mediated by a reductase) to an aldehyde, which disproportionates to a fatty acid and an alcohol. The alcohol is a metabolic end product and becomes a component of the waxes that are formed (Figure 2).

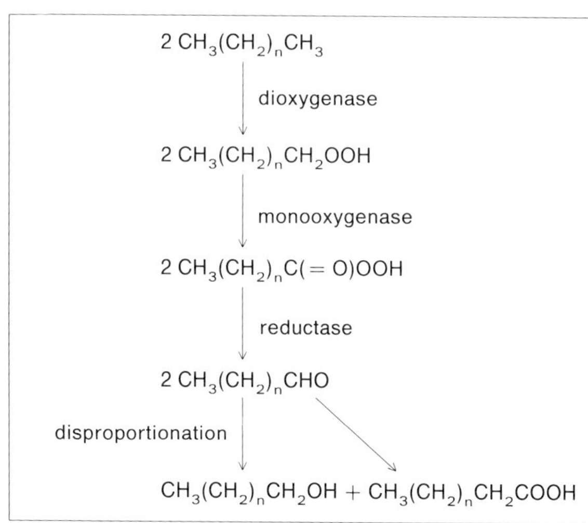

Fig. 2 Monoterminal oxidation by *Acinetobacter* H01-N

2.2 Biterminal oxidation

Much rarer than monoterminal conversion is the biterminal oxidation of the two terminal methyl groups. There is evidence for the simultaneous oxidation of both groups [13] and also for a consecutive oxidation of the methyl group in the ω position of the monocarboxylic acid formed. AVETISOVA [14] showed that two P_{450} cytochrome systems (see Section 3) exist in *Candida maltosa* (Figure 3). One system is induced by long-chain alkanes and primarily initiates alkane oxidation by end-group hydroxylation. The other P_{450} system is induced by *n*-decane, and as its chief function catalyzes the hydroxylation of fatty acid groups in the ω position. Monoterminal oxidation is, however, the usual mechanism in *Candida maltosa* [15].

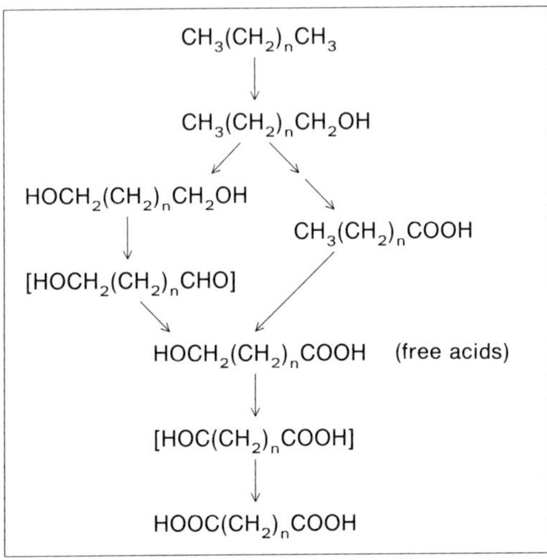

Fig. 3 Biterminal oxidation of *n*-alkanes

2.3 Subterminal oxidation

Many pseudomonads oxidize alkanes with three to eight carbon atoms subterminally at the β carbon atom, forming a ketone, before they oxidize the terminal methyl group and split off the carboxyl group of the α-oxocarboxylic acid. Other pseudomonads, and fungi of the genus *Fusarium*, perform a variation of this: the enzymatic conversion proceeds according to the BAYER-VILLIGER reaction, which is a known organic chemistry reaction mediated by hydrogen peroxide (Figure 4).

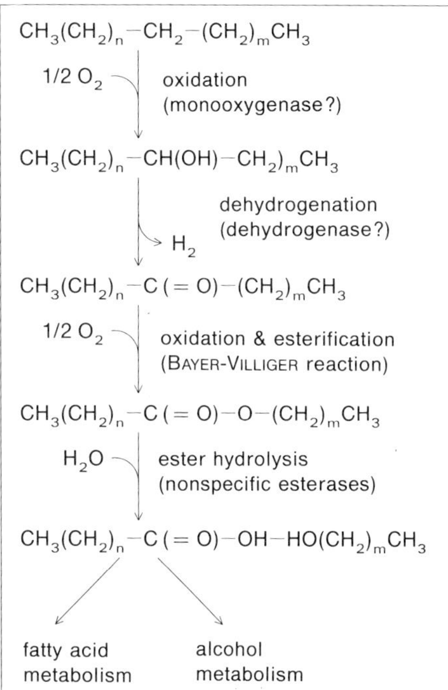

Fig. 4 Subterminal oxidation of long-chain alkanes

3 Enzymes of the degradative pathways

3.1 Alkane monooxygenases

The primary oxidation proceeds by consuming reducing equivalents; while exergonic reactions that release up to 24–26 kJ enzymatically can be used for energy production, those with 70–75 kJ, such as the primary oxidation step, cannot. Multicoupled systems such as the redox systems with rubredoxin (Fe^{2+}/Fe^{3+}), or cytochrome P_{450}, nonheme-bound iron, and a flavoprotein have been demonstrated (Figure 5).

3.2 Alcohol dehydrogenases and alcohol oxidases

Bacterial alcohol dehydrogenases have been well characterized with regard to their substrate specificity, their inducibility, and their inducers [6]. On the other hand, several research groups [16–19] have provided evidence that hydrogen peroxide is generated, we know that yeasts may also have inducible alcohol oxidases. The cellular activity of

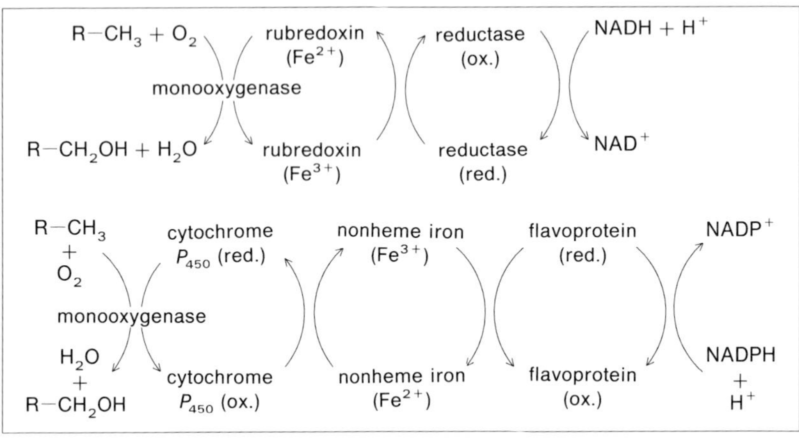

Fig. 5 Redox systems in alkane oxidation (with or without cytochrome P_{450})

these oxidases is orders of magnitude greater than that of alcohol and aldehyde dehydrogenase [19]; thus, alcohol dehydrogenases may play only a secondary role in alkane oxidation by yeasts.

4 Chlorinated aliphatic hydrocarbons

There are volatile short-chain chlorinated alkanes in the environment because they are used industrially as solvents that pose no threat of explosion or combustion; the ones most likely to be found are dichloromethane, 1,1,1-trichloroethane, trichloroethylene, and perchloroethylene. Since they have densities of 1,300 to 1,600 kg/m^{-3}, they are more likely to contaminate ground water than unsubstituted hydrocarbons are. Trichloroethylene is the contaminant most frequently found in U.S. waste sites that need to be cleaned up [20].

The aerobic degradation of dichloromethane as sole carbon source was first described in 1980 by BRUNNER [21]. Four principal metabolic pathways, described in books and review articles on biological dehalogenation, are primarily responsible for the enzymatic dehalogenation of chlorinated aromatic hydrocarbons and biocides and their conversion to unhalogenated compounds [22–25].

a. During *reductive dehalogenation* (Figure 6), the halogen atom is replaced by a hydrogen atom; reducing equivalents are simultaneously consumed; for example, under anaerobic conditions dechlorination proceeds in a chain reaction from tetrachloroethylene via trichloroethylene, *cis*-dichloroethylene, and vinyl chloride [26].
b. During *biodehydrodehalogenation* (Figure 7), a hydrogen and a halogen atom are simultaneously removed from the same molecule and a C–C double bond is formed.

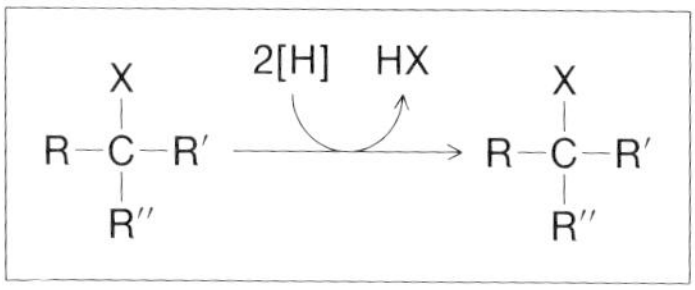

$$R{-}C(X)(R''){-}R' \xrightarrow{2[H] \quad HX} R{-}C(X)(R''){-}R'$$

Fig. 6 Reductive dehalogenation

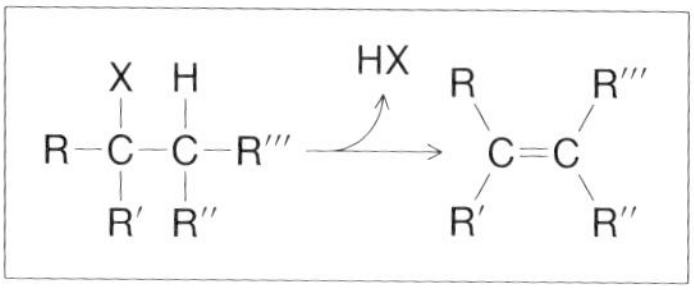

$$R{-}C(X)(R'){-}C(H)(R''){-}R''' \xrightarrow{HX} (R)(R')C{=}C(R''')(R'')$$

Fig. 7 Dehydrodehalogenation

$$R{-}C(X)(R''){-}R' \xrightarrow{H_2O \quad HX} R{-}C(OH)(R''){-}R'$$

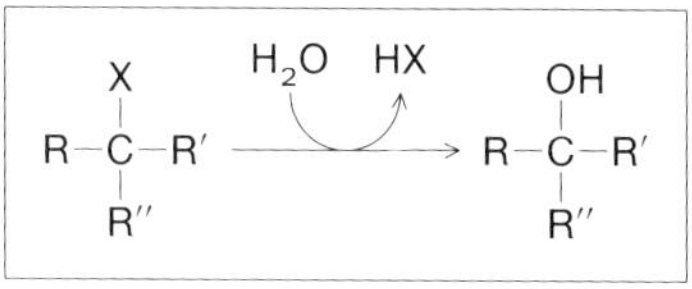

Fig. 8 Hydrolytic dehalogenation

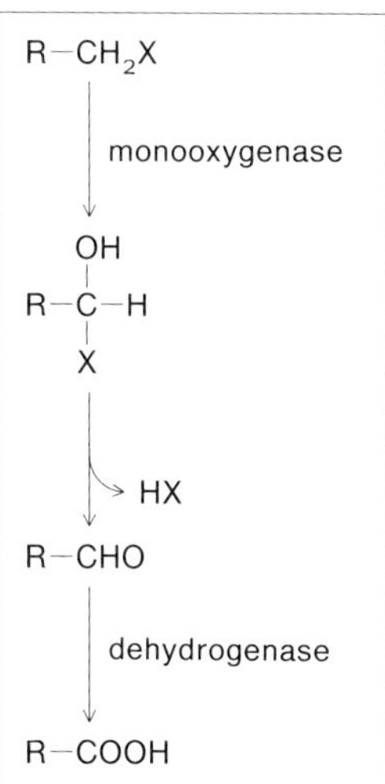

Fig. 9 Oxidative dehalogenation

For example, dibromoethane is quantitatively converted to ethylene by soil microorganisms; only one of the two dehalogenation steps may involve dehydrodehalogenation [27].

c. The nucleophilic substitution of the halogen atom by the hydroxyl ion is called *hydrolytic dehalogenation* (Figure 8). This mechanism has been suggested for the oxidation of short-chain halogenated hydrocarbons such as dichloroethane and dichloromethane [28, 29].

d. During *oxidative dehalogenation* (Figure 9), chlorine is removed by oxidation with consumption of oxygen. For example, *Methylococcus capsulatus* oxidatively dehalogenates chloromethane in its bioconversion to formaldehyde [30].

In most instances of degradation of short-chain chlorinated hydrocarbons, cometabolism is involved; *Pseudomonas fluorescens* PFL 12 does aerobic degradation best when glucose or yeast extract is the carbon source [31]. The obligate methanotrophic bacteria, such as *Methylomonas methanica* NCIB 11130 and *Methylosinus trichosporium* OB 36, biodegrade *trans*-1,2-dichloroethylene cometabolically in the presence of methane [32]. Microorganisms that metabolize chlorinated hydrocarbons do not perforce biodegrade trichloroethylene via inducers of the aromatic pathway, as does a strain of *Acinetobacter* [33,34]; a counterexample is *Pseudomonas fluorescens* PFL 12 [31].

Bachmann *et al.* [35, 36] studied the degradation of α-hexachlorocyclohexane in a soil slurry as a function of redox potential. Under aerobic conditions, the α-hexachlorocyclohexane was mineralized, but under anaerobic conditions, intermediates were the degradation products. Furthermore, anaerobic degradation was strongly repressed in the presence of carbon sources that were more readily biodegradable. Even lindane, a γ-hexachlorocyclohexane, can be metabolized aerobically as the sole carbon source [37].

5 Yields of metabolites in the biodegradation of alkanes

Because the substrates are expensive, the only biodegradation products of alkanes that are worth considering for commercial purposes are those with high value. A mutant of *Candida tropicalis* accumulates long-chain dicarboxylic acids from β-monocarboxylic acids because it cannot β-oxidize them, but the dicarboxylic acids undergo a separate β-oxidation; various short-chain dicarboxylic acids accumulate simultaneously, so that a single dicarboxylic acid of choice cannot be obtained in pure form [38]. It follows that many chromosomally encoded enzymes may be taken advantage of expediently only by cloning the required enzymes. Long-chain alcohols, short-chain fatty acids, and optically active epoxides are products of potential interest. For example, *Pseudomonas oleovorans* forms epoxides from 1-alkenes with chain lengths of 5 to 12 carbon atoms; the (*R*)-1,2-epoxyalkane is preferentially produced [39]. As a further example, chiral resolution by *Xanthobacter* Py2 can be used to produce certain optically active epoxides: the bacteria metabolize only the 2*S* forms of *trans*-2,3-epoxybutane, *trans*-2,3-epoxypentane, and *cis*-2,3-epoxypentane [40], but biodegrade both enantiomers of the 1,2-epoxyalkanes. Other interesting products of alkane degradation are the optically pure polyesters of poly-(*R*)-3-hydroxyalkanoates and poly-(*R*)-3-hydroxyalkenoates, which, like polyhydroxybutyric acids (PHB), could be used as biodegradable synthetic molecules. PHB can be up to 80% of the dry weight of bacterial cells; in contrast, the cell mass of *Pseudomonas oleovorans* contains no more than 25% polyesters of long-chain monomers, because this stored metabolite, unlike PHB, is again degraded [39, 41].

6 Microbial degradation in the environment

Alkanes contaminate marine ecosystems, and, to a lesser extent, fresh water, soils, and ground water. In sewage treatment plants large quantities of oil can be separated by clarification and then incinerated, but purification is the usual strategy of choice. It has been estimated that at least 70,000 toxic waste sites exist in what was formerly known as West Germany (or the Federal Republic of Germany, FRG); the cost of investigating, monitoring, and reclaiming these sites has been estimated as high as DM 20 billion [42]. Microbial methods have been explored and used in the former FRG since 1984, in addition to the usual methods for handling oil-contaminated soil such as transporting it to special confined dump sites, washing, incinerating, or solidifying it by physical/chemical means [43]. Because of problems in location, on-site purification must be done in bottom-sealed clamps, in a more confined manner than are in-ground processes in the USA. On-site purification is usually combined with a washing cycle and containment.

The microflora at a site are often able to decontaminate the area unassisted, after a period of adaptation that is sometimes decades long. The farm field that we studied, near an interstate highway, had a high biodegradative capacity but had only 10^6 (mean value) viable hydrocarbon decomposers per gram of dried soil [44]. Soils not previously contaminated with oil contained 10^5 to 10^6 such microorganisms; those with a history of oil contamination had 10^6 to 10^8 microorganisms per gram of dried soil [45]. The microbial soil population of farm fields metabolized trimethylcyclohexane (as a cyclic alkane) and pristane (as a branched alkane) as rapidly as tetradecane; an increase in the ratio of pristane to tetradecane should thus not be used as a measure of microbial degradation. Biodegradation did not proceed cometabolically when 91–97% of the tetradecane decomposers grew on cyclohexane or pristane as sole source of carbon, although cycloalkanes are primarily biodegraded cometabolically [8].

Adequate oxygen supply is important for the rate of biodegradation *in situ* in clamps or in landfarming processes; the oxygen may be supplied as diatomic molecules, hydrogen peroxide [46], or ozone. Nitrate is not adequate as the sole oxygen source because alkane monooxygenases and alcohol oxidases depend on molecular oxygen. In addition, the rapid metabolism of 1 g of oil with rapid microbial growth under nonlimited mineral nutrient conditions requires about 120 mg of nitrogen and 20 mg of phosphorus; therefore, fertilization of contaminated sites enhances the biodegradation of the oil [45]. Because the biodegradation of hydrophobic hydrocarbons is associated with the formation of surfactants, there were several attempts to use chemical surfactants to enhance biodegradation. In some cases they were inhibitory, and in others they were without effect or stimulated the biodegradation of the oil [47]. Although surfactants are effective in increasing oil recovery in tertiary oil production [48], the extent of degradation could not be increased [49]. In oil-contaminated mudflats, chemical surfactants decreased the amount of oil removed, but microbiologically produced surfactants applied at the same time increased it (Figure 10), although the two surfactants had comparable HLB (hydrophilic lipid balance) values, that is, the surfactant molecules were equally hydrophilic [50]. Our group has demonstrated the enhancing effect of microbially generated surfactants on the degradation of a model oil, in a well-stirred bioreactor with 10% soil, and in a fixed-bed reactor in which water percolates through soil

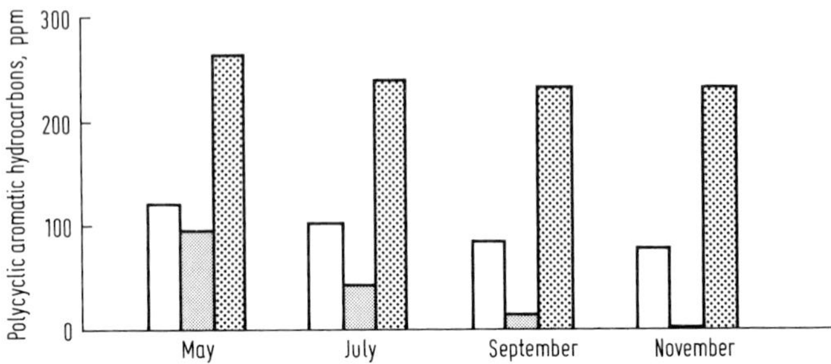

Fig. 10 Persistence of crude-oil hydrocarbons in mudflats (after [50])

□ ARAMCO, 1 liter oil applied to 2 m^2, 10 times
▨ ARAMCO plus biosurfactant, 1 liter oil and 1 gram trehalose lipid applied to 2 m^2, 10 times
▩ ARAMCO plus chemical surfactant, 1 liter oil and 100 ml OSR-5 in 1 liter seawater applied to 2 m^2, 10 times

Table 1 Effect of added glycolipids on the efficiency of oil degradation in submerged culture

Culture conditions	Removal of model oil		HLB	Degradation rate, g oil per kg of soil dry weight per day
	Time, h	Extent, %		
partial oxygen limitation				
no added surfactant	114	81	–	16.3
trehalose 6,6′-corynomycolate	71	93	4.0	37.2
sophorose lipids	75	97	6.9	39.0
cellobiose lipids	79	99	8.0	32.3
rhamnose lipids	77	94	9.5	28.6
trehalose 2,3,4,2′-tetraester	94	95	10.0	23.8
no oxygen limitation				
no added surfactant	79	89	–	25.7
sophorose lipids	57	95	6.9	46.5

The glycolipids were produced microbially at our institute

[51, 52]. The surfactants can also be produced in larger amounts [53]. Surfactants increase the extent of biodegradation in submerged cultures from 81–91% to 95% and above 99% (Table 1).

Degradation of a mixture of hydrocarbons begins earlier because the soil population is able to biodegrade the aliphatic hydrocarbons directly without first forming biosurfactants. Biosurfactants, although they are indeed biodegradable, are metabolized after the slightly soluble aliphatic hydrocarbons [51]. If higher clamps are used, or if hydrogen peroxide can be stabilized successfully with compounds such as pyrophosphate [46], sufficiently stable biosurfactants should improve the extent and rate of biodegradation in circumstances where they are introduced.

7 Concluding remarks

We have learned in numerous experiments that the microbial ecosystem is often quite capable of degrading alkanes *in situ*; the removal of limitations to growth promotes the biodegradation of alkanes much more effectively than does the introduction of metabolically more efficient but foreign strains. This is true of marine ecosystems as well as of soils. In particular, the release to the environment of genetically altered hybrid strains that are very efficient in the laboratory should be conducted under tight restrictions as long as the effects have not been studied adequately.

It is of particular importance to analyze the microbiological profile of an ecosystem once it has been contaminated with chemical substrates such as short-chain chlorinated hydrocarbons; for example, dichloromethane dehalogenases at a single site can differ 5- to 6-fold in activity. An active strain therefore does not need to have 15–20% of its total cell protein in the form of this one enzyme, in contrast to the 11 strains presently characterized [54]. Evolution by gene transfer is a possibility in mixed culture populations, and could contribute to improving existing degradative pathways and to accelerating the selection of novel degradative pathways for metabolizing new substrates.

8 Acknowledgements

We thank the Ministry of Research and Development (BMFT) for supporting our work on improved degradation of oil with biosurfactants.

9 References

[1] LINGENS, F.: Microbial Degradation of Aromatic Compounds. In: FINN, R. K.; PRÄVE, P.; SCHLINGMANN, M.; CRUEGER, W.; ESSER, K.; THAUER, R.; WAGNER, F., Eds.; *Biotechnology Focus 2*, Hanser Publishers, Munich, Vienna, New York, (1988/89); pp 285–305 (in English). LINGENS, F.: Mikrobieller Abbau von aromatischen Verbindungen. In: PRÄVE, P.; SCHLINGMANN, M.; CRUEGER, W.; ESSER, K.; THAUER, R.; WAGNER, F., Eds.; *Jahrbuch Biotechnologie Band 2*, Carl Hanser Verlag, München, Wien (1988/89); pp 297–318 (in German).

[2] HORTMANN, L.; REHM, H. J.: Inhibitory Effect of Undecanoic Acid on the Biosynthesis of Long-Chain Fatty Acids in *Mortellia isabellina*; *Appl. Microbiol. Biotechnol.* **20** (1984) 139–145.

[3] MCKENNA, E. J.; KALLIO, R. E.: Biology of Hydrocarbons; *Ann. Rev. Microbiol.* **19** (1965) 183–206.

[4] VAN DEN LINDEN, A. C.; THIJSSE, G. J. E.: The Mechanism of Microbial Oxidation of Petroleum Hydrocarbons; *Adv. Enzymol.* **27** (1965) 469–546.

[5] RATLEDGE, C.: Degradation of Aliphatic Hydrocarbons. In: WATKINSON, R. J., Ed.: *Developments of Biodegradation of Hydrocarbons*; Applied Science Publishers, London, (1978); pp 1–45.

[6] Rehm, H. J.; Reiff, I.: Mechanisms and Occurrence of Microbial Oxidation of Long-Chain Alkanes; *Adv. Biochem. Eng.* **19** (1981) 173–215.

[7] Singer, M. E.; Finnerty, W. R.: Microbial Metabolism of Straight-Chain and Branched Alkanes. In: Atlas, R. M., Ed.; *Petroleum microbiology*, Macmillan Publishing Co., New York (1984); pp 1–59.

[8] Perry, J. J.: Microbial Metabolism of Cyclic Alkanes. In: Atlas, R. M., Ed.; *Petroleum Microbiology*, Macmillan Publishing Co., New York (1984); pp 61–97.

[9] Rehm, H. J.: Mikrobiologie und Biochemie der Kohlenwasserstoffe. In: Schweisfurth, R., *et al.*, Eds.; *Angewandte Mikrobiologie der Kohlenwasserstoffe in Industrie und Umwelt*, Exper. Publ., Ehingen, FRG (1988); pp 1–16.

[10] Zajic, J. E.; Seffens, W.: Biosurfactants; *CRC Crit. Rev. Microbiol.* **5** (1984) 87–107.

[11] Reddy, P. G.; Singh, H. D.; Roy, P. K.; Baruah, J. N.: Predominant Role of Hydrocarbon Solubilization in the Microbial Uptake of Hydrocarbons; *Biotech. Bioeng.* **24** (1982) 1241–1269.

[12] Finnerty, W. R.: Lipids of *Acinetobacter*. In: Applewhite, T. H., Ed.: *Proceedings of the World Conference on Biotechnology for the Fats and Oils Industry*, Kraft, Inc., Glenview, IL, (1988); pp 184–188.

[13] Yi, Z. H.; Rehm, H. J.: Degradation Pathways from *n*-Tridecane to α,ω-Tridecandioic Acid in a Mutant of *Candida tropicalis*; *Appl. Microbiol. Biotechnol.* **15** (1982) 144–146.

[14] Avetisova, S. M.; Sokolov, Y.; Kozlov, V. I.; Davydov, R. M.; Davydov, E. R.: The Induction of Cytochrome P_{450} Two Forms in *Candida* Yeast by *n*-Alkanes of Different Chain Length. In: Vereczky, L.; Magyar, K., Eds.; *Cytochrome P_{450} Biochemistry, Biophysics and Induction*; Proceedings, 5th International Conference on Cytochrome P_{450}, Budapest, (1988); pp 455–458.

[15] Blasig, R.; Mauersberger, S.; Riege, P.; Schunck, W.-H.; Jockisch, W.; Franke, P.; Müller, H.-G.: Degradation of Long-Chain *n*-Alkanes by the Yeast *Candida maltosa*. II. Oxidation of *n*-Alkanes and Intermediates Using Microsomal Membrane Fractions; *Appl. Microbiol. Biotechnol.* **28** (1988) 589–597.

[16] Il'chenko, A. P.: Oxidase of Higher Alcohols in the Yeast *Torulopsis candida* Grown on Hexadecane; *Mikrobiologiya* **53** (1984) 903–906.

[17] Krauzova, V. I.; Il'chenko, A. P.; Sharyshev, A. A.; Lozinov, A. B.: Possible Pathways of the Oxidation of Higher Alcohols by Membrane Fractions of Yeasts Culture on Hexadecane and Hexadecanol; *Biokhimiya* **50** (1986) 726–732.

[18] Krauzova, V. I.; Kuvichkina, T. N.; Sharyshev, A. A.; Romanova, I. B.: Formation of Lauric Acid and NADH in the Oxidation of Dodecanol by Membrane Fractions of the Yeast *Candida maltosa*, Cultured on Hexadecane; *Biokhimiya* **50** (1986) 23–27.

[19] Kemp, G. D.; Dickinson, F. M.; Ratledge, C.: Inducible Long Chain Alcohol Oxidase from Alkane-Grown *Candida tropicalis*; *Appl. Microbiol. Biotechnol.* **29** (1988) 370–374.

[20] Wilson, J. T.; Wilson, B. H.: Biotransformation of Trichloroethylene in Soil; *Appl. Environ. Microbiol.* **49** (1985) 242–243.

[21] Brunner, W.; Staub, D.; Leisinger, T.: Bacterial Degradation of Dichloromethane; *Appl. Environ. Microbiol.* **40** (1980) 950–958.

[22] Neidleman, S. L.; Geigert, J.: Biological Dehalogenation. In: *Biohalogenation: Principles, Basic Roles and Applications*, Ellis Horwood, Ltd., Chichester, U.K. (1986); pp 156–171.

[23] Müller, R.; Lingens, F.: Mikrobieller Abbau halogenierter Kohlenwasserstoffe: Ein Beitrag zur Lösung vieler Umweltprobleme? *Angw. Chem.* **98** (1986) 778–787.

[24] Müller, R.; Lingens, F.: Mechanismen der mikrobiellen Dehalogenierung von Chlorkohlenwasserstoffen; *GIT Supplement* 5/1987, pp 4–9.

[25] Rochkind-Dubinsky, M. L.; Sayler, G. S.; Blackburn, J. W.: *Microbial Decomposition of Chlorinated Aromatic Compounds, Microbiology Series,* **Vol. 18**, Marcel Dekker, Inc., New York, Basel, (1986).

[26] Bosma, T. N. P.; Holliger, C.; van Neerven, A. R. W.; Schraa, G.; Zehnder, A. J. B.: Reduzierende Dechlorierung chlorierter Kohlenwasserstoffe in anaeroben Sedimentsäulen. In: *2nd International TNO/BMFT-Congress*, Hamburg (1988); pp 745–746.

[27] Castro, C. E.: Biodehalogenation; *Environ. Health Perspectives* **21** (1977) 2335–2339.

[28] Janssen, D. B.; Scheper, L.; Dijkhuizen, L.; Witholt, B.: Degradation of Halogenated Aliphatic Compounds by *Xanthobacter autotrophicus* GJ 10; *Appl. Environ. Microbiol.* **49** (1985) 673–677.

[29] Kohler-Staub, D.; Leisinger, T.: Dichloromethane Dehalogenase of *Hyphomicrobium* sp. Strain DM2; *J. Bacteriol.* **162** (1985) 676–681.

[30] Dalton, H.: Methane Monooxygenases from a Variety of Microbes. In: Dalton, H., Ed.; *Microbial Growth on C1 Compounds*, Proceedings, Third International Symposium, Heyen, London, (1980); pp 1–10.

[31] Vandenbergh, P. A.; Kunka, B. S.: Metabolism of Volatile Chlorinated Aliphatic Hydrocarbons by *Pseudomonas fluorescens*; *Appl. Environ. Microbiol.* **54** (1988) 2578–2579.

[32] Janssen, D. B.; Grobben, G.; Hoekstra, R.; Oldenhuis, R.; Witholt, B.: Degradation of *trans*-1,2–Dichloroethene by Mixed and Pure Cultures of Methanotrophic Bacteria; *Appl. Microbiol. Biotechnol.* **29** (1988) 392–399.

[33] Nelson, M. J.; Montgomery, S. O.; Mahaffey, W. R.; Pritchard, P. H.: Biodegradation of Trichlorethylene and Involvement of an Aromatic Biodegradative Pathway; *Appl. Environ. Microbiol.* **53** (1987) 949–954.

[34] Nelson, M. J.; Montgomery, S. O.; Pritchard, P. H.: Trichlorethylene Metabolism by Microorganisms That Degrade Aromatic Compounds; *Appl. Environ. Microbiol.* **54** (1988) 604–606.

[35] Bachmann, A.; Walet, P.; Wijnen, P.; de Bruin, W.; Huntjens, J. L. M.; Roeloefsen, W.; Zehnder, A. J. B.: Biodegradation of Alpha- and Beta-Hexachlorocyclohexane in a Soil Slurry under Different Redox Conditions; *Appl. Environ. Microbiol.* **54** (1988) 143–149.

[36] Bachmann, A.; De Bruin, W.; Jumelet, J. C.; Rijnaarts, H. H. N.; Zehnder, A. J. B.: Aerobic Biomineralization of Alpha-Hexachlorocyclohexane in Contaminated Soil; *Appl. Environ. Microbiol.* **54** (1988) 548–554.

[37] Tu, C. M.: Utilization and Degradation of Lindane by Soil Microorganisms; *Arch. Microbiol.* **108** (1976) 259–263.

[38] Hill, F. F.; Venn, I.; Lukas, K. L.: Studies on the Formation of Long-Chain Dicarboxylic Acids from Pure *n*-Alkanes by a Mutant of *Candida tropicalis*; *Appl. Microbiol. Biotechnol.* **24** (1986) 168–174.

[39] de Smet, M. J.; Eggink, G.; Witholt, B.; Kingma, J.; Wynberg, H.: Characterization of Intracellular Inclusions Formed by *Pseudomonas oleovorans* during Growth on Octane; *J. Bacteriol.* **154** (1983) 870–878.

[40] Weijers, C. A. G. M.; de Haan, A.; de Bont, J. A. M.: Chiral Resolution of 2,3-Epoxyalkanes by *Xanthobacter* Py2; *Appl. Microbiol. Biotechnol.* **27** (1988) 337–340.

[41] Lageveen, R. G.; Huisman, G. W.; Preusting, H.; Ketelaar, P.; Eggink, G.; Witholt, B.: Formation of Polyesters by *Pseudomonas oleovorans*: Effect of Substrates on Formation and Composition of Poly-(*R*)-3-Hydroxyalkanoates and Poly-(*R*)-Hydroxyalkenoates; *Appl. Environ. Microbiol.* **54** (1988) 2924–2932.

[42] Deckwer, W.-D.; Weppen, P.: Technologien zur Sanierung von Bodenkontaminationen und Altlasten; *Chem.-Ing.-Tech.* **59** (1987) 457–464.

[43] Battermann, G.: Beseitigung einer Untergrund-Kontamination mit Kohlenwasserstoffen durch mikrobiellen Abbau; *Chem.-Ing.-Tech.* **56** (1984) 926–928.

[44] Oberbremer, A.; Müller-Hurtig, R.: Aerobic Stepwise Hydrocarbon Degradation and Formation of Biosurfactants by an Original Soil Population in a Stirred Reactor; *Appl. Microbiol. Biotechnol.* **31** (1989) 582–586.

[45] Bossert, I.; Bartha, R.: The Fate of Petroleum in Soil Ecosystems. In: Atlas, R. M., Ed.; *Petroleum Microbiology*, Macmillan Publ. Co., New York (1984); pp 355–397.

[46] Stimulation of Biooxidation Processes in Subterranean Formations; European Patent Appl. 85 308094.3 and 85 308096.3; FMC Corp., Philadelphia (1985).

[47] Cooney, J.: The Fate of Petroleum Pollutants in Freshwater Ecosystems. In: Atlas, R. M., Ed.; *Petroleum Microbiology*, Macmillan Publ. Co., New York (1984); pp 435–473.

[48] Wagner, F.: Gewinnung mikrobieller Produkte für die Tertiärförderung von Erdöl. In: Schweisfurth, R., *et al.*, Eds.; *Angewandte Mikrobiologie der Kohlenwasserstoffe in Industrie und Umwelt*, Expert Publ., Ehningen, (1988),; pp 65–83.

[49] Atlas, R. M.; Bartha, R.: Effects of Some Commercial Oil Herders, Dispersants and Bacterial Inocula on Biodegradation of Oil in Seawater. In: Ahearn, D. G; Meyers, S. P., Eds.; *The Microbial Degradation of Oil Pollutants*, Louisiana State University publ. no. LSU-SG-73-01, Baton Rouge, LA (1973); pp 283–289.

[50] van Bernem, K.-H.: Experimentelle Untersuchungen zur Wirkung von Rohöl und Rohöl/Tensid-Gemischen im Ökosystem Wattenmeer. II. Eindringverhalten und Persistenz von Rohölkohlenwasserstoffen in Sedimenten nach experimenteller Kontamination; *Senckenbergiana marit.* **16** (1984) 13–23.

[51] Oberbremer, A.; Müller-Hurtig, R.; Wagner, F.: Effect of the Addition of Microbial Surfactants on Hydrocarbon Degradation in a Soil Population in a Stirred Reactor; *Appl. Microbiol. Biotechnol.* **32** (1990) 485–489.

[52] Müller-Hurtig, R.; Meier, R.; Kindervater, R.; Wagner, F.: Einfluß von Biotensiden auf den Ölabbau in wasserdurchströmten Bodenfestbettreaktoren; *Dechema-Jahrestagung der Biotechnologen*, Frankfurt (1989); pp 825–826.

[53] Syldatk, C.; Wagner, F.: Production of Biosurfactants. In: Kosatic, N.; Cairns, W. L.; Gray, N. C. C., Eds.; *Biosurfactants and Biotechnology*, Marcel Dekker, Inc., New York (1987); pp 89–120.

[54] Scholtz, R.; Wackett, L. P.; Egli, C.; Cook, A. M.; Leisinger, T.: Dichloromethane Dehalogenase with Improved Catalytic Activity Isolated from a Fast-Growing Dichloromethane-Utilizing Bacteria; *J. Bacteriol.* **170** (1988) 5698–5704.

Industrial Biotechnology

Lipases: Selected Applications in Preparative Organic Chemistry

by H. ERDMANN, K. FRITSCHE, M. KORDEL, S. LANG, W. LOKOTSCH, M. MARKWEG, M. SCHNEIDER, C. SYLDATK, and F. WAGNER

Contents

Dr. H. Erdmann and Dr. M. Kordel,
Gesellschaft für Biotechnologische Forschung mbH,
Mascheroder Weg 1,
D-3300 Braunschweig, Fed. Rep. Germany;

Dr. K. Fritsche, Dr. S. Lang, Dipl.-Chem. W. Lokotsch, Dipl.-Biol. M. Markweg, Dr. C. Syldatk, and Prof. Dr. F. Wagner,
Institut für Biochemie und Biotechnologie,
Technische Universität Braunschweig,
Konstantin-Uhde-Str. 5,
D-3300 Braunschweig, Fed. Rep. Germany;

Prof. Dr. M. Schneider,
Fachbereich 9,
Bergische Universität GH-Wuppertal,
D-5600 Wuppertal 1, Fed. Rep. Germany

1 Introduction

In *Focus 1* we had a chapter on techniques for the production of chiral compounds by bioreduction, and in *Focus 2* we had a chapter on the preparative enzymatic hydrolysis of optically active compounds. We continue this series of chapters with this one on preparative organic chemistry with lipase-catalyzed reactions.

Lipases, a subgroup of the esterases, catalyze four different reactions:

1. ester hydrolysis,
2. ester synthesis,
3. transesterification, and
4. acyl transfer.

M. SCHNEIDER refers to the catalytic mechanisms of ester synthesis as irreversible acyl transfer. With a water-insoluble substrate the lipase-catalyzed reaction occurs at the interface of the aqueous two-phase system. The specific enzyme activity depends strongly on the substrate concentration and on the interfacial area, which itself depends on the dispersion of the second phase in the first.

Since the change in free energy in lipase-catalyzed ester hydrolysis is greater than 1 kJ/mol, we can take advantage of the equilibrium for preparative purposes, with reaction conditions suited specifically for ester synthesis or for transesterification. An appropriate preparative ester synthesis uses nonpolar organic solvents; the lipases are highly stable and also are stereoselective as chiral catalysts.

2 Fermentation and purification of a lipase from *Pseudomonas* sp. ATCC 21808 (References [1–4])

Dr. Marianne KORDEL

2.1 Introduction

Pseudomonas sp. ATCC 21808 produces an active and stable lipase (temperature optimum, 40–50 °C; pH optimum, 7–10; temperature and pH stability, 60 minutes at 37 °C and pH 4 without loss of activity) [1] that is excreted into the medium and subsequently can be isolated from the medium [2]. On indicator plates containing Tween [*Tween 20* is polyoxyethylene (20) sorbitan monolaurate, *Tween 40* is the monopalmitate, *Tween 80* is the monooleate] liberated fatty acids appear as white calcium salt precipitates around the colonies after incubation at 30° for 24 to 48 hours. The yield of lipase produced during a fermentation can be increased by adding an inducer such as olive oil.

2.2 Objectives

We describe

- the production of microbial lipase, and
- the purification of the lipase.

2.3 Chemicals and equipment required

The bacterium *Pseudomonas* sp. ATCC 21808 can be ordered from the American Type Culture Collection (ATCC), 12301 Parklawn Drive, Rockville, MD 20852, USA. The necessary chemicals and equipment can be obtained from various sources.

Bacto-Peptone
Yeast extract
Agar
Glucose
Olive oil (commercial)
Tween 20, 40, or 80
Sodium glutamate
$CaCl_2$
KH_2PO_4
$MgSO_4 \cdot H_2O$
KCl
NaCl
NaOH
Glycine
n-Octyl-β-D-glucopyranoside
Isopropanol
Q SEPHAROSE™ fast-flow chromatography columns, 5 cm diameter, 30 cm long
pH meter
Gradient mixer
Fraction collector
Gel electrophoresis system and power supply
Octyl SEPHAROSE™ CL-4B
Petri dishes
Test tubes
Kapsenberg caps
Transfer loop
Sterile pipettes
Volumetric flask and cylinder
1-l Baffled Erlenmeyer flasks
Incubator/shaker
Incubator
Refrigerated centrifuge
Bench-top centrifuge
Titration equipment
Crossflow ultrafiltration unit
Ultrafiltration filters NMWL 10,000
Two-pen recorder
Vacuum pump
Flow photometer

Time required: the experiment requires approximately 5 days, for fermentation and purification.

2.4 Experimental procedure

Preparation of media

The following media are made up for culturing the bacteria and are autoclaved at 121 °C for 20 minutes. Agar media are poured into petri dishes when cooled to 60 °C.

(a) *Indicator plates*
5 g/l Bacto-Peptone™; 3 g/l yeast extract; 0.1 g/l $CaCl_2$; 10 ml/l of Tween 20, 40, or 80; 15 g/l agar.
(b) *M 820 Stock culture medium* [3]
5 g/l sodium glutamate; 1 g/l KH_2PO_4; 0.2 g/l $MgSO_4 \cdot H_2O$; 0.1 g/l KCl; 10 g/l glucose; 15 g/l agar; final pH 6.5.
(c) *Inoculum medium*
5 g/l peptone; 10 ml/l olive oil.
(d) *Main culture medium*
5 g/l peptone; 5 g/l yeast extract; 0.5 g/l NaCl; 10 ml/l olive oil; final pH 7.

Microbial culture procedure (Figure 1)

Two milliliters of inoculum medium is inoculated with a bacterial colony from an M 820 stock culture and incubated overnight at 30 °C. The main culture medium is then inoculated with this, and the bacteria are cultured in baffled 1-liter shake flasks (with a maximum of 300 ml in a flask) at 20 °C (or room temperature), with shaking at 150 rpm. If large volumes are to be cultured, the ratio of inoculum volume to main culture medium volume should be 1:100. Lipase production begins after about 40 hours, and its course is monitored by sampling at 4-hour intervals: the samples are centrifuged and the lipase activity in the supernatant is measured by titration/pH-stat (Section 4.4) with esterase standards. As soon as the lipase productivity decreases, the culture is centrifuged at $10{,}000 \times g$.

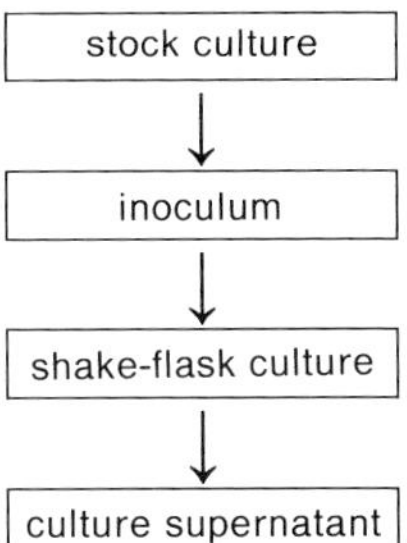

Fig. 1 Culture of *Pseudomonas* sp. ATCC 21808 to produce extracellular lipase

Enzyme purification (Figure 2)

The culture supernatant is concentrated by crossflow ultrafiltration using membrane filters with a nominal molecular weight cutoff of 10,000; it can be stored at −20 °C for further purification. Because the lipase adheres strongly to the membrane filters, the filters are rinsed thoroughly with glycine buffer (20 mM glycine, pH 9).

The ultrafiltration retentate is solubilized, before it is purified chromatographically, by titrating it with 2 N NaOH to pH 9 and adding *n*-octyl-*β*-D-glucopyranoside (OG) at 0.4% (final concentration) as needed. About 100,000 U (μmol/min) of lipase is loaded on a Q Sepharose™ column (300 ml bed volume, equilibrated with 20 mM glycine buffer at pH 9). The column is then washed with 2 bed volumes of the glycine buffer.

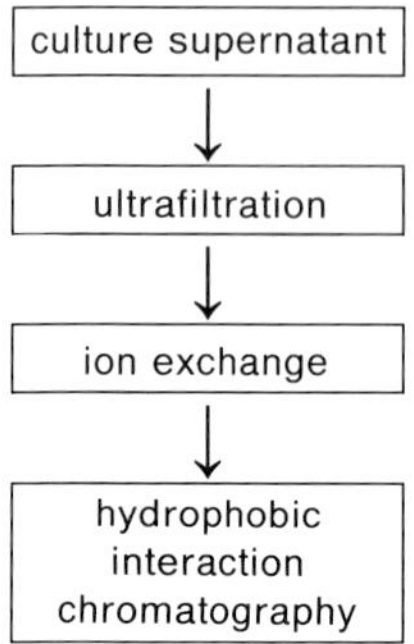

Fig. 2 Purification of lipase from *Pseudomonas* sp. 21808

Table 1 Purification of lipase from *Pseudomonas* sp. ATCC 21808

Processing step	V	Protein	Activity*			Overall yield	Purification factor
	ml	mg	U	U/ml	U/mg	%	
ultrafiltration retentate	1655	7282	92,183	56	12.5	–	–
Q SEPHAROSE column	972	418	91,746	130	220	99	17.3
octyl SEPHAROSE column	640	26	51,776	81	2,023	56	159
Q SEPHAROSE column	292	9.6	32,208	110	3,309	35	260

* Activity was determined with the *p*-nitrophenyl palmitate assay (Section 4); U = μmol/min

The lipase is eluted with a gradient in 6 bed volumes, from 20 mM glycine buffer of pH 9 to 1 M NaCl / 0.4% OG in glycine buffer. Because of interactions with the SEPHAROSE matrix the lipase cannot be eluted without detergent. The yield after this step is about 90% (Table 1).

The active fractions are combined and 0.5-volume aliquots of 0.4 M $CaCl_2$ in 20 mM glycine buffer, pH 9, are added with stirring. A precipitate of Ca^{2+}-lipid complexes is produced and is separated by centrifugation at $11{,}000 \times g$. The clear supernatant is loaded on an octyl-SEPHAROSE column (bed volume 350 ml) that has been equilibrated with 20 mM glycine / 0.4 M $CaCl_2$, pH 9 buffer. The column is washed with 2 bed volumes of the same buffer, then by 2 bed volumes of 20 mM glycine, pH 9. The lipase is eluted with a gradient over 6 bed volumes to 55% isopropanol in 20 mM glycine buffer followed by 2 bed volumes of this buffer. The lipase is stable in 55% isopropanol at 4 °C. The yield is about 60 to 70% (Table 1).

Isopropanol can be removed from the samples by adsorbing the combined active fractions on a Q SEPHAROSE™ column and eluting as described above. This step produces yields between 60 and 100% (Table 1).

The purity of the enzyme is such that only one foreign protein can be detected if the preparation is subjected to SDS-polyacrylamide gel electrophoresis and the gel is stained with $AgNO_3$ [4].

3 Immobilization of lipases (References [5–11])

Dr. Helmut ERDMANN

3.1 Introduction

Immobilized enzymes have commercial advantages over free enzymes in industrial applications, particularly in food technology and in the production of reagent-grade chemicals [8, 9, 11]. Carrier-bound enzymes are more stable with respect to pH fluctuations, temperature stress, solvent effects, and bacterial contamination; other advantages are the high catalyst concentration in the reaction vessel and the ease with which the catalyst can be separated from the reaction mixture and reused several times [5, 6, 7, 10].

These advantages are limited by the loss of activity upon immobilization and multiple use, and by the cost of the carrier. Glass, steel, aluminum, activated charcoal, cellulose, starch, and a variety of synthetic polymers have all been used as carriers for immobilized enzymes. None of these carriers is suitable for immobilizing all enzymes. The important characteristics of a carrier are particle size, porosity, electrical charge, and hydrophobicity or hydrophilicity. The enzymes can be fixed onto the carrier by inclusion, adsorption, or covalent chemical binding.

Lipases are particularly suitable for immobilization because of their great stability under adverse reaction conditions such as high temperatures and the presence of organic solvents.

Two materials are particularly useful in modern biotechnology for immobilizing enzymes: these are the polymer carrier VA-epoxy BIOSYNTH™ (from Riedel-de Haen in Germany) and DUOLITE™ ion-exchange resin (from Rohm and Haas in the USA).

The binding of the enzyme depends on

- the pH value,
- the concentration of buffer,
- the concentration of salt,
- the concentration of protein,
- the temperature,
- time,
- the rate at which the reaction mixture is shaken, and
- the viscosity of the reaction mixture.

In the following we present the requirements for producing commercially available extracellular lipase from *Pseudomonas fluorescens*. The reaction parameters must be optimized for each situation if other lipases are used.

Two methods are available for immobilizing the lipase on a carrier.

Column method

The carrier is packed in a column, and a pump circulates the enzyme solution through the column. The binding of the enzyme to the carrier is monitored optically by a photometer. When there is no further change in the extinction, binding is complete.

Batch method

The carrier is mixed with the enzyme solution until the enzyme is maximally bound to the carrier, as determined by photometric or enzymatic monitoring of samples.

The following describes the method for binding enzymes to carrier using the batch technique.

3.2 Objective

- We describe the immobilization of lipase from *Pseudomonas fluorescens* on various carriers.

3.3 Chemicals and equipment required

The lipase from *Pseudomonas fluorescens* can be obtained from Fluka Co. (# 62312). The carrier polymer VA-epoxy BIOSYNTH™ is from Riedel-de Haen and the ion-exchange resin DUOLITE A 568™ from Rohm and Haas. Other chemicals and equipment can be obtained from various suppliers.

K_2HPO_4
KH_2PO_4
4-Morpholinoethanesulfonic acid (MES)
Sodium azide
100-ml Volumetric flasks
Tris(hydroxymethyl)aminomethane (Tris)
250-ml Fritted-glass filter flask
100-ml Unbaffled Erlenmeyer flasks
Test tubes
Vortex mixer
Refrigerated centrifuge
Rotary shaker

The required buffer solutions are as follows:

Buffer A: 0.5 M potassium phosphate buffer, pH 7, + 0.02% sodium azide
Buffer B: 0.1 M potassium phosphate buffer, pH 7, + 0.02% sodium azide
Buffer C: 0.05 M MES buffer, pH 5.5, + 0.02% sodium azide
Buffer D: 0.05 M Tris buffer, pH 7, + 0.02% sodium azide

Time required: the experiment can be completed within 2 days.

3.4 Immobilizing lipase on VA-epoxy polymer

The epoxy-activated carrier polymer VA-BIOSYNTH™ is a copolymer of vinyl acetate (VA) and divinylethylene urea, fabricated into porous spherical particles with diameters between 50 and 200 μm. The amino groups of proteins bind covalently to the epoxide groups on the carrier surface (Figure 3).

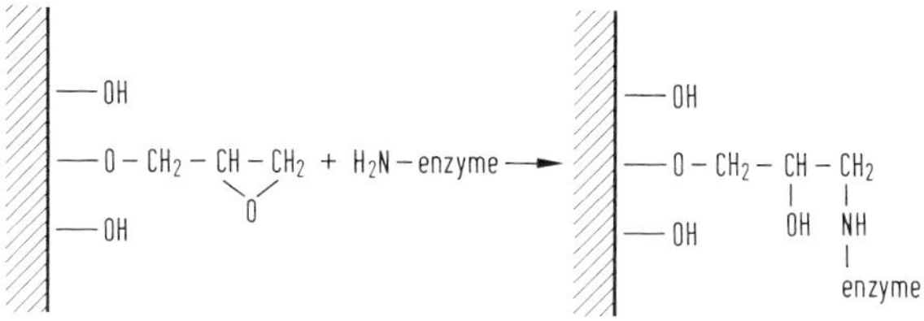

—OH
—O – CH_2 – CH – CH_2 + H_2N – enzyme → —O – CH_2 – CH(OH) – CH_2 – NH – enzyme
—OH

Fig. 3 Binding of lipase enzyme to epoxide groups on the surface of the carrier polymer VA-BIOSYNTH

Experimental procedure

In a 15-ml test tube that can be sealed, 50 mg of lipase is solubilized in 10 ml of buffer A at room temperature by careful use of the vortex mixer, taking care to prevent foaming. The enzyme solution is centrifuged at $30{,}000 \times g$ for 20 minutes at 5 °C; the clear supernatant is stored in the refrigerator.

The sediment is solubilized in 10 ml of buffer A, as just described, and centrifuged. If a centrifuge is not at hand, the enzyme solution may be filtered (on a membrane filter with a pore size of 0.5 μm). The volume of the combined supernatants is adjusted to 30 ml with buffer A. A 0.5-ml aliquot is removed and stored at 5 °C for later determination of recovery.

Ten mg of VA-BIOSYNTH™ is washed on the fritted-glass filter with 500 ml of buffer A. The washed carrier is shaken for 20 minutes with the enzyme solution in a sealed 100-ml unbaffled Erlenmeyer flask on a rotary shaker (90–100 rpm) in the cold (5 °C).

The reaction mixture is passed through a fritted-glass filter, and the filtrate is kept refrigerated for later determination of yield. The carrier is washed with 500 ml of buffer B, resuspended in 20 ml of buffer B, and kept refrigerated at 4 °C until needed.

3.5 Immobilizing lipase on DUOLITE™ ion-exchange resin

DUOLITE A 568™ is a porous, granular, weakly basic anion-exchange resin composed of a polycondensate of phenol-formaldehyde.

Experimental procedure

The procedure for immobilizing lipase on DUOLITE A 568 differs from immobilization on VA-epoxy in several ways.

- The lipase solution is prepared as described above, but with buffer C;
- 10 mg of DUOLITE A 568 is washed on the fritted-glass filter with 500 ml of buffer C;
- the enzyme solution and carrier are shaken together for 2 hours to immobilize the enzyme, under the same conditions as previously described;
- the carrier is washed with 500 ml of buffer D on the fritted-glass filter, resuspended in 20 ml of buffer D, and kept at 4 °C until needed.

3.6 Determining the yield of immobilized enzyme

To determine the yield on immobilization, the protein content and the enzyme activity of the reaction mixture before and after binding are detected.

3.7 Storing immobilized lipases

Immobilized lipases can be stored at 4 °C for up to a month without loss of activity. Freeze-drying is recommended for long-term storage.

To prepare the enzyme for freeze-drying, the carrier is washed with 500 ml of demineralized water on the fritted-glass filter. After freeze-drying it is stored in a desiccator with silica gel at 4 °C. The lyophilized immobilized lipases can be stored in this manner for extended periods of time without loss of activity or microbial contamination.

4 Quantitative determination of lipase activity in microbial culture supernatants (References [12, 13])

Dr. Siegmund LANG and Dipl.-Biol. Manuela MARKWEG

4.1 Introduction

Lipolytic enzymes are extracellular proteins widely distributed among microorganisms. Lipases are formally called triacylglycerol acylhydrolases (EC 3.1.1.3) because they hydrolyze triglycerides. They can be detected in culture supernatants because of their special property of splitting the ester bonds of triglycerides, which they do at an aqueous/lipid interface. Their action releases glycerol and fatty acids (Figure 4), although various lipases have regiospecificities that render the situation more complicated and that we will ignore here. The fatty acids can be determined by alkaline titration (Section 4.4).

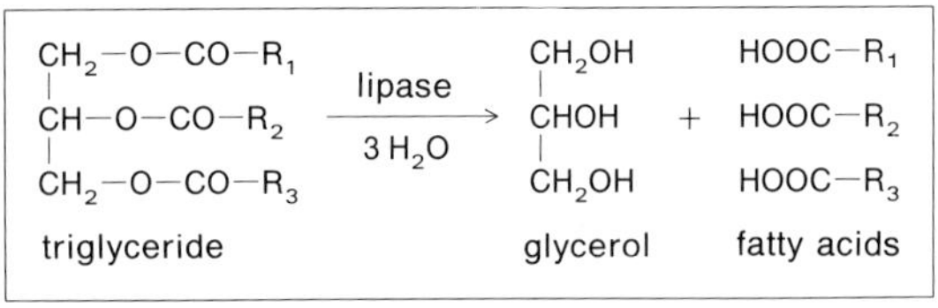

$$\begin{array}{lcll} CH_2-O-CO-R_1 & & CH_2OH & HOOC-R_1 \\ | & \text{lipase} & | & \\ CH-O-CO-R_2 & \xrightarrow[3\,H_2O]{} & CHOH \;\; + & HOOC-R_2 \\ | & & | & \\ CH_2-O-CO-R_3 & & CH_2OH & HOOC-R_3 \\ \text{triglyceride} & & \text{glycerol} & \text{fatty acids} \end{array}$$

Fig. 4 Lipase-catalyzed hydrolysis of triglycerides; R_1, R_2, and R_3 are aliphatic hydrocarbon chains

Lipases also accept water-insoluble long-chain fatty acid esters other than triglycerides as substrates. Therefore, *p*-nitrophenyl palmitate can be used as a model substrate for quantitative determination of activity; *p*-nitrophenyl anions and palmitic acid are liberated (Figure 5). Since *p*-nitrophenol is strongly colored, it can be photometrically determined in alkaline solution at its absorption maximum of 410 nm. This assay is more sensitive than the one mentioned in the preceding paragraph.

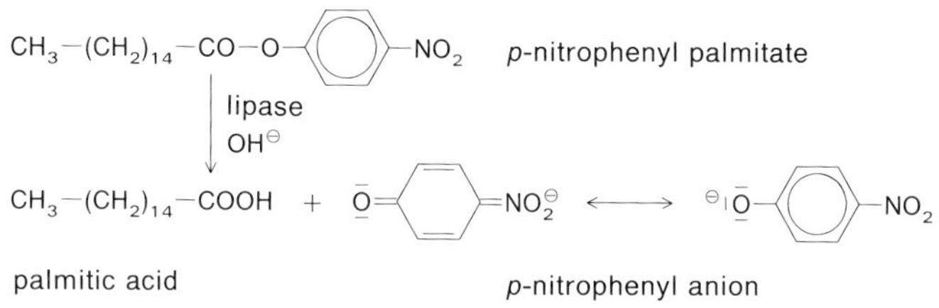

$CH_3-(CH_2)_{14}-CO-O-C_6H_4-NO_2$ *p*-nitrophenyl palmitate

lipase
$OH^{\ominus}$

$CH_3-(CH_2)_{14}-COOH + O{=}C_6H_4{=}NO_2^{\ominus} \longleftrightarrow {}^{\ominus}O-C_6H_4-NO_2$

palmitic acid *p*-nitrophenyl anion

Fig. 5 Lipase-catalyzed hydrolysis of *p*-nitrophenyl palmitate

4.2 Objectives

We describe here

- the determination of lipases and
- the measurement of lipase activity by enzymatic hydrolysis of triglycerides and fatty acid esters.

4.3 Chemicals and equipment required

The bacterium *Pseudomonas cepacia* DSM 50181 can be obtained from the German Microbial Culture Collection (Deutsche Sammlung von Mikroorganismen, DSM) in Göttingen. Other chemicals can be obtained from various sources.

$Na_2HPO_4 \cdot 2\,H_2O$
$NaH_2PO_4 \cdot H_2O$
Bacto-Yeast Nitrogen Base medium (YNB)
Olive oil
Bacto-Agar
Nutrient agar
0.05 N Tris-HCl buffer
Ethanol
Acetone
0.05 N NaOH
Chloroform
Anisaldehyde
Sulfuric acid
Glacial acetic acid
Oleic acid
p-Nitrophenyl palmitate
Sodium deoxycholate

Gum arabic
Sterilizing filters (0.2 μm)
Syringes (20 ml)
500-ml Erlenmeyer flasks with 2 baffles
Shaker
Centrifuge (20,000 × *g*)
Test tubes
50-ml Erlenmeyer flasks with 2 baffles
Rubber stoppers
pH meter
Silica gel 60 F_{254} TLC plates, on aluminum foil
Developing cabinet
Drying cabinet
Water bath
Photometer

Time required: the experiment requires approximately 4 days, including time for culturing *Pseudomonas cepacia* DSM 50181.

4.4 Experimental procedure

Preparation of culture media

The following medium is prepared, adjusted to pH 7, and autoclaved 20 minutes at 121 °C.

Yeast Nitrogen Base medium (YNB):

$Na_2HPO_4 \cdot 2H_2O$	21.72 g/l
$NaH_2PO_4 \cdot H_2O$	10.76 g/l
YNB	6.70 g/l
Olive oil	5.00 g/l

The YNB is made up as a 10× solution, sterilized by passing through a 0.2 μm membrane filter, and added to the other components aseptically.

The incubation temperature is 30 °C.

Nutrient agar is used for the agar slants.

Cultivation of Pseudomonas cepacia and recovery of the culture supernatant

A loopful of material from a fresh agar slant culture is inoculated into 100 ml of medium in a 500-ml Erlenmeyer flask with baffles and is shaken at 100 rpm for 48 hours at 30 °C. The cell suspension is then centrifuged for 15 minutes at 20,000 × *g* at room temperature. The supernatant is used as such in the enzyme tests.

Quantitative determination of lipase activity by the method of Liu *et al. [12]*

a. *Reagents*
   The reaction mixture contains 2 ml of 0.05 N Tris-HCl buffer, pH 7.7, 1 ml of olive oil, and 2 ml of culture supernatant.
b. *Enzyme assay procedure*
   The enzyme assays are carried out in rubber-stoppered narrow-neck 50-ml Erlenmeyer flasks with two baffles. The vessels are shaken (240 rpm) for 1 hour on a rotary shaker at 37 °C.
   After incubation, the reaction is stopped by adding 20 ml of an ethanol : acetone mixture (1 : 1, v : v) and chilling the flasks in an ice bath. All mixtures are titrated to pH 10.4 with 0.05 N NaOH to obtain the experimental values, *R2*. The values for the negative controls, *R1*, are obtained by adding enzyme after the reaction mixtures are in the ice bath.
c. *Calculation of the volumetric lipase activity (LA)*
   The volumetric lipase activity (*LA*) can be calculated as follows:

$$\mathrm{LA} = \frac{\text{NaOH required}}{\text{Amount of enzyme} \cdot \text{incubation time}} \left[\frac{\mu\text{mol}}{\text{ml} \cdot \text{min}}\right]$$

For the quantities in the example in this section, this is:

$$\mathrm{LA} = \frac{(\text{NaOH/R2} - \text{NaOH/R1}) \cdot 50}{\text{Lipase solution} \cdot \text{incubation time}} \left[\frac{\text{ml} \cdot \mu\text{mol/ml}}{\text{ml} \cdot \text{min}}\right]$$

d. *Definition of enzyme activity*
We define 1 unit (U) of enzyme activity as the amount of enzyme that liberates 1 micromole of fatty acid per milliliter per minute (μmol/ml · min) from the triglyceride.

e. *TLC of the free fatty acids*
The titration data are qualitatively confirmed by thin-layer chromatographic (TLC) analysis of the free fatty acids. All samples are adjusted to the same volume with distilled water, and 20 μl of each is spotted onto silica gel 60 F_{254} plates. A chloroform : acetone mixture (96 : 4, v : v) is used as the mobile phase, developing vertically. Free fatty acids and their derivatives can be detected as blue-violet spots after the plates are sprayed with an anisaldehyde : sulfuric acid : glacial acetic acid reagent (1 : 2 : 100, v : v : v) and heated at 150 °C in a drying cabinet. The intensity of color of the fatty acid spots is a visible measure of the lipolytic activity.

Quantitative determination of lipase activity by the method of W*INKLER et al.* [13]

a. *Reagents*
Solution 1: 30 mg of *p*-nitrophenyl palmitate, dissolved in 10 ml of isopropanol.
Solution 2: 207 mg of sodium deoxycholate and 100 mg of gum arabic, dissolved in 90 ml of 0.05 N sodium phosphate buffer, pH 8.
Substrate solution: Solution 1 and solution 2 are mixed together and stirred to form a stable emulsion; this solution must always be prepared fresh.

b. *Enzyme assay procedure*
The substrate solution, 2.4 ml per assay, is equilibrated for 5 minutes in a 37 °C water bath, and 100 μl of culture supernatant (enzyme suspension) is added; the mixture is shaken briefly and then incubated for 15 minutes. The sample is read photometrically at 410 nm against an enzyme-free control. If the results do not fall on a straight line when plotted, the enzyme mixture must be diluted.

c. *Calculation of the volumetric lipase activity (LA)*
The molar extinction coefficient, ε, of *p*-nitrophenol in solution 2 is 14.9 $cm^2/\mu mol$ and *DF* is the dilution factor. Therefore,

$$LA = \frac{E_{410}}{14.9 \cdot 1 \cdot 15 \cdot 1/25} \left[\frac{1}{cm^2 \cdot \mu mol^{-1} \cdot cm \cdot min \cdot DF}\right]$$

$$= 0.11 \cdot E_{410}\ (\mu mol/ml \cdot min).$$

d. *Definition of enzyme activity*
We define 1 unit (U) as the amount of enzyme that converts 1 micromole of *p*-nitrophenyl palmitate per milliliter per minute (μmol/ml · min).

5 Enzymatic ester syntheses by irreversible acyl transfer

(References [14–20])

Prof. Dr. Manfred SCHNEIDER

5.1 Introduction

Ester hydrolases (esterases and lipases) catalyze four different reactions involving carboxylic acid esters (Figure 6):

(a) ester hydrolysis;
(b) esterification;
(c) transesterification; and
(d) acyl transfer.

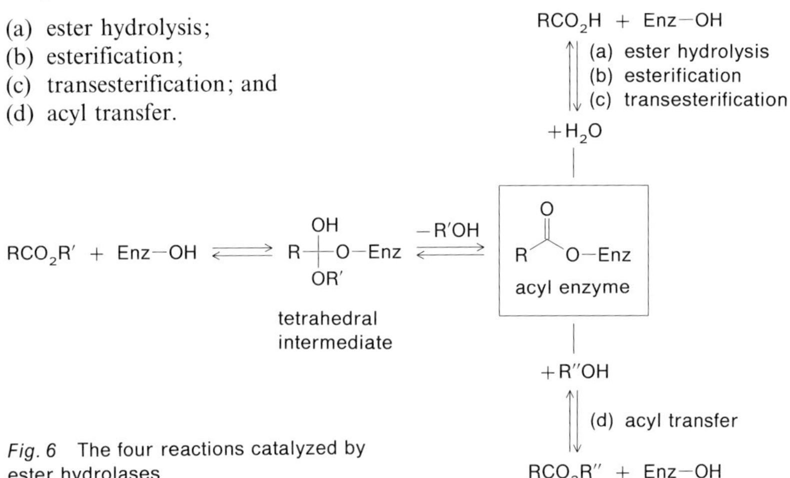

Fig. 6 The four reactions catalyzed by ester hydrolases

Although not in all cases proven beyond doubt, it is generally accepted that the catalytic mechanisms are similar to those of serine proteases which have been studied in great detail. The reaction, shown simplified in Figure 6, proceeds via a nucleophilic attack on the carbonyl group of the substrate ester, RCO_2R', by the activated primary alcohol of the serine group in the amino acid sequence of the enzyme protein. The tetrahedral intermediate is stabilized by elimination of R'OH thus leading to an activated ester, termed acyl enzyme. This is the central intermediate in all four hydrolytic esterase-catalyzed reactions, (a) to (d).

In aqueous systems the acyl group from the acyl enzyme is always transferred to the nucleophilic H_2O, present in large excess; the equilibrium is far toward the hydrolysis products. In nonaqueous systems, the reverse reaction is favored and ester synthesis occurs readily [reactions (b) and (c)].

Esterifications by enzymatic acyl transfer [reaction (d)] are particularly attractive for syntheses because no H_2O is involved. If a donor ester, which often serves also as a solvent, is used in excess, the acyl groups from acyl enzyme intermediates can be transferred to other nucleophiles such as R″OH, and new RCO_2R'' esters can be produced. The success of such acyl transfer reactions depends largely on the nucleophilicity of the acceptor alcohol R″OH. Upon inspection of reaction (d) it is evident that the desired transfer can be achieved only if the initially eliminated alcohol R′OH is less nucleophilic than R″OH or at least of comparable nucleophilicity. In other words, enzymatic esterifications by acyl transfer proceed at reasonable reaction rates only when the reverse reaction of the acyl enzyme with R′OH is insignificant. Classical acyl donors (such as EtOAc, MeOAc, tributyrin, etc.) can esterify primary alcohols with no difficulty, albeit at low reaction rates. However, little or no esterification takes place with secondary alcohols or substrates of low nucleophilicity.

Significantly higher rates of esterification and the conversion of almost any alcohol are possible if the reverse reaction of the acyl enzyme is inhibited, that is, if acyl enzyme formation is irreversible. As illustrated in Figure 7, the enzymatic hydrolysis of vinyl or isopropenyl esters produces acyl enzymes by liberating the formed enols as aldehydes or ketones [14]. In the absence of other nucleophiles (such as H_2O) reverse reactions are thus prevented. As shown in Fig. 6, the intermediate can react only with the acceptor alcohols R″OH.

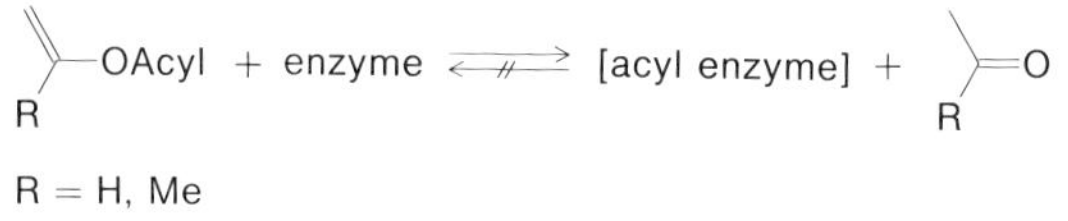

$$\text{CH}_2{=}\text{C(R)}{-}\text{OAcyl} + \text{enzyme} \rightleftharpoons\!\!\!\!\!\!/\;\; [\text{acyl enzyme}] + \text{CH}_3{-}\text{C(R)}{=}\text{O}$$

R = H, Me

Fig. 7 Irreversible enzymatic hydrolysis of vinyl or isopropenyl esters

Enzymes are chiral catalysts; many of their reactions proceed stereoselectively. The ability of ester hydrolases both to differentiate between (a), enantiomers of racemic esters, and (b), enantiotopic estergroups in achiral compounds having prochiral centers has been well documented [15]. These properties also apply to esterifications by acyl transfer. It is therefore possible to convert enantioselectively not only racemic alcohols but also achiral diols with enantiotopic hydroxyl groups into enantiomerically pure products. In the following examples we discuss enzymatic esterifications by irreversible acyl transfer (Fig. 7) in three structurally different systems:

(1) primary alcohols,
(2) racemic secondary alcohols, and
(3) achiral diols.

In all three cases, vinyl acetate is the acyl donor and lipase SAM-II (*Pseudomonas* sp.) is the biocatalyst [16].

5.2 Experimental procedure

(1) Esterification of a primary alcohol: conversion of benzyl alcohol into benzyl acetate (Figure 8) [17]

Benzyl alcohol (10.3 g, 0.1 mol) and vinyl acetate (21.5 g, 0.25 mol) are dissolved in 150 ml of *n*-hexane contained in a 250-ml round-bottomed flask. After the addition of 200 mg of lipase SAM-II (1,600 U, measured with a tributyrin standard), the flask is

$$PhCH_2OH + CH_2{=}CHOAc \xrightarrow{\text{Lipase SAM-II}} PhCH_2OAc + CH_3CHO$$

Fig. 8 Esterification of a primary alcohol: conversion of benzyl alcohol into benzyl acetate

sealed with a polyethylene (PE) stopper and the mixture is stirred at room temperature. The course of the reaction is followed by TLC (SiO_2; Et_2O : *n*-hexane at 1 : 4; R_f of benzyl alcohol, 0.12; R_f of benzyl acetate, 0.41). The reaction is complete after 20 hours (benzyl alcohol is no longer detectable by TLC). The enzyme is filtered off, washed with *n*-hexane, and stored in a refrigerator for further use. The filtrate is concentrated on a rotary evaporator and the residue is distilled *in vacuo*. The yield of benzyl acetate is 14.7 g (98%); b.p.$_{18}$ is 100 °C, and purity is 99.2% (by gas chromatography, GC).

(2a) Enantioselective esterification of a racemic secondary alcohol such as (R, S)-1-phenylethanol (Figure 9) [18]

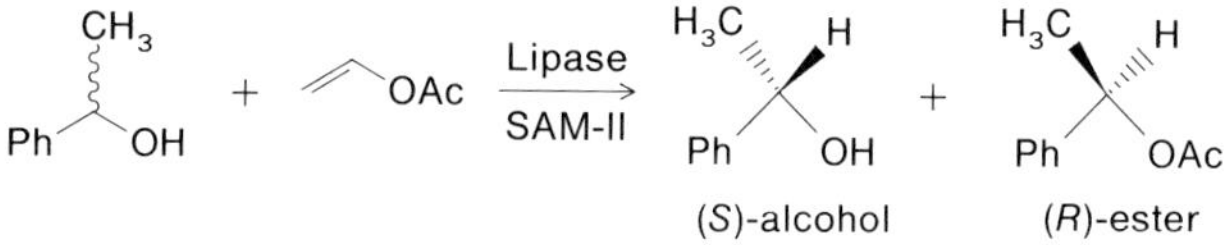

CH3 / Ph OH + OAc —Lipase SAM-II→ H3C H / Ph OH (*S*)-alcohol + H3C H / Ph OAc (*R*)-ester

Fig. 9 Enantioselective esterification of a racemic secondary alcohol such as (*R*, *S*)-1-phenylethanol

In a sealable sample vial, (*R*, *S*)-1-phenylethanol (1.22 g, 10 mmol; distilled; free of acetophenone) and vinyl acetate (2.6 g, 30 mmol) are dissolved in 15 ml of *t*-BuOMe. After the addition of 200 mg of lipase SAM-II (1,600 U, measured with a tributyrin standard), the mixture is stirred at room temperature; samples are analyzed by GC to follow the course of the reaction. The desired conversion of 50% is usually reached after 48 hours.

The enzyme is filtered off (with a #3 fritted-glass filter) and stored in a refrigerator for further use. After the solvent and excess vinyl acetate are removed, the remaining mixture consists of (*S*)-1-phenylethanol and (*R*)-1-acetoxy-1-phenylethane. These two are separated by column chromatography (SiO_2; Et_2O : petroleum ether at 1:4). We recovered 0.58 g (4.8 mmol, 48%) of (*S*)-1-phenylethanol {R_f, 0.08; $[\alpha]_D^{20}$: $-45°$ (c = 5, MeOH), 99% enantiomeric excess, e.e.} and 0.78 g (4.7 mmol, 47%) of (*R*)-1-acetoxy-1-phenylethane {R_f, 0.48; $[\alpha]_D^{20}$: 114° (c = 2, MeOH), 99% e.e.). (*R*)-1-acetoxy-1-phenylethane was hydrolyzed chemically (0.3 g of K_2CO_3 in 10 ml of MeOH for 1 hour) leading, after conventional work-up, to 0.5 g of (*R*)-1-phenylethanol {$[\alpha]_D^{20}$: 45.5° (c = 5, MeOH), 99% e.e.}.

(2b) Enantioselective esterification of a racemic secondary alcohol such as trans-2-benzylcyclohexanol (Figure 10) [19]

In a sealable sample vial, (±)-*trans*-2-benzylcyclohexanol (1.9 g, 10 mmol) and vinyl acetate (2.6 g, 30 mmol) are dissolved in 15 ml of *t*-BuOMe. After the addition of 400 mg of lipase SAM-II (3,200 U, measured with a tributyrin standard), the mixture is stirred

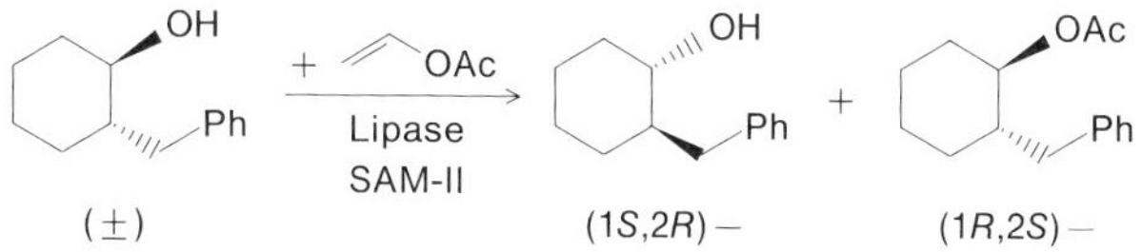

Fig. 10 Enantioselective esterification of a racemic secondary alcohol such as *trans*-2-benzylcyclohexanol

for 48 hours at room temperature. The enzyme is filtered off, the solvent and excess vinyl acetate are removed *in vacuo*, and the residue [consisting of (1*S*, 2*R*)-2-benzylcyclohexanol and (1*R*, 2*S*)-1-acetoxy-2-benzylcyclohexane] is separated by column chromatography (SiO_2; Et_2OH : *n*-hexane at 1 : 2). We recovered 0.76 g (40%) of (1*S*, 2*R*)-2-benzylcyclohexanol $\{[\alpha]_D^{20}$: 55.5° (c = 1, MeOH); ≥ 95% enantiomeric excess, e.e.$\}$ and 0.95 g (41%) of (1*R*, 2*S*)-1-acetoxy-2-benzylcyclohexane $\{[\alpha]_D^{20}$: − 19.7° (c = 1.2, MeOH); ≥ 95% e.e.$\}$.

(3) Enantioselective esterification of an achiral 1,3-diol such as 2-O-benzylglycerol (Figure 11) [20]

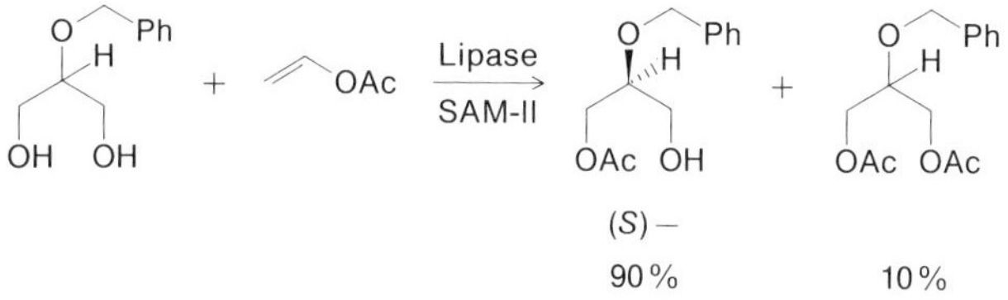

Fig. 11 Enantioselective esterification of an achiral 1,3-diol such as 2-*O*-benzylglycerol

In a flask, 2-*O*-benzylglycerol (52 g, 0.285 mol) and vinyl acetate (30.7 g, 0.357 mol) are dissolved in 250 ml of *t*-BuOMe. After the addition of 1.5 g of lipase SAM-II (12,000 U, measured with a tributyrin standard), the mixture is stirred at room temperature and the course of the reaction is followed by TLC (SiO_2; Et_2O : petroleum ether at 1 : 2; R_f of the diol, 0.05; R_f of the monoacetate, 0.28; R_f of the diacetate, 0.54). After the 2-*O*-benzylglycerol is completely esterified (62 hours), the enzyme is filtered off (1.4 g) and the filtrate is concentrated on a rotary evaporator. Recovered were 64 g of a mixture of the mono- and diacetate in a 9 : 1 ratio determined by gas chromatography with a 20-m OV-17 column at 180 °C; the R_t of the monoacetate is 9.3 min, and the R_t of the diacetate is 12.4 min. After fractional distillation 57 g (90%) of (*S*)-1-acetoxy-2-*O*-benzylglycerol was obtained $\{[\alpha]_D^{20}$: − 16.3° (c = 1.7, $CHCl_3$ stabilized with amylene); 85% enantiomeric excess$\}$.

6 Enzymatic enantioselective hydrolysis of DL-menthyl acetate with lipase from *Candida cylindracea* (References [21–23])

Dr. Christoph SYLDATK, Dipl.-Chem. Kirsten FRITSCHE,
Dipl.-Chem. Wolfram LOKOTSCH

6.1 Introduction

The commercially available lipase from the yeast *Candida cylindracea* can be used for more than just triglyceride hydrolysis; because of its enantioselective ability, it is also useful for the enzymatic separation of racemates. Therefore, this enzyme is often the catalyst of choice for producing chiral building blocks in organic synthesis [21, 23]. Lipase from *Candida cylindracea* has the advantages of being relatively inexpensive and stable; furthermore, it accepts a wide range of compounds as substrates. However, the structure of the substrates may have a significant impact on the enantioselectivity of the hydrolysis reaction.

6.2 Objective

- We discuss the enzyme-mediated reaction by which the DL-menthyl acetate racemic mixture is separated by ester hydrolysis to L-menthol and D-menthyl acetate, as shown in Figure 12.

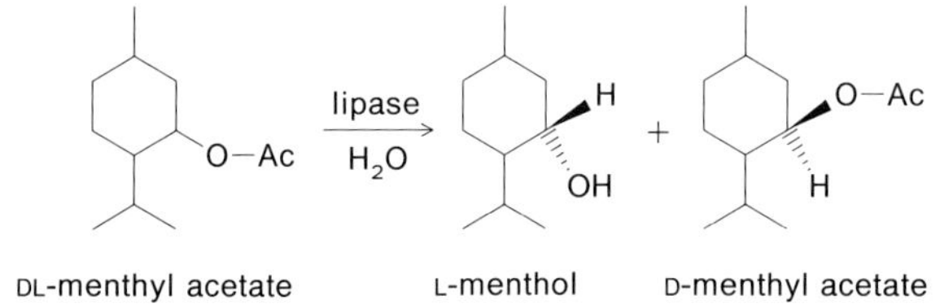

DL-menthyl acetate —(lipase, H_2O)→ L-menthol + D-menthyl acetate

Fig. 12 Resolution of DL-menthyl acetate into L-menthol and D-menthyl acetate using lipase

6.3 Chemicals and equipment required

Candida cylindracea lipase, ca. 600,000 U/g solid, may be purchased from Sigma (St. Louis, MO, USA or Deisenhofen, FRG)

DL-Menthyl acetate
L-Menthol
D-Menthol
Na_2HPO_4
KH_2PO_4
Isopropyl isocyanate

Pipettes and pipetting bulb
Thermostated waterbath shaker, or shaker
Test tubes or sealable sample vials
Round-bottomed flasks

n-Hexane
Dichloromethane
100-ml Erlenmeyer flasks
Capillary gas chromatograph
Capillary column Cp-Sil 5 CB, 10 m (Chrompack, Frankfurt, FRG)
Capillary column XE-60-L-VAL-L-PEA (Chrompack, Frankfurt, FRG)
Vortex mixer
Rotary evaporator

Time required: the experiment can be performed in two days.

6.4 Experimental procedure

Candida cylindracea lipase (200 mg, corresponding to 120,000 units) in 20 ml of Sörensen phosphate buffer, pH 6.8, is added to a 100-ml Erlenmeyer flask and preincubated with shaking at 30 °C. The reaction is started with the addition of 100 µl (460 µmol) of DL-menthyl acetate. At intervals (such as 0, 20, 40, and 60 min; 2, 3, and 4 hours), 1-ml samples are removed, added to test tubes, diluted with 1 ml of *n*-hexane, and mixed on the vortex mixer to extract the organics. After the phases have separated (centrifuging briefly if an emulsion has formed), the menthyl acetate and menthol are in the organic phase. The concentrations of each compound can be determined by gas chromatography on the Cp-Sil 5 CB capillary column (with a temperature gradient from 50 °C to 110 °C, 10 °C/min; carrier gas H_2 at 0.5 bar; split injection; the retention time of menthol is 3.1 min and of menthyl acetate is 4.5 min). If a capillary gas chromatograph is not available, the compounds can be determined quantitatively using packed columns such as SE 30 or OV-17.

After the reaction is over, the enantiomeric purity of the menthol formed can be determined using the method of KÖNIG [22]. The solvent is first removed *in vacuo*, then approximately 1 mg of pure enantiomer is taken up in 150 µl of dichloromethane and diluted with 100 µl of isopropyl isocyanate (this MUST done in a hood, with appropriate precautions). The mixture is incubated for 30 minutes at 100 °C in a closed derivatization reaction vessel. After the solvent and excess isopropyl isocyanate have been removed (for example, in a stream of dry nitrogen) the residue is again solubilized in dichloromethane, and is analyzed by GC on an XE-60-L-VAL-L-PEA capillary column (with a column length of 50 m; carrier gas is H_2 at 0.5 bar; temperature is isothermal, at 140 °C; split injection; the retention time of the L-menthol derivative is 29.3 min, and of the D-menthol derivative is 30.3 min).

After 50 hours of incubation, 115 µmol of L-menthol, corresponding to 25% of the initial DL-menthyl acetate, is produced at an enantiomeric excess of 60%.

7 Enzymatic enantioselective synthesis of L-menthyl esters from DL-menthol and organic acids, using lipase from *Candida cylindracea* in nonpolar organic solvents

(References [24–27])

Dr. Christoph SYLDATK, Dipl.-Chem. Kirsten FRITSCHE,
Dipl.-Chem. Wolfram LOKOTSCH

7.1 Introduction

The lipase from *Candida cylindracea* has been used successfully in recent years for enantioselective ester synthesis in nonpolar, water-free organic solvents [24–26], as well as for the enantioselective ester hydrolysis in aqueous solutions described in Section 6. The nonaqueous method is frequently employed in asymmetrical organic synthesis because it has several advantages over the aqueous method.

- Most of the reactants and the eventual products are relatively insoluble in aqueous systems, and they may also be subject to unspecific chemical hydrolysis.
- Processing the reaction mixture usually requires extracting it with organic solvents.
- When soluble free lipases are used in aqueous systems, they must subsequently be discarded.

The lipase from *Candida cylindracea* is highly stable in nonpolar organic solvents. Because the enzyme does not dissolve but is only suspended in the organic solvent, it can be separated easily after the reaction; the enzyme is filtered or sedimented from the solvent containing the substrates and products, and is reused.

A disadvantage of the nonaqueous method is that the reaction of ester synthesis is considerably slower than the corresponding ester hydrolysis in an aqueous system. Furthermore, water is a product in equimolar amounts along with the ester; it can shift the equilibrium in the direction of ester hydrolysis.

As in the case of ester hydrolysis of DL-menthyl acetate (Section 6), the structure of the initial substrates for ester synthesis in organic solvents has a significant influence on the enantioselectivity of the reaction.

7.2 Objectives

- In this section we describe enantioselective ester syntheses that produce various menthyl esters (Figure 13).
- We report on the effect of size of the acyl donor on the enantioselectivity of the reactions.
- We have studied the effect of water content on the reaction equilibrium.

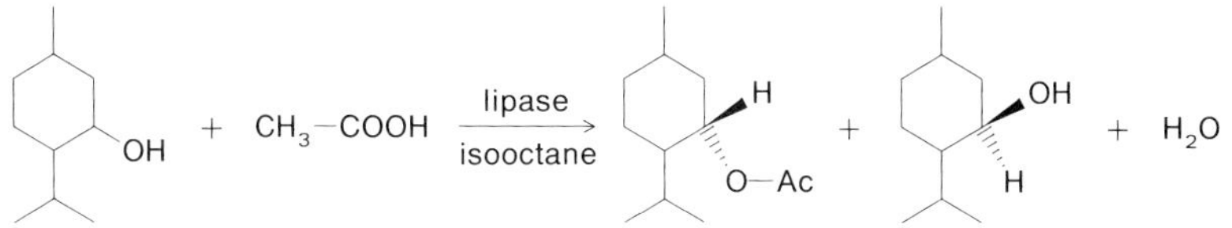

DL-menthol + CH_3-COOH $\xrightarrow[\text{isooctane}]{\text{lipase}}$ L-menthyl acetate + D-menthol + H_2O

DL-menthol | acetic acid | L-menthyl acetate | D-menthol | water

Fig. 13 Enantioselective ester synthesis to produce menthyl esters

7.3 Chemicals and equipment required

Candida cylindracea lipase, ca. 600,000 U/g solid, may be purchased from Sigma (St. Louis, MO, USA, or Deisenhofen, FRG)

L-menthol
D-menthol
(DL-menthol)
Acetic acid
Butyric acid
Lauric acid
Phenylvaleric acid
Oleic acid
n-Heptane or isooctane
Capillary gas chromatograph
Capillary column Cp-Sil 5 CB, 10 m (Chrompack, Frankfurt, FRG)
Capillary column XE-60-L-Val-L-PEA (Chrompack, Frankfurt, FRG)

Thermostated waterbath shaker, or shaker
Pipettes and pipetting bulb
Test tubes or sealable sample vials
Round-bottomed flasks
Sealable 100-ml Erlenmeyer flasks

Time required: ester synthesis takes approximately one week, but close attention is required only on the first and last days of the reaction.

7.4 Experimental procedure

The experiment can be conducted in two ways. The first is to use DL-menthol as the substrate for the ester synthesis with an appropriate acid as acyl donor. When conversion is 50%, the enantiomeric purity of the unconverted menthol is determined. This method corresponds to the ester hydrolysis of DL-menthyl acetate described in detail in Section 6.4. The second is to conduct parallel reactions with L-menthol and D-menthol and compare the specific activities of the *Candida cylindracea* lipase on the different enantiomers. The purity of the enantiomers need not be established by GC.

For ester synthesis, 20 ml of isooctane or *n*-heptane is added to each 100-ml Erlenmeyer flask, which already contains 60 µmol of the appropriate menthol enantiomer and 60 µmol of the corresponding acid. To start the reaction, 200 mg (120,000 U) of *Candida cylindracea* lipase is added to each flask; the flasks are sealed and incubated with shaking at 30 °C. Aliquots of 0.5 ml are removed from each reaction vessel at intervals (such as 2, 4, 6, and 8 hours; 1, 2, 3, and 4 days). After the lipase has been removed by sedimentation or filtration, the quantitative analysis by GC can be performed directly as in the ester hydrolysis of DL-menthyl acetate (Section 6.4).

If the first method is being followed, the reaction is stopped when approximately 50% of the DL-menthol is converted; the purity of the unconverted menthol enantiomers can be determined with GC after processing and derivatization, as described in detail in Section 6.4 for the ester hydrolysis of DL-menthyl acetate. The rate of ester synthesis is a function of the acyl donor (Table 2).

Table 2 Specific activity and enantioselectivity of *Candida cylindracea* lipase with various acyl donor substrates in esterification [27]

Substrate	Specific activity, μmol · g catalyst^{-1} · h^{-1}	
	L-Menthol	D-Menthol
acetic acid	7.2	0
butyric acid	4.2	0
lauric acid	25.6	4.2
linoleic acid	27.2	24.8

To study the effect of water on the course of the reaction, ester synthesis is conducted as described previously, with acetic acid as the acyl donor. Water in aliquots of 2, 4, 8, 12, 20, 40, 60, and 100 μl is added to a reaction series and the effect of the H_2O on the course of the reaction is followed by GC; the reaction rate becomes slower as water content increases.

8 Enzymatic enantioselective production of L-menthyl esters by transesterification of DL-menthol and triglycerides with lipase from *Candida cylindracea* in nonpolar organic solvents (References [23, 27])

Dr. Christoph SYLDATK, Dipl.-Chem. Kirsten FRITSCHE,
Dipl.-Chem. Wolfram LOKOTSCH

8.1 Introduction

Transesterification is another method of separating racemates of DL-menthol using lipase from *Candida cylindracea* [23, 27]. This type of esterification, like the direct ester synthesis described in Section 7, is performed in nonpolar organic solvents; since triglycerides are the acyl donors, water is not formed and the esterification is exceedingly efficient, because there is no ester hydrolysis. Another result of maintaining a low water content in the reaction mixture is that the lipase is much more stable and can be reused (the commercial lipase preparation used in the reaction contains about 6% water), or immobilized lipase can be used. As in the ester hydrolysis and ester synthesis described above, the enantioselectivity in alcoholysis is strongly dependent on substrate structure.

8.2 Objectives

- We describe the enantioselective transesterification for producing L-menthyl esters with various triglycerides as acyl donors (Figure 14).
- We report on the effect of chain length of the triglyceride on the rate of reaction and the enantioselectivity.
- We have studied substrate and product inhibition of the reaction.
  This experiment can be carried out along with the ester syntheses and can be evaluated with them.

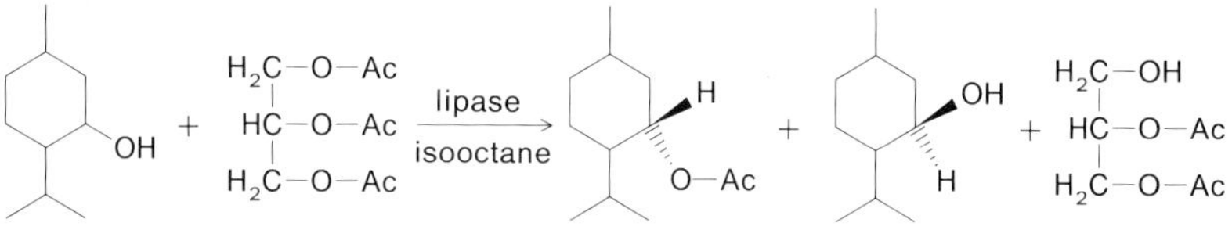

DL-menthol + triacetin —(lipase, isooctane)→ L-menthyl acetate + D-menthol + diacetin

Fig. 14 Enantioselective transesterification for producing L-menthyl esters with a triglyceride acyl donor

8.3 Chemicals and equipment required

Candida cylindracea lipase, ca. 600,000 U/g solid, may be purchased from Sigma (St. Louis, MO, USA, or Deisenhofen, FRG)

L-Menthol
D-Menthol
(DL-Menthol)
DL-Menthyl acetate
Triacetin
Tributyrin
Trilaurin
Triolein
Diacetin
n-Heptane or isooctane
Capillary gas chromatograph
Capillary column Cp-Sil 5 CB, 10 m (Chrompack, Frankfurt, FRG)
Capillary column XE-60-L-Val-L-PEA (Chrompack, Frankfurt, FRG)

Thermostated waterbath shaker, or shaker
Pipettes and pipetting bulb
Test tubes or sealable sample vials
Round-bottom flasks
Sealable 100-ml Erlenmeyer flasks

Time required: transesterification reactions take approximately a week, but close attention is required only on the first and last days of the reaction.

8.4 Experimental procedure

This experiment, like the ester synthesis described in Section 7, can be conducted in two ways. The first involves transesterification starting with DL-menthol and using appropriate triglycerides as acyl donors; once a recovery of 50% has been achieved, the purity of the unconverted menthol enantiomers is determined as described in Section 6.4. In the second method, parallel reactions with pure D- or L-menthol are conducted and the specific activities of *Candida cylindracea* lipase for each of the pure enantiomers are compared. The purity of the enantiomers need not be established by GC.

For the transesterification, 20 ml of isooctane or *n*-heptane is added to each 100-ml Erlenmeyer flask, which already contains 60 μmol of the appropriate menthol enantiomer and 60 μmol of the corresponding triglyceride. To start the reaction, 200 mg (120,000 U) of *Candida cylindracea* lipase is added to each flask; the flasks are sealed and incubated with shaking at 30 °C. Aliquots of 0.5 ml are removed from each reaction vessel at intervals (such as 2, 4, 6, and 8 hours; 1, 2, 3, and 4 days); after the lipase is removed by sedimentation or filtration, the quantitative analysis by GC can be conducted directly as described in Section 6.4 for the ester hydrolysis of DL-menthyl acetate. For example, if triacetin is reacting under the conditions just described, the triacetin has a retention time of 6.3 min on the GC, the menthol 3.1 min, and the menthyl acetate 4.5 min.

If the first method is being used, the reaction is stopped when approximately 50% of the DL-menthol is converted. The reaction rates differ markedly for different triglycerides (Table 3). The purity of the enantiomers of the unconverted DL-menthol is determined by GC after processing and derivatization, as was described in detail in Section 6.4 for the ester hydrolysis of DL-menthyl acetate. As we found for ester synthesis, enantiomeric purities also differ significantly.

Table 3 Specific activity and enantioselectivity of *Candida cylindracea* lipase with various acyl donor substrates in transesterification [27]

Substrate	Specific activity, $\mu mol \cdot g\ catalyst^{-1} \cdot h^{-1}$	
	L-Menthol	D-Menthol
triacetin	7.7	0
tributyrin	26.1	1.0
trilaurin	28.7	1.6
trilinolein	30.4	22.2

In order to evaluate product or substrate inhibition in the alcoholysis of L-menthol with triacetin, a series of reactions is carried out over a 24-hour period according to the method just described; in these reactions, we can

- vary the triacetin concentration between 1 and 200 μl;

- add diacetin in concentrations between 1 and 200 µl to the standard reaction; or
- add menthyl acetate in appropriate concentrations.

By determining the decrease in menthol concentration after 24 hours with the GC, we can estimate the inhibition of alcoholysis by the various components.

9 References

Section 2

[1] Kobayashi, M.: Lipid Metabolism Improving and Antiatheromatic Agent; US Patent 3,875,007 (1975).

[2] Kordel, M.; Hofmann, B.; Schomburg, D.; Schmid, R. D.: Extracellular lipase of *Pseudomonas* sp. ATCC 21808: Purification, characterization, crystallization, and preliminary X-ray diffraction data; in preparation.

[3] *Catalogue of Strains I*, American Type Culture Collection, Rockville, MD, 15th edition (1982).

[4] Cooper, T. G.: *Biochemische Arbeitsmethoden*, Walter de Gruyter, Berlin, New York (1981).

Section 3

[5] Burg, K.; Mauz, O.; Noetzel, S.; Sauber, K.: Neue synthetische Träger zur Fixierung von Enzymen; *Die Angewandte Makromolekulare Chemie* **157** (1988) 105–121.

[6] Kawamoto, T.; Sonomoto, K.; Tanaka, A.: Esterification in Organic Solvents: Selection of Hydrolases and Effects of Reaction Conditions; *Biocatalysis* **1** (1987) 137–145.

[7] Keller, R.; Schlingmann, M.: European Patent Application 0,141,224 A1 (1985).

[8] Marlot, C.; Lagrand, G.; Triantaphylides, C.; Baratti, J.: Ester Synthesis in Organic Solvent Catalyzed by Lipases Immobilized on Hydrophilic Supports; *Biotechnol. Lett.* **7** (1985) 647–650.

[9] Nielsen, T.: Industrial Application Possibilities for Lipase; *Fette Seifen Anstrichmittel* **87** (1985) 15–19.

[10] Eigtved, P.: European Patent Application 0,140,542 A1 (1985).

[11] Yokozeki, K.; Yamanaka, S.; Takinami, K.; Hirose, Y.; Tanaka, A.; Snomoto, K.; Fukui, S.: Application of Immobilized Lipase to Regio-Specific Interesterification of Triglyceride in Organic Solvent; *Eur. J. Appl. Microbiol. Biotechnol.* **14** (1982) 1–5.

Section 4

[12] Liu, W. H.; Beppu, T.; Arima, K.: Effect of Various Inhibitors on Lipase Action of Thermophilic Fungus *Humicola lanuginosa* S-38; *Agr. Biol. Chem.* **37** (1973) 2487–2492.

[13] Winkler, U. K.; Stuckmann, M.: Glycogen, Hyaluronate and Some Other Polysaccharides Greatly Enhance the Formation of Exolipases by *Serratia marcescens*; *J. Bacteriol.* **138** (1979) 663–670.

Section 5

[14] Degueil-Castaing, H.; DeJeso, B.; Drouillard, S.; Maillard, B.: Enzymatic Reactions in Organic Synthesis: 2-Ester Interchange of Vinyl Esters; *Tetrahedron Lett.* **28** (1987) 953–954.

[15] SCHNEIDER, M.; ENGEL, N.; BOENSMANN, H.: Enzymatische Synthesen chiraler Bausteine aus Racematen: Herstellung von (1*R*,3*R*)-Chrysanthemum-, Permethrin- und Caronsäure aus racemischen Diastereomerengemischen; *Angew. Chem.* **96** (1984) 52.
SCHNEIDER, M.; ENGEL, N.; HÖNICKE, P.; HEINEMANN, G.; GÖRISCH, H.: Enzymatische Synthesen chiraler Bausteine aus prochiralen *meso*-Substraten: Herstellung von Methyl(hydrogen)-1,2-cycloalkandicarboxylaten; *ibid.* **96** (1984) 55.
SCHNEIDER, M.; ENGEL, N.; BOENSMANN, H.: Enzymatische Synthesen chiraler Bausteine aus prochiralen Substraten: Herstellung von Malonsäuremonoalkylestern; *ibid.* **96** (1984) 54, and references cited therein.

[16] Lipase SAM-II by Amano Pharmaceutical Co., from Fluka Chemie AG, CH-9470 Buchs, Switzerland (Cat. # 62312), and Mitsubishi Int. GmbH, D-4000 Düsseldorf 1, FRG.

[17] BERGER, M.; SCHNEIDER, M.: Lab Manual „*Organisch-chemisches Grundpraktikum*", Bergische Universität-GH-Wuppertal (1989); unpublished.

[18] LAUMEN, K.; BREITGOFF, D.; SCHNEIDER, M.: Enzymic Preparation of Enantiomerically Pure Secondary Alcohols. Ester Synthesis by Irreversible Acyl Transfer Using a Highly Selective Ester Hydrolase from *Pseudomonas* sp.; An Attractive Alternative to Ester Hydrolysis. Improved method; *J. Chem. Soc., Chem. Commun.* (1988) 1459.

[19] Unpublished, but see [18] and also see LAUMEN, K.; BREITGOFF, D.; SEEMAYER, R.; SCHNEIDER, M. P.: Enantiomerically Pure Cyclohexanols and Cyclohexane-1,2-diol Derivatives; Chiral Auxiliaries and Substitutes for (−)-8-Phenylmenthol; A Facile Enzymatic Route; *J. Chem. Soc., Chem. Commun.* (1989) 148.

[20] Patent DE P 3,624,703 (22.7.86); see also BREITGOFF, D.; LAUMEN, K.; SCHNEIDER, M. P.: Enzymatic Differentiation of the Enantiotopic Hydroxymethyl Groups of Glycerol; Synthesis of Chiral Building Blocks; *J. Chem. Soc., Chem. Commun.* (1986) 1523.

Section 6

[21] CAMBOU, B.; KLIBANOV, A. M.: Comparison of Different Strategies for the Lipase-Catalyzed Preparative Resolution of Racemic Acids and Alcohols: Asymmetric Hydrolysis, Esterification and Transesterification; *Biotechnol. Bioeng.* **26** (1984) 1449.

[22] KÖNIG, W. A.; FRANCKE, W.; BENECKE, I.: Gas Chromatographic Enantiomer Separation of Chiral Alcohols; *J. Chromatogr.* **239** (1982) 227.

[23] LANGRAND, G.; BARATTI, J.; BUONO, G.; TRIANTAPHYLIDES, C.: Lipase-Catalyzed Reactions and Strategy for Alcohol Resolution; *Tetrahedron Lett.* **27** (1986) 29.

Section 7

[24] KLIBANOV, A. M.: Enzymes That Work in Organic Solvents; *Chemtech* **16** (1986) 354.

[25] KOSHIRO, S.; SONOMOTO, K.; TANAKA, A.; FUKUI, S.: Stereoselective Esterification of DL-Menthol by Polyurethane Entrapped Lipase in Organic Solvent; *J. Biotechnol.* **2** (1985) 47.

[26] LANGRAND, G.; SECCHI, M.; BUONO, G.; BARATTI, J.; TRIANTAPHYLIDES, C.: Lipase-Catalyzed Ester Formation in Organic Solvents – An Easy Preparative Resolution of α-Substituted Cyclohexanols; *Tetrahedron Lett.* **26** (1985) 1857.

[27] LOKOTSCH, W.; FRITSCHE, K.; SYLDATK, C.: Resolution of D,L-Menthol Using Free and Immobilized Lipase of *Candida cylindracea*; *Appl. Microbiol. Biotechnol.* **31** (1989) 467.

Section 8

[23], [27] above.

Chemical Screening as Applied to the Discovery and Isolation of Microbial Secondary Metabolites

by S. GRABLEY, J. WINK and A. ZEECK

Contents

Dr. Susanne GRABLEY,
Hoechst AG, ZF I/HL III,
D-6230 Frankfurt am Main 80, Fed. Rep. Germany

Dr. Joachim WINK,
Hoechst AG, ZF I/HL III,
D-6230 Frankfurt am Main 80, Fed. Rep. Germany

Prof. Dr. Axel ZEECK,
Institute for Organic Chemistry,
University of Göttingen,
Tammannstrasse 2,
D-3400 Göttingen, Fed. Rep. Germany

1 Introduction

A new chapter in the chemistry of natural products began when penicillin was discovered over 60 years ago and everybody recognized that secondary metabolites of microbial origin offered new possibilities for the treatment of infectious diseases. In the last few decades about 8000 compounds, usually called "antibiotics", have been isolated from cultures of a great variety of microorganisms. Although extensive chemical, biological, and pharmacological studies of many substances have been published, fewer than 100 of the substances have found use on a large scale, and even then only after being converted into chemical or biological derivatives.

It soon became evident that the structural variety of microbial secondary metabolites is based on a chemistry for which there are no parallels in the plant and animal kingdoms. The wealth of novel compounds and structural types is astonishing and is, at the same time, a clear indication that microorganisms have a very widely diversified secondary metabolism. These novel low molecular weight natural products are all derived from the universal metabolites of primary metabolism by one-way biosynthetic pathways. We can speculate that the enzymes that are necessary for their elaboration were possibly genetically determined in primeval times.

Nevertheless, to date there has been no satisfactory answer to the question *why?* An extraordinary variety of new metabolites is produced after enzyme activity is initiated in suitable nutrient solutions. A classification that attempts to trace each compound back to its biogenetic source can help us systematize the profusion [1].

Two factors are of particular importance in assessing and planning what we can expect in the area of microbial secondary metabolites in the future.

(1) Biologically active microbial metabolites are useful in considerably broader areas than heretofore. Antibiotics used to be considered only for antibacterial purposes. Today, there are other indications by pathogens (fungi, protozoa, helminths, viruses) as well as in pharmacological areas (immunomodulation, inhibition of blood clotting, lipid metabolism, carbohydrate metabolism, tumor therapy), with additional fields of application in veterinary medicine and plant protection. Biological screening is now possible in a large number of test systems; routine testing can only be done in a limited number. Furthermore, relatively few of the metabolites are eventually used for the indications that caused them to be selected in the initial screening of culture broths. For example, cyclosporin was discovered because it had antifungal activity, but now it is used clinically for its immunosuppressive effect, which was detected only later.

(2) Only a few genera and species of microorganisms have a secondary metabolism worth noting. Even strains of a genus often show considerable differences in this respect. Talented producers can make *different* secondary metabolites at the *same time*, and these very frequently differ from each other in both, their structure and their spectrum of activity.

Obviously, we are likely to overlook interesting secondary *metabolites* if our biological screening has a single focus. Since every new test system that is brought into use selects other active substances from the pool of microbial secondary metabolites, we could, as a working hypothesis, regard all microbial secondary metabolites as potentially active.

Therefore, a plan to supplement a purely biological screening with a chemical screening strategy [2–4] would make good sense. The whole panoply of strain-specific secondary metabolites in cultures of selected and, if possible, newly isolated microorganisms would be detected, isolated, and characterized. Only characterized compounds, not crude extracts or culture filtrates, would be allowed to proceed to the subsequent biological testing. This would result in a pool of both new and known natural products whose profile of activity could now be determined according to different criteria in wide-ranging biological tests.

Chemical screening resembles classical antibiotic or modified target-oriented screening in that it is based on microbiological research, both in the isolation and culture of strains and in the fermentation process. The successful isolation and preparation of products depend on the reproducible accumulation of metabolites selected in the initial screening; the results obtained with 100-ml shake flask cultures must first be reproduced in 10-liter laboratory fermenters and subsequently on the pilot plant scale (up to 500 liters). The work-up of cultures is of critical importance to eventual success: a culture is of no value until it can be used to obtain pure substances for structural elucidation, testing, and, possibly, the preparation of derivatives. The reproducibility of the biological test results is inextricably linked with the reproducibility, and hence the standardization, of the test substances themselves.

Since the work of product isolation is usually relegated to the background during discussion of new natural products and their chemistry and effects, it is easy to come to the mistaken conclusion that separation is a purely routine operation.

However, each substance, or at least each class of substances, requires its own analytical techniques; in the case of new compounds these must first be established. Only when the analysis is available can effective preparative separation steps be developed [5–8]; these, after setting up on the laboratory scale, must be adapted to the pilot-plant scale. The large number of possible separation methods must be rapidly narrowed down to the appropriate optimum process. The most important criterion for the chosen process must be its suitability for quantities of 10 to 20 liters, but aspects such as environmental protection and safety in the workplace should also be considered.

In this chapter we will outline a general procedure for discovering and isolating microbial secondary metabolites using chemical screening as an example. Following this, we will explain in detail the stages in the purification of a few substances.

2 General procedures in chemical screening

If secondary metabolites are formed by a microbial isolate in adequate amounts (at least 1 mg/l under defined culture conditions), we can usually detect them by staining thin-layer chromatograms of the crude extracts with suitable spray reagents. Furthermore, if we use several different spray reagents, which respond to different classes of substances, we can learn something about the pattern of metabolites a particular strain produces. All stages in the chemical screening must be standardized, if we are to be sure

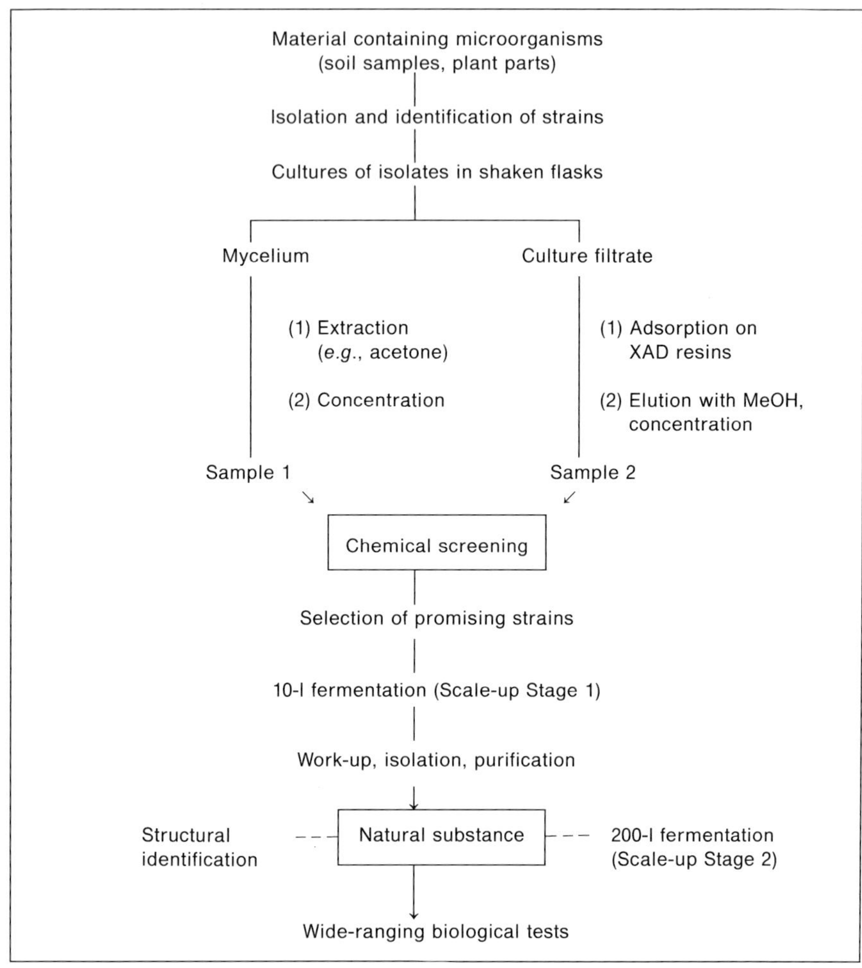

Fig. 1 General procedure for chemical screening

that the crude extracts of different strains are comparable and if we are to succeed in classifying any recurring spots on the chromatograms. The working procedure usually followed is shown in Figure 1.

This method uses readily cultured microorganisms and all the common nutrient solutions, and works best in reproducibly screening for substances with certain lipophilic properties. The keys to finding secondary metabolites are *preparation of samples* and *detection*.

The first step in preparing a sample is to separate the mycelium from the culture supernatant. The mycelium can be extracted with various solvents (such as acetone or ethyl acetate). The most satisfactory way to separate interesting substances from culture filtrates is with polystyrene adsorber resins (XAD type): the mostly strongly polar constituents of the nutrient solution pass through the resin, but the bulk of the secondary metabolites is adsorbed. The adsorbed substances can be eluted with an organic solvent such as methanol. The mycelial extract and the XAD eluate are each concentrated, yielding Sample 1 and Sample 2 (Figure 1).

The materials of interest are detected by an analytical separation method such as thin-layer chromatography. This is an effective technique that uses two mobile phase systems (polar and nonpolar) on silica gel plates. Some invisible constituents can be seen when the plates are examined under UV light, however it is more helpful to use several different spray reagents. Ehrlich's reagent, orcinol, anisaldehyde/sulfuric acid, tetrazolium blue, phosphomolybdic acid, ninhydrin, and others are particularly effective spray reagents for staining the reactive functional groups of natural products. The discovery of *new* natural products thus depends very strongly on the standardization of methods, on the visual recognition of substances that have been isolated before, and on experience in recognizing particular spots as strain-specific. This means that one must decide whether a spot is a constituent of the nutrient medium, is frequently formed as a widely distributed microbial metabolite, or is specific to the screened isolate and therefore merits further attention.

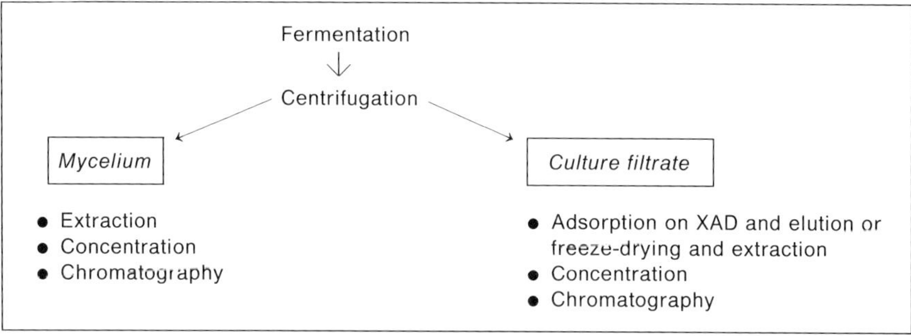

Fig. 2 Work-up of natural substances at the 10-liter scale

As soon as we leave the screening stage and embark upon isolating the natural products, we have to match the processing method to each substance individually. When doing a work-up at the 10- or 20-liter scale (Figure 2) we can manage with a few standardized methods that are based mainly on column chromatography and are more or less direct descendants of the screening methods. Separating the mycelium and concentrating the solutions cause few problems, since the volumes are relatively small; the difficulties come mainly in optimizing the steps by which the secondary metabolites are chromatographically separated and purified. A laboratory-scale production run of 10–20 liters of culture medium usually yields enough material for chemical characterization.

The picture changes markedly when gram or kilogram quantities of the natural products are required for biological testing or the preparation of derivatives. A procedure that has proven satisfactory at the pilot-plant scale of 100–500 liters is outlined in

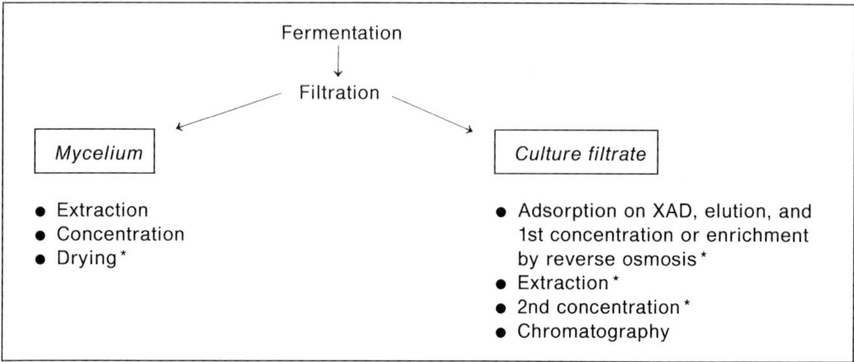

Fig. 3 Work-up of natural products on the pilot-plant scale (200 liters). Asterisks mark the stages of work-up at pilot-scale that must be carried out in addition to those on the laboratory scale

Figure 3. Some basic problems are the large volume of fermentation solution, its nutrient constituents, and the added antifoam agent, such as Desmophen (polyurethane), which is often concentrated with the natural products during work-up. Therefore, additional stages of extraction during work-up are necessary. Another difficulty is the large amount of water involved if the desired substance is in the culture filtrate and the volume is more than 100 liters. Additional stages of concentration, such as reverse osmosis, become necessary.

3 Examples of chemical screening methods

Structures of some of the compounds we have isolated are shown in Figure 4. Information about the microbial sources, culture media, constituent substances, R_f values, biological activity, and behavior on staining, is given in Tables 1, 2, and 3.

3.1 Polyethers and macrolides[1] from *Streptomyces* sp. DSM 3816

After fermentation is complete, the mycelium and supernatant have to be separated. In the laboratory they are usually centrifuged, or they are filtered if volumes are larger. The supernatant or filtrate can be discarded because 90% of the substance to be isolated is in the mycelium or attached to the mycelial wall.

[1] A *macrolide* is a macrocyclic lactone; if there are two lactone moieties in one macrocyclic system, it is called a *macrodiolide*.

Table 1 Producer strains and nutrient media for secondary products isolated by chemical screening

Strain	Medium	Secondary products
Streptomyces sp. DSM 3816	Soybean meal/mannitol	SM 54, SM 62, SM 77
Streptomyces cyaneus DSM 4351	Oat flakes	SM 104
Streptomyces sp. DSM 3813	Soybean meal/mannitol	SM 35, SM 43, SM 76
Streptomyces diastaticus DSM 4201	Oat flakes	SM 98, SM 99
Penicillium sp. DSM 4210	Malt medium	SM 131, SM 133

Table 2 Name, class, biological activity, and formula number ref. Fig. 4 of the isolated secondary products

Secondary product	Substance class	Biological activity	Formula ref.
SM 54 Elaiophylin	Macrodiolide	Anthelminthic	(1)
SM 77 Nigericin	Polyether	Antibacterial/antiviral	(2)
SM 62 A to C Niphimycin, amycins	Macrolide	Antifungal	(3)
SM 104	Pseudosugar	–	(4)
SM 35, SM 43, SM 76 Umycins A to C	Phenoxazinone	Anthelminthic	(5)
SM 98, SM 99 Elmycins A and B	Angucyclinone	Antiprotozoal	(6)
SM 131, SM 133	10-Membered lactone	–	(7)

Table 3 R_f values and characteristic reactions to spray reagents of the natural products described

Substance	R_f values, mobile phase		Spray reagent	
	I	II	Anisaldehyde	Orcinol
SM 54 (Elaiophylin)	0.77	0.32	violet	green
SM 77 (Nigericin)	0.97	0.62	yellow	ocher
SM 62 A (Niphimycin)	0.40	0	violet	–
SM 62 B (Amycin A)	0.30	0	violet	–
SM 62 C (Amycin B)	0.45	0	violet	–
SM 104	0.45	0.30	brown	brown
SM 35 (Umycin A)	0.8	0.15	brown	gray
SM 43 (Umycin B)	0.55	0	orange	orange
SM 76 (Umycin C)	0.4	0.2	orange	gray
SM 98 (Elmycin A)	0.90	0.55	gray	gray
SM 99 (Elmycin B)	0.85	0.35	green	green
SM 131	0.6	0.3	brown	brown
SM 133	0.6	0.3	brown	brown

I = Butanol/glacial acetic acid/water (4 : 1 : 5) (upper phase), II = Chloroform/methanol (9 : 1)

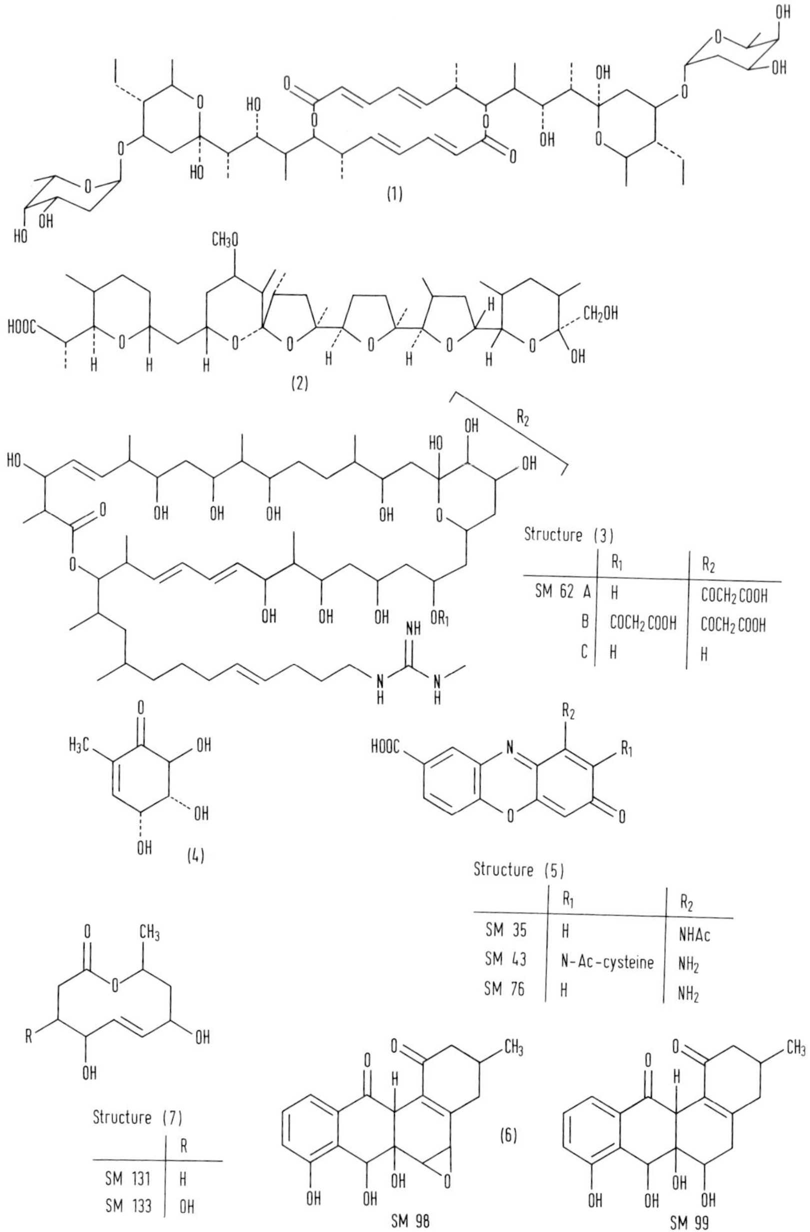

Fig. 4 Structures of the natural products described

Mycelium from a lab-scale fermentation is lyophilized. The macrodiolide *elaiophylin* (Figure 4, structure 1) and the polyether *nigericin* (Figure 4, structure 2) are isolated by extraction with ethyl acetate and separated by chromatography on a silica gel column. The macrolides *niphimycin* (SM 62 A) and *amycin* (SM 62 B and C) (Figure 4, structure 3), are extracted from the residue with methanol and obtained in pure form by preparative HPLC [9–11].

This procedure is not practical on a larger scale, so we did the following instead. The mycelium was extracted with hexane, and the organic phase was concentrated so that the nigericin crystallized out. The mycelial residue was extracted with ethyl acetate; again, the organic phase was concentrated and the desired product (elaiophylin) crystallized out. The macrolides could be obtained in pure form if a methanol extract of the mycelium was chromatographed on silica gel. The silica gel chromatography turned out to be difficult to develop into an effective method of purification, so even 200-liter fermentations, which produced yields of 10–20 g of macrolide, had first to be processed by preparative HPLC followed by DCCC (droplet counter-current chromatography) so that material for biological testing could be obtained as quickly as possible.

3.2 Pseudosugar from *Streptomyces cyaneus* DSM 4351

The *pseudosugar* SM 104 (Figure 4, structure 4) accumulates exclusively in the fermentation broth, so we had to design a work-up completely different from the one we used for the metabolites of DSM 3816. In the laboratory, the broth can be lyophilized, the dried product extracted with methanol, and the pseudosugar purified by chromatography on a silica gel column. On a larger scale we found that adsorption on XAD was the method of choice. Pseudosugar SM 104 binds to Amberlite XAD-16 in 90% yield, and could be eluted with methanol. Pure SM 104 is obtained by concentration and crystallization.

3.3 Phenoxazinones from *Streptomyces* sp. DSM 3813

Separation of the phenoxazinones called *umycins A*, *B*, and *C* (Figure 4, structure 5) is like that of many other microbial metabolites in that it is impossible to avoid tedious chromatographic purification on both the laboratory and pilot-plant scales. Only with time-consuming, systematic research were we able to simplify the work-up. Efforts like that are not justified until a natural product has been granted the status of a developmental product. *Umycins A*, *B*, and *C* accumulate mainly in the fermentation broth. On the laboratory scale, therefore, the culture filtrate can be lyophilized, and the resuspended lyophilizate chromatographed on silica gel and SEPHADEX LH-20 columns. If the volume of culture is larger, it should be concentrated by reverse osmosis before drying. Subsequent extraction with methanol produces the starting material for column chromatography [12, 13].

3.4 Angucyclinones from *Streptomyces diastaticus* DSM 4201

The angucyclinones *elmycin A* and *B* (Figure 4, structure 6) are found both in the broth and in the mycelium [14]. This means that isolation is a lot of work, on either the industrial or the laboratory scale. The procedure is summarized in Figure 5.

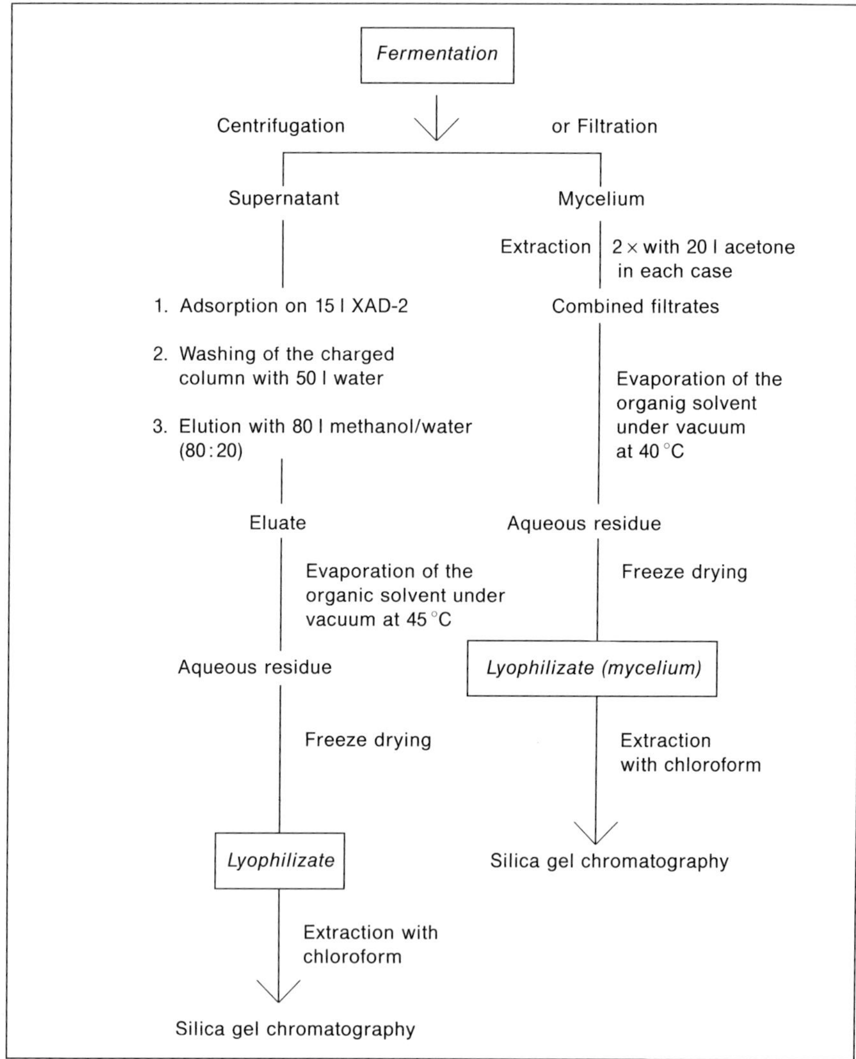

Fig. 5 Isolation and purification of the angucyclinones elmycin A and elmycin B

3.5 Ten-membered lactones from *Penicillium* sp. DSM 4210

The isolation of low molecular weight lactones (Figure 4, structure 7) from fungi is our final example. The metabolites SM 131 and SM 133 are found in both the culture supernatant and the mycelium. Laboratory work-up is analogous to initial screening. The supernatant is adsorbed on XAD; the mycelium is extracted, and the extract chromatographed on silica gel.

Larger volumes of fermentation broth must first be concentrated by reverse osmosis. Preliminary separation of the concentrate is by liquid/liquid extraction (for example, with a water/ethyl acetate system in a mixer-settler [15]). The extracts are combined and chromatographed on silica gel. The mycelium is extracted with acetone, the organic solvent is removed, and the extract lyophilized. This is followed by further extraction with ethyl acetate, evaporation, and final chromatographic purification.

4 Acknowledgements

We thank the Federal Ministry for Research and Technology (BMFT) for its support of the project "Biological-Chemical Screening", during which we carried out the work described in this chapter.

5 References

[1] ROHR, J.; ZEECK, A.: Biogenetisch-chemische Klassifizierung fermentativ hergestellter Sekundärstoffe. In: PRÄVE, P.; SCHLINGMANN, M.; CRUEGER, W.; ESSER, K.; THAUER, R.; WAGNER, F., Eds.; *Jahrbuch Biotechnologie Band 2*, Carl Hanser Verlag, München, Wien (1988/89); pp 263–295 (in German). ROHR, J.; ZEECK, A.: Biogenetic-Chemical Classification of Secondary Metabolites Produced by Fermentation. In: FINN, R.K.; PRÄVE, P.; SCHLINGMANN, M.; CRUEGER, W.; ESSER, K.; THAUER, R.; WAGNER, F., Eds.; *Biotechnology Focus 2*, Hanser Publishers, Munich, Vienna, New York (1988/89); pp 251–283 (in English).

[2] UMEZAWA, S.; TSUCHIYA, K.; UMEZAWA, H.; *et al.*: A New Antibiotic, Dienomycin. I. Screening Method, Isolation and Chemical Studies; *J. Antibiotics* **23** (1970) 20–27.

[3] ZÄHNER, H.; DRAUTZ, H.; WEBER, W.: Novel Approaches to Metabolite Screening. In BU'LOCK, J. D.; *et al.*, Eds.: *Bioactive Microbial Products; Search and Discovery*, Academic Press, New York (1982); pp 51–70.

[4] BREIDING-MACK, S.; ZEECK, A.: Secondary Metabolites by Chemical Screening. I. Calcium 3-Hydroxyquinoline-2-carboxylate from a *Streptomyces*; *J. Antibiotics* **40** (1987) 953–960.

[5] WEINSTEIN, M. J.; WAGMAN, G. H., Eds.; *Antibiotics Isolation, Separation and Purification*. J. Chromatogr. Library, Vol. 15, Elsevier, Amsterdam (1978).

[6] WAGMAN, G. H.; WEINSTEIN, M. J.: *Chromatography of Antibiotics*, 2nd ed. J. Chromatogr. Library, Vol. 26, Elsevier, Amsterdam (1984).

[7] ASZALOS, A., Ed.: *Modern Analysis of Antibiotics*, Marcel Dekker, New York, Basel (1986).

[8] WAGMAN, G. H.; COOPER, R.: *Natural Products Isolation.* J. Chromatogr. Library, Vol. 43, Elsevier, Amsterdam (1989).
[9] Verfahren zur Herstellung von Elaiophylin und dessen Verwendung (A Process for Manufacturing Elaiophylin and the Use of the Latter); Eur. Pat. Appl. (EPOS) 02 97 523.
[10] Amycin und dessen Derivate, Verfahren zu deren Herstellung und deren Verwendung (Amycin and its Derivatives, a Process for Their Manufacture and Their Use); Eur. Pat. Appl. (EPOS) 02 72 668.
[11] GRABLEY, S.; HAMMANN, P.; RAETHER, W.; WINK, J.; ZEECK, A.: Secondary Metabolites by Chemical Screening. 2: Amycin A and B, Two Novel Niphimycin Analogs Isolated from a High Producer Strain of Elaiophylin and Nigericin; *J. Antibiotics* **43** (1990) 639–647.
[12] Neue pharmakologisch wirksame Phenoxazinone und deren Derivate, Verfahren zu ihrer Herstellung und ihre Verwendung (New Pharmacologically Active Phenoxazinones and Their Derivatives, a Process for Their Manufacture and Their Use); Eur. Pat. Appl. (EPOS) 02 60 486.
[13] GRABLEY, S.; VOELSKOW, H.; WINK, J.; BREIDING-MACK, S.; ZEECK, A.: Secondary Metabolites by Chemical Screening. 9: Umycins, Novel Phenoxazinones from *Streptomyces*; *J. Antibiotics*, in preparation.
[14] DOBREFF, S.; ZEECK, A.; GRABLEY, S.; WINK, J.; SEIBERT, G.; SCHMIDT-BÄSE, K.; EGERT, E.: Secondary Metabolites by Chemical Screening. 11: Elmycin A and B, New Angulol Antibiotics from *Streptomyces*; *J. Antibiotics*, in preparation.
[15] HANSON, C.: *Neuere Fortschritte der Flüssig-Flüssig-Extraktion (Recent Advances in Liquid/Liquid Extraction)*, Verlag Sauerländer Aarau, Frankfurt/M. (1974).

Synthesis of Nucleic Acid Fragments and Complete Structural Genes

by H. G. GASSEN, S. BERTRAM, and J. STOLLWERK

Contents

Prof. Dr. H. G. GASSEN, Dr. S. BERTRAM, and Dipl.-Ing., FH, J. STOLLWERK,
Institut für Biochemie,
Technische Hochschule Darmstadt,
Petersenstr. 22,
D-6100 Darmstadt, Fed. Rep. Germany

1 Introduction

Nucleic acid fragments of defined sequence were first realized in Germany, from two essentially different starting points in 1960: the isolation of nucleic acid as a natural substance, and its total chemical synthesis. In Marburg, H. Witzel [1] isolated and characterized short fragments of ribonucleic acid (RNA) of a single sequence from yeast RNA using hydrolysis with bismuth II hydroxide. These fragments served as substrates for elucidating the mechanism of pancreatic ribonuclease [2]. The function of ribonucleic acid as a template was still unknown at that time. Preparative nucleic acid chemistry also focused primarily on RNA, although the double helix structure of deoxyribonucleic acid (DNA) had been known since 1953.

If procedures with recombinant DNA and cloning had not been developed, the synthesis of DNA would have continued to have a low priority. Gene technology and the sequencing of DNA and protein have raised DNA synthesis to the position it currently occupies in molecular biology. Today, when we speak of the chemical synthesis of genes, we think of the multistep process of chemical oligonucleotide synthesis, enzyme-mediated coupling of fragments, insertion of DNA into a plasmid, and plasmid multiplication in a bacterium. Only when the combination of chemistry and microbiology was brought to bear upon the problem was success in DNA synthesis possible.

2 Synthesis of the phosphoric acid diester bond

All chemical oligonucleotide synthesis is based on the principle of a condensation reaction between a phosphoric acid monoester group of the first nucleotide and a primary or secondary alcohol group of the second nucleoside or nucleotide (Figure 1). The reactants must be arranged so that formation of a $3' \rightarrow 5'$ phosphodiester linkage is the only possible reaction. This is more easily done in DNA synthesis than in RNA synthesis because the 2′-hydroxyl group found in RNA is absent in DNA. Therefore the synthesis of each oligonucleotide includes four reaction steps.

- Reactive groups on the heterocyclic base and on the deoxyribose or ribose are protected;
- the phosphoric acid monoester group is activated;
- the condensation reaction is performed; and
- the protecting groups are removed.

Numerous chemical procedures have been described for each reaction step and have been summarized in useful review articles [3–9].

Undesired side reactions are avoided by protecting the hydroxyl groups of the sugar moiety and the exocyclic amino groups of the nucleic acid bases; protecting groups for both these purposes are now standardized.

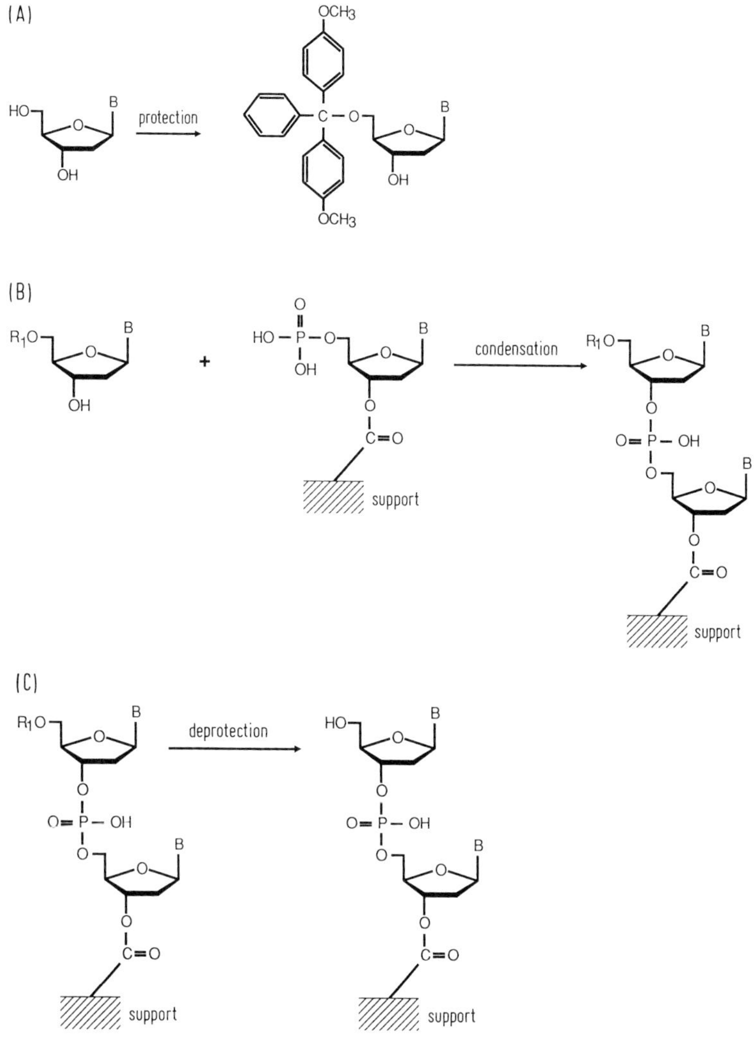

Fig. 1 Formation of the 3′→5′ phosphoric acid diester bond using the phosphoric acid diester procedure. Starting materials are a nucleoside and a nucleotide.
A: The 5′ position of the deoxyribose moiety of the nucleoside is protected by the dimethoxytrityl (DMTr) group; the exocyclic amino groups of the bases of both components must also be blocked.
B: The monomers react under the influence of a coupling agent such as dicyclohexylcarbodiimide to form a dinucleotide.
C: After the DMTr group is removed, the dinucleotide can react with the next monomer.
Key to symbols: R_1 = DMTr group; B = protected nucleic acid bases

2.1 The phosphoric acid triester, phosphoramidite, and H-phosphonate methods

The first syntheses of the phosphodiester bond used the *phosphodiester method.* Three additional methods are in current use. They are

- the *phosphoric acid triester method,*
- the *phosphoramidite method,* and, most recently,
- the *H-phosphonate method.*

The diester method (Figure 1), used mainly by KHORANA and his co-workers [10, 11], is now of historical significance only. This method involves condensing a nucleoside phosphoric acid monoester and a nucleoside with its 5′ position protected, using dicyclohexylcarbodiimide as the coupling agent; KHORANA synthesized the gene for tyrosine suppressor tRNA using this method [12–24]. The synthesis of the complete gene, however, did not rely exclusively on chemical reactions: assembly was by enzymatic ligation of oligonucleotide units. The diester method is hardly ever used nowadays.

The triester method of synthesis (Figure 2) was pioneered by LETSINGER [25], and has been perfected over the years to such a degree that it can serve as a standard in oligonucleotide synthesis [26–34]. It is used primarily in the so-called *filter paper (segmental solid support) method* developed by FRANK and BLÖCKER [35]. In the triester method, the growing oligonucleotide chain is nonpolar, because no negative charge is introduced by the added nucleotide. Thus, even long-chain oligonucleotides remain soluble in nonpolar solvents. The yield and the reaction velocity in this process depend most strongly on the coupling agent and the catalyst. Commonly used catalysts include *N*-methylimidazole, 4-dimethylaminopyridine, and pyridine *N*-oxide.

If large quantities of an oligonucleotide of a defined sequence are to be synthesized, the techniques developed by VAN BOOM [36, 37] using bifunctional phosphorylating agents are most suitable. No coupling agents are required when using *N*-methylimidazole as the catalyst.

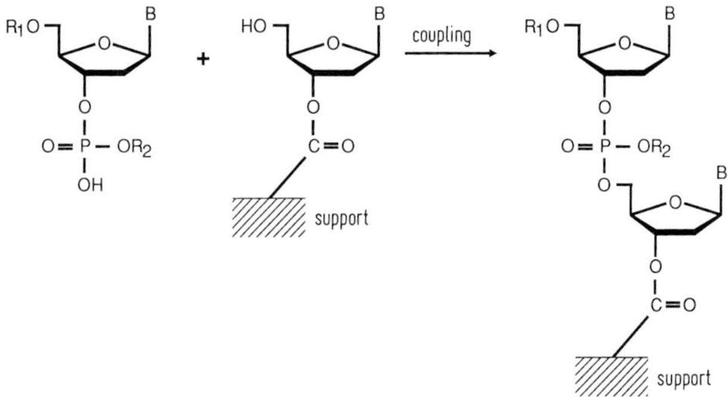

Fig. 2 Formation of the phosphoric acid diester bond using the phosphoric acid triester method. Key to symbols: R_1 = deoxyribose protecting group; R_2 = phosphate protecting group; B = protected nucleic acid bases

In the phosphoramidite method [38–40], the esters of phosphorous acid (trivalent phosphorus) are used. The trivalent phosphorus is oxidized to pentavalent phosphorus (phosphate) after condensation with iodine. The chemical reaction is shown in Figure 3. The phosphitylating agent is *β*-cyanoethyl-*N*,*N'*-diisopropylaminophosphine, and the catalyst is tetrazole. The phosphoramidite method is used in all solid-phase syntheses, especially in systems that employ automated processes.

The most recent development in DNA chemistry is the hydrogen phosphonate (H-phosphonate) method [41–43] (Figure 4). Its chief drawback is the limited stability of the H-phosphonate diester in the presence of pivaloyl chloride [44].

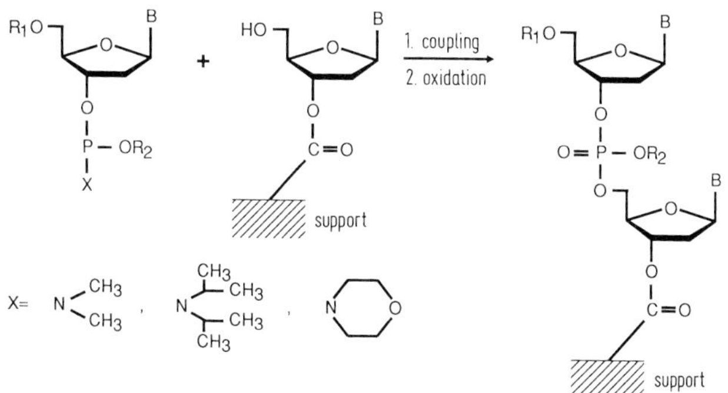

Fig. 3 Formation of the phosphoric acid diester bond using the phosphoramidite method. Key to symbols: R_1 = deoxyribose protecting group; R_2 = phosphite protecting group; B = protected nucleic acid bases; X = amidites

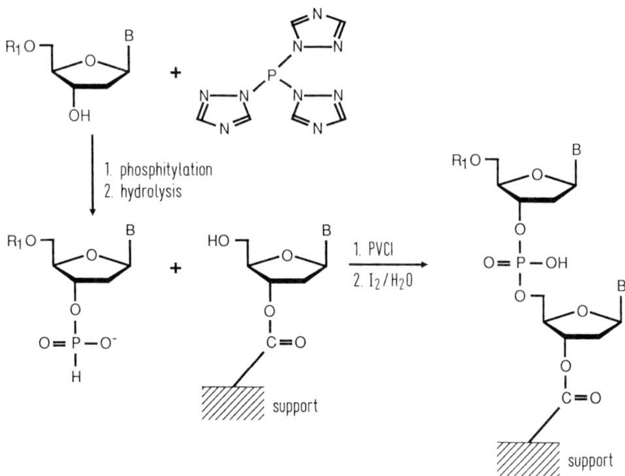

Fig. 4 Formation of the phosphoric acid diester bond using the H-phosphonate method. The oxidation with I_2/H_2O is done only after the last nucleotide is added. Key to symbols: R_1 = deoxyribose protecting group; B = protected nucleic acid bases; PVCl = pivaloyl chloride

2.2 Solid-phase synthesis

The synthesis of long-chain deoxyribonucleotides (with about 100 bases) became a routine task for technicians in many molecular biology laboratories with the advent of the solid-phase process. Polymer synthesis on a solid phase was first developed for peptide chemistry: the MERRIFIELD synthesis produced proteins more than 100 amino acids long.

Similar success eluded DNA chemists for a long time because the chain became steadily more polar the longer it grew, as a negative charge was added at each step in the diester process. Only when the triester method was introduced and phosphite diesters began to be used was solid-phase DNA synthesis successful. The yield and the velocity of this process have now been optimized to the point where additional improvements are unlikely. Rarely has a method of synthesis taken over the world's molecular biology laboratories as rapidly as has solid-phase oligonucleotide synthesis.

2.3 The filter paper method of FRANK and BLÖCKER

FRANK and BLÖCKER [35] developed the so-called *filter paper* or *segmental solid support method*, which permits the simultaneous synthesis of up to 50 oligonucleotides. The usual supports in this method are filter paper discs 2 to 3 cm in diameter. The synthesis goes on simultaneously in four reaction vessels, one each for A, C, G, and T. All oligonucleotides that are to have a particular one of the four nucleotide units added on are put together in the appropriate reaction vessel. After a standard synthesis cycle the filters are relocated: each oligonucleotide is moved to the vessel that contains the next nucleotide to be added. Although this method yields excellent results in the hands of experts – FRANK and BLÖCKER have synthesized a great many genes in this manner – it has not come into wide general use.

2.4 Automated synthesizers for deoxyoligonucleotides

The synthesis of deoxyoligonucleotides occurs in a series of identical reaction steps, some of which are conducted under inert gas because the chemical reagents employed hydrolyze quite readily. The mechanization or automation of the reaction sequence was therefore attempted very early. Here again, peptide synthesis showed the way, with the automation of solid-phase peptide synthesis by MERRIFIELD [45] and by LETSINGER and KORNET [46] serving as a model for oligonucleotide production. The easy and efficient automated control available with microprocessors has also pushed development forward. The first automated synthesizer for peptides was developed in 1965 by STEWART [47]; the cycle time for each amino acid incorporated was 4 hours. GUTTE and MERRIFIELD [48] mechanically synthesized ribonuclease A with this machine, in a process that required 369 reactions or 11,391 individual steps.

By 1980, the first machines for oligonucleotide synthesis were introduced. They consisted of four fundamental elements:

- a control unit,
- metering systems for solvents and reagents,
- mechanical valves, and
- a reaction vessel.

Various models of this equipment differed only in the quality of the precision pumps and valves. With experience we can now say that pneumatic systems with precision valves are the most successful. Argon is used as the compressed gas. Figure 5 is a schematic illustration of an automated DNA synthesizer.

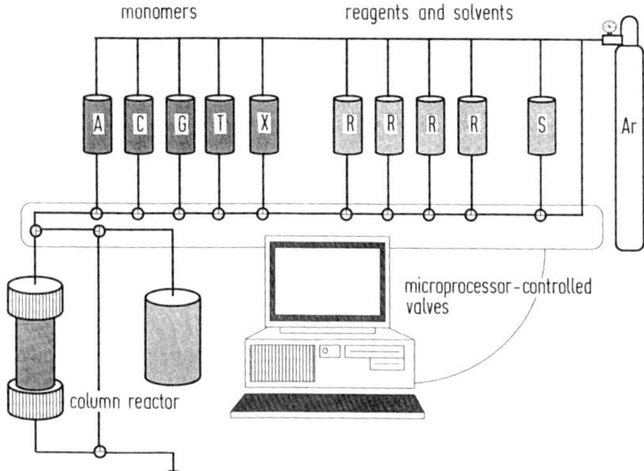

Fig. 5 Schematic illustration of an automated DNA synthesizer. A microprocessor controls the process. Key to symbols: A, C, G, T, X = nucleotide solutions; S = solvent; R = reagents; Ar = argon supply cylinder

Table 1 Commercially available automated DNA/RNA synthesizers[1]

Company (USA)	Automated DNA/RNA synthesizer	Approx. cost of equipment (US$)
Applied Biosystems	Model 391 (1 col) Model 392 (2 col) Model 394 (4 col) – w. built-in microprocessors	12,500 to 42,000
Cruachem	PS 150 (1 col)	1,495
	(2 col)	1,995
	(3 col) manual	2,495
	(4 col)	2,995
	PS 250 (2 col) – w. IBM comp. computer	13,995
MilliGen/Biosearch	Cyclone Plus (1 col)	14,950
	Cyclone Plus (2 col)	17,450
	Model 8700 (4 col)	34,000
	Model 8750 (4 col) – w/o computer	35,000
	Model 8750 (4 col) – w. NEC/AT computer	38,000
	Model 7500 (3 col) – w. NEC/AT computer	45,000
	Model 8800 (1 col, large scale) – w. NEC/AT computer	75,000
Pharmacia LKB Biotechnology	Gene Assembler-Plus (2 col)	
	– w/o computer	14,000
	– w. computer	17,600

[1] Company information, Oct. 1990

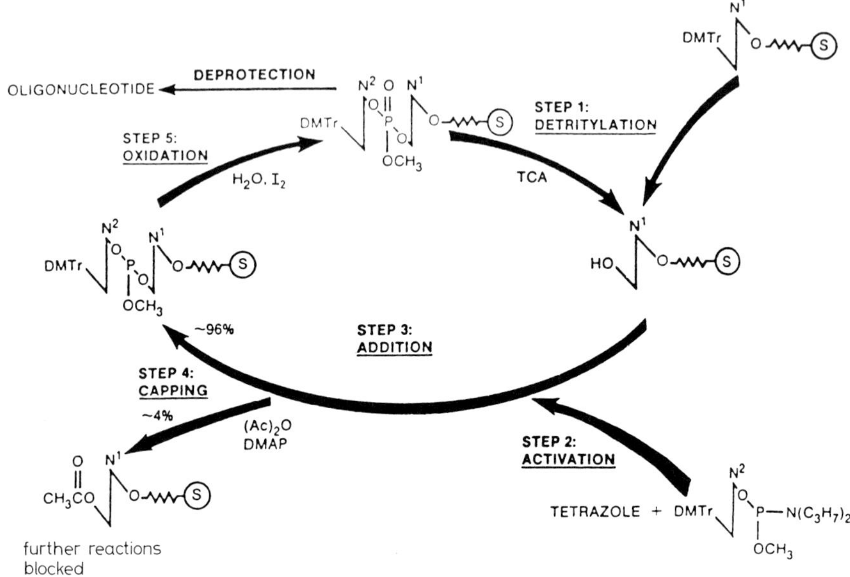

Fig. 6 Diagram of one cycle of the solid-phase synthesis of oligonucleotides. (This illustration has been made available through the kindness of Applied Biosystems, USA.) Key to symbols: TCA = trichloroacetic acid; DMTr = dimethoxytrityl chloride; DMAP = dimethylaminopyridine

Synthesizers that are currently commercially available are listed in Table 1; they have been perfected to such a degree that deoxyoligonucleotides of chain length in excess of 150 nucleotides can be synthesized. The first nucleoside (primer) is attached by its 3′-hydroxyl group to the support, which is silica gel or controlled-pore glass (CPG). The synthesis is thus directed from 3′ to 5′. Each reaction cycle includes the following steps (Figure 6).

- The protecting group at the 5′ position of the primer is split off (detritylated);
- the synthesis monomer is activated, and added or coupled;
- free unconverted 5′-hydroxyl groups are blocked from participating in further reactions (capped); and
- the phosphite compound is oxidized to a phosphate compound.

The chemistry has been so standardized that the yield is about 99% for each cycle.

There have been numerous suggestions for improvements to individual chemical steps in the oligonucleotide synthesis process, but to date none has superseded an existing step. The best evidence for the efficiency of the process is that 100 structural genes have been synthesized chemically at the time of writing. Some of them are named in Table 2.

Table 2 Some chemically synthesized structural genes of important proteins [58]

Amylase	Eglin	Myosin
Angiotensin	Endorphin	Ribonuclease
Bradykinin	Galactosidase	Rhodopsin
Calcitonin	Globin	Secretin
Casein	Glucagon	Somatostatin
Chymosin	Hygromycin	Substance P
E Growth Factor	Hirudin	Thaumatin
Interferon	HLA Antigen	Tetragastrin
Interleukin	Lysozyme	Thymosin
Insulin	Myoglobin	Ubiquitin

(30 examples of approximately 100 chemically synthesized structural genes)

3 Purification methods

Although oligonucleotide synthesis can attain yields of 99% per reaction step, the end products after removal of protecting groups (*deprotection*) must be purified; in certain cases, such as the synthesis of short-chain oligonucleotides used as primers in sequencing, purification is not really necessary. Gel electrophoresis and high-performance liquid chromatography (HPLC) are the two purification methods used.

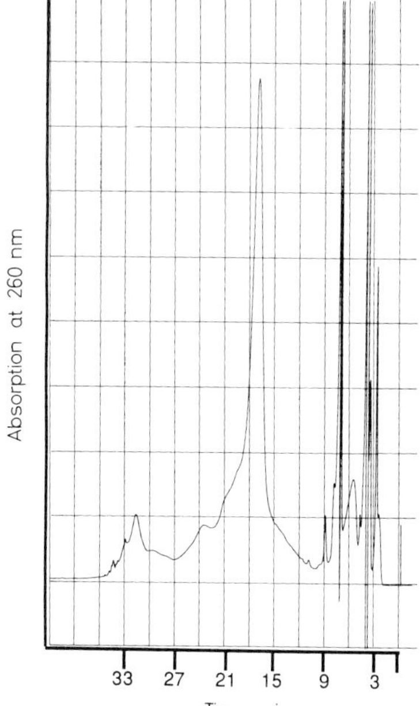

Fig. 7 Profile of a reversed-phase HPLC purification of an oligonucleotide. The oligonucleotide chain is 90 nucleotides long. The peaks at times between 2 and 12 minutes correspond to oligonucleotide fragments. The desired oligonucleotide is eluted between 15 and 21 minutes (the product peak). The peak between about 27 and 33 minutes corresponds to faulty sequences of longer chain length

Electrophoretic separations usually take advantage of polyacrylamide gels, which contain 6–20% acrylamide [25, 26], 1–2% *N,N′*-methylenebisacrylamide, and 7 M urea. The position of the oligonucleotide on the gel can be visualized by UV shadowing or with a fluorescent dye such as ethidium bromide. The appropriate gel sections are cut out and the oligonucleotides are eluted with water or buffer [49]. The solution is desalted with a molecular sieve. Although this purification procedure sounds involved, it has the advantage that a gel can hold up to 12 samples that can be purified simultaneously.

High-performance liquid chromatography is more widely used. The oligonucleotides can be separated according to charge, size, or hydrophobic nature according to the eluent (mobile phase) in the column. Disadvantages of this technique are the equipment involved and the cost per separation. C-18 reversed-phase columns are particularly useful for routine tasks such as the separation of still-tritylated oligonucleotides (Figure 7), in which the terminal nucleotide of the oligonucleotide still bears the trityl protecting group. The hydrophobic trityl group effects partitioning from the desired product of all those sequences that are shorter and therefore not tritylated. After detritylation, the oligonucleotide can be purified to over 95% by repeating the separation step. When oligonucleotides with chain lengths of more than 100 nucleotides are being purified, the formation of stable secondary structures may cause problems.

4 Analysis of oligonucleotides of defined sequence

It is usual nowadays to analyze oligonucleotides only for their chain length, by labeling them with ^{32}P-phosphate at the 5′-hydroxyl group and subjecting them to polyacrylamide gel electrophoresis (*PAGE*) (Figure 8). The gel is autoradiographed, and sequencing markers of known chain length are used to determine the length of the unknown sequence. If the exposure is prolonged, a series of weaker secondary bands becomes visible, indicating contamination with shorter oligonucleotides that result from sequence fragmentation.

Three methods are available for sequencing analyses of DNA fragments with fewer than 100 nucleotides.

- The oligonucleotide can be cleaved enzymatically into nucleotides and the nucleotide composition can be determined by HPLC [50].
- The MAXAM-GILBERT technique of chemical DNA sequencing [49] can be adapted for oligonucleotides.
- FRANK and BLÖCKER [51] have developed the "wandering spot" sequence analysis technique. This involves two-dimensional fingerprinting, so that the sequences of oligonucleotides up to 20 nucleotides in length can be read on a single autoradiograph.

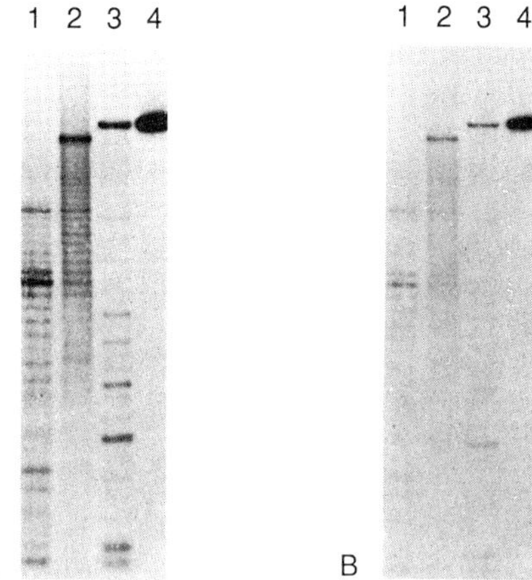

Fig. 8 Autoradiograph of an analysis by gel electrophoresis of chemically synthesized oligonucleotides of intermediate chain length. The oligonucleotides after synthesis were enzymatically labeled by a DNA kinase at the 5′ end with radioactive [γ-^{32}P] ATP and subsequently analyzed on a 12% polyacrylamide gel (sequencing gel), either as a crude mixture or after undergoing purification by reversed-phase HPLC. The film for autoradiograph A was exposed for 24 hours, and that for autoradiograph B for 1 hour.

Track 1 consists of a crude mixture of 44–mers;
track 2 consists of 44–mers after reversed-phase HPLC purification;
track 3 consists of a crude mixture of 46–mers;
track 4 consists of 46–mers after reversed-phase HPLC purification.

In both mixtures, the product after purification by HPLC was more concentrated.

In autoradiograph A, compare the uppermost (product) bands in tracks 2 and 4 with those in tracks 1 and 3. The product bands in track 1 are almost indistinguishable and there are many bands from nucleotide fragments; this is indicative of a poor synthesis. However, a product band does emerge after purification (track 2). Track 3 has the strong product bands and few fragment bands indicative of a good synthesis, and purification results in a pure product (track 4 has product bands only).

This kind of analysis is possible only with a lengthy (24-hour) exposure of the film (autoradiograph A). If the exposure is short (1 hour), erroneous results are displayed (autoradiograph B)

5 Chemical synthesis of RNA molecules of defined sequence

Synthesis of ribonucleic acids of defined sequence is chemically more difficult than the production of corresponding DNA fragments; the 2′-hydroxyl group of the ribose must be blocked in addition to the 5′- or 3′-hydroxyl group. Removing the protecting group at acid pH is difficult because the phosphate group can attach to either the 2′ or the 3′ position.

The molecule of choice for RNA chemists and biochemists has always been transfer ribonucleic acid (tRNA). Since it is of manageable size – 60 to 90 nucleotides – and because it has sequences with many modified nucleotides, the synthesis of this molecule presents a special challenge to the chemist. Short-chain RNA molecules can be used as primers in DNA sequencing analysis and for strand replication. Ribose derivatives of phosphoramidites have recently become available, so now RNA synthesis can be conducted with automated synthesizers (Table 1).

RNA fragments with a chain length of about 20 nucleotides can also be synthesized enzymatically using polynucleotide phosphorylase and RNA ligase [52]; this method is best suited for the synthesis of small quantities, and especially for preparing radioactively labeled ribonucleotides.

6 Synthesis of structural genes by a method combining chemical and enzymatic techniques

Khorana and his co-workers synthesized the first structural gene in about 1970; this gene did not encode for a protein, but rather for a tyrosine suppressor tRNA in yeast [11]. Although this may seem a puzzling objective today, one must keep in mind that DNA could not be sequenced at that time, so the chemists had few options of what they should synthesize. The picture changed immediately once the DNA sequencing techniques of Maxam and Gilbert and of Sanger became known. However, the principles of gene synthesis had already been established by Khorana. The synthesis of structural genes by a combination of chemical and enzymatic methods can be outlined as follows.

- Deoxyoligonucleotides of maximum chain length are synthesized chemically and purified (*synthesis*);
- *DNA ligase* is used to join the oligonucleotides with phosphodiester bonds enzymatically (*ligation*);
- the DNA fragments are inserted into a *vector*, the bacterial host of the vector is *transformed*, and *clones* are selected that contain the vector with proper insert (*transformation*);
- the DNA in the recombinant vector is produced in milligram quantities (*production*);
- the gene is verified by sequencing (*verification*).

The gene for angiotensin II, the first structural gene to be produced that coded for a protein, could not be expressed in bacteria [53]. Shortly thereafter the genes coding for the peptide hormones somatostatin [54] and the A and B chains of insulin [55] were produced; bacterial expression was possible for them. With additional improvements in technique, the genes coding for human leukocyte interferon (containing 514 nucleotide pairs) [56] and for bovine rhodopsin (with more than 1000 nucleotide pairs) [57] were synthesized. Seliger and his co-workers [58] compiled a comprehensive list of all chemically synthesized genes and published it in *Nucleic Acids Research* in 1988; the list contained more than 100 genes, so that now the structural gene and the regulatory gene sequence of the host cell of almost every medically important protein have been produced chemically.

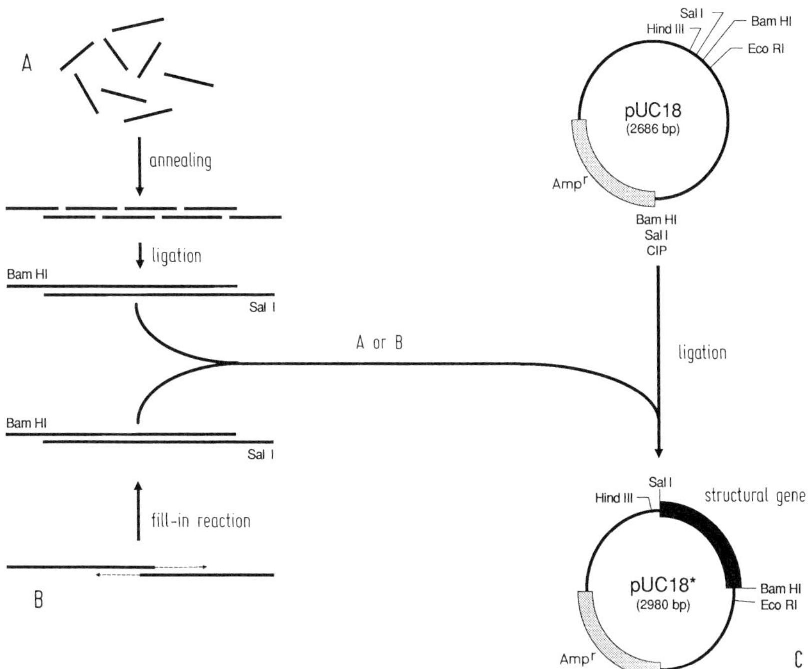

Fig. 9 Gene synthesis, (A) according to the technique of KHORANA and (B) the fill-in technique, and (C) insertion of a gene into vector pUC18. Key to symbols: *Sal* I, *Hind* III, *Bam* HI, *Eco* RI are designations of restriction enzyme cleavage sites or the enzymes themselves; Amp^r is the gene for ampicillin resistance; CIP = calf intestinal phosphatase (alkaline phosphatase)

Two different approaches to gene synthesis are in use, the method of KHORANA [22] (Figure 9 A) and the *fill-in* technique [59–61] (Figure 9 B). With the first method, the entirety of the double strand of DNA must be synthesized with complementary and overlapping oligonucleotides. In the fill-in method for synthesizing the two single strands, only half of each strand, in alternating overlapping regions containing about 10 nucleotides, need be synthesized; filling in the gaps in the strands is done enzymatically. This technique has advantages for oligonucleotides 150–200 nucleotide pairs long.

Most laboratories use KHORANA's method, however. It permits gene cloning without great effort. The gene is synthesized by starting with completely overlapping DNA fragments, each consisting of about 60 nucleotides. The two terminal units must each be complementary to the terminal ends of linearized vector DNA. The oligonucleotides associate with each other, at maximum complementarity, with WATSON-CRICK hydrogen bonding to form the double strand (this step is called *annealing*), in a *single-pot reaction* [62], and are subsequently incubated with a linearized vector. Hydrogen bonds are also formed between the terminal ends of the vector DNA and the complementary ends of the annealed oligonucleotide product. The result is a linear DNA molecule composed of vector DNA and gene DNA. Plasmids pUC18 or pUC19 are the most usual vectors. The vector is linearized by sequence-specific hydrolysis with restriction endonucleases, producing two different single strands, each terminating with four nucleotides; this is done so that the orientation during hybridization with the terminal ends of the annealed product is correct. Prior to this, the terminal 5′-phosphate groups at the ends of the vector are removed enzymatically with alkaline phosphatase. Finally, the oligonucleotides are covalently ligated to each other and to the vector DNA by DNA ligase (the *ligation* step). The resultant recombinant plasmid DNA is then inserted into *Escherichia coli* (*E. coli*) cells (the *transformation* step) and subsequently replicates. This synthesis and cloning strategy is successful, with about 10^6 to 10^7 *transformants* (transformed cells) produced per microgram of DNA. About half of the transformants bear the plasmid with the inserted gene; the remaining cells contain faulty products: because each oligonucleotide is only about 90% pure, and because of the single-pot ligation, many faulty sequences arise. Therefore, the nucleotide sequence of the DNA insert must be verified by sequencing.

Chemical DNA synthesis could not have come this far without cloning methods. If the amino acid sequence is known, we can obtain a gene more rapidly by synthesizing it chemically than by isolating it from a cDNA library or from a genomic DNA library [63]. Gene synthesis also has the advantages that the gene can be tailored to the codon requirements of the host in which it is to be expressed and that the ligating sequences can be optimized for each vector. Similar considerations are involved in the selection of cleavage sites for restriction enzymes and in the replacement of gene sequences in site-specific *in vitro* mutagenesis [64].

Those who have followed the advances in chemical gene synthesis and the successful expression of synthetic genes in microorganisms over the past 10 years are enthusiastic about the technical potential of these techniques.

7 Applications of deoxyoligonucleotides in molecular biology

The chemical synthesis of oligonucleotides has expanded enormously the opportunities for gene technology in the production of biological substances.

- Oligonucleotides of defined sequence can be used in analysis, for example, to sequence a 2- to 10-kilobase (kb) DNA fragment by a so-called *walk* [64].
- Oligonucleotides are used as primers in the synthesis of cDNA from an RNA template, because neither reverse transcriptase (that transcribes RNA into DNA) nor DNA polymerase (that synthesizes DNA from a DNA template) is capable of catalyzing the formation of the first phosphoric acid diester bond.
- Oligonucleotides serve as *linkers* and *adapters*. These are sequences (*constructs*) that are required for joining a structural gene to a vector: frequently during cDNA synthesis, a DNA fragment is obtained with a 5′ terminus missing a few nucleotides; if the amino acid sequence is known, one can replace missing DNA sequences by deoxyoligonucleotides.

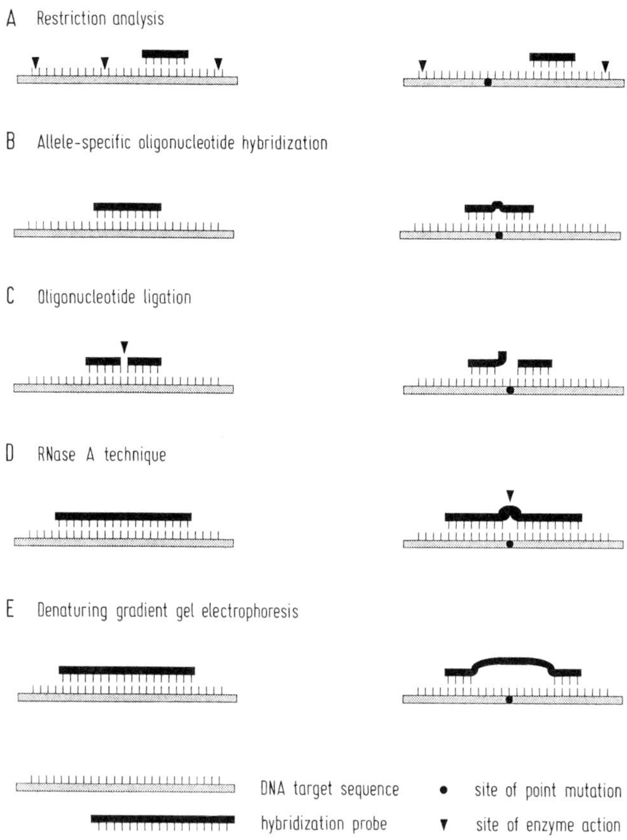

Fig. 10 Various ways in which oligonucleotides of defined sequence can be used as gene probes (hybridization probes). Often, a change in a single base (a point mutation) suffices to block a permanent hybridization under defined reaction conditions.
A: DNA is cleaved into specific fragments by restriction enzymes. A fragment can be detected with a matching hybridization probe (left). If a restriction enzyme cleavage site is destroyed by a point mutation, the fragment detected by the probe is correspondingly longer (right).
B: Different forms (alleles) of a gene can be distinguished with hybridization probes. Under appropriate reaction conditions, the hybridization of the normal gene (left) is more stable than that of the gene with a point mutation (right).
C: A point mutation (right) blocks the ligation of two oligonucleotides that are complementary to the target DNA. In the DNA without the mutation (left), the reaction can proceed.
D: The probe in this technique is of RNA. If the probe is not completely complementary to the DNA, it can be excised at the locus of the mismatch by the enzyme RNase A.
E: If DNA is not completely complementary to the hybridization probe because of a point mutation, it can be detected with gradient gel electrophoresis under special conditions of denaturation. The incompletely matched double strand on the right dissociates at lower detergent concentrations than the one on the left, which is a complete complementary double strand

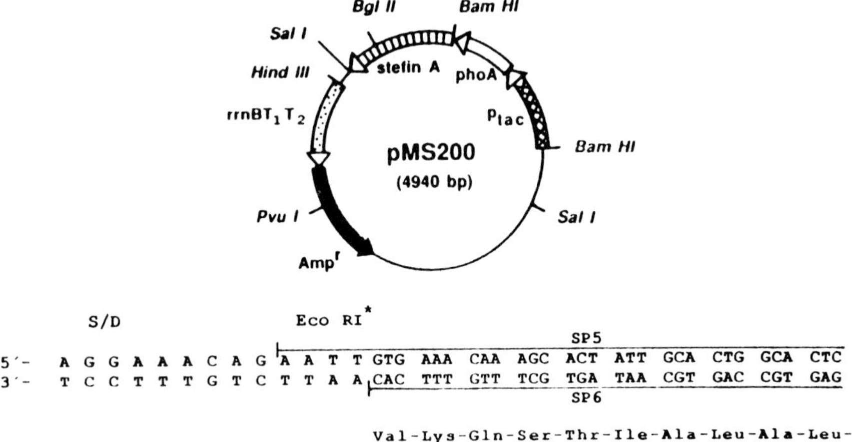

Bam HI

TTA CCG TTA CTG TTT ACC CCT GTG ACA AAA GCA GGA TCC ATG ATC CCG GGT- 3´
AAT GGC AAT GAC AAA TGG GGA CAC TGT TTT CGT CCT AGG TAC TAG GGC CCA- 5´

-Leu-Pro-Leu-Leu-Phe-Thr-Pro-Val-Thr-Lys-Ala-Gly-Ser-Met-Ile-Pro-Gly-

Fig. 11 The use of oligonucleotides of defined sequence as elements in the construct of expression vector pMS200: The stefin A gene was periplasmically expressed with vector pMS200 [71]. In order for periplasmic expression to occur, a *signal sequence* had to be inserted upstream from the gene; in this case the signal sequence was for the enzyme alkaline phosphatase, and was constructed from two chemically synthesized oligonucleotides (SP5 and SP6). The vector pMS200 is illustrated in the upper portion of the figure. The lower portion is the sequences (left) at the transition between the Shine-Dalgarno (*S/D*) sequence and the signal sequence, and (right) at the transition between the signal sequence and the stefin A gene. Nucleotides as well as amino acids are indicated. The amino acids for the signal sequence of alkaline phosphatase and the nucleotides of oligonucleotides SP5 and SP6 are underlined. The first four codons of the stefin A gene and the first four amino acids of the stefin A protein (Met-Ile-Pro-Gly-) are at the bottom right. During construction, an *Eco* RI cleavage site was destroyed (indicated by *Eco* RI*) and a *Bam* HI cleavage site was introduced *de novo*. With the addition of the new cleavage site came the incorporation of two additional amino acids (Gly and Ser). Key to symbols: *Pvu* I, *Hind* III, *Sal* I, *Bgl* II, *Bam* HI are designations for restriction enzyme cleavage sites, or for the enzymes themselves; Amp^r is the gene for ampicillin resistance; $rrnBT_1T_2$ is the transcription terminator

- Regulatory sequences such as promoters, ribosomal binding sites, and terminators, all of which are always host-specific, can be chemically synthesized and inserted into an expression plasmid.
- Point mutations can be introduced into a structural gene using an oligonucleotide modified by site-specific mutagenesis [64], with the result that a given amino acid is selectively replaced within the protein.
- An additional important application of short-sequence oligonucleotides is their use as probes for identifying certain DNA sequences. Oligonucleotides labeled radioactively or with a dye are used as probes for screening gene libraries (colony screen), verifying DNA fragments (Southern blotting), or detecting point mutations.
- The most important area in which we expect to use oligonucleotides is DNA sequencing. So far it has been customary to subclone long DNA sequences (about 10 to 20 kb) obtained as inserts from genomic DNA libraries into smaller fragments of about 600 to 800 base pairs (bp) [49]. The effort this would require with a microbial system can be reduced significantly by using oligonucleotides for the DNA sequencing. Even with automated gene sequencers, the technique is still a little complicated and time-intensive.
- Yet another area of application is medical research [65, 66], in which modified oligonucleotides are becoming increasingly important.

Because the use of oligonucleotides has been described extensively in manuals for gene technology [49, 67, 68], we will limit ourselves to oligonucleotides as gene probes (*hybridization probes*) (see Figure 10) and as elements for engineering constructs such as expression vectors (see Figure 11).

Very large quantities of oligonucleotides will be required in the future for the test-tube replication of DNA. This relatively new technique is called the *polymerase chain reaction* (PCR). If two primer oligonucleotides are used for each of the DNA strands (the *plus* strand and the *minus* strand) the template can be replicated enzymatically up to 10^6-fold [69, 70].

8 Summary

The synthesis of oligonucleotides has become extremely sophisticated, thanks to commercially available automated synthesizers and the excellent chemical techniques they incorporate. The DNA sequences of some 150 nucleotides can be synthesized without difficulty even by nonspecialists. Efforts with RNA have yet to meet with similar success.

Short DNA fragments with sequences of 15 to 20 nucleotides will be required in quantity in the future. Their chief applications will be for DNA sequencing, for uses as linkers/adapters for expressible genes, and for DNA identification. Certainly the gene technology of the future could not proceed efficiently without chemical oligonucleotide synthesis.

9 Acknowledgements

Figures 1 through 4 and 6 through 10 were prepared by Dipl.-Biol. Holger ZINKE.

10 References

[1] WITZEL, H.: Einfluß der Nucleotidbasen auf die nichtenzymatische Spaltung der Ribonucleinsäure-Diesterbindung; *Ann. Chemie* **635** (1960) 182–191.

[2] WITZEL, H.: Zur Katalyse bei der Ribonuclease-Reaktion. In: *Mechanismen enzymatischer Reaktionen*, Springer Verlag, Berlin, GÖttingen, Heidelberg (1964); pp 123–125.

[3] ENGELS, J.; UHLMANN, E.: Gene Synthesis. In: *Advances in Biochemical Engineering/Biotechnology*, **Vol. 37**, Springer Verlag, Berlin, Heidelberg (1988); pp 73–127.

[4] GASSEN, H. G.; LANG, A., Eds.; *Chemical and Enzymatic Synthesis of Gene Fragments*, VCH Publ. Co., Weinheim (1982).

[5] US Congress, Office of Technology Assessment: *Mapping Our Genes – The Genome Projects: How Big, How Fast?* OTA-BA-373, US Government Printing Office, Washington, DC.

[6] GAIT, M. J.: *Oligonucleotide Synthesis – A Practical Approach*, IRL Press, Oxford (1984).

[7] MIZUNO, Y.: The Organic Chemistry of Nucleic Acids. *Studies in Organic Chemistry*, **Vol. 24**, Kodansha, Tokyo (1986).

[8] KIEFER, H.; BANNWART, W.: Strategies of Oligonucleotide Synthesis; *Immunol. Methods* **3** (1985) 69–83.

[9] SELIGER, H.; GUPTA, K. C.; KOTSCHI, U.; SPANEY, T.; ZEH, D.: New Preparative Methods in Oligonucleotide Chemistry and Their Application to Gene Synthesis. I. Improvements in the Chemistry of Solid-Phase Oligonucleotide Synthesis; *Chem. Scr.* **26** (1986) 561–567.

[10] KHORANA, H. G.: Studies on Nucleic Acids: Total Synthesis of a Biologically Functional Gene; *Bioorg. Chem.* **7** (1978) 351–393.

[11] KHORANA, H. G.: Total Synthesis of a Gene; *Science* **203** (1979) 614–625.

[12] KHORANA, H. G.; AGARWAL, K. L.; BESMER, P.; BUCHI, H.; CARUTHERS, M. H.; CASHION, P. J.; FRIDKIN, M.; JAY, E.; KLEPPE, K.; KLEPPE, R.; KUMAR, A.; LOEWEN, P. C.; MILLER, R. C.; MINAMOTO, K.; PANET, A.; RAJBHANDARY, U. L.; RAMAMOORTHY, B.; SEKIYA, T.; TAKEYA, T.; VAN DE SANDE, J. H.: Total Synthesis of the Structural Gene for the Precursor of a Tyrosine Suppressor Transfer RNA from *Escherichia coli.* 1. General Introduction; *J. Biol. Chem.* **251** (1976) 565–570.

[13] VAN DE SANDE, J. H.; CARUTHERS, M. H.; KUMAR, A.; KHORANA, H. G.: Total Synthesis of the Structural Gene for the Precursor of a Tyrosine Suppressor Transfer RNA from *Escherichia coli.* 2. Chemical Synthesis of the Deoxypolynucleotide Segments Corresponding to the Nucleotide Sequence 1–31; *J. Biol. Chem.* **251** (1976) 571–586.

[14] MINAMOTO, K.; CARUTHERS, M. H.; RAMAMOORTHY, B.; VAN DE SANDE, J. H.; SIDOROVA, N.; KHORANA, H. G.: Total Synthesis of the Structural Gene for the Precursor of a Tyrosine Suppressor Transfer RNA from *Escherichia coli.* 3. Synthesis of Deoxyribopolynucleotide Corresponding to the Nucleotide Sequence 27–51; *J. Biol. Chem.* **251** (1976) 587–598.

[15] Agarwal, K. L.; Caruthers, M. H.; Fridkin, M.; Kumar, A.; van de Sande, J. H.; Khorana, H. G.: Total Synthesis of the Structural Gene for the Precursor of a Tyrosine Suppressor Transfer RNA from *Escherichia coli*. 4. Synthesis of Deoxyribopolynucleotide Segments Corresponding to the Nucleotide Sequence 47–78; *J. Biol. Chem.* **251** (1976) 599–608.

[16] Jay, E.; Cashion, P. J.; Fridkin, M.; Ramamoorthy, B.; Agarwal, K. L.; Caruthers, M. H.; Khorana, H. G.: Total Synthesis of the Structural Gene for the Precursor of a Tyrosine Suppressor Transfer RNA from *Escherichia coli*. 5. Synthesis of the Deoxyribopolynucleotide Segments Representing the Nucleotide Sequence 71–103; *J. Biol. Chem.* **251** (1976) 609–623.

[17] Agarwal, K. L.; Caruthers, M. H.; Buchi, H.; van de Sande, J. H.; Khorana, H. G.: Total Synthesis of the Structural Gene for the Precursor of a Tyrosine Suppressor Transfer RNA from *Escherichia coli*. 6. Synthesis of the Deoxyribopolynucleotide Segments Corresponding to the Nucleotide Sequence 100–126; *J. Biol. Chem.* **251** (1976) 624–633.

[18] Sekiya, T.; Besmer, P.; Takeya, T.; Khorana, H. G.: Total Synthesis of the Structural Gene for the Precursor of a Tyrosine Suppressor Transfer RNA from *Escherichia coli*. 7. Enzymatic Joining of the Chemically Synthesized Segments to Form a DNA Duplex Corresponding to the Nucleotide Sequence 1–26; *J. Biol. Chem.* **251** (1976) 634–641.

[19] Loewen, P. C.; Miller, R. C.; Panet, A.; Sekiya, T.; Khorana, H. G.: Total Synthesis of the Structural Gene for the Precursor of a Tyrosine Suppressor Transfer RNA from *Escherichia coli*. 8. Enzymatic Joining of the Chemically Synthesized Segments to Form DNA Duplexes Corresponding to Nucleotide Sequences 23–60 and 23–66; *J. Biol. Chem.* **251** (1976) 642–650.

[20] Panet, A.; Kleppe, K.; Khorana, H. G.: Total Synthesis of the Structural Gene for the Precursor of a Tyrosine Suppressor Transfer RNA from *Escherichia coli*. 9. Enzymatic Joining of Chemically Synthesized Deoxyribopolynucleotide Segments Corresponding to Nucleotide Sequence 57–94; *J. Biol. Chem.* **251** (1976) 651–657.

[21] Caruthers, M. H.; Kleppe, R.; Kleppe, K.; Khorana, H. G.: Total Synthesis of the Structural Gene for the Precursor of a Tyrosine Suppressor Transfer RNA from *Escherichia coli*. 10. Enzymatic Joining of Chemically Synthesized Segments to Form the DNA Duplex Corresponding to the Nucleotide Sequence 86–126; *J. Biol. Chem.* **251** (1976) 658–666.

[22] Kleppe, R.; Sekiya, T.; Loewen, C.; Kleppe, K.; Agarwal, K. L.; Buchi, H.; Besmer, P.; Caruthers, M. H.; Cashion, P. J.; Fridkin, M.; Jay, E.; Kumar, A.; Miller, R. C.; Minamoto, K.; Panet, A.; RajBhandary, U. L.; Ramamoorthy, B.; Sidorova, N.; Takeya, T.; van de Sande, J. H.; Khorana, H. G.: Total Synthesis of the Structural Gene for the Precursor of a Tyrosine Suppressor Transfer RNA from *Escherichia coli*. 11. Enzymatic Joining to Form the Total DNA Duplex; *J. Biol. Chem.* **251** (1976) 667–675.

[23] Ramamoorthy, B.; Lees, R. G.; Kleid, D. G.; Khorana, H. G.: Total Synthesis of the Structural Gene for the Precursor of a Tyrosine Suppressor Transfer RNA from *Escherichia coli*. 12. Synthesis of a DNA Duplex Corresponding to a Sequence of 23 Nucleotide Units Adjoining the C-C-A End; *J. Biol. Chem.* **251** (1976) 676–694.

[24] Brown, E. L.; Ramamoorthy, B.; Ryan, M. J.; Khorana, H. G.: Chemical Synthesis and Cloning of a Tyrosine tRNA Gene; *Methods Enzymol.* **68** (1979) 109–151.

[25] Letsinger, R. L.; Mahadevan, V.: Stepwise Synthesis of Oligodeoxyribonucleotides on an Insoluble Polymer Support; *J. Am. Chem. Soc.* **88** (1966) 5319–5324.

[26] Eckstein, F.; Scheit, K. H.: Synthese von Nucleosidphosphaten mit Phosphorsäurebis-(β,β,β-trichlorethylester)-chlorid; *Angew. Chem.* **7** (1967) 317–318.

[27] Eckstein, F.; Rizk, I.: Synthese von Oligodesoxynucleotiden über Phosphorsäuretriester; *Chem. Ber.* **102** (1969) 2362–2377.

[28] Reese, C. B.; Saffhill, R.: Oligonucleotide Synthesis via Phosphotriester Intermediates: The Phenyl-Protecting Group; *J. Chem. Soc., Chem. Commun.* **767** (1968) 767–768.

[29] Catlin, J. C.; Cramer, F.: Deoxyoligonucleotide Synthesis via the Triester Method; *J. Org. Chem.* **38** (1973) 245–250.

[30] Itakura, K.; Bahl, C. P.; Katagiri, N.; Michniewicz, J. J.; Wightman, R. H.; Narang, S. A.: A Modified Triester Method for the Synthesis of Deoxyribopolynucleotides; *Canad. J. Chem.* **51** (1973) 3649–3651.

[31] Itakura, K.; Katagin, N.; Bahl, C. P.; Wightman, R. H.; Narang, S. A.: Improved Triester Approach for the Synthesis of Pentadecathymidylic acid; *J. Am. Chem. Soc.* **97** (1975) 7327–7332.

[32] Miyoshi, K.; Itakura, K.: Solid-phase Synthesis of Nonadecathymidylic Acid by the Phosphotriester Approach; *Tetrahedron Lett.* **20** (1979) 3625–3638.

[33] Efimov, V. A.; Buryakova, A. A.; Reverdatto, S. V.; Chakkmakhcheva, O. G.; Ovchinnikov, Yu. A.: Rapid Synthesis of Long-Chain Deoxyribooligonucleotides by the *N*-Methylimidazolide Phosphotriester Method; *Nucleic Acids Res.* **11** (1983) 8369–8387.

[34] Froehler, B. C.; Matteucci, M. D.: 1-Methyl-2(2-hydroxyphenyl)imidazole: A Catalytic Phosphate Protecting Group in Deoxyoligonucleotide Synthesis; *J. Am. Chem. Soc.* **197** (1985) 278–279.

[35] Frank, R.; Heikens, W.; Heisterberg-Moutsis, G.; Blöcker, H.: A New General Approach for the Simultaneous Chemical Synthesis of Large Numbers of Oligonucleotides: Segmental Solid Supports; *Nucleic Acids Res.* **11** (1983) 4365–4377.

[36] Marugg, J. E.; Piel, N.; McLaughlin, L. W.; Tromp, M.; Veeneman, G. H.; van der Marel, G. A.; van Boom, J. H.: Polymer Supported DNA Synthesis Using Hydroxybenzotriazole Activated Phosphotriester Intermediates; *Nucleic Acids Res.* **12** (1984) 8639–8651.

[37] van Boom, J. H.; Burgers, P. M. J.; van der Marel, G.; Verdegaal, C. H. M.; Wille, G.: Synthesis of Oligonucleotides with Sequences Identical with or Analogous to the 3′-End of 16S Ribosomal RNA of *Escherichia coli*: Preparation of A-C-C-U-C-C via the Modified Phosphotriester Method; *Nucleic Acids Res.* **4** (1977) 1047–1063.

[38] Adams, S. P.; Kavka, K. S.; Wykes, E. J.; Holder, S. B.; Galluppi, G. R.: Hindered Dialkylamino Nucleoside Phosphite Reagents in the Synthesis of Two DNA 51–mers; *J. Am. Chem. Soc.* **105** (1983) 661–663.

[39] Caruthers, M. H.; Beaucage, S. L.; Becker, C.; Efcavitch, W.; Fisher, E. F.; Galluppi, G.; Goldman, R.; deHaseth, P.; Martin, F.: New Methods for Synthesizing Deoxyoligonucleotides. In: *Genetic Engineering*, **Vol. 4**, Plenum Publishing Company (1982); pp 1–17.

[40] Sinha, N. D.; Biernat, J.; Köster, H.: *β*-Cyanoethyl-*N*,-*N*-dialkylamino-*N*-morpholinomonochloro Phosphoramidites, New Phosphitylating Agents Facilitating Ease of Deprotection and Work-up of Synthesized Oligonucleotides; *Tetrahedron Lett.* **24** (1983) 5843–5846.

[41] Garegg, P. J.; Regberg, T.; Stawinksi, J.; Strömberg, R.: Formation of Internucleotidic Bonds via Phosphonate Intermediates; *Chem. Scr.* **28** (1985) 280–282.

[42] Froehler, B. C.; Ng, P. G.; Matteucci, M. D.: Synthesis of DNA via Deoxynucleoside H-phosphonate Intermediates; *Nucleic Acids Res.* **14** (1986) 5399–5407.

[43] Garegg, P. J.; Lindh, I.; Regberg, T.; Stawinski, J.; Strömberg, R.: Nucleoside H-phosphonates. III. Chemical Synthesis of Oligodeoxyribonucleotides by the Hydrogenphosphonate Approach; *Tetrahedron Lett.* **27** (1986) 4051–4054.

[44] Kuyl-Yeheskiely, E.; Spierenburg, M.; van den Elst, H.; Tromp, M.; van der Marel, G. A.; van Boom, J. H.: Reaction of Pivaloyl Chloride with Internucleosidic H-phosphonate Diester; *Recueil Trav. Chim. Pays-Bas* **105** (1986) 505–506.

[45] Merrifield, R. B.: Solid-phase Peptide Synthesis. I. The Synthesis of a Tetrapeptide; *J. Am. Chem. Soc.* **85** (1963) 2149–2154.

[46] Letsinger, R. L.; Kornet, M. J.: Popcorn Polymer as a Support in Multistep Synthesis; *J. Am. Chem. Soc.* **85** (1963) 3045–3046.

[47] Stewart, J. M.; Young, J. D.: *Solid-phase Peptide Synthesis*, W. H. Freeman, San Francisco (1969).

[48] Gutte, B.; Merrifield, R. B.: The Synthesis of Ribonuclease A; *J. Biol. Chem.* **246** (1971) 1922–1941.

[49] Maniatis, T.; Fritsch, E. F.; Sambrook, J.: *Molecular Cloning. A Laboratory Manual*, Cold Spring Harbor Laboratory Press, Cold Spring Harbor, NY (1982).

[50] Fritz, H. J.; Eick, D.; Werr, W.: Analysis of Synthetic Oligodeoxyribonucleotides. In: Gassen, H. G.; Lang, A., Eds.; *Chemical and Enzymatic Synthesis of Gene Fragments*, VCH Publishing Co., Weinheim (1982); pp 199–223.

[51] Frank, R.; Blöcker, H.: The "Wandering Spot" Sequence Analysis of Oligodeoxyribonucleotides. In: Gassen, H. G.; Lang, A., Eds.; *Chemical and Enzymatic Synthesis of Gene Fragments*, VCH Publishing Co., Weinheim (1982); pp 225–246.

[52] Uhlenbeck, O. C.: Enzymatic Synthesis of ^{32}P-Oligonucleotides. In: Gassen, H. G.; Lang, A., Eds.; *Chemical and Enzymatic Synthesis of Gene Fragments*, VCH Publishing Co., Weinheim (1982); pp 161–168.

[53] Köster, H.; Blöcker, H.; Frank, R.; Geussenhainer, S.; Kaiser, W.: Total Synthesis of a Structural Gene for the Human Peptide Hormone Angiotensin II; *Hoppe-Seyler's Z. Physiol. Chem.* **356** (1975) 1585–1593.

[54] Itakura, K.; Hirose, T.; Crea, R.; Riggs, A. D.: Expression in *Escherichia coli* of a Chemically Synthesized Gene for the Hormone Somatostatin; *Science* **198** (1977) 1056–1063.

[55] Crea, R.; Kraszewski, A.; Hirose, T.; Itakura, K.: Chemical Synthesis of Genes for Human Insulin; *PNAS USA* **75** (1978) 5765–5769.

[56] Edge, M. D.; Greene, A. R.; Heathcliffe, G. R.; Meacock, P. A.; Schuch, W.; Scanlon, D. B.; Atkinson, T. C.; Newton, C. R.; Markham, A. F.: Total Synthesis of a Human Leukocyte Interferon Gene; *Nature (London)* **292** (1981) 756–762.

[57] Feretti, L.; Karnik, S. S.; Khorana, H. G.; Nassal, M.; Operian, D. D.: Total Synthesis of a Gene for Bovine Rhodopsin; *PNAS USA* **83** (1986) 599–603.

[58] Gröger, G.; Ramalho-Ortigao, F.; Steil, H.; Seliger, H.: A Comprehensive List of Chemically Synthesized Genes; *Nucleic Acids Res.* **16** (1988) 7763–7771.

[59] Bergmann, C.; Dodt, J.; Köhler, S.; Fink, E.; Gassen, H. G.: Chemical Synthesis and Expression of a Gene Coding for Hirudin, the Thrombin-Specific Inhibitor from the Leech *Hirudo medicinales*; *Biol. Chem. Hoppe-Seyler* **367** (1986) 731–740.

[60] Rossi, J. J.; Kierzek, R.; Huang, T.; Walker, P. A.; Itakura, K.: An Alternate Method for Synthesis of Double-Stranded DNA Segments; *J. Biol. Chem.* **257** (1982) 9226–9229.

[61] Scarpulla, R. C.; Narang, S.; Wu, R.: Use of a New Retrieving Adaptor in the Cloning of a Synthetic Human Insulin A-Chain Gene; *Analyt. Biochem.* **121** (1982) 356–365.

[62] Strauss, M.; Bartsch, O. F.; Stollwerk, J.; Trstenjak, M.; Böhning, A.; Gassen, H. G.; Machleidt, W.; Turk, V.: Chemical Synthesis of a Gene for Human Cystatin C and Its Expression in *E. coli*; *Biol. Chem. Hoppe-Seyler* **369** (1988) 209–218.

[63] Appelhans, H.: Anlage von Genbibliotheken. In: Gassen, H. G.; Martin, A.; Bertram, S., Eds.; *Gentechnik*, Gustav Fischer Verlag, Stuttgart (1987); pp 202–215.

[64] GASSEN, H. G.; KÖHLER, S.: Chemische Synthese von DNA-Fragmenten. In: GASSEN, H. G.; MARTIN, A.; BERTRAM, S., Eds.; *Gentechnik*, Gustav Fischer Verlag, Stuttgart (1987); pp 270–293.

[65] MATSUKURA, M.; SHINOZUKA, K.; ZON, G.; MITSUYA, H.; REITZ, M.; COHEN, J. S.; BRODER, S.: Phosphorothioate Analogs of Oligodeoxynucleotides: Inhibitors of Replication and Cytopathic Effects of Human Immunodeficiency Virus; *PNAS USA* **84** (1987) 7706–7710.

[66] ZON, G.: Synthesis of Backbone-Modified DNA Analogs for Biological Application; *J. Prot. Chem.* **6** (1987) 131–145.

[67] GLOVER, D.M., Ed.; *DNA Cloning – A Practical Approach*, **Vols. I & II** (1985), **Vol. III** (1987), IRL Press, Oxford, Washington, DC.

[68] ADAMS, S. P.; GALLUPPI, G. R.: DNA Synthesis and Applications to Molecular Biology; *Med. Res. Rev.* (1986) 135–170.

[69] MULLIS, K. B.; FALOONA, F. A.: Specific Synthesis of DNA *in vitro* via a Polymerase-Catalyzed Chain Reaction; *Methods Enzymol.* **155** (1987) 335–350.

[70] HIGUCHI, R.; KRUMMEL, B.; SAIKI, R. K.: A General Method of *in vitro* Preparation and Specific Mutagenesis of DNA Fragments: Study of Protein and DNA Interactions; *Nucleic Acids Res.* **16** (1988) 7351–7367.

[71] STRAUSS, M.; STOLLWERK, J.; LENARCIC, B.; TURK, V.; JANY, K.-D.; GASSEN, H. G.: Chemical Synthesis of a Gene for Human Stefin A and Its Expression in *E. coli*; *Biol. Chem. Hoppe-Seyler* **369** (1988) 1019–1030.

Use of Mammalian Cell Cultures in Biotechnology

by M. WRANN, W. SCHEIRER, E. WASSERBAUER

Contents

Dr. M. WRANN, W. SCHEIRER, E. WASSERBAUER,
Sandoz Forschunginstitut GmbH,
Brunner Strasse 59,
A-1235 Vienna, Austria

1 Introduction

Until recently, the possibility of using mammalian and human cell lines as vehicles for the production of pharmaceutically important proteins has been viewed with skepticism. This was because, on the one hand, people felt that any protein could be produced microbially, after genetic manipulation; and, on the other hand, they considered the large-scale culture of mammalian cells either difficult to handle or too expensive.

Today we recognize advantages in mammalian cell culture, especially in the post-translational modification of protein molecules. In some cases, the process methodology can result in acceptable, reproducible, and economically satisfactory yields. The accepted guidelines in gene technology and the process development in this technology have frequently received too little attention, with the ultimate result that some products that could be better obtained from mammalian cells are produced by other means, with a loss in quality or at much greater cost.

In this chapter we analyze the problems just mentioned and present what is technically feasible at this time, and we introduce and discuss the biological as well as the industrial factors for the utilization of mammalian cell systems for biotechnology processes.

2 Biological considerations

2.1 Overview

As a result of recent successes in cell biology and molecular biology, biologically active proteins are now being produced in quantities several orders of magnitude larger than the amounts produced naturally. Cultures of normal or continuously growing animal or human cells can be induced by chemicals, viruses, or plasmid transfection to produce certain proteins at a very high level. The proteins are then isolated from the culture medium or from the cells.

At the present state of development, the highest yields on a large scale are obtained from microorganisms transformed by expression plasmids. The plasmids carry the gene for a given protein and also regulate its selective production. Proteins consist of an uninterrupted string of linked amino acids and can take on various three-dimensional conformations. Obviously, the larger is the molecular size of the protein, the more difficult is the task of using evolutionarily distant organisms to fold it into the correct three-dimensional conformation and get the right disulfide bond formation. The biological activity and immunogenicity of a protein are usually strongly influenced by correct molecular folding and some other possible modifications. This is then the overriding problem that has to be considered for each step in the selected process, be it cell growth, protein biosynthesis, isolation from the culture medium, or purification.

The biotechnology of proteins has many potential applications; in this chapter, however, we limit our discussion to the pharmaceutical sector.

2.2 Biosynthesis of proteins

A protein to be synthesized in a host organism may undergo various possible modifications. Before a project on protein production and purification is undertaken, the problems that could arise at the molecular level or during production must be evaluated. They can involve the genome, the process of translation, or the posttranslational modification, and are summarized in Table 1.

Table 1 Factors to be considered and problems that may arise at the molecular level

DNA, transcription	plasmid stability, mutation, promotion, repression
mRNA, translation	stability nontranslatable structures no splicing or faulty splicing initiation of translation Metpeptide-Metpropeptide-Metprepropeptide processing of signal peptidase stuttering
Posttranslational modifications	folding, formation of disulfide bridges methylation, acetylation, amidation phosphorylation, glycosylation proteolytic modification (necessary/unnecessary) secretion assembly of subunits

The methods required to analyze the physical and chemical properties of proteins are presented in Table 2. The properties to be assayed include biological activity, composition, molecular configuration, and product purity.

2.3 Biosynthesis of proteins in cultured mammalian cells

Some of the problems just referred to can be avoided by using animal or human cells as host cells for protein production. This technique is particularly advantageous in the production of large proteins, and, in certain cases, is the only way to produce those proteins that are posttranslationally modified in the host cell. If cells are to produce sufficient quantities of a desired protein, they must be transfected by stable plasmids, fused with other cells, or have incorporated multiple copies of a particular gene into chromosomal DNA. Most mammalian cell cultures must have serum added to the culture medium; the presence of serum proteins can complicate the purification of protein from the culture supernatant.

Since the costs of animal cell culture are still higher than the costs of a microbial system, the production rate per cell in an animal cell culture project should be high, and the pharmaceutical dose of the protein should be appropriately low. This could be the case for hormones, immunostimulants, enzymes, monoclonal antibodies, and anti-idiotypic antibodies, as well as for vaccines.

Table 2 Analytical methods applicable to the analysis of molecular biological problems

Biological activity	*in vitro*: bioassay *in vivo*: animal model
Biochemical analysis	amino acid analysis amino acid sequencing peptide mapping characterization of peptides determination of carbohydrates or methylation mass spectrometry
Molecular weight	SDS-PAGE, gel filtration ultracentrifugation, mass spectrometry
Purity	(SDS)-PAGE, combined with special staining (silver, Coomassie, immunoblotting), IEF reversed-phase HPLC, ion-exchange HPLC *N*-terminal sequence capillary electrophoresis
Aggregation	gel filtration, ultracentrifugation
Charge and microheterogeneity	IEF
Oxidation state of cysteines and localization of disulfide bridges	alkylation and peptide mapping
Folding	X-ray diffraction analysis, circular dichroism, NMR
Detection of process-specific contaminants	proteins: immunoblotting with dummy antiserum carbohydrates: carbohydrate analysis fatty acids: fatty acid analysis DNA: chemical analysis, hybridization LPS: *Limulus* lysate test

2.4 Biosynthesis of plasmid-encoded proteins in microorganisms

Because it costs little to produce microorganisms by fermentation, and because plasmid-encoded microorganisms are very capable, it is possible to produce large quantities of human proteins in *Escherichia coli* (*E. coli*), yeast, *Bacillus subtilis* (*B. subtilis*), and other host organisms. Human insulin was the first human protein to pass the approval process of the FDA (Food and Drug Administration) in the USA, and it is now being marketed successfully. Human interferons and human growth hormone are also being marketed as well as Interleukin 2, tissue plasminogen activator, Erythropoietin and colony stimulating factors; other hormones and lymphokines are in clinical trials.

Only in clinical trials can the human efficacy, toxicity, and immunogenicity be judged reliably, because the human body may recognize modifications of the protein that might escape detection by even the most up-to-date analytical methods. Because the problems listed in Table 1 cannot be predicted, experiments are imperative. Some predictions are possible, however; for example, *E. coli* is unable to synthesize complex glycoproteins, and yeast glycosylates proteins differently than do animal cells. A list of the advantages and disadvantage of various host cell systems is presented in Table 3.

Table 3 Advantages and disadvantages of various host cell systems

Host	Advantages	Disadvantages
E. coli	uniform molecular weight; no glycosylation; low cost of production (product reliable)	additional methionine at the *N*-terminus; high probability of incorrect folding and of faulty disulfide bridges
Mammalian cells	secreted proteins; correct folding and disulfide bridges; correct glycosylation	faulty glycosylation?; possibly incomplete glycosylation; heterogeneous product; contamination by DNA; cell proteins and viruses critical; relatively high cost of production
Yeasts	glycosylation faulty or absent; potential for secretion and correct folding; reliable products; inexpensive products	same disadvantages as *E. coli* for intracellular proteins; for secreted proteins: possibly incorrect folding and faulty disulfide bridges, totally different glycosylation

2.5 Methods of isolation of proteins

The method of protein isolation must be carefully tailored to the situation, depending on the cell fraction where the desired protein is being produced: the culture supernatant, the cytoplasm, or inside cellular organelles (Figure 1). The individual steps of isolation are performed on one of these fractions, using various of the methods listed in Table 4.

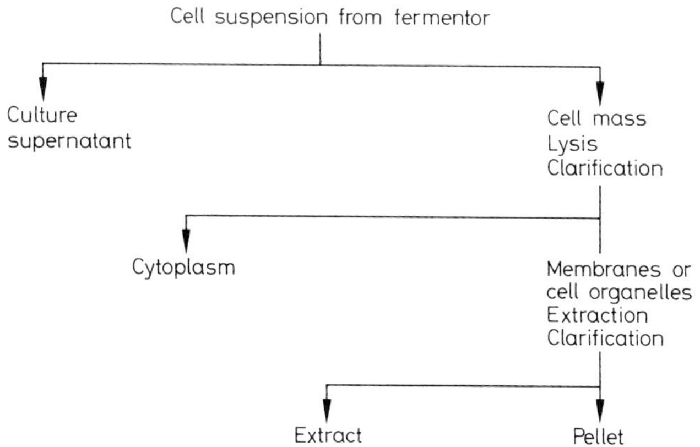

Fig. 1 Processing of fermentation liquor for the isolation of protein

Table 4 Methods and agents used in protein isolation

Cell harvesting	centrifugation, cross-flow filtration
Cell lysis	mechanical agents (bead mill, high-pressure homogenization) chemical, enzymatic agents
Clarification	centrifugation cross-flow filtration precipitation
Extraction	detergents chaotropic agents alcohol salt

2.6 Purification of proteins

2.6.1 Concentrating the protein

In most cases, the starting volume of liquid at the beginning of the purification process is large; therefore, appropriate methods to concentrate it are required. Precipitation, molecular-sieve ultrafiltration, or ion-exchange chromatography may be used at this stage to reduce the volume and to concentrate the desired protein. Moreover, the multitude of other, unwanted, proteins should be eliminated during this first purification step; proteases are of particular concern because they are a constituent of cells and are capable of digesting the desired protein. Consequently, one must be careful not to enrich the desired protein and the proteases concurrently.

After this initial purification, the material can undergo final purification with high-performance liquid chromatography (HPLC) (Table 5).

Table 5 Methods of protein purification

Concentration of protein	precipitation ultrafiltration ion-exchange chromatography affinity chromatography
Final purification	affinity ligand binding preparative HPLC or FPLC, depending on criteria molecular size – gel filtration hydrophobicity – reversed-phase chromatography charge – ion exchange chromatofocusing preparative electrophoretic methods (*e.g.*, free-flow electrophoresis)

2.6.2 Final purification

Various types of liquid chromatography are currently used for the final purification of complex proteins. Methods based on different selection criteria such as size, ionic charge, or hydrophobic nature are used together or in series to purify the protein. In many cases, affinity chromatography, the most elegant method of protein purification, is used with appropriate ligands.

Taking advantage of HPLC facilities during the preparative step will achieve highly efficient separation in the shortest time. Methods for process scale-up are already established. Electrophoretic methods, especially free-flow electrophoresis, will be of increasing importance in future.

2.7 Criteria for initiating a development project for the pharmaceutical production of a protein

Several criteria have to be met before a project passes from research into development. Cell biologists and gene technologists will initially have their data evaluated and the patent situation explored. Furthermore, the expression system to be used must be productive and stable; it must yield a stable product that can withstand fermentation and purification and will cause no problems in pharmaceutical formulation. In most cases, these are tasks for gene technology.

There must be enough initial material to conduct pilot studies on isolation, purification, and characterization. Only then can a good evaluation be made of process efficiency, pharmacological effectiveness, and the toxicity of the preparation. In addition, the process must pass all official regulatory requirements for proteins for human use. Only then can the project enter the clinical trial phase. Table 6 is a summary by scientific discipline of the activities involved in the individual project phases.

Table 6 Participation of different disciplines in specific phases of a project

	Scientific discipline	Project-associated tasks
Research	cell biology gene technology	selecting a project; selecting and testing systems for expression; doing primary characterization
	biotechnology	developing the process; producing research material
	pharmacology toxicology	establishing *in vivo* effectiveness and nontoxicity
Development	fermentation purification	producing pilot compounds; optimizing the process; producing reference compounds; writing documentation for the production facility, the analysis of the process, and the analysis of the product

3 Industrial considerations

3.1 Overview

The generation of a product that is safe and economically feasible, using either native or genetically altered mammalian cell cultures for the industrial production of proteins, requires that a series of technological, economic, and approval (regulatory) matters be considered.

At the biological level, those requisites that are involved in the industrial aspect of the process need to be defined. Because the variables and interactions of the two aspects are rather complex, all essential parameters must be approached systematically. In this manner it is possible to equalize the lesser technological expertise still inherent in mammalian culture technology with state-of-the-art microbial technology, thereby giving the project a high chance of success and permitting a meaningful comparison to microbial technology.

When all available possibilities are exploited, the production yields can be increased markedly during process development; this decreases product cost by an order of magnitude or two in comparison to the cost of production using common laboratory procedures. Mouse monoclonal antibodies are a case in point; five years ago, the yield was about 50 milligrams per liter of culture medium; today, in some cases, the yield is in the range of a gram per liter and is obtained with a culture medium that is cheap and low in protein [1, 2, 3].

3.2 Organizational requirements

If process development is to achieve the most satisfactory results, certain organizational requirements have to be in place. In addition to appropriate facilities and instrumentation, technical support staff familiar with the specialized science should be available [4]. The organization must have product standards as well as a standardized methodology for all culture and product analyses, with known accuracy and reproducibility. Furthermore, a well-characterized cell culture stock of sufficient size is necessary.

Before process development begins, a product-related development plan must be drawn up that illustrates all required steps, their sequence, their interrelationships, and the points at which decisions must be made. Based on this chart (Figure 2), personnel requirements and the time required for the development phase can be estimated quite precisely.

The characterization of the cell line and of the product are central to all subsequent steps. Once these are established, a promising system and methods for process technology can be tentatively selected.

Preliminary fermentation runs are conducted to establish the physical and chemical limits, to optimize the conditions, and to determine the nutrient requirements of the system. After a series of optimization cycles, the process can move into the pilot stage, where, in addition to being further optimized, it produces both engineering data and sufficient experimental product.

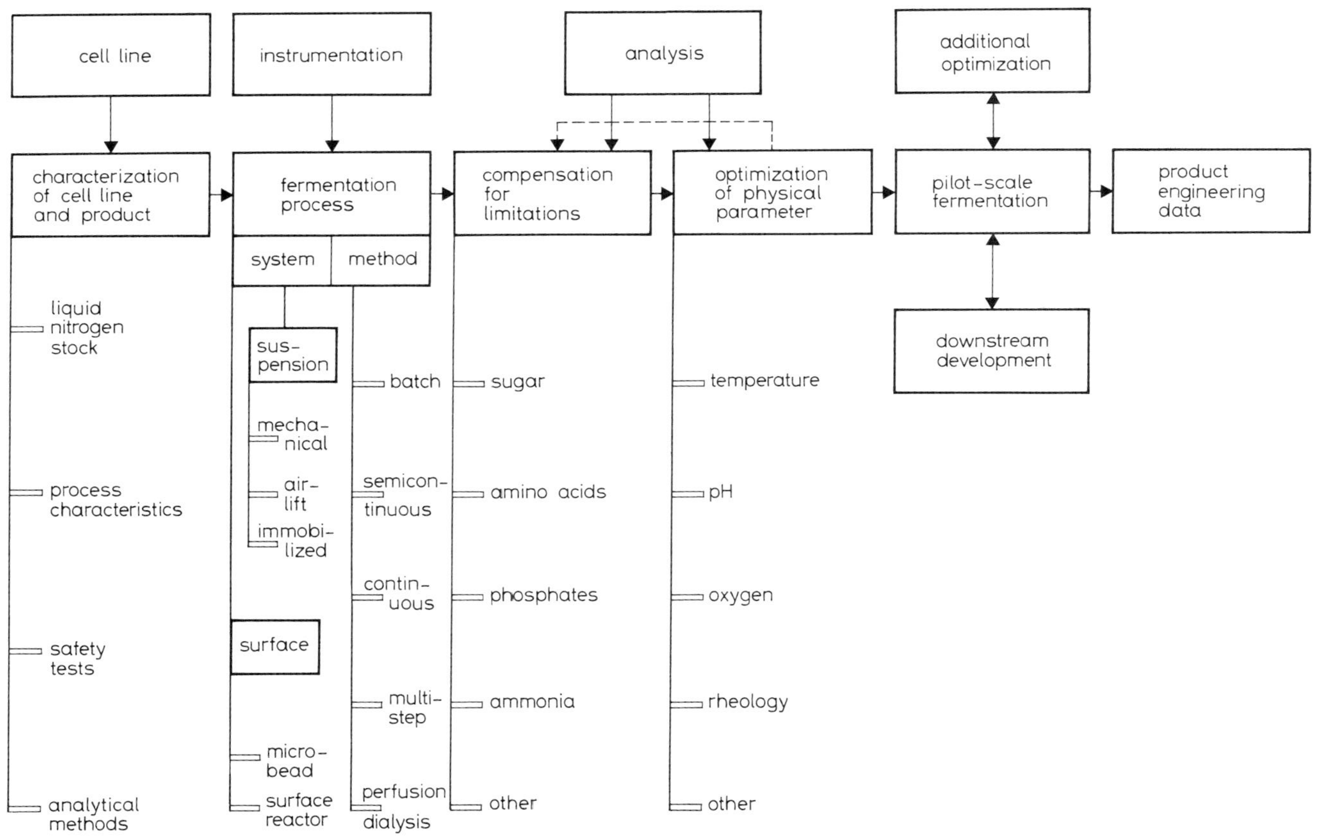

Fig. 2 General organizational plan for developing a production process using animal cell culture

The last step involves extracting product from the culture medium and purifying it, applying the procedures tested and optimized in bench-scale experiments. Optimization of product recovery has to be coordinated with process development of the fermentation, because the purification procedures and the fermentation process influence each other with respect to costs and yields. If these interactions are not taken into account, the gains in yield achieved during fermentation optimization can be forfeited by a poorly performing purification process.

3.3 Product characteristics associated with the process

Table 7 is a list of the essential data necessary for process development. In addition to the characterization of the product and the test systems mentioned earlier, the product stability under the various conditions arising during the process must be carefully established. Of particular concern in this regard are the sensitivity of the product towards cellular enzymes and its stability at different process temperatures in the culture medium.

Once these data are known, the limiting values for operationally conducting the fermentation can be derived; these include retention time, physical conditions in the fermentor, maximum cycling time in the fermentor, use of membrane filters, and possible protective substances added to the medium. Incremental improvements can be calculated from deviations and from data on reproducibility of the test system; this has to be accomplished with a single optimization step in order to be clearly detectable in the test system.

Table 7 Product characteristics associated with the process

Standard analytical system(s)
relevance
precision
reproducibility
Biochemical characteristics
Biological characteristics
Stability
culture requirements
culture medium
storage requirements
Standards (reference products)
Special properties, such as
special sensitivity
special hazards, etc.

Table 8 Cell characteristics that depend on the culture system

Surface dependence
Mechanical sensitivity
Genetic stability
Cell size and size distribution
Requirement for essential process steps

Table 9 Cell characteristics that depend on the method of culture

Phenotypic uniformity
Maximum culture duration
Minimum doubling time
Possible key culture parameters
Product generation characteristics
Maximum product production rate

3.4 Cell characteristics that depend on the culture system

The properties of a cell line significantly affect the selection of the culture system as well as the culture method employed. The most important properties that influence the choice of the culture system are discussed in the following and are also listed in Table 8.

The essential criterion for choosing a suitable culture system is the degree to which the culture depends on surfaces; the culture habit can range from strongly surface dependent growth to a suspension of single cells, with several stages in between. However, because surface-culture systems have significant disadvantages in operation and scale-up, a surface-adjusted cell line should be examined for adaptability to suspension growth. A change in growth habit could, however, entail a detrimental change in production behavior; this must not be ignored when evaluating the adaptability of a cell line to suspension growth.

The mechanical sensitivity of different cell lines, the sedimentation behavior, and the size of cell aggregates all vary; they all influence the hydrodynamic design, the aeration systems of a reactor, and the scale-up. There is little information on how to determine the mechanical sensitivity of cells [5, 6, 7] and about the dynamic conditions within different cell culture reactors [8, 9]. Because the scale-up behavior varies from one culture system to another, each cell line needs to be experimentally evaluated for a given system [10].

Genetically altered cells may be subject to chromosomal instability, resulting in the loss of a chromosome or part of a chromosome, and in either case the possible loss of the genes for the desired product. In the absence of selection pressure, the culture is usually rapidly overgrown by nonproductive cell populations. It is clearly preferable to improve the cell material *per se* and thus eliminate this difficulty, but it may also be possible to minimize the problem through process technology. If this is the approach, the maximum production cycle time must be determined before start-up of a new culture run. Carefully selected retention systems can prolong the productive time; on the one hand, extreme culture conditions can constitute an added selection pressure in favor of a productive population; on the other hand, the reduction in growth rate in a retention system can be regulated without culture washout by simply lowering the temperature or other measures. Since loss of chromosomes is a function of growth rate and total number of cell divisions, an acceptable production process can be established even for labile cell lines by selecting an appropriate culture system.

Cell size and cell size distribution are parameters essential to the design of a retention-type facility. Cell size can also be used as a check during fermentation, since cell size, mechanical influences, osmotic pressure, and physiological condition are interrelated [11].

If the product is constitutively secreted, it can be recovered from the cell culture supernatant directly, somtimes secretion has to be induced, or alternatively it is recovered by processing the cell mass; monoclonal antibodies are a typical example of a secretion product, and tumor antigens are primarily cell-bound. Depending on the type of production desired, a single-step process, a multistep procedure, or a continuous system can be selected.

3.5 Cell characteristics that depend on the method of culture

Following the selection of a culture system, a review of fermentation methods must be undertaken by analyzing the cell characteristics outlined in Table 9.

When we analyze the phenotypic uniformity of a given cell population for productivity, we get clues for improving the productivity within the framework of physical and chemical process optimization, and for determining the production cycle. If cells are not uniform, we may find a relationship between cell cycle, growth rate, and physiological parameters, and we may then use a fermentation method that compensates for the nonuniformity.

Maximum culture periods must be established in order to pass pharmaceutical approval processes; they are established from the changes in karyotype and phenotype determined during continuous culture. As a result, limits on the operating time of the process are set according to accepted safety guidelines.

The ratio between the actual growth rate and the maximum possible growth rate measured experimentally for a given culture can serve as a conservative criterion for evaluating the quality of the system and the method. If a reduction in growth rate appears useful it can also be employed as a regulatory parameter. Since the specific content of ATP is correlated with the growth rate, it can serve as the unit of measure [12].

Existing parameters that may be suitable to serve as key process parameters may be integrated into the methodology once their specificity has been established. One example is lactic dehydrogenase (LDH) liberated during replication of foot-and-mouth disease viruses in host cells, an effect used in determining the optimal harvest time of the virus during vaccine production [13]; another is the rate of glucose uptake by hybridoma cells, which can, in a few cases, be correlated with the specific production rate of monoclonal antibodies.

Yet another parameter of process mechanics is the specific production characteristic of each cell line. This characteristic is a function of productivity and growth rate. According to MERTEN [14], there are three principal types of production characteristics:

A. production primarily during the stationary phase,
B. production as a function of growth rate (best production at maximum growth rate, no production in stationary phase), and, lastly,
C. production during both the growth and the stationary phases.

Obviously, culture conditions must be tailored as closely as possible to these biological characteristics. If we look at culture methods in light of their generally accepted differences [15, 16], the individual requirements are as follows.

- A batch culture will be suitable for all three types of production because all growth rates are expressed (Figure 3a). It will not be optimal for any of the production characteristics, however, because the culture conditions are optimal only during a relatively short phase, and subsequently nutrient limitations as well as feedback regulation became noticeable.
- A continuous culture (Figure 3b) is not appropriate for type A production because the growth rate remains relatively high in chemostat cultures. However, continuous

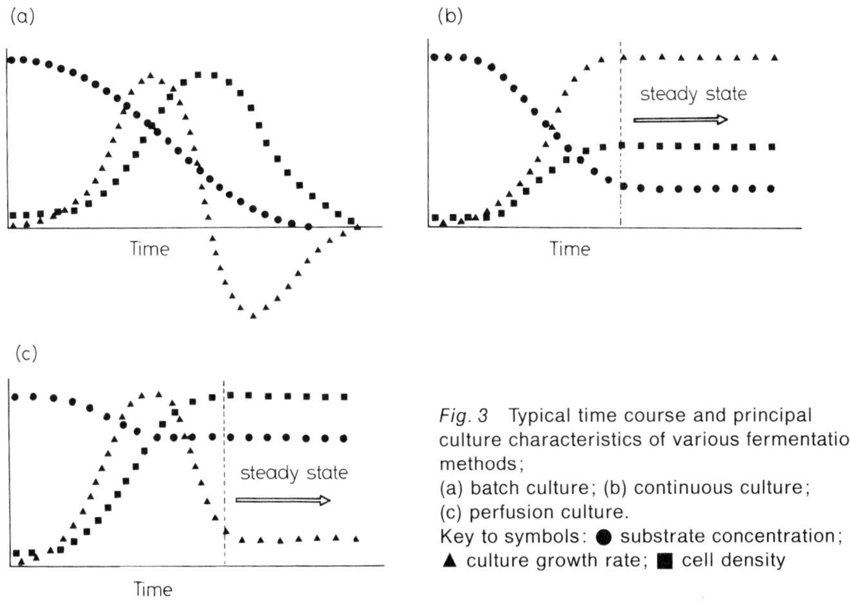

Fig. 3 Typical time course and principal culture characteristics of various fermentation methods;
(a) batch culture; (b) continuous culture;
(c) perfusion culture.
Key to symbols: ● substrate concentration; ▲ culture growth rate; ■ cell density

culture seems very appropriate for production types B and C, since they could be carried out in a turbidostat or chemostat with maximum productivity.

- In the perfusion system (Figure 3c) there are various ways by which growth rate and cell density can be freely monitored, without the need for chemostat control. As a result, optimal adaptability to a given production characteristic is assured (with the exception of type A production, which requires that the growth rate be near zero). This method can function at very high cell densities, which reduces costs for nutrient media [17]; also, because retention time in the reactor is inversely proportional to cell density, there is less product inactivation in the reactor.

4 Conclusion

Of course, mammalian cell cultures are not the sole or optimal production system for pharmaceutically useful proteins. Contrariwise, they should not be dismissed *a priori* as an impractical and expensive system that must be avoided under any circumstances. Rather, an evaluation must be undertaken at various levels of decision-making with full consideration of all parameters; the evaluation will subsequently serve as the basis for a final decision regarding the choice of a production system.

Some criteria, such as the need for glycosylation, particularly the need for correct glycosylation, or the possibility of incorrect disulfide bondings can sometimes be predicted. In most cases, however, including molecular biology, the possible differences between microbial and mammalian expression systems must be investigated experimentally in order to reach a decision. Each system intended for industrial production must be analyzed through experiments, because cells change their characteristics as a function of the product and the expression system. A clear grasp of the fundamentals and the commercial systems and processes will be obtained if all steps described in Section 3 receive full observation and evaluation.

We have recently seen how efficient fermentation and industrial protein purification can be controlled. Current product yields in the range of grams per liter are being reported [1, 2, 3], including costs of culture media comparable to those of microbial fermentations (a few German marks per liter) when low-serum or serum-free media are used. Because yields from cell culture material after purification are usually much greater than those obtained with microbial systems, typically between 20% and 50%, a product generated with eukaryotic cell culture systems can be less expensive than one produced with microorganisms.

It is therefore advantageous, when choosing a new product, to define the molecular biological aspects experimentally and also to adapt the appropriate strains to the technological process. A final decision on the production system that yields a product of pharmacological value equal to the native material must be based on feasibility and market analyses of the purified product produced under optimal process conditions.

5 References

[1] TAKAZAWA, Y.; TOKASHIKI, M.; HAMAMOTO, K.; MURAKAMI, H.: High Density Perfusion Culture of Hybridomas Recycling High Molecular Weight Components; *Cytotechnology* **1** (1988) 171.

[2] VELEZ, D.; REUVENY, S.; MILLER, L.; MACMILLAN, J. D.: Effect of Feeding Rate on Monoclonal Antibody Production in a Modified Perfusion-Fed Fermentor; *J. Immunol. Methods* **102** (1987) 275.

[3] SCHOENHERR, O. T.; VAN GELDER, P. T. J. A.: Culture of Animal Cells in Hollow-Fibre Dialysis Systems. In: SPIER, R. E.; GRIFFITHS, J. B., Eds.; *Animal Cell Biotechnology*, **Vol. 3**, Academic Press, London, Orlando (1988); p 337.

[4] SCHEIRER, W.: Laboratory Management of Animal Cell Culture Processes; *Trends in Biotechnology* **5** (1987) 261.

[5] NEREM, R. M.: Shear Stress Effects on Anchorage-Dependent Mammalian Cells; *Engineering Foundation Conference on Cell Culture Engineering*, Palm Coast, FL, Feb. 1988.

[6] PETERSEN, J. F.; KUNAS, K. T.; CHERRY, R. S; PAPOUTSAKIS, E. T.: Shear and Other Hydrodynamic Effects on Cultured Animal Cells; *J. Biotechnol.* **7** (1988) 229.

[7] KRETZMER, G.; JÄMMRICH, U.; SCHÜGERL, K.: Shear Stress Effects on Anchorage-Dependent and Suspended Mammalian Cells. In: SPIER, R. E.; GRIFFITHS, J. B.; STEPHENNE, J.; CROOY, P. J.; *Advances in Animal Cell Biology and Technology for Bioprocesses;* 9th Meeting of the European Society of Animal Cell Technology (ESACT), (Knokke, Belgium, 1988), Butterworth & Co., UK (1989); pp 172–174.

[8] SMITH, C. G.; GREENFIELD, P. F.; RANDERSON, D. H.: Growth and Productivity of Hybridomas in Turbulent Shear Flow. Oral Presentation at 9th Meeting of the European Society of Animal Cell Technology, "Advances in Animal Cell Biology and Technology for Bioprocesses", Knokke, Belgium, Sept. 1988.

[9] Moo-Young, M.; Christy, Y.: Bioreactor Design for Aeration of Shear-Sensitive Fermentation Cultures. In: Durand, G.; Bobichon, L.; Florent, J., Eds.; *Proceedings of 8th International Biotechnology Symposium*, Paris, 1988, Société Française de Microbiologie.

[10] Bliem, R.; Katinger, H.: Scale-Up Engineering in Animal Cell Technology; *Trends in Biotechnology* **6** (1988) Part I: pp. 190–195, Part II: pp. 224–230.

[11] Oyaas, K.; Berg, T. M.; Bakke, O.; Levine, D. W.: Hybridoma Growth and Antibody Production under Conditions of Hyperosmotic Stress. In: Spier, R. E.; Griffiths, J. B.; Stephenne, J.; Crooy, P. J.; *Advances in Animal Cell Biology and Technology for Bioprocesses;* 9th Meeting of the European Society of Animal Cell Technology (ESACT), (Knokke, Belgium, 1988), Butterworth & Co., UK (1989); pp 212–220.

[12] Scheirer, W.; Merten, O. W.: Instrumentation of Animal Cell Bioreactors. In: Ho, C.; Wang, D. I. C., Eds.; *Animal Cell Bioreactors*, Butterworths, Sevenoaks, Kent, UK (1991); p 409.

[13] Spier, R. E.: Determination of the Time to Harvest Foot-and-Mouth Disease Virus Cultures by Measurements of the Supernatant Concentration of Lactic Dehydrogenase; *Biotechnol. Bioeng.* **19** (1977) 929.

[14] Merten, O. W.: Batch Production and Growth Kinetics of Hybridomas; *Cytotechnology* **1** (1988) 113.

[15] Katinger, H.; Scheirer, W.: Mass Cultivation and Production of Animal Cells. In: Spier, R. E.; Griffiths, J. B., Eds.; *Animal Cell Biotechnology*, **Vol. 1**, Academic Press, London, Orlando (1985); p 167.

[16] Scheirer, W.: High Density Growth of Animal Cells within Cell Retention Fermenters Equipped with Membranes. In: Spier, R. E.; Griffiths, J. B., Eds.; *Animal Cell Biotechnology,* **Vol. 3**, Academic Press, London, Orlando (1988); p 263.

[17] van der Velden-De Groot, T.; Witterland, W.; Beavery, E. C.; van Wezel, T. L.: Evaluation of the Continuous Perfusion Culture System for the Production of Monoclonal Antibodies. In: Spier, R. E.; Griffiths, J. B., Eds.; *Modern Approaches to Animal Cell Technology*, Butterworths, Sevenoaks, Kent, UK (1987); p 513.

Industrial and Technical Information

Codes and Regulations

Genetic Engineering in Japan: Regulations, Expenditures, and Results*

1 General situation

Japan's efforts to develop genetic engineering began later than did those of the USA. In 1980, 75% of all the genetic engineering patents registered in Japan originated from abroad, predominantly from the USA [1].

Since then, however, genetic engineering methods have been given high priority and have been actively promoted. The Research Association for Biotechnology, to which 14 Japanese companies and some major Research Institutes belong, was founded in 1981 under the aegis of the Ministry of International Trade and Industry (MITI); it has, for example, an ambitious program ten-year for the industrial exploitation of recombinant microorganisms, and of animal cell lines in mass culture; this plan has largely been implemented (Table 1).

Table 1 Genetic engineering portion of the program of the Research Association for Biotechnology

Company/Institute	Goals
Sumitomo Chemical Co.	Hydroxylation of steroids by hepatic P450 monooxygenase, cloned in yeast
Mitsui Toatsu Co.	*Bacillus* host-vector system to produce neutral protease
Mitsubishi Kasei Institute of Life Sciences	Secretion vectors for yeasts (*Saccharomyces* and *Kluyveromyces*); expression of human nerve growth factor (NGF)
Fermentation Research Institute	Host-vector system for thermophiles (*Thermus*)
Research Institute for Polymers and Textiles	Constructs with the dihydrofolate reductase gene

The number of genetic engineering experiments in industry and in the universities continues to grow [2], as shown in Table 2.

Japan's activities in this sector far surpass those in the Federal Republic of Germany both in number and in scope [3]. According to official statistics for 1987, 4934 Japanese scientists are working on genetic engineering experiments (Table 3). Comparable figures for Germany are not available, but must be substantially less than this number [4].

The result of this concerted effort is that by 1985 half of all recombinant patents registered in Japan were of Japanese origin [5].

* Prof. Dr. Rolf D. Schmid,
Gesellschaft für Biotechnologische Forschung mbH (GBF),
3300 Braunschweig, Fed. Rep. Germany

Table 2 Genetic engineering experiments in Japan

Number of experiments	1984 <201	1984 >201	1985 <201	1985 >201	1986 <201	1986 >201
National institutes	87	–	125	–	154	–
Other institutes	82	–	79	–	98	–
Technical experts	75	3	111	1	104	3
Patent agents	29	–	54	–	50	–
Industry	456	19	650	34	832	44
Universities	1858	–	2398	–	3324	–
Total	2578	22	3417	35	4562	47
Number involving animal and plant cell cultures					537	3

Table 3 Genetic engineering in Japan: expenditures (in billions of yen), and numbers of researchers

Year	Total	Industry	Research institutes	Universities
Expenditure (billions of yen)				
1981	9.4	5.8	1.2	2.4
1982	12.8	7.9	1.9	3.0
1983	19.9	12.9	2.6	4.3
1984	24.7	13.7	4.5	6.5
1985	36.2	19.3	8.8	8.1
1986	43.7	25.8	8.7	9.8
1987	50.4	26.8	11.3	12.3
Number of people working in research				
1987	4.934	1.768	657	2.509
1988	5,388	1,742	799	2,847

2 Regulations relating to genetic engineering

The structures of science and of the economy are not the same in centrally governed Japan and the federal structure of Germany. Accordingly, there are more directives regulating genetic engineering experiments in Japan than Germany; these are, however, well coordinated. The most important of these regulations, listed according to the Ministry responsible, are the following:

1. Science Technology Agency (STA) (Department in the Prime Minister's office responsible for central research projects and coordination of research among the other ministries)
   Guideline for the application of recombinant DNA to experiments (first issued 28. 8. 1979; the current version is dated 16. 9. 1987)
   The directives contain provisions relating both to materials and methods. There are no penalty provisions or regulations in the event of an infringement. Some types of experiment are subject to special application or approval procedures with more stringent requirements. The STA has the immediate and final decision on such applications, without the right to legal appeal. Only in doubtful cases is the decision-making power vested in a special committee of the Council for Science and Technology, which reports directly to the Prime Minister. This special committee (10 permanent members and 5 delegates from various ministries) issues and amends the guidelines for the application of rDNA techniques, implements approval procedures for special applications, and coordinates special tasks.
2. Ministry of Education (responsible for 150 state and approximately 350 private universities)
   Guideline for the application of recombinant DNA to experiments at research institutes of universities, etc. (dated 1979; the most recent version is dated 24. 8. 1985)
   The guidelines relate to university research institutes, and correspond closely to the STA guidelines in terms of content and procedure.
3. Ministry of Health and Welfare (responsible for the release of pharmaceutical products, important areas in government research in the area of medicine and health, and the hospitals in Japan)
   Guideline for the production of medication based on recombinant DNA technology (most recent version dated 11. 12. 1986)
   Although the regulations are not legally binding, they have the *de facto* quality of administrative orders in applications of the Pharmaceutical Law.
   The Ministry of Health is generally considered a strict supervisory authority. It has legal and control procedures, regulated in detail, which are also applied to genetic engineering on the basis of the Pharmaceutical Law of 1960 in its 1983 version.
4. Ministry of International Trade and Industry (MITI) (responsible for trade and industry in general)
   Guidelines for industrial applications of recombinant technology (dated July 1986; most recent version dated 11. 12. 1986)
   This guideline will be discussed in more detail below. It has the following outline:

1 General
1.1 Purpose
1.2 Definitions
2 Evaluation of safety of the recombinant strain
2.1 Regulations
2.2 Details
2.2.1 Host organism
(taxonomy, genetic characterization, pathogenic and physiological properties, prior history)
2.2.2 rDNA
(structure, design methods)

2.2.3 Properties of DNA donor and vector
(taxonomy, pathogenic and physiological properties, recombinants)
2.3 Safety evaluation and classification
2.3.1 Safety evaluation and classification of the recipient (GILSP, Category 1, Category 2, Category 3)
2.3.2 Safety evaluation and classification of the recombinant organism
3 Equipment, apparatus, operation, and management
3.1 Regulations
3.2 Equipment and apparatus
3.3 Operation and management

In conformity with the 1986 Recommendations of the OECD, four levels of safety are distinguished ("OECD Recombinant DNA Safety Considerations"):

1. Good Industrial Large-Scale Practice (GILSP)
   A recipient organism should be nonpathogenic; it should contain no adventitious agents such as pathogenic viruses, phages, and plasmids; and it should have either a long history of safe industrial use, or built-in environmental limitations that permit optimum growth in an industrial setting but limit survival without adverse consequences in the environment.
2. Category 1
   This category includes nonpathogenic recipient organisms not included in the above GILSP.
3. Category 2
   A recipient organism in this category has undeniable pathogenicity for humans and might cause infection if handled directly. However, the infection will probably not result in a serious outbreak in those instances where effective preventive and therapeutic methods are known.
4. Category 3
   A recipient organism in this category is capable of causing disease and is not included in Category 2 above. It is to be handled with care, but there are known effective preventive and therapeutic methods for the disease. A recipient organism which, whether directly handled or not, might be significantly harmful to human health and result in a disease for which no effective preventive nor therapeutic method is known, shall be assigned a classification separate from Category 3 and be treated in a special manner.

In the GILSP category, a "semi-closed system" is the only requirement for protection of the environment, but releases should be kept to a minimum.

The MITI is also preparing guidelines for the safety of equipment used in genetic engineering processes. Some of these will become industrial standards (JIS Standards), with financial incentives to encourage their observance. Methods for evaluating the safety of industrial plants will also be developed in accordance with the OECD recommendations (on such matters as aerosol formation, tightness of filters). The guidelines are expected to be issued in 1990.

Since the autumn of 1987, uniform application procedures to the MITI and the Health Ministry have been in place for the industrial production of recombinant materials to be marketed as either industrial reagents or medicaments. Responsibility is determined in accordance with a catalog.

5. Ministry of Agriculture, Forestry, and Fishery (MAFF)
   (responsible for the sectors mentioned, the food industry, and the release of recombinant plants)
   Guideline for the application of recombinant DNA to agriculture, forestry and fishery (announced on 18. 12. 1986; projected date of publication was autumn 1988, postponed to beginning of 1989)
   The National Institute for Agro-Environmental Studies in Tsukuba has built a special greenhouse for experiments with recombinant plants, in which the plants must be studied from the standpoint of safety; the results must be presented to the Ministry for its approval before the plants may be released.

Overall, the safety requirements for genetic engineering in Japan are comparable to the standard of the USA and less stringent than those in the Federal Republic of Germany. Essentially, a legal licensing and control procedure exists only within the framework of the "Pharmaceutical Law" operated by the Health Ministry. There are no plans for a uniform monitoring law. The existing guidelines have been revised a number of times, always in the direction of becoming more relaxed. The Japanese regulations are, however, complex in that the method by which approval is obtained depends on the use to which the recombinant product is to be put.

For example, approval from the MITI is sufficient for marketing recombinant reagents and industrial reagents, but application must be made to the Health Ministry for the release of recombinant pharmaceutical products.

Approval by the Agricultural Ministry (MAFF) alone is not sufficient for foodstuffs manufactured with recombinant microorganisms; the consumer organizations must also declare their agreement.

3 P-3 and P-4 laboratories

There are currently around twenty P-3 laboratories operating in the public domain in Japan. The RIKEN Institute, which belongs to the STA, maintains a laboratory in Tsukuba in which all four safety levels have been installed (Figure 1 [7]). The two-story building has a total floor area of 2500 m^2 and houses two P-4, two P-3, four P-2, and three P-1 laboratories. The two P-4 laboratories are equipped with special safety extractor hoods containing incubators, centrifuges, animal cages, and so forth (Figure 2).

4 Recombinant products and their markets

Table 4 is a list of the recombinant methods currently permitted in Japan. The number of approved methods rises continually, and the number of products has increased even further since the middle of 1988.

The market for recombinant and cell culture products in Japan has reached some 77 billion yen (approximately DM 1.1 billion), categorized as in Table 5.

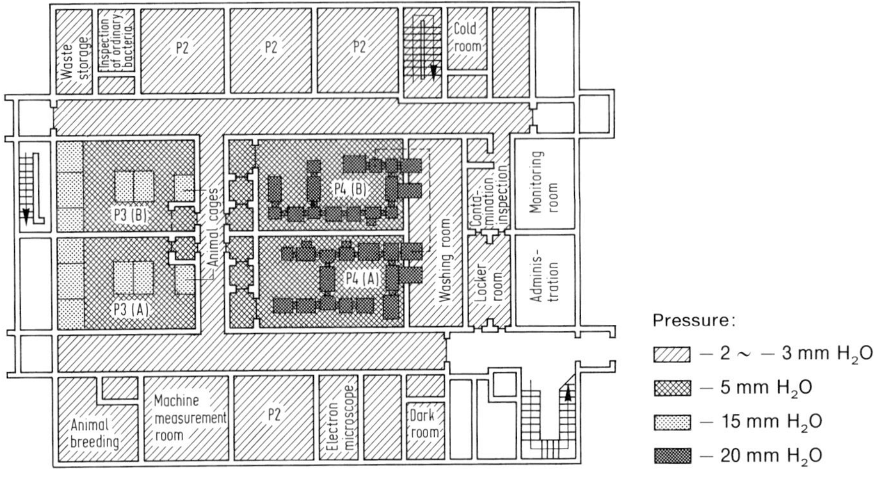

Fig. 1 Institute for Physical and Chemical Research (RIKEN), Life Science Institute in Tsukuba (Japan): floor plan with 4 P-2, 2 P-3, and 2 P-4 laboratories

Fig. 2 Institute for Physical and Chemical Research (RIKEN), Life Science Institute in Tsukuba (Japan): view inside one of the two P-4 laboratories

Table 4 Recombinant DNA methods approved for industrial use in Japan (up to May 1988)

Company	Product	Use	Host	Source of the DNA	Safety level
Approval on 25. 3. 1987					
Toyo Jozo	pyruvate oxidase	diagnostics	*E. coli* K 12	aerobic, Gram + cocci	GILSP
Kirin Beer	creatinase	diagnostics	*E. coli* K 12	aerobic, Gram + cocci	GILSP
	erythropoietin	therapeutics	CHOC	h-liver cells	GILSP
Chugai Pharma	erythropoietin	therapeutics	CHOC	h-liver cells	GILSP
Toyobo Biotech	tPA	therapeutics	MRIII	h-uterus cells	Categ. 1
Approval on 15. 5. 1987					
Chugai Pharma	granulocyte CSF	therapeutics	CHOC	h-oral epithelium	GILSP
Kyowa Hakko	h-IFN-β	therapeutics	*E. coli* K 12	h-fibroblasts	GILSP
	h-IFN-γ	therapeutics	*E. coli* K 12	h-lymphocytes	GILSP
Asahi Chemicals	h-TNF	therapeutics	*E. coli* K 12	h-liver cells	GILSP
Meiji Seika	h-INF-γ	therapeutics	*E. coli* K 12	h-lymphocytes	GILSP
Dainippon Pharma	h-TNF	therapeutics	*E. coli* K 12	h-macrophages	GILSP
Approval on 11. 8. 1987					
Fujisawa Pharma	somatomedin C	therapeutics	*E. coli* K 12	chem. synthesis	Categ. 1
Suntory	h-IFN-γ	therapeutics	*E. coli* K 12	chem. synthesis	GILSP
Nippon Roche	h-IFN-γ	therapeutics	*E. coli* K 12	h-periph. lymphocytes	GILSP
	h-IFN-α-2α	therapeutics	*E. coli* K 12	h-bone marrow cells	
Takeda	h-IFN-	therapeutics	*E. coli* K 12	h-periph. lymphocytes	GILSP
	h-IFN-α-α	therapeutics	*E. coli* K 12	h-bone marrow cells	GILSP
	h-IL-2	therapeutics	*E. coli* K 12	h-periph. lymphocytes	GILSP
Mitsubishi Chem.	HB vaccine	vaccine	CHOC	HB virus	GILSP
Wakunaga Pharm	h-EGF	therapeutics	*E. coli* K 12	chem. synthesis	GILSP
Kirin Beer	granulocyte CSF	therapeutics	*E. coli* K 12	chem. synthesis	GILSP

(continued on next page)

Table 4 (cont.)

Approval on 17. 11. 1987					
Earth Pharm.	h-IL-1 derivative	therapeutics	*E. coli* K 12	h-periph. lymphocytes	GILSP
Ohtsuka Pharm.	h-IL-1 derivative	therapeutics	*E. coli* K 12	h-periph. lymphocytes	GILSP
Toyobo Biotech	erythropoietin	therapeutics	MRIII	h-liver cells	GILSP
Toyo Jozo	sarcosine oxidase	diagnostics	*E. coli* K 12	aerobic Gram + rods	GILSP
Sumitomo Pharm	tPA	therapeutics	CHOC	h-epidermal cells	GILSP
Nihon Chemicals	h-superoxide dismutase	therapeutics	*E. coli* K 12	h-placenta	GILSP
Chemical and Serum Therapy Center	HB antibodies	diagnostics	bakers' yeast	HB virus	GILSP
Takeda	HB vaccine	vaccine	bakers' yeast	HB virus	GILSP
Chemical and Serum Therapy Center	HB vaccine	vaccine	bakers' yeast	HB virus	GILSP
Approval on 16. 2. 1988					
Midori Juji	HB vaccine	vaccine	bakers' yeast	HB virus	GILSP
Unitika	leucine dehydrogenase	diagnostics	*E. coli* K 12	aerobic, Gram + rods	GILSP
Kikkoman Soy Sauce	creatinase	diagnostics	*E. coli* K 12	aerobic, Gram + rods	GILSP
	sarcosine oxidase	diagnostics	*E. coli* K 12	aerobic, Gram + rods	GILSP
Shionogi Pharma	h-IL-2	therapeutics	*E. coli* K 12	h-spleen cells	GILSP
	h-PSTI	therapeutics	bakers' yeast	h-pancreas cells	GILSP
Kyowa Hakko	CSF	therapeutics	*E. coli* K 12	h-macrophages	GILSP
Approval on 16. 5. 1988					
Asahi Chemicals	cytidine 5′-diphosphat-idylcholine	therapeutics	bakers' yeast	bakers' yeast	GILSP
Fujisawa Pharm	somatomedin C	therapeutics	*E. coli* K 12	chem. synthesis	GILSP
Sumitomo Pharm	tPA	therapeutics	CHOC	h-epidermal cells	GILSP
Suntory	h-sodium diuretic peptide	therapeutics	*E. coli* K 12	h-heart cells	GILSP
	h-IFN-γ	therapeutics	*E. coli* K 12	chem. synthesis	GILSP

KEY: CHOC, Chinese Hamster Ovary Cells; CSF, Colony-Stimulating Factor; EGF, Epidermal Growth Factor; GILSP, Good Industrial Large-Scale Practice; HB, Hepatitis B; IL, Interleukin; IFN, Interferon; MRIII, Mouse-R-III derived cell line; PSTI, Pancreatic Serum Trypsin Inhibitor; TNF, Tumor Necrosis Factor; tPA, Tissue Plasminogen Activator; h, human

Table 5 Market figures for recombinant and cell culture products in Japan (in billions of yen)

Technology	Sales, billions of yen	
	1987	1988
Recombinant products	11.0	27.6
Products from fused cells	25.0	34.4
Products from cell cultures	13.3	15.0
Total	49.3	77.0

[1] At the end of 1988, the exchange rate for 100 yen was DM 1.40

Acknowledgment

I thank Mrs. A. KAWASHIMA for her tireless help in obtaining documents in Japan, and the Fonds der Chemischen Industrie (FCI) for financial support.

5 References

[1] *Biotechnology in Japan, The Reference Source*, Science Forum Inc., Palo Alto (1984).

[2] *Nikkei Biotech.*, 24.5.87, p 14.

[3] Statistics Office, Management and Coordination Agency, Tokyo, December 1988.

[4] RauCon, Heidelberg, personal communication.

[5] SCHMID, R. D.: *Appl. Microbiol. Biotechnol.* **24** (1986) 355–365.

[6] Guideline for the Industrial Application of Recombinant DNA Technology, provisional translation, Ministry of International Trade and Industry, Tokyo.

[7] *Science and Technology in Japan*, Tokyo, February 1989, p 37.

[8] *BIDEC Annual 1988 (Jap.)*, Biotechnology Development Corporation, Tokyo, pp 214–216.

[9] Bio-Intelligence; *Nikkei Biotech.*, 16. 1. 1989, p 1.

[10] SCHMID, R. D.: *Biotechnology in Japan – A Comprehensive Guide*, Springer, Heidelberg, New York (1991).

Opportunities and Risks of Gene Technology[1]

Explanatory comments

This position paper, issued by the Biotechnology Study Group, responds to the report ***Opportunities and Risks of Gene Technology****, issued by the Enquete-Kommission (Inquiry Commission) of the German Parliament. The group has limited itself to statements it considered to be of special importance. It has not only provided critical comments, but has also willingly offered to cooperate on matters of safety in gene technology. The study group and its member associations are ready to cooperate and to act on the suggestions of the Inquiry Commission. Various measures have been suggested; appropriate responses are in progress.*

Maximum possible safety in the use of genetic engineering methods in biotechnology can be guaranteed only when scientific experts cooperate closely with legislative and regulatory bodies. The following positions should be understood in light of the above.

Scientific societies whose areas of interest include parts of the interdisciplinary field of biotechnology have joined to form the *Biotechnology Study Group* (*AG BioT* is the German abbreviation). AG BioT includes the following member associations (*e.V.* is the German abbreviation for "registered society"):

- **Study group on Diagnostics, Düsseldorf** (AGD – Arbeitsgemeinschaft für Gen-Diagnostik e.V., Düsseldorf);
- **German Society for Clinical Apparatus, Clinical Technique and Biotechnology, Frankfurt am Main** (DECHEMA – Deutsche Gesellschaft für Chemisches Apparatewesen, Chemische Technik und Biotechnologie e.V., Frankfurt am Main);
- **German Bunsen Society for Physical Chemistry, Frankfurt am Main** (Deutsche Bunsen-Gesellschaft für Physikalische Chemie, Frankfurt am Main);
- **German Society for Hygiene and Microbiology, Bonn** (Deutsche Gesellschaft für Hygiene und Mikrobiologie, Bonn);
- **Society for Biological Chemistry, Frankfurt am Main** (Gesellschaft für Biologische Chemie, Frankfurt am Main);
- **Society of German Chemists (SGC), Microbiology Study Group of the SGC Division of Food Chemistry and Forensic Chemistry, Frankfurt am Main** (Gesellschaft Deutscher Chemiker, AG Mikrobiologie der GDCh-Fachgruppe "Lebensmittelchemie und gerichtliche Chemie", Frankfurt am Main);
- **Society of German Engineers, Division of Process Engineering and Chemical Engineering, Düsseldorf** (VDI, Gesellschaft Verfahrenstechnik und Chemieingenieurwesen, Düsseldorf)[2];
- **Coal Mining Association, Essen** (Steinkohlenbergbauverein, Essen);
- **Association for General and Applied Microbiology, Göttingen** (VAAM-Vereinigung für Allgemeine und Angewandte Mikrobiologie, Göttingen);

---

[1] Comments and constructive suggestions of the Biotechnology Study Group on the subject of the report of the Enquete (Inquiry) Commission of the German Parliament

[2] The Society of German Engineers (VDI), Division of Process Engineering and Chemical Engineering, has waived the right to send two representatives to the commission, because questions on bioprocess technology have not been addressed

- **Research and Teaching College for Alcohol Production and Fermentation Technology, Berlin** (*corresponding member*) (Versuchs- und Lehranstalt für Spiritusfabrikation und Fermentationstechnologie, Berlin).

A committee of two representatives from each member association, in addition to four personally invited colleagues from a given field, has reviewed the report of the Inquiry Commission of the German Federal Parliament on **Opportunities and Risks of Gene Technology**. This committee kept two considerations in mind:

- Its comments on the recommendations of the report may be of great importance to the future development of gene technology, and to biotechnology as a whole.
- Both constructive suggestions and newly introduced knowledge will enable the member associations to enhance the potential of their members for the positive development of biotechnology.

AG BioT feels it is not qualified to address problems associated with human reproductive technology, and consequently advocates a scientific and a legal distinction between, firstly, human reproductive technology (which is included in the guidelines for embryo protection) and, secondly, the scientific and industrial application of gene technology and the accompanying safety considerations, as presently regulated by guidelines. The following comments and suggestions are therefore restricted to topics the latter area.

The report of the Inquiry Commission has been put together with great care. In its introduction it states that some positions are the result of compromises, made possible by agreements between the various members. The AG BioT agrees with many of the Commission's recommendations. However, AG BioT regrets that the balance found in the presentation of the main section (fully addressing the great opportunities in gene technology) is often lacking in the recommendations. Since most readers will look only at the recommendations, the faultfinding aspects (risks) may receive undue emphasis in public discussions.

The following is an attempt to clarify those points that, in the opinion of AG BioT, are most important for the development of the science and technology of gene technology in the Federal Republic of Germany.

Introduction

Applications of gene technology in the basic and applied biosciences are burgeoning. This will predictably result in new processes and products for the field of biotechnology. The report of the Inquiry Commission is a preliminary review of this area; it makes a significant contribution by stressing increased safety in this research and its future applications, and the AG BioT strongly supports the Commission's position.

The Inquiry Commission report goes into considerable detail on the benefits to humankind that will accrue in many areas as a result of gene technology. This new technology is already indispensable for the understanding of widespread diseases and infections, and possible treatments for them. A polarized atmosphere has developed, however, with increased safety requirements on the one hand and the opportunities for gene technology in research, development, and production on the other. The AG BioT hopes to mediate with constructive contributions to risk detection and risk reduction in biotechnology.

Fundamentals

AG BioT notes that the potential dangers from genetically engineered cells can readily be assessed by the molecular biology that went into their design. This specifically means that if microorganisms have been altered by genetic engineering techniques, they do not necessarily become pathogens just because recombinant DNA technology was used in their creation. As a rule, cells with DNA brought in from other cells should not present a greater danger than that inherent in the donor organisms.

The fact is that since the early seventies, with recombinant DNA in use in many hundreds of laboratories, not a single accident has been reported. As a result of this experience, each revision of the guidelines, issued in a number of countries over the past decade for protection against dangers posed by *in vitro* recombinant nucleic acids, has been less restrictive than the one before. Furthermore, we know more about the properties of organisms or proteins modified by genetic engineering than we do about organisms obtained by classical mutation techniques or about the continuing procession of spontaneous natural mutants. Therefore, AG BioT believes that all guidelines for the handling of organisms containing recombinant DNA must start with an objectively accurate assessment of the potential dangers. We urgently recommend that "genetically engineered" and "pathogenic" microorganisms, cells, etc., be differentiated. Statements such as those on page XVII, line 1 and 2, can lead to an equality which technically is not justified.

Safety guidelines of the ZKBS (Central Commission for Biological Safety) and Advisory Board

The AG BioT supports all efforts to strengthen the responsibilities and decision-making capabilities of the ZKBS (Central Commission for Biological Safety) in scientific and technical matters, and to give the ZKBS clear and explicit responsibilities.

Advisory Board (Section E 5)
The AG BioT recognizes the efforts required, starting with the work of the Inquiry Commission, to establish an Advisory Board that would conduct concurrent discussions on science, technology, and society. It is the nature of the problem that fundamental discussion on medium- and long-term developments takes precedence over immediate decisions in individual cases. Therefore, the parallel existence of the ZKBS and the Advisory Board makes sense, provided that their responsibilities are clearly defined and the various relationships are appropriate to the problems under consideration, and that the Advisory Board is so constituted that the participation of experts in science and technology is guaranteed.

Proposal for action
Therefore, the AG BioT proposes, in accordance with Section E 5.3, that representatives from the AG BioT membership in molecular biology, microbiology, biochemistry, and bioprocess technology be appointed as members of the Advisory Board.

Release of genetically altered microorganisms into the environment (Section C 4)

AG BioT shares the opinion that a potential ecological risk may exist if certain modified microorganisms – including genetically engineered ones – are released into the environment. However, in large part it is possible to assess and define the risks and limit them to an acceptable level with measures appropriate to each case. We therefore reject a moratorium on the release of genetically modified microorganisms into the environment; on the contrary, each case requires a risk assessment and an individual decision based upon that assessment.

Proposal for action
An *ecological risk study group* (headed by Prof. Dr. K. H. DOMSCH) was formed in 1985 within the framework of the working committee on *Safety in Biotechnology* (headed by Prof. Dr. W. FROMMER). In August 1987 this study group was to present its position paper, **Considerations for the Release of Genetically Engineered Microorganisms into the Environment**, and publish it subsequent to its discussion by the full committee (Session 10/16/87). This analysis by the study group is intended to develop clear criteria for defining the risks in decision-making, thus providing a way to investigate, both theoretically and experimentally, the risks of a release to the environment in any given case. The analysis will also identify where research is needed, so that matters currently in doubt can be resolved as technology improves.

Genome analysis and gene therapy (Section C 6)

Now that it has been established that errors in methodology as well as mistaken genetic diagnoses can be avoided, hereditary disease can be accurately detected with genetically engineered DNA probes.

Proposal for action
A solution to the problems facing us includes both technical and human genetic considerations.

- Methodological safety can be improved under the following circumstances: if new knowledge in basic research is regularly exchanged; if the use of generally accepted methods is promoted; and if certain diagnostic tasks are centralized.
- Diagnostic results can be correctly interpreted only when the individual's background is kept in mind; if this is done within a genetic counseling relationship, the procedure has sufficient genetic reliability. Diagnostic genetics, therefore, cannot be justified without genetic counseling.

Handling of animal cell cultures (Sections D 1, D 2)

Although accepted techniques for safe handling of animal cell cultures exist, a summary of good cell culture practices needs to be issued.

Proposal for action
The DECHEMA Safety in Biotechnology working committee will establish a study group to propose research in areas requiring clarification.

Handling and use of genetically engineered microorganisms in food biotechnology

Proposal for action
Athough the report by the Inquiry Commission found no potential danger here, VAAM (Association of General and Applied Microbiology, one of the members of AG BioT) offered to examine this question again carefully.

Since the public has natural emotional objections to the use in food biotechnology of microorganisms containing recombinant DNA, an ill-conceived use of such microorganisms could lead to significant sales losses and could discredit a product for years. Therefore, informing and educating the public will be most helpful. Minimum standards for the introduction of microorganisms containing recombinant DNA into food should be established (examples include documenting the absence of pathogenic factors and toxins by making known the base sequence analysis of the inserted gene, or not releasing genes for resistance to antibiotics). An AG BioT expert panel should authorize VAAM to draft its recommendations on this subject.

Quality standards for laboratories and industries that use genetic engineering technology (Section D 4)

It is appropriate to require certain qualifications of all persons dealing with genetically engineered microorganisms. AG BioT suggests that large numbers of individuals be trained and tested in the necessary theory and practice; professionals should encounter no special restrictions, or, alternatively, they may be granted proficiency qualifications automatically.

Proposal for action
DECHEMA, a member association of AG BioT, in agreement with the chemical professional trade association, took on the task of developing the required basic and advanced training. At the time this is being written, a grant application for the project has been submitted to the BMFT (Federal Ministry for Research and Technology). A system of courses consisting of several modules is being established with the active participation of experts from academia and industry. This should begin to satisfy the immediate need for education for laboratory and factory; the system is tailored as much to the level of scientist and engineer – as supervisory and management personnel – as to the level of foreman and laborer. Graduates of these courses will receive a certificate that attests to the knowledge and expertise they acquired.

Simultaneously, a group of competent university and college teachers will work on implementing safety protocols in the university and college curricula, so that graduates in the relevant disciplines will have acquired the knowledge necessary for the safe handling of biological material during their studies at those schools.

The DECHEMA Safety in Biotechnology working committee is providing the projects and the planning with support and critical guidance. The member associations of AG BioT have offered to support the incorporation of the results into the respective college and university curricula.

Concordance (Sections E 2, E 5)

AG BioT emphatically supports the Inquiry Commission's demand for concordance in the guidelines for biological safety. Members of the AG BioT have offered to serve as advisors on all matters.

Because of the active participation of AG BioT in the group on Safety in Biotechnology established by the *European Biotechnology Federation*, we are reaching a state where know-how at the scientific and technological levels can be exchanged internationally. AG BioT, however, points out that if the laws or guidelines for biotechnology in general, and for gene technology in particular, are to be made internationally harmonious, compromises are necessary; an example is the work done by the Inquiry Commission. Therefore, experience has taught us that bureaucratic requirements and maximum restrictions, if promulgated by only a few countries, are no more than a hindrance. Unfortunately, it is already apparent that the Federal Republic of Germany is largely being bypassed when new biotechnology companies are being established, because of its very restrictive handling of the necessary permits; consequently, the economic results of significant research initiatives will be lost to our country.

Legislation (Sections D 2.4.1.3 and D 2.4.1.4)

AG BioT agrees that we do not currently need to push for general legislation in biological safety. The safety of the workplace, both in laboratories and industry, is assured by the Accident Prevention Specifications (UVV) in biotechnology of the Workmen's Compensation Board. The UVV (specifications) in biotechnology are presently being completed, and safety is also furthered by the UVV instruction sheets, also in preparation. The binding nature of the UVV biotechnology requirements will hold, provided that the recommendations of the Inquiry Commission are included. Because of these UVV biotechnology specifications, the registration requirement applies to all laboratories and industrial settings that conduct work with recombinant DNA, whether *in vivo* or *in vitro*. The UVV is a set of overall guidelines; detailed regulations are given in the instruction sheets.

Work with pathogenic microorganisms is regulated by the federal law on infectious diseases (BSG).

The problems of release to the environment are covered by federal laws on emission prevention, water conservation, plant protection, etc., and by the power that these laws confer on the judicial branch. The recommendations of the ZKBS (Central Committee on Biological Safety) should be implemented within the framework of these laws. It appears to be neither necessary nor advisable to incorporate this area into the federal law on infectious diseases (BSG).

Frankfurt am Main, 3 August 1987

Prof. H.-J. Rehm
(Chairman, Biotechnology Study Group)

Members of the Commission's Biotechnology Study Group:

Dr. M. Bahn; Prof. Dr. D. Behrens; PD[3], Dr. A. J. Driesel; Prof. Dr. C. C. Emeis; Prof. Dr. W. Hillen; Prof. Dr. W. Frommer; PD[3], Dr. J. Koempf; Prof. Dr. P. Praeve; Prof. Dr. H.-J. Rehm; Dr. J. Schabronath; Prof. Dr. H. Simon; Prof. Dr. U. Stahl; Prof. Dr. M. Teuber; Prof. Dr. E. L. Winnacker

[3] PD = Privat Dozent, and corresponds roughly to Lecturer or Assistant Professor

The Release of Genetically Engineered Organisms to the Environment

The Royal Commission on Environmental Pollution, United Kingdom

The Rt. Hon. the Lord Lewis of Newnham, Chairman

Thirteenth Report (abbreviated version)

Presented to Parliament by Command of Her Majesty
July 1989

Recommendations for the regulation of release

Statutory control of releases of genetically engineered organisms (GEOs) to the environment must be put in place.

Both the Secretary of State for the Environment and the Health and Safety Commission (HSC) (acting on behalf of the Secretary of State for Employment) should be involved in decisions on release.

The Secretary of State for the Environment should take primary responsibility for control with respect to the environmental consequences of such releases.

The control of releases of genetically engineered organisms should be governed by a statute establishing controls in respect of environmental protection and providing a framework within which the Secretary of State would be empowered to make regulations including a system for licensing. The statute should, in addition, impose a duty of care obliging all those responsible for the release of a GEO, whether for experimental or commercial purposes, to take all reasonable steps for the protection not only of human health and safety but also of the environment.

A release licence should be required before the release of a genetically engineered organism may take place. It should be an offence, carrying a substantial penalty, to release a GEO without having first obtained a release licence or to fail to comply with any conditions attached to the licence.

Any release licence should be granted by the Secretary of State for the Environment and the HSC (referred to as the licensing authorities) acting jointly. They should also have the power to revoke a release licence or to amend its terms if they had reason to believe that the continuation of the licence was inadvisable.

Anyone proposing that a GEO be released into the environment should be required to notify the licensing authorities of his intention and to furnish them with details of the organism concerned and the method of release, including the results of an assessment of safety carried out by a local safety assessment committee.

The new Genetic Manipulation Regulations should be revised to provide that the HSC's approval to release be given in the form of a licence.

In the light of experience the licensing authorities may consider it to be safe to issue a release licence for a class or category of related GEOs. Persons or organizations wishing to make releases under such a licence should, however, be required to submit their proposals to the licensing authorities who would decide whether they fall within the scope of that licence. The authorities should have the power to require that any proposal with features which gave rise to concern should be the subject of an application for a specific release licence, even though it appeared to be covered by a licence for a category.

Each stage of release in the development of a GEO should be the subject of a licence. The organism may then be proposed for use as or in a product. It should be assessed once more at that stage and be subject to licensing by the licensing authorities for sale, supply or use as or in a particular product. If no other product control applies to that product the licence should be issued directly by them.

Where other product controls apply, the product control authority should be required to inform the licensing authorities of any application for approval of a product which is or which contains a genetically engineered organism. They in response would inform the product control authority whether they were willing to issue a licence for the product. This applies both to products developed in this country and to those imported. Anyone applying for approval to the sale or supply of a GEO as or in a product should therefore be required to state in the application that it is genetically engineered.

The Secretary of State for the Environment should be given additional powers including the power to: set up advisory committees; draw up and publish codes of practice; maintain a register of people and organizations approved to carry out releases; make information available to the public and to other authorities; deal with emergencies and impose obligation on others to establish emergency arrangements; carry out to require others to carry out appropriate monitoring; require the provision of information about releases; require the proper disposal of waste products and, if necessary, cleaning up of release sites; inspect premises; and recover the costs of regulation.

The Government should consider whether the powers of the HSC need to be extended, in respect of the release of GEOs, to cover some or all of the powers listed in [the preceding paragraph] which it does not already exercise.

The first consideration in the proper control of releases of GEOs is a thorough, expert scrutiny of every proposed release. At this stage of the development of the technology we consider that each case needs to be scrutinized by a national committee of experts. Prior to such scrutiny a local committee based within the organization developing the GEO should screen the proposal to ensure that only well thought out proposals come forward for national scrutiny.

The Secretary of State for the Environment and the HSC should refer each application for a release licence, or for product approval, in respect of a genetically engineered organism, to a committee of experts and should take account of its recommendations. The primary function of the committee should be the assessment of such proposals with regard to environmental protection and human health and safety.

The present Intentional Introduction Sub-Committee of the Advisory Committee on Genetic Manipulation (ACGM) should be constituted as a committee in its own right, distinct from the ACGM. It should be charged with giving advice to both the HSC and the Secretary of State for the Environment. We refer to it as the Release Committee.

The Release Committee should have close links with the ACGM. This may be achieved through common membership and a joint secretariat.

Members of the Release Committee should have expert knowledge of genetic engineering techniques, microbiology, theoretical or field ecology or other relevant disciplines. They should be drawn from universities, other institutions, industry and workers' representatives. Persons engaged in the development and release of genetically engineered organisms should not be debarred from membership of the Committee but interests should be declared appropriately. Experts from the UK and from other countries should be invited to join the Committee on an *ad hoc* basis when needed for the assessment of particular proposals. There should also be representation from relevant Government departments and agencies and from local authority environmental health officers.

In addition to advising on proposals for release, including any conditions which should be attached to licences, the Release Committee should have other functions including: development of codes of practice and guidance for applicants; advising on the scope for categorizing releases; advising on the need for research especially on matters relating to release; reviewing the outcomes of releases; licensing with overseas organizations in relevant fields; and advising on possible needs for changes in legislation or procedures. The Committee should be asked to produce an annual report on its activities, on developments in the subject and on lessons learned. Adequate resources should be provided for its effective operation.

The functions of the Department of the Environment's Interim Advisory Committee on Introductions (IACI) should be taken on by the Release Committee, so that there will be no continuing need for IACI.

The Secretary of State for the Environment and the HSC, acting on advice from the Release Committee, should compile and maintain a register of persons authorized to release GEOs. It would be an offence for a person not so registered to be responsible for carrying out a trial release. A registered person would be held personally responsible by the registration authorities for the use of appropriately qualified and trained staff for every aspect of the release and for the issuing of adequate instructions for them. He or she should be required to record the names of all staff engaged in the release and to make the names available to the registration authorities if requested.

Appropriate arrangements should be made for the registration of companies or other organizations which carry out trial releases. Criteria for their entry to the register should include the employment of suitably qualified personnel, the provision of appropriate training, designation of safety officers and the establishment of a local safety assessment committee. Registered organizations should be required to identify one or more registered persons who would be responsible for releases.

Registration, either of persons or of organizations, could be made in respect of a single release, a specified series of releases or any release of a specified class or classes of organism. In addition to trial releases, it might occasionally be appropriate to require the registration of releasers of a licensed product; provision for this should be made in the legislation.

The new legislation should provide that any person, or the directors of any company or other organization, responsible for carrying out the release of a genetically engineered organism without the necessary licence and registration, will be subject to strict liability for any damage arising. It should also provide that neither the licensing and registration authorities, nor members of the Committee on whose advice they or either of them acted in granting the licence or registration, should be liable in respect of the consequences of the release.

The Secretary of State for the Environment should have the power to impose, in the release licence, a condition that the licence holder monitors the spread and fate of the organisms and of any introduced genes, the environmental impact of the release and any unexpected ecological event. The licence holder should be required to report the results of the monitoring to the licensing authorities, with immediate reporting of any significant untoward occurrence. There should be provision for monitoring to be required, on a temporary basis, in the case of licensed products where necessary.

There should be a public register of applications for release licences and of licences granted. This should contain the names and addresses of the persons or organizations making the applications, particulars of the organisms, the purposes of the releases and descriptions of the release sites. The register should be maintained nationally. Relevant sections of it should be kept in the localities of releases.

Information about release of GEOs, concerning foreseeable effects and arrangements for monitoring and dealing with emergencies, should be made available by the DOE or the HSE on request.

The national register should contain details of applications and licences granted for the sale or supply of GEOs as or in products. The register of authorized releasers should also be made public.

Persons or organisations applying for licences to carry out trial releases of GEOs should be required to place advertisements, in the local press serving the areas of intended releases, announcing their proposals. Anyone applying for a licence for the sale or supply of a GEO as or in a product should be required to place a notice in the London Gazette and an advertisement in an appropriate national newspaper.

The legislation should empower the licensing authorities to allow public access to the information on the basis of which the Release Committee has made its recommendation. It should also enable them, if they considered it appropriate before allowing access, to invite the applicant to comment on the request for information and to take account of the applicant's views on commercial confidentiality.

The licensing authorities will need to be able to communicate information about release proposals to the European Commission (EC) and competent authorities in other EC member states and other countries; if a specific power is necessary for that, it should be given to them. The UK authorities should also, if necessary, be empowered to make information available to the OECD, UNEP and other international organisations, and should do so to the fullest extent possible.

Members of the public should have the opportunity to make representations to the licensing authorities in respect of any application for a release license within 30 days of the appearance of the local or national advertisement. The applicant, and anyone who has made such representations, should subsequently receive a copy of the recommendation made by the Release Committee and be given the opportunity to make representations about that recommendation before the decision of the licensing authorities is taken.

The powers over the release of genetically engineered organisms which are to be exercised by the Secretary of State for the Environment and the HSC should apply to the marine environment within UK territorial waters. They should exercise these powers in consultation with the Minister responsible for fisheries.

There will need to be an extension of controls over contained work on genetically engineered organisms to minimize the risk of damage to the environment. These will include powers to require the proper disposal of waste products and to regulate storage, transport and import for contained use. The powers should be given to the Ministers already having responsibilities in each area.

The Secretary of State for the Environment should be given power in respect of waste disposal from contained work on genetically engineered organisms. In exercising his power, including the issue of advice by HER MAJESTY's Inspectorate of Pollution to the waste disposal authorities immediately responsible, he should receive advice from the ACGM and, as appropriate, from the Release Committee and elsewhere.

The Secretary of State for the Environment, together with agriculture and other Ministers, should conduct a review of issues arising over the selection and use of naturally occurring organisms. They should consider the possibility of enacting more comprehensive controls than those afforded by the Wildlife and Countryside Act and other present legislation.

Recommendations other than for the regulation of release

It is important that any definition of genetic engineering should be kept under review by experts and amended as necessary both to clarify if necessary the position of new techniques and to modify the coverage in the light of experience.

International measures are called for in relation to commercial releases of plants. Viable samples of current commercially-used plant varieties should be conserved so that it will be possible to return to these in order to eliminate an undesirable trait if necessary. There should be lineage registers which record the history of plant varieties including information on any introduced genes. In addition, before organisms with introduced genes are released the introduced DNA sequence for the new genes should be characterized for future reference.

Research on selective, readily degradable chemical pesticides leaving no objectionable residues and which are non-toxic to humans should not be abandoned in the enthusiasm for biological control. The development of agricultural practices such as integrated pest management, which may help to reduce the scale of the problem with which pesticides are trying to deal, should also continue to receive attention.

Local safety assessment committees may not need the same range of expertise as the national committee but should contain ecologists as well as experts in genetic engineering. Other members with relevant local knowledge and expertise should be appointed where possible.

Local authority environmental health officers should be invited to serve on local safety assessment committees. In order to make an informed contribution on a subject which is somewhat outside the range of current EHO responsibilities, training and advice will be needed.

It is clearly desirable that there should be international agreement on the information to be required of releasers and the procedures for assessment. We hope that the Government will use the final version of the ACGM's revised guidelines for information and risk assessment as a model in international discussions on this subject.

The progression from laboratory to widespread release should go through a series of stages gradually relaxing the degree of containment at each, for example from laboratory, to greenhouse, to single field trial, to wider trials, to full marketing.

As products move through stages of release, responsibility for scrutiny may fall progressively to various bodies. Close links are needed between these bodies together with arrangements for exchange of information about assessments and about the results of releases that have taken place.

There should be a step-by-step approach to innovation in the releases that take place so that the modifications made at each step do not introduce an unacceptable degree of uncertainty.

The use of debilitating mechanisms should always be considered when genetically engineered micro-organisms are proposed for release.

The potential for clean-up and decontamination of a release site should always be considered but it would nevertheless be prudent to work on the assumption that, once released, it may not be possible totally to eradicate an organism, particularly a micro-organism, from the environment.

Releasers should be given clear advice by the Release Committee, both in general guidance on good practice and in specific comments on their releases, about the manner in which releases should be carried out, including arrangements for security, for monitoring, for clean-up and for dealing with contingencies. Compliance with these arrangements should be checked by appropriately trained inspectors with authority to take action where necessary.

At least until more knowledge is gained and confidence acquired about the behavior of GEOs in the environment, releasers should be required to carry out monitoring. When assessing a proposal, the Release Committee should consider the extent, methods and arrangements for the monitoring that should be carried out.

The monitoring of the release of a GEO should normally continue after completion of the experiment for an appropriate period depending on the nature of the release, with agreed arrangements for reporting the outcome.

There is scope for coordinating the general monitoring of the environment to develop a systematic approach. The DOE should take the lead in promoting and funding this coordination work as part of its responsibilities for the protection of the environment.

The Release Committee should carry out regular reviews of the information it has obtained about the outcome of releases. Consideration should be given to publishing the results of the reviews.

International exchanges of information between assessment bodies could provide valuable material to assist in assessing release proposals. The European Commission has proposed regular exchanges of information on this subject between member states. We support this initiative.

We recommend that the UK authorities should, in appropriate circumstances, notify proposed releases of GEOs to the competent authorities not only in other EC member states but also in other countries and should take full account of their views.

The relationship between living organisms and their environment is such that proposed releases of GEOs must be considered in the appropriate environmental context. This aspect of the draft ED Directive on the release of GEOs needs further thought and should be the subject of careful discussion between the European Commission and member states.

The list of exclusions from the products section of the draft ED Directive on the release of GEOs considerably weakens the value of the proposals. Where product controls exist, those responsible for them must, before they authorize release of a product which is or which contains a genetically engineered organism, receive expert advice on those features which differentiate it from, for example, a chemical product. For products which are subject to no control, it is essential that controls should be established in respect of those which are or which contain GEOs.

We support proposals by the European Commission for regular meetings of officials from member states to discuss and exchange information on release proposals, for an expanded research program on biotechnology, in particular on risk assessment of releases, and for the creation of a database of releases.

The ACGM, in consultation with the Release Committee, HSE, DOE and MAFF, should revise its containment guidelines to take into account potential harm to the environment from the escape of GEOs.

The risk of accidents in the use or storage of commercially produced GEOs needs to be considered when proposals for products which are or which contain GEOs are put forward for assessment. Clear labelling, including instructions for storage, use, disposal and action to be taken in the event of an accident, should be considered where potential hazards exist.

Well-designed protocols for procedures at the laboratory, field trial site and production process plant are very important in reducing the risk of accidents occurring. Staff should be appropriately trained so that they understand how to handle the GEOs and associated equipment safely. Response plans should be drawn up to deal with the consequences of an accident and staff should be trained to implement them.

HER MAJESTY's Inspectorate of Pollution should consider the waste disposal issues raised by the development of genetic engineering techniques and, in consultation with the appropriate authorities, issue advice on the selection of BPEOs for the disposal of the wastes.

When a proposed product which is or which contains a GEO is submitted for assessment, a licence should be granted only if any waste or residue can be disposed of safely and if appropriate advice on waste disposal appears on the product label.

Guidance for the disposal of GEOs in biological and biotechnological waste, and its enforcement, should be kept under review to ensure that it remains appropriate. The waste disposal procedures recommended for field trials should also be kept under review.

We have recommended a degree of public access to information about releases which goes beyond the access allowed in respect of most products. Some of this information could be of value to other companies. A regime of intellectual property rights should be developed which provides sufficient protection to enable the release of adequate information to the public without undermining the commercial viability of the development and thereby damaging the incentive for innovation.

Knowledge of genetics and ecology should be included in the curriculum in schools. Students should be aware of the factors involved in judging the impact on the environment of a proposed release.

There is a need for a substantially enhanced research base in the basic sciences underpinning the release of genetically engineered organisms to the environment. It should be located in the universities and research institutes and should receive adequate funding. Such research should be in three major areas: the molecular biology of organisms in the environment, interactions between organisms and the environment and basic studies on ecology and population biology.

Basic research should be supplemented by projects related to specific environmental issues commissioned by the relevant Government departments.

Guidelines for the Characterization of Immobilized Biocatalysts

(Reprinted from *Enzyme Microb. Technol.* **5** (July) (1983) 304–307.)

The Working Party on Immobilized Biocatalysts within The European Federation of Biotechnology

Approved at the EFB Working Party Meeting in Basel on 22–23 November 1982.

I. Introduction

A. Purpose and scope

The recommendations relate to particulate catalysts with emphasis on preparative and industrial applications. Additional guidelines relevant to analytical, biomedical, photochemical, and other special applications will have to be drawn up by experts in these fields. However, some of the general considerations are relevant also to these areas.

General guidelines for characterizing immobilized biocatalysts were drawn up at the first Enzyme Engineering Conference in 1971 [1]. Since then, not only the problems but also the perspectives have changed. Furthermore, examination of the extensive literature shows that adequate characterization of the catalysts is an exception rather than a rule with the result that a large part of the published information cannot be evaluated and classified.

Recognizing the need for better information, a group of scientists reviewed the problem, compiling lists of fundamental parameters and describing or proposing methods to measure them [2]. However, no general recommendations were made.

The present Working Party, appointed by the European Federation of Biotechnology, has considered the subject and drawn up new guidelines.

For reasons outlined in the next section, we propose that investigations dealing with immobilized biocatalysts should contain answers to questions like the following:

What is the quantity of the free enzyme (organelle, cell) preparation needed to prepare a unit volume of wet catalyst?

What are the dimensions of the wet particles?

Are the observed reaction rates diffusion-limited?

In what way are the reaction rates affected by changes in concentrations of the reactants in the concentration range of interest and how do these rates compare with those catalysed by the same quantity of free catalyst?

Does the catalyst hold out promise for practical applications in terms of its mechanical (and other relevant) properties and also its stability under conditions of its intended use?

The goal of the Working Party has been to define the minimum parameters to give a satisfactory description of an industrial immobilized biocatalyst. We realise that our recommendations should be read as general minimum requirements and general guidelines. Special situations give rise to further demands.

B. Reasons underlying the guidelines

Immobilized biocatalysts represent a particularly complex area of heterogeneous catalysis involving physical, chemical, biological and technical disciplines. This complexity arises not only from the nature of the subject, but also from the practical motivation which explicitly or implicitly underlies most of the current work. The relevant criteria for applications are technical or technological feasibility and cost. These have to be met usually by making a compromise between a number of desirable properties and contradicting requirements. The catalysts therefore represent a practically unlimited range of compositions and morphologies, from structures like controlled-pore glass to cell homogenates cross-linked with multi-functional reagents.

There are no *a priori* requirements on the absolute value of the characteristics, such as rate per unit volume of catalyst, concentration of the reagent, particle size, dimensional stability, or on the values of the experimental variables, such as concentrations, temperature and flow rates. These values are meaningful only in relation to each other and in relation to the objective of the investigation. However, some of the quantities – such as dimensions of the wet particles – are an essential part of the description of the system because other important quantities – such as reaction rates, the compression behaviour of packed beds, and the conditions prevailing within the catalyst particles in the course of their use – depend on them. For this reason, only a reasonably complete set of data (including precise information about the preparation of the catalyst and adequate specification of the conditions of the measurements) can be evaluated and reproduced.

We still know very little about the changes of molecular catalytic properties associated with the chemical modification of proteins and effects of the microenvironment other than those due to electrostatic forces. On the other hand, measurements with particles of simple, defined structure have established the applicability of diffusion laws to immobilized enzymes, and the implications have been successfully used to design carriers for applications requiring high reaction rates per unit catalyst volume.

However, the mathematical equations relating the fundamental physical parameters and external variables to the properties of interest become very complex unless the structure of the carrier matrix and the distribution of catalyst within the matrix are simple or unless diffusion limitations are negligible. Thus, direct measurement of the desired quantities in the experimental range of interest will generally require less work, and give more reliable results than calculations. In particular, estimates of the useful life of catalysts based on wide extrapolation of experimental data are unreliable because many unrelated factors, including fouling by microbial matter, may change the properties of the system abruptly.

For these reasons, we recommend a pragmatic and phenomenological characterization of the catalysts. It is the only practical way to obtain adequate information for sensible comparisons and design purposes with a reasonable amount of effort. Once this information is available, further measurements will obviously be carried out with those types of preparations which are interesting in a practical or theoretical context. In no way do we wish to discourage detailed studies on particular systems. We do wish to raise the minimum level of characterization of all immobilized biocatalysts to an acceptable level.

C. Definition of immobilized biocatalysts

Immobilized biocatalysts are enzymes, cells or organelles (or combinations of these) which are in a state that permits their reuse.

Example:

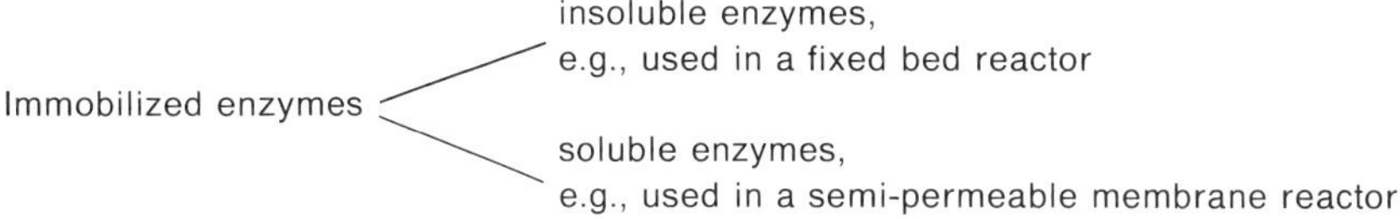

II. Recommendations

List of desirable minimum requirements for characterization of an immobilized biocatalyst

0. *General description*
   - 0.1 Reaction scheme
   - 0.2 Enzyme and microorganism
   - 0.3 Carrier type
   - 0.4 Method of immobilization
1. *Preparation of the immobilized biocatalyst*
   - 1.1 Method of immobilization, reaction conditions
   - 1.2 Dry weight yield, activity left in supernatant
2. *Physical/chemical characterization*
   - 2.1 Biocatalyst shape, mean wet particle diameter, swelling behaviour
   - 2.2 a. compression behaviour in column systems, or
     b. abrasion in stirred vessels, or
     c. minimum fluidization velocity and abrasion in fluidized beds.
3. *Immobilization biocatalyst kinetics*
   - 3.1 Initial rates vs. substrate concentration for free and immobilized biocatalyst, effect of pH and buffer
   - 3.2 Diffusional limitations in the immobilized biocatalyst system (effect of particle size or enzyme load on activity)
   - 3.3 Degree of conversion vs. residence time (points on a curve)
   - 3.4 Storage stability (residual initial rate after storage for different periods)
   - 3.5 Operational stability [residual initial rate (or transforming capacity of reactor system) after operation for different periods].

III. Guidelines for the characterization of an immobilized biocatalyst

0. *General description*

This section should be a summary giving a short description of the enzyme, microorganism or organelle, the carrier type, the method of immobilization, and the reaction scheme for the process in which the biocatalyst is to be used. The reaction scheme should also state possible side reactions at the actual operating conditions and indicate (if relevant) information on equilibria/thermodynamics, etc.

1. Preparation of the immobilized biocatalyst

1.1 Method of immobilization, reaction conditions

It is very important that the information given here is so detailed that *anyone can reproduce the procedure*. Amount and concentration of *each component used* must be specified (also the activity of the soluble enzyme). The method of activity determination must be described in detail (see section 3.1). The rest of the characterization will be of little value if the information about preparation of the biocatalyst is inadequate and does not allow other scientists to reproduce the procedure.

Comments like FDA acceptance of the carrier/system could also be included here.

1.2 Dry weight yield, activity left in supernatant

The yield is an important parameter in the evaluation of the immobilization procedure. Always state the *dry weight* of the biocatalyst preparation obtained when the immobilization procedure described in 1.1 is followed. This must also be done in the cases where drying is omitted in the preparation procedure. The activity left in the supernatant indicates the amount of the originally used enzyme activity which has not been incorporated in the immobilized biocatalyst. If the immobilization conditions cause activity loss of free enzyme in solution, this should be mentioned. The method of activity determination must be described in detail and the method should preferably be identical to the method used in the determination of the activity of free enzyme used in the preparation of the immobilized biocatalyst. If it is impossible to use the same assay for activity, it should be stated how to compare the activity obtained with the different methods.

Furthermore, enzyme and or cell leakage from the immobilized biocatalyst in the reactor should be stated.

2. Physical/chemical characterization

2.1 Biocatalyst shape, mean wet particle diameter, swelling behaviour

The *shape* of the wet biocatalyst particles and *mean wet particle size distribution* (or other characteristic particle dimensions) must be given here. The method of determination has to be described in detail (*e.g.*, wet sieving or microscopic measurement). The *swelling behaviour* can be described by taking *x* ml or gram of the immobilized biocatalyst prepared by the procedure described in 1.1 and determining how much *settled* (and eventually total) *volume* it will occupy in the reactor at operating conditions. It must be stated how much dry weight the *x* ml or gram corresponds to. The method of determining the biocatalyst settled volume in the reactor must be described in detail, including equilibration time. Preferably, the drying procedure should also be described.

2.2a Compression behaviour in column systems

Compression behaviour in column systems should be determined with at least 20 g dry material (*e.g.*, according to [3]). If insufficient material is available, the information will necessarily be of limited value; however, the minimum requirement is that the method used for pressure drop determination, including dimensions of the system, be stated. Specification of the starting-up procedure should be included.

It is essential to measure the extent of compression over a prolonged period of time and the ideality of the flow.

2.2b Abrasion in stirred vessels

The minimum requirement here is to state the particle size distribution at the start and at the end of the experiment and – if the system is continuous – also the biocatalyst weight at the start and at the end of the experiment.

The experimental conditions have to be described in detail, including geometry and dimensions of both vessel and stirrer as well as stirrer speed.

Additional ways of providing information about abrasion are to measure enzyme activity in the supernatant or to determine settling time after different periods of operation.

2.2c Minimum fluidization velocity and abrasion in fluidized bed

At least 20 gram dry material (and preferably more) must be used to obtain reliable values for minimum fluidization velocities. The information will be of very limited value if too little material is available.

The experimental conditions and the system construction must be stated in detail.

Measure at least the fluid velocity at which the bed starts to expand, preferably several points on the curve bed volume vs. fluid velocity should be given. (Be aware that the stationary phase is followed by an expanded phase which again is followed by the fluidized phase). Abrasion is determined as in 2.2b.

If other types of reactors than those mentioned above are used (*e.g.*, tubular reactors or filter press reactors), give some indication of the pressure drop. Always describe experimental conditions, system design, and dimensions in detail.

3. Immobilized biocatalyst kinetics

3.1 Initial rates vs. substrate concentration for free and immobilized biocatalyst, effect of pH (buffer)

The initial rates vs. substrate concentration for both the free and the immobilized biocatalyst must be presented showing the original data for as many substrate concentrations as possible and *not* as regression lines. If curves are also shown, then specify the equation representing the curves.

The measurement should be performed on the biocatalyst preparations as they are. Whenever possible, in solution and suspension in batch analysis, respectively, *at identical experimental conditions for the free and immobilized biocatalyst.*

Always describe the exact method of analysis in sufficient detail so that anyone can reproduce it. *If* the method for free biocatalyst analysis differs from that of the immobilized biocatalyst, do not forget to state the differences.

The time elapsed from the preparation of the immobilized biocatalyst to the determination of initial rates must be shown, and the storage conditions must be described.

For reactions generating or consuming hydrogen ions, the buffers of the system act as co-reagents, and the buffer concentration may be rate-determining. For such reactions, the measurement should be carried out with two different buffers (differing by about 1 p*K* unit), each at two concentrations.

If temperature profiles are described, the conditions should be thoroughly described.

Describe the effect of growth factors when the immobilized biocatalyst contains live cells as catalysts.

3.2 Diffusion limitations in the immobilized biocatalyst system

The best way of determining if diffusion phenomena can occur in an immobilized biocatalyst system depends very much on the system type. In comparisons between immobilized enzymes and cells, diffusional limitations both by the carrier and the cell material should be taken into consideration.

If it is possible to change the activity loading of the biocatalyst particles, measure initial rates at two (preferably more) different activity loadings of particles with an identical size distribution. In some systems, it may be possible to change activity loading by introducing some denatured material in the immobilization procedure.

If the method of immobilization is reversible, measure the initial rate after resolubilizing the catalyst.

Determination of the effect of particle size on initial rate can be used in many cases. Determine initial rates for at least two (preferably more) particle sizes.

In any case, state in detail how the different loadings or the resolubilization of the particles have been achieved and how the initial rates have been measured.

3.3 Degree of conversion vs. residence time

Determine the degree of conversion for as many different values of residence time as possible.

Describe the size and design of your system thoroughly and give all details about experimental conditions, including the equilibration times between changes of flow rate and the corresponding measurements of conversion (it is often necessary to allow the throughput of several bed volumes before a new steady state is attained).

3.4 Storage stability

Measure initial activities of the immobilized biocatalyst after different periods of storage. State the storage conditions in detail. *The actual data points must be presented. Never make conclusions that are based on extrapolations.*

3.5 Operational stability

Stability data from experiments that have been run for less than 50% of expected operational time are in general of little value.

Always present the exact data from your stability measurements. Extrapolation of stability data is completely unacceptable and of no use at all.

Stability data should be presented as immobilized activity (determined by initial rate measurements) vs. time where the biocatalyst has been running at application operating conditions during the whole experimental period (except when initial rates are determined).

Catalytic efficiency is a property of the whole reaction system and cannot therefore be defined in terms of a single parameter like the initial rate. Furthermore, in packed beds, different rates of inactivation can occur in different sections of the bed.

If runs must be made at a constant flow rate, at least measure initial rates of catalyst samples at the beginning and at the end of the experiment. Preferably, initial rates should be determined several times during the experiment.

Perform trials both at optimal analytical conditions and at the conditions used in actual practice. Most activity data are determined and reported at conditions far from those in actual practice, especially with regard to substrate concentration.

Furthermore, stability data are often reported for a period of time, without stating if the reactor was actually run continuously or intermittently with storage periods at reduced temperature in between.

The same degree of fluidization should be maintained in fluidized beds.

For batch reuse systems, specify exactly how the biocatalyst is treated between uses and also the storage time and conditions between uses. For analytical systems, state the number of assays and the total time involved in performing all the assays.

In work with whole cell preparations, any formation of by-products should be stated.

Specify the experimental conditions in detail.

Appendix

Members of the Working Party on Immobilized Biocatalysts (November 1982)

Address of the Working Party: Susanne RUGH (secretary) or Poul B. POULSEN (chairman), both at Novo Industri A/S, Novo Allé, DK-2880 Bagsvaerd, Denmark.

A. BALLESTEROS (Spain)
G. VAN BEYNUM (Netherlands)
O. BORUD (Norway)
K. BUCHHOLZ (Federal Republic of Germany)
A. ÇAGLAR (Turkey)
P. S. J. CHEETHAM (United Kingdom)
J. DUNNE (Ireland)
L. FOGARTY (Ireland)
J. L. IBORRA (Spain)
J. KONECNY (Switzerland)
E. LASZLO (Hungary)
M. D. LILLY (United Kingdom)
P. LINKO (Finland)
G. MANECKE (Federal Republic of Germany)
W. MARCONI (Italy)
B. MATTIASSON (Sweden)
K. MÅRTENSSON (Sweden)
P. B. POULSEN (Denmark)
K. ROERING (Finland)
S. RUGH (Denmark)
V. SCARDI (Italy)
G. SCHMIDT-KASTNER (Federal Republic of Germany)
P. J. SICARD (France)
G. SMITS (Belgium)
U. VON STOCKAR (Switzerland)

D. A. SZEWCZUK (Poland)
A. SZENTIRMAY (Hungary)
D. THOMAS (France)
J. TRAMPER (Netherlands)
E. VANDAMME (Belgium)

In order to illustrate the Guidelines for the Characterization of Immobilized Biocatalysts, the article in reference [4] has been worked out.

The general description will be found under **Introduction** and the section **Materials and methods**.

The preparation of the immobilized biocatalyst is described under **Materials and methods**.

Physical and chemical characterization is described in the first subsection under **Results and discussion** and finally the description of the kinetics is in the last part of the article.

References

[1] WINGARD, JR., L. B., Ed.; *Enzyme Engineering*, Interscience Publishers, New York (1972) pp 15–18.
[2] BUCHHOLZ, K., Ed.; *Characterization of Immobilized Biocatalysts*, Dechema Monograph, Vol. 84, Verlag Chemie, Weinheim (1979).
[3] NORSKER, L.: *Starch* **31** (1979) 13–16.
[4] VAN GINKEL, C. G.; *et al.*: Characterization of *Nitrosomonas europaea* Immobilized in Calcium Alginate; *Enzyme Microb. Technol.* **5** (1983) 297–303.

Gene Technology and Its Management in Germany*

A recommendation for a resolution (Pamphlet 11/5320) on the Report of the Inquiry Commission (Enquete-Kommission) on Opportunities and Risks of Gene Technology has been before the Parliament of the Federal Republic of Germany (FRG) since 1989.

The report, after a few preliminary comments, emphasizes the ambivalent nature of this modern science; not only does biotechnology offer new technological possibilities for solving numerous scientific and practical problems, but in certain instances it also causes those practicing it to express ethically motivated reservations.

The recommendation of the Parliament is undoubtedly an attempt to approach an unknown area of science with caution, but on the whole it must be regarded as overly cautious and too wide-ranging.

The report of the Inquiry Commission covers 25 individual areas of application of gene technology. For example, it recommends special support for basic research in Biological Material Conversion and Raw Material Supplies, Plant Production, Animal Production, Environment, Health, Ecological Genetics, and Pharmacokinetics, as well as Environmental Release. Later sections deal with the more protean aspects, such as Genetic Counseling, Genome Analysis for Employees, Somatic Gene Therapy, Germ Line Manipulations, Legal Aspects, Law for Gene Technology, and other questions.

Of course, one can attempt no more than a short summary and interpretation of the Inquiry Commission's resolution here. Generally, the Commission's recommendations are reasonable; they point out that gene technology, like other technological developments, can present a potential hazard, along with numerous positive aspects; interference with the human germline is opposed by all.

The order of the Administrative Court of Hesse on 6 November 1989, to immediately halt construction and pilot operation of the Hoechst Human Insulin Facility, was based upon the reasoning that the unit in question does indeed need approval, but cannot be approved. No evaluation of the facility with regard to its technical safety and the biotechnology to be used was conducted, because this would – in the Court's opinion – be meaningless for the ruling. Rather, a clear legislative statute is required because the current "partial laws" offer insufficient protection.

The ruling was fundamentally based on the assumption that technology is intrinsically dangerous and must be regulated by laws. Thus, the ruling affects not only technical facilities in which certain operations are being conducted, but also has bearing on the use of gene technology methods per se. Central to the Court's decision is that new technologies may be developed and applied only if special legislation provides approval. No further comment on the consequences of this reasoning is needed. It is in this context that a bill for a gene technology law has been brought before the German Parliament for decision.

This line of thought is rather difficult to follow, since Germany is a highly civilized country that depends on the use of the most modern technologies in its numerous

* by Prof. Dr. Paul Präve,
Hoechst AG, Zentralforschung II, Biotechnik, Gebäude G 836, Pf. 80 03 20,
6000 Frankfurt/Main 80, Germany.

educational institutions (universities, etc.) and manufacturing facilities; if this decision is compared with developments in other countries, the reasoning is even harder to grasp.

Experience in science research and teaching has shown that work with highly pathogenic microorganisms can be safely controlled; furthermore, what we know of traditional genetics, as it is handled worldwide, paints a picture that is significantly different from the one currently dominating public discussion about gene technology. Particularly because recombinant DNA technology can be precisely targeted, risks are minimized and safety in biological investigations is increased.

Often throughout human history, situations have developed that required a change in thought. Galileo, Copernicus, and Darwin undoubtedly destroyed the worldview of those who believed that people and the earth were at the center of the universe; similar problems arise in assessing and evaluating modern science and, in particular, gene technology. The fact that most people are inadequately educated in science and technology, or, rather, that they have an inaccurate conception of the scientific method, particularly of the trial and error necessary for conducting research, adds to the difficulty. The trial-and-error approach, which renders the results of scientific research falsifiable, is exactly what can affect the outcome of any scientific endeavor, whether the product is a diphtheria vaccine, a microchip, or a bridge. That the public seriously misunderstands scientific interrelationships also results in the unrealistic demand for 100% safety in numerous scientific and technological applications, a standard that, according to the criteria of natural science, can never be attained.

Gene technology today is practiced worldwide in thousands of laboratories and production facilities. Biological research today without gene technology is no longer conceivable; thoughtful management, in concert with numerous other key technologies, will be indispensable.

At first it seems surprising that even at the end of the 20th century, opposition to technology is at all possible at current levels of public discussion, given the high degree of informed, civilized awareness. The debate, however, is evidence that when science and technology remain incomprehensible to the population at large, intangible anxieties find vent.

Therefore, every scientist and technologist must take on the task of making his or her world of work and thought more accessible to the populace.

Research Initiatives and Funding

BRIDGE (Biotechnology Research for Innovation and Development Growth in Europe): Decision of the Council on a Specific Program for Research and Technological Development in Biotechnology (1990 to 1994)

Dated 27 November 1989; abbreviated version

The Council of European Communities (CEC) –

supported by the Treaty founding the European Economic Community (EEC), and in particular by Article 130q, Paragraph 2, with the advice of the Commission (*Commission on Biotechnology Research and Development*), and in collaboration with the European Parliament, including the Economic and Social Committee, makes the following statements:

According to Article 130k of the Treaty, the program is to be executed by means of specific programs that are developed for each venture.

With Resolution 87/516/Euratom (European Atomic Energy Association)/EEC, as altered by Resolution 88/193/EEC/Euratom, the Council adopted a common program in research and technological development (1987 to 1991) that, among other things, provides for ventures in the exploration and optimal utilization of biological resources.

The previously cited resolution aims, particularly in this joint research effort, to strengthen the scientific and technological base of European industry, especially in strategic fields of high technology, as well as to support industry to make it more competitive in the international arena. According to the same resolution, a joint venture is justified when it can contribute to strengthening the economic and social unity of the Community, and to promoting harmonious development in all fields, while striving for scientific and technological quality. The BRIDGE Program for biotechnology research promotes innovation, development, and growth in Europe, and is intended to contribute to attaining these goals. The program specifically includes the following ventures.

- A joint research and development network is to be established, so that the efforts of individual states may transcend national borders. Participants should take advantage of this network to promote basic technology, risk analysis, and the transfer of technology to industry and agriculture within the infrastructure.
- The strategic significance of new developments in biotechnology will be continually evaluated, and the requisite coherence between the different areas of Community politics that concern biotechnology will be promoted.

The outcomes of several recent resolutions have, unequivocally, proven the usefulness of joint ventures in biology and the need to expand such ventures. Resolution 81/1032/EEC, on a multi-year research and training program sponsored by the EEC in biomolecular engineering; Resolution 85/195/EEC, on a research venture program, projected over several years (1985–1989) by the EEC, to address biotechnology; and Resolution 88/420/EEC, revising the previous two, are cases in point.

Special attention must be given to ethical and social questions that may arise in connection with this program.

The total or partial participation of EEC nonmember countries in the projects of this program is desirable.

Furthermore, small and medium-sized companies should be integrated into the biotechnology research and development program to the greatest extent possible.

Executing the research and training program within *COST* (*Cooperation in Science and Technology*) significantly complements the research and development endeavors within biotechnology.

The *Committee on Research in Science and Technology Research* (*CREST*) has delivered a position paper –

has made the following stipulations:

Article 1

A specific EEC program for research and development in biotechnology (BRIDGE) is to be established according to Appendix I for a period of four years beginning 1 January 1990.

Article 2

The budget for the program should be 100 million ECU (European Currency Unit), including expenses for 28 employees. Preliminary budget allocations are listed in Appendix II.

Article 3

Details on conducting the program and the extent of financial participation of the Community are presented in Appendix I.

Article 4

(1) The Commission is to review the program in the third year of operation and present a report on the result of this review to the European Parliament and the Council. The report will be amended as necessary with suggestions for change or expansion of the program.

(2) After the program period is ended, the Commission is to evaluate the results and report its findings to the European Parliament and the Council.

(3) The above reports are to be prepared on the basis of the goals and criteria outlined in Appendix III, in accordance with Article 2, Paragraph 2 of Resolution 87/516/Euratom/EEC.

Article 5

The Commission is in charge of conducting these programs. It is to be supported by a committee with an advisory function (subsequently referred to as the Committee) that consists of representatives from member states presided over by a representative of the Commission.

The treaties concluded by the Commission regulate the laws and obligations of all parties and, especially, the dissemination, protection, and utilization of research results.

Article 6

(1) The Commission representative is to present to the Committee a draft of actions to be taken. The Committee will offer comment – based on a vote, if necessary – within a time that can be set by the chairman, based on the urgency of the question at hand.

(2) The comment is to be incorporated into the Committee protocol; in addition, each member state has the right to demand that its point of view be recorded in the protocol.

(3) The Commission is to take the comments of the Committee into consideration as much as possible. It will inform the Committee of the extent of consideration given to its comments.

Article 7

The procedures outlined in Article 6 apply especially to:

- the content of proposals;
- the evaluation of the proposals presented, and the budget estimate for the Community's participation in these ventures;
- deviations from the general rules for the Community's participation as outlined in Appendix I;
- the participation in a given venture by organizations and enterprises, identified in Article 8, Paragraph 2, that reside outside the Community;
- adjustments to details of the preliminary budget estimate shown in Appendix II;
- actions required to evaluate the program;
- agreements on the dissemination, protection, and utilization of the research results obtained within the framework of the program.

Article 8

(1) The Commission is empowered, in accordance with Article 130n of the Treaty, to negotiate agreements with international organizations, with third countries that are participating in the European collaborative effort in scientific and technological research (COST), and with European countries that have agreements with the Community on scientific and technological cooperation, with the goal of partially or fully including them in the program.

(2) Organizations and enterprises in nonmember states can participate as partners in ventures initiated within the framework of the program by meeting the criterion of mutual benefit, inasmuch as agreements have been reached between European nonmember and Community member states on scientific and technological collaboration. A contractual party residing outside the Community, participating in a venture within the program framework, cannot benefit from the Community's financing of the program. This party must contribute to the general administrative costs.

Article 9

This Decision is intended for the member states.

Brussels, 27 November 1989.
For the Council
The President
R. Dumas

Appendix I

Program content and execution, and extent of the Community's financial participation

Action I: Research and training

Content

1 Informational infrastructure

1.1 Culture collections

- developing a communications system that permits the easy and speedy access to the most important service-oriented culture collections within the Community. To this end, the following institutions should be supported:
  - (a) a support center for culture collections that specifically supply European users, with factually correct information about the available know-how and services at European culture collections (distribution of catalogs, patent regulations, printed and visual material);
  - (b) a centralized European data bank, mainly for microorganisms, which later will be extended to other cultures (animal and plant cells, viruses, plasmids). The first step towards this organization should be to make consistent the format and data of the most important culture collections within the European Community.

1.2 Processing and analyzing of bio(techno)logical data

- using computer technology (special software and hardware) required for conducting activities in protein technology and gene sequencing (see §2.1 and §2.3);
- implementing and designing databases for storage and classification of bio(techno)-logical data such as sequences, genetic maps, protein and biopolymer structures, risk-assessment data;
- utilizing current or newly developed computer technology for rapid access to European databases and sequencing networks through an electronic network, including data input, on-line catalogs, electronic order processing, and so forth.

2 Support technologies

2.1 Protein technology/molecular models

- refining interdisciplinary concepts, including genetic engineering and modern structure analyses, that target improvements of properties (such as stability, optimal pH, substrate specificity) of interesting proteins and their conjugates (including glycoproteins);
- developing methods for understanding and predicting the structure/function relationships of proteins, especially those that contribute to the folding, stability, and crystallization of proteins, including theoretical models for simulating these properties, and their interactions with other associated molecules.

2.2 Biotransformation

- developing biological reactions, using new cell lines or novel enzymes, for the synthesis of important intermediary products necessary to producing compounds with high

added value (with special attention to the bioconversion of agricultural surpluses) and to converting toxic material into nontoxic compounds;
- conducting research projects on problems in the genetic and physiologic stability of free or immobilized genetically engineered microorganisms or biotransformed cells;
- conducting research projects on problems of enzymatic activity under conditions of environmental stress (organic solutions, pH values, temperatures, immobilization);
- developing methods for isolating and purifying products generated through biotransformation (pre- and post-storage processing);
- developing special software and mathematical models for controlling and analyzing biotechnological processes.

2.3 Gene mapping, gene sequencing, and novel cloning methods

- sequencing the genome of yeast (*Saccharomyces cerevisiae*) or parts thereof, and of *Bacillus subtilis*;
- developing molecular genetic techniques for identifying important novel plant genes using the *Arabidopsis* genome as a source; characterizing identified genes;
- developing modern sequencing procedures and technologies (see §1.2) and incorporating them into sequencing projects.

3 Cell biology

3.1 Physiology and molecular genetics of industrial microorganisms

- studying gene stability and expression, postatranslational processes, genetic and metabolic regulation of excess production, transport, and secretion. These studies must be individually adapted to the state of scientific know-how, be focused on a few microorganisms of industrial interest, such as the genera *Streptomyces*, *Pseudomonas*, *Bacillus*, *Clostridium*, and *Corynebacterium*, and, later, on the large groups of lactic acid bacteria, extremophiles, yeasts, and filamentous fungi.

3.2 Basic biology of plants and associated organisms

- elucidating the central processes of sexual propagation; clarifying mechanisms of flowering, germ cell differentiation, and the molecular basis for the characterization of gametes and selection systems;
- working out the fundamentals of plant cell regeneration (the genetics and molecular biology of embryogenesis of somatic and germ cells); detecting and modifying growth-promoting signals;
- discovering the molecular restriction patterns of plants and associated organisms and the molecular fundamentals of host spectrum and virulence; describing plant defense mechanisms; developing gene technology for pathogenic fungi or mycorrhizae; regulating the expression signals of microbe-plant genes; identifying both structurally and functionally the genes that are involved in symbiotic nitrogen fixation;
- compiling the physiological characteristics of crops (their storage processes, stress physiology, nitrogen utilization).

3.3 Biotechnology of animal cells

- developing animal cell techniques and culture techniques that can lead to new or improved production of important substances for use in industry and in zoological technology;

- extending what is known of animal genetics; analyzing the genome, investigating important genes, finding methods of gene transfer, studying gene expression and regulation in cell cultures;
- improving animal husbandry; enhancing immunity through genetically engineered second-generation vaccines.

4 Preliminary research

Preliminary research in biotechnology lies at both ends of the Research-Development-Utilization continuum.

4.1 Safety tests in conjunction with the release of genetically engineered organisms into the environment

- establishing monitoring and control techniques; taking test samples of genetically engineered organisms with newly incorporated DNA sequences; devising methods and instrumentation for automated high-resolution microbial identification and establishing suitable databases; developing a reference collection of specific probes and chemical markers for a large number of specific microorganisms; finding methods for eradication;
- developing trial techniques for biological encapsulation, gene stability and gene transfer; developing micro-models and simulation methods for environmental impact analysis;
- acquiring the basic know-how on gene behavior (horizontal transfer between species, rearrangements of engineered genes within the host organism) and on the survival and adaptation of organisms that have been introduced into the environment, specifically soil bacteria; furthermore, recording changes in the host specificity spectrum and the range of the economic usefulness of genetically engineered viruses;
- inventing novel constructs, such as biologically encapsulated organisms, suicide vectors, or constructs that cannot develop outside a host organism, or genetically engineered organisms that can be destroyed in the environment by currently available specific techniques.

4.2 *In vitro* evaluation of toxicity and pharmacological activity of molecules

- developing cellular and multicellular assay systems as substitutes for *in vivo* tissues and organs;
- conducting research on the preparation, storage, viability, and growth of human cell cultures;
- developing cell lines that show stable functional properties.

Execution

This part of the program must be carried out with training programs, contracted cost-sharing research projects, and participation in certain COST activities (identified as Category A). The training programs are to be conducted by making training contracts and setting up courses for each of the above topics. Costs of these actions are to be assumed by the Community.

Industrial companies, including small and medium-sized ones, research institutions, and universities can participate on a cost-sharing basis in projects, individually or jointly, provided they are located within the Community or in European nonmember countries that have signed a general agreement for scientific and technological cooperation. Until the terms of the eventual Council guidelines on the planned release of genetically engineered organisms into the environment are set, the proposals selected must conform to the pertinent safety rules and guidelines of the country in which the field trials will take place; in countries that lack such rules or guidelines, the initiators of field trial projects must obtain the written approval of the appropriate administrative authority.

Research projects and cost-sharing arrangements in which industry participates with research institutions or universities (or both) are vigorously supported. Within the framework of the program, the participation of industry should constitute an important criterion for selection.

The Community's contribution to research contracts with a cost-sharing basis will be 50% of the total costs. However, for universities and research institutes, the Community will absorb up to 100% of the cost of projects within this program.

Two kinds of intra-EEC research programs have been planned, in which participants of more than one member country will contribute (regardless of participants from nonmember countries).

- *N projects*, for initiating research in suitable Community institutions (European Laboratories Without Walls, or ELWW) in areas in which the most important bottlenecks result from deficiencies in basic knowledge. The financial contribution from the Community to such projects shall not exceed 400,000 ECU per year;
- *T projects*, for investing significant expertise and funds to eliminate important bottlenecks caused by constraints of organization or size. The contribution of the Community to these projects will range between 1 and 3 million ECU per year for any given project.

Calls for research contracts on a cost-sharing basis are published in the official bulletin of the European Communities, and contracts are granted after a selection process.

In accordance with the Community's rules and conforming to the contractual agreements, special attention is given to disseminating the program's results. Thus will the impact of the work be maximized, and all enterprises within the entire Community, including small and medium-sized companies and those most in need, will benefit.

COST activities (Category A) associated with Action I: Research and training

Content

- primary marine biomass;
- *in vitro* cultures for maintaining pure varietal characteristics and propagating horticultural plants;
- methods for early diagnosis and identification of plant diseases;
- significance of the VA (vascular-arbuscular) mycorrhizae in soil biochemistry and in plant nutrition;
- biotechnological development of a coccidioidomycosis vaccine.

Execution

Sessions, consultation with experts, publications, exchange of researchers between laboratories, and coordination contracts are planned for implementing these **Action I** projects.

Action II: Coordination

Content

The following tasks are conducted in cooperation with the pertinent official bodies of the Commission and the member states:

(a) observing the developments in biotechnology, especially in research and development, evaluating their impact, and subsequently informing the official bodies of the Commission and the states' administrative units that have related responsibilities;
(b) determining the feasibilities of improving conditions incidental to the useful development of biotechnology in Europe, expanding the efficacy and coherence of biotechnological programs within the existing politics of the member states and the Community, and involving international cooperation;
(c) disseminating knowledge and improving public awareness and understanding of the nature, the potential, and the possible risks associated with biotechnology;
(d) determining the need and help that is necessary if increased activity of smaller companies in biotechnology is to be promoted within the Community.

Execution

This Action will extend the initiative (begun in the Biotechnology Action Program, BAP) of *ad hoc* collaborations among groups and individuals who have interests and capabilities in the biosciences and biotechnology; informal and flexible networks will thus be established, and will promote coordination by the exchange of information among participants, thereby supporting the wider spread of information required by the above tasks.

These tasks within a given institution will essentially involve analyzing, developing, and operating a clear means of communication, including job-related travel, and may, if required, also include taking requests for study reports, conducting workshops and sessions, supporting the preparation of reports, and disseminating information.

An appropriate portion of the funds for **Action II: Coordination** is allocated for the future impact of research and development within the **Action I** biotechnology program, for example, the impact on the consumer, society, environment, and development.

Appendix II

Preliminary Budget Details

	Millions of ECU	
Action I: Research and Training		
• Contract research, (allocated equally to N projects, 38.25 M ECU, and T projects, 38.25 M ECU), and broken down as follows:		
Preliminary (preregulatory) research	15.5	
Cell biology	27.0	
Support technologies	27.0	
Communication infrastructure	7.0	
Total for contract research:		76.5
• Training activities		12.0
• COST actions		2.0
Action II: Coordination		9.5
	Total	100.0[1]

Appendix III

Program goals and criteria for evaluation

The Commission's report to the Council about a common action plan that evaluates the research and development activities between 1987 and 1991 [Commission on Biotechnology Research and Development, KOM (86) 660, end] stipulates that the plans and intermediate stages of each research program must be submitted in a format that can be evaluated. The program's plan and intermediate stages are outlined as follows:

Action I: Research and Training

1 The long-range objective is to contribute to the exploitation and optimal usage of the Community's biological resources, and, consequently, to improve the research activities required for the competitiveness of the European agriculture and biotechnology industry, as well as to protect the environment. This goal should be attained by eliminating the scientific and technical bottlenecks stemming either from deficiencies in knowledge or constraints due to size or organization. Research projects are conducted in a reciprocal fashion using interdisciplinary integration that merges the requirements and options of the various member states and fuses diverse expertise in basic and applied fields.

2 Primary short-term objectives, therefore, are to prepare proposals for research and training activities at a level that corresponds to the proposed Community budget; subsequently to carry out these activities (N projects, T projects, training, and cooperation with nonmember countries) so that they contribute to a significant advance in

[1] including 9% personnel costs

interdisciplinary cooperation between countries as well as to scientific mobility. These objectives should be reviewable by 1992 or 1993.

3 Specific N-project objectives, to be reviewed in 1995, are as follows.

3.1 Networks of international collaboration should be established in each of the four programs (Communication infrastructure, Support technologies, Cell biology, and Preregulatory research).

3.2 International collaboration should be demonstrated by examining scientific publications (each individual network or ELWW produces at least one publication, either with an international authorship or by noting that material or methods have been furnished by other contractual partners).

3.3 The high quality of scientific advances should be demonstrated by an evaluation by scientific experts, and by citation analysis of the references in review papers that summarize research results.

3.4 Industrial interest should be expressed either by industrial participation in at least 20% of the active projects, or in the utilization of research results outside the legal framework of BRIDGE.

4 Special objectives that should be attained by setting up and carrying out T projects include the following.

4.1 A description should be prepared that details the research efforts and the expected benefits of specific research objectives, such as sequencing the yeast genome, automating high-resolution microbe identification, or identifying novel plant genes at the molecular level.

4.2 Two years after initiating a given T project, the advances towards the objectives and the means of attaining the goals within the allotted time should be documented.

4.3 Important contributions should be cited, indicating that the objectives were indeed achieved and that the special needs of industry, agriculture, and environmental protection were served.

5 The training program should award stipends for about 2 years to young scientists, and for 1 to 2 years to senior scientists, to work in laboratories with high scientific standards. The following actions are of special concern.

5.1 A publicity campaign should be conducted in all Community member states.

5.2 Most awardees should return to another member state of the Community, if not to their country of origin, to work in biotechnology after their training period is completed.

5.3 Training courses, summer schools, and workshops supported by the program should be organized, with contributions by industrial participants whenever possible. Where indicated, representatives of other disciplines are welcome to participate.

The above criteria may be reexamined in 1993, in part, and a further review should be done in 1998.

Action II: Coordination

An evaluation of this Action will indicate whether or not the program has in fact fulfilled the tasks outlined in the Council decision, and whether or not the tasks' execution has indeed contributed to the stated objectives. More specific criteria for evaluation are as follows.

1 With regard to coordination with member states, this Action should support the personnel responsible for biotechnology in the administrations of the member states;

1.1 They should receive information about the current and projected Commission initiatives in the relevant areas of biotechnology;
1.2 They should receive information about biotechnology projects and planning in other member states;
1.3 They should thereby have the means to consider joint or independent ventures of other member states in biotechnology in their own national plans or initiatives.
2 With regard to the impact on the conditions for biotechnology in Europe, this Action must itself be evaluated to establish whether and to what extent it has contributed to the improvement of the basic conditions for the safe development and the profitable application of biotechnology in Europe, taking into particular consideration Europe's ability to compete internationally, the creation and growth of small companies, and the climate of public opinion on biotechnology.
3 Account should be taken of the impact on the development of international cooperation in biotechnology, particularly in the areas of research and development, including the developing countries.
4 In view of the results obtained from the biotechnology research sponsored by the Community, individual nations, or by the private sector, two things need to be demonstrated. They are:
4.1 Has the BRIDGE program contributed to the implementation of results (generated by the research activities outlined previously) in those regions of the Community where no research was conducted?
4.2 Has the BRIDGE program appropriately taken into account all selection criteria listed in Appendix III of Resolution 87/516/Euratom/EEC? This resolution also includes the criterion that a program is expected to contribute to strengthening the economic and social unity of the Community, while making every effort to achieve scientific and technological quality.

ECLAIR (European Collaborative Linkage of Agriculture and Industry through Research) [1]

Research and development at the interface between agriculture and industry can help solve some problems presently faced by the agricultural sector of the Community.

Consequently, this type of research activity should be added to the Community's existing research programs in biotechnology and agriculture, and the application of its results to the social and economic objectives of the Community should be promoted. The most urgently important developments required are:

- new means of agricultural production that are adapted to industrial needs,
- technologies for transformation, and
- other tools for agriculture, such as means of pest control and fertilizers, that are less of a burden to the environment.

It is important to involve small and medium-sized agricultural and industrial enterprises as much as possible in this biotechnology-based program of agro/industrial development.

Of the articles in the Decision of the European Community Council, four are particularly worthy of mention here.

Article 1

It was resolved that a research and development program for agro/industrial research and technology development, based on biotechnology as described in the Appendix, would begin on 1 July 1988 and continue for a 5-year period.

The program includes contractual research, and grants for coordination, training, and travel. Participants may be industrial or agricultural operations run individually, by associations, or by cooperative organizations, research institutions, universities, or combinations of these that are located within the Community.

Article 2

(1) In projects conducted under cost-sharing contracts, the Community's contribution as a rule will be a maximum of 50% of the total expense; the remainder must be borne by the partners, and especially by industrial and agricultural concerns.

In projects carried out by universities and research institutes, the Community may absorb up to 100% of the expense.

(2) Usually at least two independent partners from different member states should participate in the projects. Generally, research institutes and universities should collaborate with one or more industrial or agricultural organizations.

Article 3

The budget for conducting this program is estimated to be 80 million ECU (European Currency Units).

---

[1] An initial 5-year program (1988–1993) for agro/industrial research and technological development in biotechnology

Article 6

The Commission will review the program in the third year of operation and provide a report on its findings to the European Parliament and the Council. It will also make suggestions for changes in content, budget, and time limits for the program.

Appendix

Program Content

The program will have the following components:

1 *Generation and testing of new varieties and novel organisms* Experiments with new or modified species or organisms (plants, domestic animals) will be conducted in appropriate numbers and under varying conditions. The following aspects will be studied: performance/yield; disease resistance; agricultural/industrial need; suitability of the organisms for industrial processing; and animal nutrition and market acceptance of the organism, its composition, and the resulting products. Particularly important is the application of new biotechnology methods in the identification, characterization, selection, variability, dissemination, cultivation, or other considerations to the development and evaluation of interesting organisms.

2 *Industrial products and services*

2.1 *More specific and efficient agricultural products resulting from basic scientific and biotechnological research and development*
Success in this area will involve creating products and services for use in agriculture that bring advantages in precision, cost efficiency, or improved plant and animal performance. These products or services must have no undesirable side effects, and be suitable for processing and marketing.

2.2 *More specific and efficient extraction, processing, and production procedures, as a result of research and development in these or other methodologies, that increase the usefulness and value of the products of agriculture, industry, and other enterprises*
Thus, these procedures, aided by biotechnology and other measures, should create or impart more value to characteristics intrinsic to the products.

3 *Integrated projects*

3.1 *Experiments to improve utilization of the entire harvest with development of systems that harvest the total plant biomass, store it, and separate it into its components that are suitable for subsequent use*
Such experimental systems must be of sufficient scale to provide for an economic evaluation and invite a significant participation of industry and agriculture. Priority is assigned to those projects that employ new biotechnology.

3.2 *Studies and development projects for the integrated application of new technologies should primarily target the general utilization of advances in biological knowledge and procedures for new agricultural systems based on technology.*

Biological Hydrogen Generation: Scientific Framework of the Program Supported by the Federal Ministry of Research and Technology (BMFT)

Part of the "Applied Biology and Biotechnology" Program of the German Federal Government

Preface

The text of the Preface is from Chapter 1, Initial Position of the BMFT Support Concept "Biological Hydrogen Generation", with minor modifications, and with an added final paragraph

Hydrogen as an energy source has figured in public and political discussion during the past two years, with the vision of solar hydrogen generation as an inexhaustible and environmentally safe energy supply being a central feature of those discussions.

One concern is that of an adequate energy supply, in view of the world's increasing population; another is the environmental burden caused by the existing systems of energy supply involved in conversion, distribution, use, and waste disposal.

The increasing levels of emitted sulfur dioxide, nitrogen oxides, and carbon dioxide are causing substantial environmental problems. The increase in carbon dioxide concentration in the atmosphere has become so significant that today – in spite of many uncertainties in interpretation – we can no longer ignore the possibility that in the next few decades we may see substantial changes in the earth's climate.

The situation is aggravated by the burning of biomass, particularly the clearing of forests by burning, and the progressive soil erosion resulting from intensive agricultural use of cultivated land. The results of these practices may force us to find substitutes for fossil fuels even before the fuel sources are depleted.

Therefore, in strategies for our energy future we must consider not only the limits on the availability of fossil energy sources (shortages of resources), but also the extent to which we can burden the atmosphere with harmful substances, especially carbon dioxide.

We therefore need to make strenuous efforts to develop the technology of generating hydrogen from solar energy. The matter is so important that even processes that, in current opinion, show little promise must be explored. Biological hydrogen generation is one of these; although its principles are known, options for its application have not been thoroughly explored yet.

1 Introduction

In April 1988, an *ad hoc* committee, mandated by the Federal Ministry of Research and Technology (BMFT), produced a report on "The Solar Hydrogen Energy Industry". In the study, solar-powered hydrogen generation was depicted as a future worldwide option, substituting for large amounts of fossil fuel for centuries to come. The report suggested expanding our fundamental knowledge about photochemically and photobiologically generated hydrogen, as well as developing further the technology of solar-powered electricity and electrolysis. Based on these recommendations, a supporting document, "Biological Hydrogen Generation", was drafted, and research and development missions were suggested. This second report also included a general evaluation and description of the initial situation, as presented and amplified at a panel discussion on 5 December 1988. This discussion resulted in a list of future research topics, divided into four subprojects:

- Biophotolytic hydrogen generation;
- Photoproduction of hydrogen from biomass;
- Hydrogen generation from biomass by fermentative processes; and
- Structure and function of hydrogenases.

Research and development missions were outlined for these four subprojects and were coordinated with each other in a panel discussion on 13 March 1989.

The participants agreed that, from the thermodynamic standpoint, only light-mediated hydrogen generation is feasible. Nevertheless, the subproject "Hydrogen generation from biomass by fermentative processes" was included because hydrogen production in the dark contains important intermediate steps of the light-driven process.

2 Scope and evaluation

The goal of the "Biological Hydrogen Generation" projects is to expand our fundamental knowledge of light-mediated hydrogen generation, so that we may assess the feasibility of applying the knowledge industrially. There is still plenty of skepticism about the likelihood of generating hydrogen by biological means for the power industry, but our knowledge is still too incomplete to render a definitive judgement. Before we can reach a decision on potential industrial usefulness, we must have a study on the biological material in question. In this context it represents basic research, but clearly has definite connotations of applied research also.

In 1983 the BMFT sponsored a research project by Dornier System GmbH on the analysis of photoelectrical and biological methods used in solar energy conversion. They did a comparative cost study of hydrogen production, including photovoltaic and electrolytic production systems; biophotolysis was rejected. The main problems noted were in maintaining high efficiency and in separating hydrogen from oxygen.

Since the Dornier study of 1983, know-how in molecular biology and in immobilization techniques has progressed. The breakthroughs in molecular biology, in particular, make possible a reassessment of the biological potential of solar hydrogen generation.

3 Research and development missions

Four subprojects were defined because there are three energetically different kinds of H_2 generation, and hydrogenase is the common key enzyme.

Hydrogen may be formed biologically from H_2O or from biomass. In the latter case, the biomass may be oxidized to CO_2 either completely or partially.

Hydrogen generation from H_2O is a strongly endergonic process that can proceed only with energy supplied by light quanta. For this, photosystems I and II are required, and thus the biophotolytic splitting of water occurs only in organisms that generate oxygen.

The production of H_2 from biomass with complete oxidation to CO_2 is also an endergonic process that can proceed only upon absorption of light energy. However, only one photosystem is involved here, because this process has a lower energy requirement than does the splitting of H_2O; therefore, the photoproduction of H_2 from biomass can also be found in phototrophic bacteria with anaerobic photosynthesis.

On the other hand, the generation of H_2 from biomass can also proceed exergonically provided the organic material is not oxidized completely. This kind of H_2 formation occurs without input of other energy (in the dark) and is characteristic of fermenting microorganisms. The list of fermentation products generated, besides hydrogen and carbon dioxide, may include acetic acid, ethanol, and butyric acid, and grows ever longer.

H_2-generating microorganisms all have a hydrogenase that catalyzes the reversible reduction of protons to hydrogen. This enzyme, along with photosystems I and II, is therefore central to studies on structure/function relationships.

Another prominent participant in H_2 formation in phototrophic organisms is the enzyme nitrogenase. This molybdoprotein catalyzes the ATP-dependent formation of H_2 as a side reaction of N_2 fixation. Studies on the structure/function relationships of this enzyme will also be important in this program. This enzyme was not given a special subgroup of projects because research on it is viewed as an integral part of the subgroups "Biophotolytic hydrogen generation" and "Photoproduction of hydrogen from biomass".

3.1 Biophotolytic hydrogen generation

Status of research

The process of photosynthesis uses solar energy to split water into molecular oxygen, protons, and electrons. The electrons are used to reduce CO_2 to biomass. Additionally or alternatively, in cyanobacteria and green algae the protons can be reduced to molecular hydrogen. The process of photosynthesis is required if the water molecule is to be split, but H_2 formation, like CO_2 fixation, can proceed in the dark. Two photosystems, organized as so-called biological solid states in the thylakoid membrane, assure an efficient photoprocess. Two light quanta drive one electron through an electron transport system; four quanta (photons) are required to generate one molecule of hydrogen. Researchers have tried very hard to mimic the natural system with artificial model systems, with no success to date. Consequently, we continue to depend on biological systems to split water photosynthetically into H_2 and O_2.

The ultimate goal is to convert the maximum possible number of reducing equivalents, generated during the dissociation of water, into molecular hydrogen. To attain this goal, the H_2-generating system and its coupling to the photosystems needs to be optimized, stabilized, and (possibly) genetically modified. To do all this we must learn more about the biochemical and physiological functions; targeted basic research is needed. Microalgae could serve as suitable objects of study, and possibly in a later phase as production systems. Prokaryotic and eukaryotic species can be maintained in large-scale cultures; in Germany in particular there has been considerable research done on culture methods and growth conditions. We can now isolate and characterize the photosystems and the H_2-generating systems; techniques of gene cloning and gene transfer (particle gun method) have become available for cyanobacteria as well as for green algae.

Research and development missions

Biological hydrogen production using solar energy is a process of photosynthetic electron transport; water is photosynthetically split, and molecular hydrogen is produced or liberated. In green algae the latter reaction is mediated by a hydrogenase; cyanobacteria have a hydrogenase, a nitrogenase complex that fixes nitrogen, or both. In green algae, several processes that include hydrogenase participation can be distinguished. The role of hydrogenase in cyanobacteria is not always clear. In some cases, this enzyme refixes the hydrogen generated during the nitrogenase-mediated N_2 fixation. Blocking the hydrogenase will then result in hydrogen liberation, as demonstrated in recent work. The cellular system appears more promising than isolated cell-free preparations for industrial applications, if steady-state conditions for generating the hydrogen can be found. Important targets of R & D work are:

- The photosynthetic process by which cyanobacteria and green algae split water; in particular, the working out of more details and functional relationships in the photosystem II complex and photosynthetic electron transport.
- The biochemistry, physiology, and molecular biology of the nitrogenase in cyanobacteria; the reversible and irreversible regulation of activity by ammonia, metabolites, and oxygen; improvement of H_2 generation through genetic modifications; research on increasing the nitrogenase concentration.
- The biochemistry, physiology, and molecular biology of hydrogenase in cyanobacteria and green algae; the optimizing of biosynthesis; isozymes isolated and correlated with various processes of hydrogen metabolism.
- Bacterial hydrogenase genes transferred to and expressed in green algae.
- The influence of exogenous factors that limit and modify light-induced H_2 production (for example, activators, inhibitors, metal ions).
- Bioprocessing techniques.

3.2 Photoproduction of hydrogen from biomass

Status of research

While photolytic hydrogen generation is limited to cyanobacteria and green algae, photoproduction of hydrogen from biomass is done principally by phototrophic bacteria with anaerobic photosynthesis, including purple bacteria, green bacteria, and some species of *Heliobacterium*. Among the best known families that use organic compounds as electron donors are the Rhodospirillaceae and the Chloroflexaceae. Other groups, such as the Chromatiaceae and the Chlorobiaceae, use reduced sulfur compounds. Many of these bacteria contain nitrogenase and under suitable conditions produce molecular hydrogen, which may be fixed all over again by another hydrogenase.

Research and development missions

The goal in photoproducing hydrogen from biomass is to trap as hydrogen all the reducing equivalents produced during the oxidation of organic compounds to CO_2. Furthermore, we want to prevent the loss of reducing equivalents to biosynthesis and the reutilization of the H_2 by hydrogenase. All this can be achieved only by genetic and metabolic means. Three research areas can be identified: (a) genetic and molecular-genetic optimization of H_2-producing strains of phototrophic bacteria with anaerobic photosynthesis; (b) enzymatic and metabolic studies with H_2-producing phototrophic bacteria; and (c) studies on bioprocessing techniques. Suitable targets for R & D missions are as follows:

- Molecular-genetic constructs of strains in which nitrogenase is derepressed, and analysis of the regulation of expression of the *nif* operon;
- Development of mutants with defective reaction patterns that compete for reducing equivalents during H_2 formation, have a functionally defective hydrogenase, or both; analysis of the rate and efficiency of H_2 formation in these mutants;
- Enzymatic studies of nitrogenase and hydrogenase;
- Metabolic studies and optimization of H_2 formation on a variety of substrates in continuous and batch culture;
- Photoproduction of hydrogen in mixed cultures: studies on the population dynamics of competition between H_2-producing phototrophic bacteria and H_2-consuming chemotrophic anaerobic bacteria;
- Bioprocessing techniques.

3.3 Hydrogen generation from biomass by fermentative processes

This section of the project is concerned with making hydrogen and optimizing the process, and also with the way microbes exploit and use excess biomass; this latter area includes the microbial consumption of byproducts and waste products from other bioprocesses.

Status of research

The production of hydrogen from biomass by fermentation under physiological conditions (about 25 °C, pH 7.0) is limited by the unfavorable redox potentials of numerous catabolic redox reactions. Those oxidation reactions that can transfer electrons directly to a low-potential electron carrier such as ferredoxin (for example, the oxidative decarboxylation of pyruvate or the oxidation of aldehydes) pose no fundamental problem, but NAD-coupled reactions (for example, the oxidation of alcohols, glyceraldehyde phosphate, or amino compounds) hold much less promise for efficient hydrogen production because of the drop in electrochemical potential. Electrons at a potential of − 125 mV, such as those generated in oxidizing fatty acids, are particularly difficult to use. Therefore, no more than 1/3 of the electrons generated by the complete oxidation of glucose under standard conditions can be released as molecular hydrogen. However, in what are called *syntrophic mixed cultures*, the electrons from all the processes named above are released quantitatively as molecular hydrogen, but this can happen only if the organisms in the association (methane bacteria, sulfate reducing bacteria, obligate fermenters of acetate) keep the external hydrogen pressure sufficiently low (1 to 100 Pa, 10^{-5} to 10^{-3} bar). We know little about the electron transport pathways in these obligate syntrophic fermenters; the key questions are whether and how energy-dependent potential shifts (reverse electron transport) participate. Some sulfate-reducing bacteria also liberate significant amounts of hydrogen from organic substrates with unfavorable redox potentials, in a manner yet unknown.

Current thinking is that mesophilic bacteria (methanogens) that can efficiently harvest low concentrations of hydrogen are responsible for liberating redox equivalents from electron transport carriers with unfavorable redox potentials. At temperatures of 90 °C, such as are found in hot springs, there is no methanogenesis, and hydrogen is liberated without the participation of hydrogen-harvesting partners. At pH 6.0 there is no significant methane formation either; the role of the methanogenic bacteria is initially assumed by the far less efficient obligate acetate fermenters, and eventually, when the pH value has declined, hydrogen is liberated without the participation of a syntrophic association. At both of these conditions, elevated temperature and low pH, hydrogen formation is thermodynamically more efficient, but we do not have systematic comparative studies that establish whether or not appropriately adapted bacteria can take advantage of these conditions.

Research and development missions

It follows from the above discussion that three different aspects of the fermentation-dependent generation of hydrogen from biomass merit study:

- The enzyme systems that participate in the generation of hydrogen from electron carriers with unfavorable redox potentials;
- The possibility of improving the fermentation-associated generation of hydrogen by changes in physical or chemical parameters (temperature, pH, etc);
- The optimization of the hydrogen yield at low partial pressures by transferring electrons with unfavorable potentials into other desirable reduced compounds such as methane.

3.4 Structure and function of hydrogenases

Status of research

Hydrogenases catalyze the reversible reduction of protons to H_2. The active site contains a transition metal, either iron or nickel, surrounded by sulfur, selenium, nitrogen, and oxygen ligands in an arrangement still unknown. This metal reacts reciprocally with iron-sulfur clusters. Depending on whether nickel, iron, selenium, or combinations of these are present, the enzyme is an Fe-, NiFe-, or NiFeSe-hydrogenase. Different hydrogenases have different electron carrier specificities, different subunit compositions, and different locations within the cell. In spite of this diversity, we deduce from sequence homologies that in all hydrogenases the subunit containing the active site appears to spring from a common phylogenetic origin.

We know next to nothing about the structure of the active site or about the catalytic mechanisms involved; we also do not understand why certain hydrogenases are more stable than others. Nobody has yet succeeded in reconstructing hydrogenase *in vitro* from its components. Expression of an active hydrogenase in host organisms is still a very recent achievement. Furthermore, we still understand very little about the function of hydrogenases, nor do we know why many organisms contain several kinds of hydrogenases or what the physiological function of these different enzymes may be. We have not yet succeeded in relating specific functions to the three kinds of hydrogenases (Fe-, NiFe-, and NiFeSe-).

Research and development missions

Photobiological generation of hydrogen will be possible only when a stable, O_2-insensitive hydrogenase is found, or when it is possible to modify the hydrogenase genetically and express it in host organisms. In addition, we must unravel the biosynthesis of hydrogenase and the relationships among structure, function, and stability.

These problems should be approached broadly, using various hydrogenases from different organisms. Objects of study in R & D missions are as follows:

- The three-dimensional structure of hydrogenase;
- The catalytic mechanisms;
- The biosynthesis of hydrogenases and its regulation; exploration of the secondary proteins participating in hydrogenase biosynthesis, and controlling expression of these proteins; development of gene-transfer systems;
- The *in vitro* reconstitution of hydrogenases;
- Site-specific and nonspecific genetic alterations; designing proteins;
- External stabilization of hydrogenases, such as by binding to carriers or by encapsulation;
- Tests of hydrogenases for possible industrial use.

The direct biophotolytic generation of H_2 will be feasible only when the coupling of hydrogenase electrons with photosystems I and II can be accomplished; hydrogenase and photosynthesis research must therefore be carried out jointly. The R & D missions for this are as follows:

- The *in vitro* coupling of hydrogenases with the photosystems and with artificial electron transport carriers;
- The stability of hydrogenase when coupled;
- The transfer of hydrogenase genes between chemotrophic and phototrophic organisms.

We assume that it will be possible to synthesize inorganic catalysts once we have a precise knowledge of the structure of hydrogenases in hand; these *biomimetic catalysts* could lower the activation energy for the reduction of protons to H_2 as efficiently as hydrogenases do, and would be important for *in vitro* systems that generate H_2 photobiologically. Moreover, they will also be useful in the electrolysis of water by photovoltaically produced current (see the expert testimony of the *ad hoc* committee). The R & D mission in this area is the following:

- Biomimetic hydrogenases.

Commercial Biotechnology

Erythropoietin: History of the Development of a New Product*

1 Introduction

Erythropoietin, or *EPO*, is an important regulator of erythropoiesis (the production of red blood cells, as in the bone marrow) in mammals. CARNOT and DÉFLANDRE, in 1906, were the first to speculate about an endocrine regulation of erythropoiesis, but convincing evidence came only in the 1950s with the work of REISSMANN, STOHLMANN, and others. The name "erythropoietin" was coined in 1959 by GORDON. Since then, a great deal of research has been done on erythropoietin; its function has been investigated, its structure determined, and correlations with diseases established. Eventually, EPO was produced by genetic engineering and thus became available for clinical use.

Today EPO has been approved and is available for treating various diseases. This product is of particular interest to the gene technologist because it is a completely novel product and not a substitute, in contrast to all other genetically engineered pharmaceuticals approved to date, such as human insulin, human growth factor, tissue plasminogen activator, and hepatitis-B vaccine. The market for EPO is estimated at a billion dollars. This chapter describes the difficult developmental history through the year 1989, and is an attempt to show that the problems that arise in developing such a product are not solved by successful technological know-how and clinical trials. On the contrary, obstacles appear at every level: patent disputes, approval processes, questions about contractual obligations, opposition to genetic engineering, the market analysis, and so forth – even the abuse of this new drug by high-performance athletes. Navigating around each of these treacherous shoals and reefs was the only way to avoid shipwreck.

I will start by reviewing the mode of action of EPO and its use in medicine.

2 Biological fundamentals and medical considerations

2.1 In vivo production and mechanism of action

Erythropoietin is produced primarily in the kidney (the cell type responsible still remains to be identified), and to a lesser degree in the liver; it is subsequently released into the blood stream and acts upon the bone marrow, effecting the differentiation of stem cells into erythrocytes.

Workers making both clinical and experimental observations have found that the EPO content of plasma and urine depends on the oxygenation of the site of EPO induction. The relationship is an inverse one: production of EPO is stimulated by anemic anoxia (the result of bleeding or hemolytic anemia), hypobaric hypoxia (brought about by reduced barometric pressure, etc.), and histotoxic hypoxia (caused by exposure to cobalt, etc.). Conversely, production of EPO is reduced by hyperoxia or excess erythrocyte levels. The kidney seems a logical site for the production of EPO because oxygen consumption in the kidney depends on blood flow; in contrast to the case with most other organ functions,

* Dr. Norbert RAU,
RauCon Informatik and Consulting GmbH,
D-6912 Dielheim, Fed. Rep. Germany

a great deal of oxygen is required for sodium transport. Anemias with increased plasma flow also exhibit increased oxygen consumption; the consequence is hypoxia in the kidney and the initiation of EPO synthesis. In polycythemia, however, blood has a higher viscosity and therefore a lower flow rate. Because of the reduced filtration rate and the accompanying reduced consumption of oxygen, the kidney is not hypoxic and produces no EPO. The induction of erythrocyte production thus does not need any further hormonal control.

About 10% to 15% of total EPO is generated in the liver. Here again, we do not know the exact cell type responsible. Some scientists believe that Kupffer cells synthesize EPO because macrocytes are able to produce some EPO; others think that hepatocytes are the agents, because some hepatomas produce EPO *in vivo* as well as *in vitro*. It is noteworthy that in a fetus all the EPO originates in the liver. There is, therefore, a feedback loop between the red blood cell count and the amount of secreted EPO.

There are few references to any direct effect of EPO upon stem cells; even the early progenitor cells are only slightly influenced by it. As the erythrocyte progenitor cells mature into erythrocytes, their sensitivity to EPO rises significantly, probably because surface receptors emerge. Erythropoietin acts by binding onto the surface receptors of the target cells (erythroblast and progenitor cells), but we do not yet know whether it is brought to the site directly or whether second messengers participate also. The number of surface receptors is relatively small, about 800 to 1,000 per cell. (This number was obtained with iodine-labeled EPO, but we do not know whether labeling reduces biological activity.) Differentiation to erythroblasts depends primarily on the presence of erythropoietin, but the mechanism of action is not clear. The final step of maturation to erythrocytes is probably not dependent on EPO.

2.2 Medical considerations

As the fifties ended, people knew that erythropoietin is a hormone that regulates the process of erythropoiesis, and that it is produced primarily in the kidney; they assumed that anemia in chronic kidney insufficiency is due at least in part to a deficiency of this kidney hormone, and that EPO administered in sufficient quantity could help. With the advent of molecular biology, there is now enough EPO for clinical trials. The conclusion is that EPO does mitigate the symptoms of anemic hypoxia in patients with chronic renal insufficiency, and perhaps it will work in anemias with other origins also.

The first clinical trials with recombinant erythropoietin have confirmed that EPO not only increases the concentration of hemoglobin but also improves the general health of dialysis patients. No serious side effects have been observed to date. Sometimes the patient's blood pressure rises because blood volume and blood viscosity increase suddenly; the elevated blood pressure and expanded blood volume can be controlled successfully in dialysis patients, but can present a problem in anemic patients in whom the blood volume cannot be corrected as easily.

I continue by discussing briefly a few diseases that are associated with the maturation of erythrocytes or their quantity, as well as with erythropoietin itself. I will also mention some clinical trials that have used EPO to alleviate symptoms.

Renal anemia in childhood

Renal anemia is defined as a reduction in oxygen carriers, compared to an age-dependent standard; it arises within the framework of chronic renal insufficiency (CRI). CRI in childhood is characterized by an irreversible increase in the level of serum creatinine, and is caused equally by congenital nephropathies, inherited urinary tract malformations, and acquired glomerulopathies. Anemia is one of the most significant symptoms in children with CRI. As the renal anemia steadily progresses, the effects on the heart and circulatory system mount. The stress on cardiac volume, along with the chronic organ hypoxia, results in limited physical capabilities; the extent of renal anemia and the drop in physical performance parallel each other. The physical well-being of children with kidney insufficiency depends, therefore, on the severity of their renal anemia. The pediatric importance of this disease has increased with the successful establishment of routine dialysis treatment in the past decade, as well as with kidney transplantation and the rising number of reported cases.

At first, when extremely sensitive assays were not yet available, the medical community thought that an absolute deficiency of serum erythropoietin was a decisive pathogenetic factor in renal anemia; now that we have more sensitive assays, we see the importance of erythropoietin deficiency in CRI in proper perspective. With conservative treatment of CRI children, the serum erythropoietin concentration rises as the degree of anemia increases. However, this rise becomes less pronounced as uremia progresses. In the end-stage of CRI, the compensatory rise in erythropoietin is totally absent. In summary, children with CRI have a relative deficiency of EPO in the preterminal phase of their illness, and an absolute deficiency in the final phase.

We still do not have many treatments for renal anemia that are based on its causes, not merely its symptoms. Erythropoietin preparations synthesized in cultured cells will be a welcome addition.

Chronic kidney insufficiency

The causes of chronic kidney insufficiency are complex. Chronic glomerulonephritis, cystic kidney, nephropathy, and nephrolithiasis can bring it on. The consequence of chronic kidney insufficiency is often a moderate to severe normochromic normocytic anemia. Three causes have been adduced: firstly, a deficiency of EPO; secondly, the inhibition of erythropoiesis by the byproducts of uremia; and thirdly, a shortening of the lifetime of erythrocytes; frequently, iron or folate deficiency and osteitis fibrosa accompany the other symptoms. Of these causes, the chief is surely the deficiency of EPO.

Dialysis patients

Worldwide, the number of dialysis patients is estimated at 175,000. Patients with CRI and its associated anemia suffer not only from their immediate disease; they also require dialysis on a regular basis to cleanse their blood and, moreover, need frequent additional blood transfusions to raise the number of erythrocytes to the necessary minimum value.

Some dialysis patients who did not require repeat transfusions were treated in phase I and phase II studies with recombinant human erythropoietin (rHuEPO, epoietin) provided by Ortho, Cilag, and Amgen. They began with a two-week control period without treatment; they then received increasing dosages of rHuEPO by intravenous injection three times a week, after their dialysis. Initial dosages were 24 IU/kg for two weeks, then 48 IU/kg, and, after a further two weeks, 96 IU/kg. In one patient, hemoglobin rose from 5.3 to 10.0 g/dl, in another patient from 9.0 to 12.0 g/dl, and the number of reticulocytes also increased. From these clinical trials, as well as from experimental data, researchers conclude that rHuEPO, like EPO, is specific for erythroid cells. The patients did not have side effects, nor did they produce antibodies. After similar phase I and phase II studies, patients who previously had required blood transfusions on a regular basis had no further need for them after or during the treatment with rHuEPO. The risks of infection, iron overloading, and immunologic sensitization associated with transfusions were thus eliminated.

Patients being treated with EPO feel better and their appetite increases. Additional metabolites in the blood may need to be reduced by dietetic measures, more frequent dialysis, or both.

The situation after a kidney transplant

After a kidney transplant, EPO concentration increases in a significant number of patients. It drops again when the hematocrit and hemoglobin reach high values. This means that the patient's metabolism and control mechanisms are functioning again. There are some patients, however, who develop erythrocytosis (excessive number of red blood cells) 3 to 90 months after their transplant, and the condition can last from 1 to 7 years. Risk factors for developing erythrocytosis are smoking, diabetes, and the absence of rejection reactions. These are also risk factors for thromboembolism. Patients with acute tissue rejection reactions show significant drops in the level of serum EPO. In most kidney transplant patients, erythropoietically programmed stem cells with a heightened sensitivity to EPO arise.

EPO levels at high altitude and during space flight

The level of EPO increases significantly in a person who stays one to two days at an altitude above 3,500 m, and then decreases slowly, reattaining its normal value within 22 days. If the person ascends further, the EPO level rises again. The secondary polycythemia characteristic of a stay at high altitude, brought on by altitude-dependent hypoxia, remains unchanged. If a person has been at an altitude of 6,000 m, the EPO level remains above normal for two to four weeks after the stay.

Astronauts after a space flight have a 10–15% decrease in the number of red blood cells. Although a decrease like this is normally of no consequence, there can be problems if the astronaut falls ill or if there is trouble in the spacecraft's support system. The decrease in the number of erythrocytes is probably caused by weightlessness, to which other factors may be added, such as hypokinesis/hypodynamia, muscle atrophy, altered hemodynamics, stress, metabolic changes, hypoxia, reduced atmospheric pressure, radiation, and acceleration.

Polycythemia vera

Polycythemia is a myeloproliferative disorder in which all bone marrow cells participate. Characteristic symptoms are erythrocytosis and an elevated hemoglobin level. Hypervolemia and hyperviscosity and the resulting impaired blood flow are responsible for most clinical signs and symptoms.

Secondary polycythemia

Secondary erythrocytosis (without involvement of the bone marrow) can develop in connection with an abnormally high level of erythropoietin. In some cases of tumor-associated erythrocytosis and in ordinary familial erythrocytosis syndromes, the activity of EPO does not correlate with the oxygen requirement of the tissues; the physiological set point for erythropoietin production is too high. Generally, in spite of the elevated set point, the negative feedback mechanism for EPO production still works. This means that the cause of the erythrocytosis that is initiated by an increased level of EPO must be sought at various sites within the EPO-regulating system.

2.3 The medical prospects of recombinant EPO

EPO is an important therapeutic advance in the treatment of dialysis patients, who often suffer from severe anemia and therefore require repeated blood transfusions. EPO, however, also holds promise for the treatment of different kinds of anemias associated with liver disease or rheumatoid arthritis, and for diseases of the bone marrow, as well as for the anemias associated with cancer and AIDS therapies. Autologous blood transfusions in patients who need an operation is also a most promising application for EPO.

On the other hand, once we have a thorough knowledge of and clinical experience with EPO, we may find new ways to relieve primary and secondary erythrocytosis, for example, by inhibiting erythropoietin.

3 Industrial development of recombinant EPO

3.1 The competitors ...

The competitors listed in Figure 1, along with their marketing affiliates, are the ones that matter. There are a few additional companies, such as Biogen and California Biotechnology (USA), Morinaga Milk (Japan), and Celltech (England), whose level of involvement

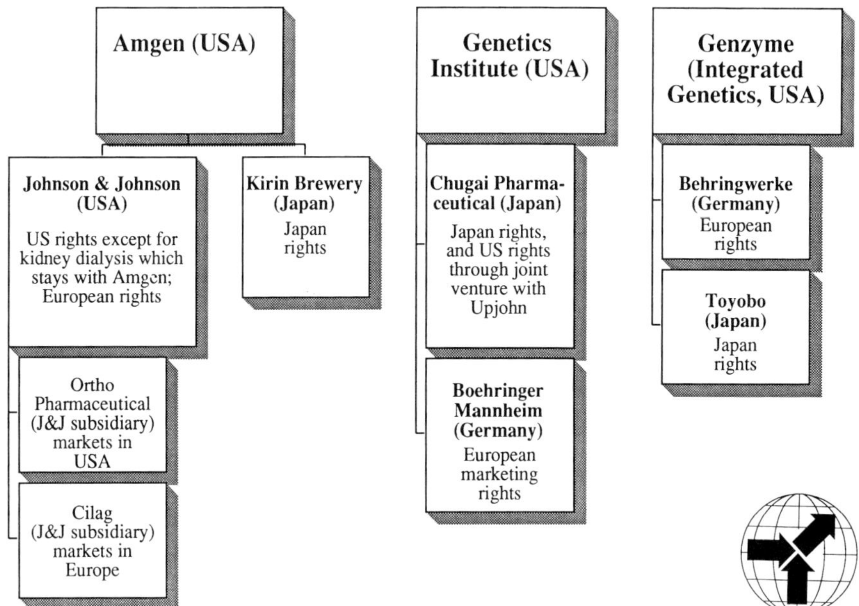

Fig. 1 Corporate competitors in the industrial development of recombinant erythropoientin

with EPO is unclear. Amgen, a company located in Thousand Oaks, CA (USA), is, without doubt, a pioneer in EPO development, producing recombinant EPO not only in CHO (Chinese hamster ovary) cells but also in bacteria, yeast, and human cells.

... in the USA

By the end of 1987, Amgen had established a lead of one to two years in the development of EPO; this lead has since shrunk to about six months. Amgen's EPO product was already in clinical testing in 1987. The company was founded as late as 1980, at a time when Genentech, Cetus, Biogen, and Genex, the "glorious pioneers of biotechnology", had already lost part of their original luster. Amgen established a partnership with Ortho Pharmaceutical, a subsidiary of Johnson & Johnson, to produce EPO. Deliveries were suspended for a time due to indecisiveness, and now Ortho is attempting to rebuild its production capacity. Furthermore, Ortho has also subcontracted production to Celltech (England). Amgen collaborates with Kirin Brewery Co., Ltd., in Japan.

Besides Amgen, Genetics Institute is the company with the greatest potential for marketing EPO. This company took a giant technological step forward when it identified the EPO receptor on the surface of red blood cells; this breakthrough could be the basis for the development of an EPO analog. In Japan, Genetics Institute cooperates with Chugai Pharmaceutical, and in Europe with Boehringer Mannheim.

Integrated Genetics, Framingham, MA (USA), is also involved, among others, in developing erythropoietin produced by molecular biology. It collaborates on development and marketing with the Behring Company, in Marburg, Germany, and with Toyobo, in Osaka, Japan. In September 1989, Integrated Genetics was purchased for $29.3 million (all dollars in this chapter are U.S.) by Genzyme (Boston) and since then operates under that name.

... in Europe

In the developmental sector, Johnson & Johnson operates with a license from Amgen; it acquired the distribution rights for Europe and a few other countries, and the codistribution rights for the North American market. In Europe, the product is distributed by the J & J subsidiary, Cilag; this has been the case since the autumn of 1988 in Switzerland and France, since the beginning of 1989 in (West) Germany, and since May 1989 in Austria.

The Behringwerke AG, Marburg, a subsidiary of Hoechst AG, had earlier signed an agreement with Integrated Genetics (now Genzyme) that established collaboration in the development, production, and marketing of EPO. Production planning was temporarily halted in September 1989, because of protests by opponents of genetic engineering and the "Grünen" (Green Party) during a public hearing.

In 1987, Boehringer Mannheim had begun clinical testing of EPO within the framework of a collaborative agreement with Genetics Institute. However, in 1989 a quarrel arose between the two companies when Boehringer Mannheim raised objections to a change in the production method introduced by Genetics Institute, reasoning that this could lead to problems in the pending drug approval process. During the first nine months of 1989, Genetics Institute delivered EPO valued at $19 million to Boehringer Mannheim.

British Bio-Technology also works in gene technology, and expects to market erythropoietin as well as other biotechnologically produced compounds.

... in Japan

By 1984, Kirin Brewery Co., Ltd., had established a joint enterprise with Amgen for $24 million that gave it the right to produce and market erythropoietin in Japan. This collaboration was favored by the fact that both these companies had achieved the same level in their research at that time. The EPO product developed in Japan is now in the final stage of clinical trials. Kirin invested $ 13 million in EPO production facilities. The product is intended for the Japanese and Chinese markets.

Chugai Pharmaceutical Co., Tokyo, is the powerful Japanese partner of Genetics Institute, and is licensed to produce genetically engineered erythropoietin on the basis of a long-term contract signed by both companies in 1983. Chugai constructed a 14,000-m^2 research facility, at a cost of 12 billion yen, for its biotechnology projects including EPO. It employs 200 people. In the meantime, it developed a method for producing erythropoietin with animal cells in a 1000-l bioreactor. In mid 1989, it became known that Chugai plans to erect what it believes will be the largest manufacturing facility for the products of gene technology in Japan.

Both Kirin and Chugai applied in Japan in early 1989 for approval of their products for the treatment of anemia.

Toyobo has developed erythropoietin jointly with Integrated Genetics (now Genzyme), and clinical trials for the treatment of anemic disorders with this product have begun.

3.2 ... and their legal disputes and fuzzy agreements

The economic base of the EPO market is characterized by an uncertain patent situation and diverse contract interpretations, which are being disputed aggressively. Two companies, Amgen and Genetics Institute, are quarreling over patent rights. Amgen filed an application for a broadly defined process for the production of EPO, and received the patent in 1987. The patent covers the EPO gene and the cells altered by gene technology that can produce EPO, but not the EPO product. A year after Amgen's application, Genetics Institute filed an application for a product patent, to cover all forms of human erythropoietin. This application was also granted.

Obviously, appropriate legal decisions could lead to one or the other group being blocked, since one company holds the patent rights to the product, and the other to the production process. An adverse decision would be especially damaging to Amgen, because Amgen is still 6 months ahead of the others. The problem could be solved with cross-licensing, but Amgen currently opposes this solution. A cross-licensing agreement would exclude all other competitors in the USA, especially Genzyme, whose product is still in preclinical trials. Amgen, at the end of 1987, filed a lawsuit claiming patent infringement by Integrated Genetics (now Genzyme), as well as by Genetics Institute and its partner, Chugai Pharmaceutical. Genetics Institute, which previously had been granted a patent that included the purified form of natural EPO, retaliated immediately with a suit against Amgen charging patent infringement. Peter Drake (analyst at Kidder, Peabody) estimated that settling these suits would take two to three years and $3 to $5 million. In December 1989 the courts decided that both patents are valid. At the time this was written, the consequences of this decision were uncertain.

To erect yet another barrier, Amgen asked the International Trade Commission (ITC) for an import ban on Chugai's EPO (obtained through Genetics Institute). Chugai could circumvent Amgen's process patent if it produced the EPO outside the USA. In February 1988, Upjohn and Chugai agreed to form a joint company named Chugai-Upjohn, which would sell EPO in the USA. Upjohn thus became involved in the lawsuit. Early in 1988 the International Trade Commission declared Amgen's claim justified. However, on 10 January 1989, the ITC judges decided against Amgen. Further, an application for an injunction against the export of EPO to Boehringer Mannheim by Genetics Institute was turned down in early 1989.

And, as if that were not enough, Amgen's partner Ortho Pharmaceuticals (Johnson & Johnson) in January 1989 turned against its partner and initiated a lawsuit that would forbid Amgen from marketing EPO in the USA. Ortho, in 1985, had bought from the financially strapped Amgen the U.S. marketing rights for EPO – with the exception of its use for dialysis patients, which market remained with Amgen. It finally occurred to Ortho that the most attractive portion of the market would still be Amgen's, so it tried its luck in court. (Angry insiders, however, maintain that Ortho started out with the belief that Amgen could not complete the costly approval process and, therefore, was willing to wait until it could inherit the dialysis business by default.) The matter was still in a court of arbitration at the time this was written.

3.3 Product approval and market expectations

On 1 June 1989, epoietin, the recombinant form of erythropoietin produced by Amgen, was approved by the American Food and Drug Administration (FDA) for the treatment of anemic disorders during hemodialysis and renal insufficiency – that is a good three quarters of a year after the Ortho/Cilag approval in Switzerland and France, and more than a half year after its approval in Germany. Amgen is marketing this product under the trade name Epogen. Estimates of the numbers of U.S. patients that might use this product range from 75,000 to 120,000, while Amgen itself considers 95,000 a preliminary estimate. Treatment costs, of great concern to the congressional committee charged with health concerns, are thought to range from $4,000 to $8,000 per patient per year; thus Amgen could potentially garner revenues of $380 to $760 million with this product in the USA alone. The price for 1,000 IU is $10.00. Four months after product approval, more than 25,000 U.S. patients have been treated with Amgen's Epogen, corresponding to $100 to $200 million per year.

The government is currently questioning the projected treatment costs. Amgen, however, maintains that the costs and consequently its earnings of this magnitude are justified by the risks it has taken – several hundred million dollars invested in research and development, as well as extensive legal costs.

The product has been assigned "Orphan Drug Status", which means that no other manufacturer is allowed to market a similar product for the same disease for seven years. (Orphan Drug Status is assigned to medications used to treat diseases or conditions affecting fewer than 200,000 U.S. patients.) In spite of the exclusive position thus granted Amgen, on 23 June 1989 the U.S. Health Care Financing Administration gave the go-ahead for an 80% reimbursement of costs. Amgen expects that sales will be $75 million in the first year after approval, and $125 million in the second year, so the cost refunds alone will add $60 million annually. Furthermore, Amgen, in contrast to Genentech, is always very conservative in its predictions, so that more, rather than less, is likely. Denise Gilbert of Montgomery Securities calculates product sales of $78 to $97 million for Amgen's fiscal year 1990 (ending 3 March 1991), but predicates these figures on the Chugai-Upjohn product not coming onto the American market before the middle of 1990.

At the end of June 1989, when the FDA gave preliminary approval for Ortho's EPO, called EPOrex, in the treatment of AIDS patients, Amgen pursued tests for the approval of EPO for home dialysis and for the treatment of childhood renal anemia.

In June 1989, Chugai-Upjohn also filed for approval under the Orphan Drug Status for its epoietin product Marogen. The FDA is currently (at the end of 1989) examining this application, because the product is supposed to be significantly different from Amgen's Epogen; however, Chugai-Upjohn has not indicated what the differences are.

What are realistic market expectations for recombinant erythropoietin? At this writing, when clinical trials and licensing applications for various diseases are not completed, projecting is difficult. Market expectations for a completely new drug depend on a whole series of imponderables; these questions, among others, need answering.

- How many patients can be reached with this drug? This turns out to be the easiest question.
- For what kinds of diseases will it be approved, when, and in what country?
- What prices are acceptable? How much of the cost is likely to be refunded?
- How long will protective legal barriers (patent protection, Orphan Drug Status) hold?
- When will competitors arrive with competing or substitute products?
- What side effects will emerge with widespread use of the drug? And how will these affect sales?

Furthermore, these questions will be answered differently in various countries because legal conditions vary, so even the predictions of experts can differ dramatically. For example, Shearson Lehman Hutton, Inc., estimates 1991 U.S. sales at $1,750 million, for the following uses:

- Surgery (autotransfusions), $550 million
- AIDS (treating anemia caused by AZT), $400 million
- Cancer (treating anemia caused by chemotherapy), $400 million
- Renal insufficiency (treating damage done by dialysis), $400 million

The same company predicts 1993 sales of $2,000 million, and $3,500 million by the mid-nineties! Amgen has warned that these predictions are overoptimistic. On the other hand, other projections are more guarded. Morgan Stanley and Co., Inc., forecasts U.S. sales of $96 million in 1990 and $104 million in 1991, which is about 6% of Shearson Lehman Hutton's estimate; according to Amgen's current market experience, these figures are too low. Some other estimates are in between, but most are below both these estimates. Some economic analysts foresee a total worldwide sales volume of $225 million in 1990 and a long-term market volume of $350 million. This too may be rather conservative. Sales in 1990 will give us a better feel for the future sales potential.

4 Outlook

The situation at the end of 1989 can be evaluated from several points of view.

EPO as a novel genetically engineered product is clearly successful: it is a new therapeutic principle, with few side effects, a solid ethical justification for its existence as a drug, and a possible billion-dollar market. A recent cloud on the horizon is its exploitation by high-performance athletes; they apparently discover every product that can be misused in their drive for an absurd increase in performance level. A performance gain of 10% is an accepted value. *Business Week* quoted a Swedish skier thus: "You feel it kick in when you are at the top of what you can do".

The coming months should bring a series of approvals, so that EPO may possibly help many patients with anemia brought on by cancer or AIDS treatments, for example.

The story of EPO leaves one feeling ambivalent about the field of gene technology; on the one hand, a novel product achieves success, but on the other hand, the competitors and partners do battle tooth and nail, with questionable methods, and thereby appear to lose sight of the well-being of patients. The tale appears to be fairly typical for the field of gene technology (or only for its early history), in that too many companies vie for the development of one product, so that some necessarily must fail. Those who believe in the philanthropic future of high technology are disappointed that a large proportion of the financial resources needed for product development has been diverted into lawyers' fees for meaningless lawsuits. This case is certain to disillusion those who believe that a scientifically brilliant principle can produce economic success on its own.

5 Sources

The following references were used in writing this chapter, as were numerous short communications from *Abstracts in BioCommerce*, *Biofutur*, *Bio/Technology*, *Genetic Engineering News*, and *Genetic Technology News* for the years 1985 to 1989, as well as company information. The sources are not cited in the text, so as not to sacrifice clarity, because the chapter is only a brief review.

6 References

[1] ANON.: Protein für Dialysepatienten; *Chem. Rundschau* (**1987**) 10.

[2] EGRIE, J. C.; BROWNE, J.; LAI, P.; LIN, F. K.: Characterization of Recombinant Monkey and Human Erythropoietin; *Prog. Clin. Biol. Res.* **191** (1985) 339–350.

[3] EGRIE, J. C.; STRICKLAND, T. W.; LANE, J.; AOKI, K.; COHEN, A. M.; SMALLING, R.; TRAIL, G.; LIN, F. K.; BROWNE, J. K.; HINES, D. K.: Characterization and Biological Effects of Recombinant Human Erythropoietin; *Immunobiol* **172** (1986) 213–224.

[4] ERSLEV, A. J.; CARO, J.: Secondary Polycythemia: A Boon or a Burden? *Blood Cells* **10** (1984) 177–191.

[5] ERSLEV, A. J.; CARO, J.: Physiologic and Molecular Biology of Erythropoietin; *Med. Oncol. Tumor Pharmacother.* **3** (1986) 159–164.

[6] ERSLEV, A. J.: Erythropoietin Coming of Age; *New Engl. J. Med.* **316** (1987) 101–103.

[7] ESCHBACH, J. W.; EGRIE, J. C.; DOWNING, M. R.; BROWNE, J. K.; ADAMSON, J. W.: Correction of the Anemia of End-Stage Renal Disease with Recombinant Human Erythropoietin; *New Engl. J. Med.* **316** (1987) 73–78.

[8] FISHER, J. W.; HAGIWARE, M.: Effects of Prostaglandins on Erythropoiesis; *Blood Cells* **10** (1984) 241–260.

[9] FRIED, W.; BARONE-VARELAS, J.; MORLEY, C.: Factors that Regulate Extrarenal Erythropoietin Production; *Blood Cells* **10** (1984) 287–304.

[10] GOLDE, D. W.; GASSON, J. C.: Blutbildende Hormone; *Spektrum der Wissenschaft* **9** (1988) 70–79.

[11] GOLDWASSER, E.: Erythropoietin and Its Mode of Action; *Blood Cells* **10** (1984) 147–162.

[12] HEILMANN, E.; GOTTSCHALK, D.; GOTTSCHALK, I.; LISON, A. E: Studies in Polycythemia after Kidney Transplantation; *Clin. Nephrol* **20** (1983) 94–97.

[13] JERESKI, L.: It Gives Athletes a Boost – Maybe Too Much; *Business Week* (11.12.1989) (1989) p 58.

[14] KRYSTAL, G.; PANKRATZ, H. R. C.; FARBER, N. M.; SMART, J. E.: Purification of Human Erythropoietin to Homogeneity by a Rapid Five-Step Procedure; *Blood* **67** (1986) 71–79.

[15] KULKARNI, V.; RITCHEY, K.; HOWARD, D.; DAINIAK, N.: Heterogeneity of Erythropoietin-Dependent Erythrocytosis: Case Report in a Child and Synopsis of Primary Erythrocytosis Syndromes; *Brit. J. Haematol.* **60** (1985) 751–758.

[16] LAMPERI, S.; CAROZZI, S.; ICARDI, A.: Polycythaemia is Erythropoietin-Independent after Renal Transplantation; *Proc. Eur. Dial. Transplant. Assoc. Eur. Re. Assoc.* **21** (1985) 928–931.

[17] LANGE, R. D.; ANDREWS, R. B.; TRENT, D. J.; REYNIERS, J. P.; DRAGANAC, P. S.; FARKAS, W. R.: Preparation of Purified Erythropoietin by High Performance Liquid Chromatography; *Blood Cells* **10** (1984) 305–314.

[18] LEE-HUANG, S.: Cloning and Expression of Human Erythropoietin cDNA in *Escherichia coli*; *Proc. Natl. Acad. Sci. USA* **81** (1984) 2708–2712.

[19] LIN, F. K.; SUGGS, S.; LIN, C. H.; BROWNE, J. K.; SMALLING, R.; EGRIE, J. C.; CHEN, K. K.; FOX, G. M.; MARTIN, F.; STABINSKY, Z.; BADRAWI, S. M.; LAI, P. H.; GOLDWASSER, E.: Cloning and Expression of Human Erythropoietin Gene; *Proc. Natl. Acad. Sci. USA* **82** (1985) 7580–7584.

[20] MCGONIGLE, R. J.; WALLIN, J. D.; SHADDUCK, R. K.; FISHER, J. W.: Erythropoietin Deficiency and Inhibition of Erythropoiesis in Renal Insufficiency; *Kidney Int.* **25** (1984) 437–444.

[21] MILLEDGE, J. S.; COTES, P. M.: Serum Erythropoietin in Humans at High Altitude and Its Relation to Plasma Renin; *J. Appl. Physiol.* **59** (1985) 360–364.

[22] MIYAKE, T.; KUNG, C. K. H.; GOLDWASSER, E.: Purification of Human Erythropoietin; *J. Biol. Chem.* **252** (1977) 5558–5564.

[23] MUELLER-WIEFEL, D. E.: Die renale Anämie – ihre Diagnostik, Pathogenese, Kompensation u. Therapie im Kindesalter; *Monatsschr. Kinderheilk* **132** (1984) 72–79.

[24] REJMANN, A. S.; GRIMES, A. J.; COTES, P. M.; MANSELL, M. A.; JOEKES, A. M.: Correction of Anaemia Following Renal Transplantation: Serial Changes in Serum Immunoreactive Erythropoietin, Absolute Reticulocyte Count and Red-Cell Creatine Levels; *Br. J. Haematol* **61** (1985) 421–431.

[25] ROCHEMAN, M.: L'Erythropoiétine; *Biofutur* (**1987**) Jan., p 47.

[26] ROSENLOEF, K.; FYHRQUIST, F.; GROENHAGEN-RISKA, C.; BOEHLING, T.; HALTIA, M.: Erythropoietin and Renin Substrate in Cerebellar Haemangioblastoma; *Acta Med. Scan.* **218** (1985) 481–485.

[27] SHERWOOD, J. B.: The Chemistry and Physiology of Erythropoietin; *Vitamins and Hormones* **41** (1984) 161–211.

[28] SUDA, J.; SUDA, T.; KUBOTA, K.; IHLE, J. N.; SAITO, M.; MIURA, Y.: Purified Interleukin-3 and Erythropoietin Support the Terminal Differentiation of Hemopoietic Progenitors in Serum-Free Culture; *Blood* **67** (1986) 1002–1006.

[29] SUMMERFIELD, G. P.; GYDE, O. H. B.; FORBES, A. M. W.; GOLDSMITH, H. J.; BELLINGHAM, A. J.: Haemoglobin Concentration and Serum Erythropoietin in Renal Dialysis and Transplant Patients; *Scand. J. Haematol.* **30** (1983) 389–400.

[30] TALBOT, J. M.; FISHER, K. D.: Influence of Space Flight on Red Blood Cells; *Fed. Proc.* **45** (1986) 2285–2290.

[31] VAN BRUNT, J.: EPO: Biotech's Next Blockbuster Drug? *Bio/Technology* **5** (1987) 199.

[32] WICKRE, C. G.; NORMAN, D. J.; BENNISON, A.; BARRY, J. M.; BENNET, W. M.: Postrenal Transplant Erythrocytosis: A Review of 53 Patients; *Kidney Int.* **23** (1983) 731–737.

[33] WINEARLS, C. G.; PIPPARD, M. J.; DOWNING, M. R.; OLIVER, D. O.; REID, C.; COTES, P. M.: Effect of Human Erythropoietin Derived from Recombinant DNA on the Anaemia of Patients Maintained by Chronic Haemodialysis; *Lancet* (**1986**) Nov. 22, 1175–1178.

[34] YANGAWA, S.; HIRADE, K.; OHNOTA, H.; SASAKI, R.; CHIBA, H.; UEDA, M.; GOTO, M.: Isolation of Human Erythropoietin with Monoclonal Antibodies; *J. Biol. Chem.* **259** (1984) 2707–2710.

[35] ZINS, B.; DRUEEKE, T.; ZINGRAFF, J.; BERERHI, L.; KREIS, H.; NARET, C.; DELONS, S.; CATAIGNE, J. P.; PETERLONGO, F.; CASADEVALL, N.; VARET, B.: Erythropoietin Treatment in Anaemic Patients on Haemodialysis; *Lancet* (**1986**) Dec. 6, p 1329.

Company Profiles I

Amgen, Inc.*

Amgen was founded in 1980 under the name "Applied Molecular Genetics". It is based in Thousand Oaks, CA, and as of October 1989 employs 650 people. Shareholders include Abbott Laboratories with 7.1% (Chairman George B. RATHMANN was previously Vice President of Research and Development at Abbott) and Smith-Kline Beecham with 3%.

Amgen was initially financed in 1981 with $19 million (U.S.) from venture and private capital. When the company went public in June 1983, the shares were offered at $19 and generated $42 million. The second issue, in March 1986, yielded a mere $39 million at $15 per share. In September 1986 Amgen experienced a sharp market rise and has since been traded at more than twice that price, thereby surpassing the majority of its biotechnology competitors. A third stock issue in 1987 produced $75 million at $34 per share. Further financing was accomplished through "Research and Development Limited Partnerships", a special cooperative of limited partnerships that, according to American law, enjoys a favorable tax rate. This provided at least $160 million.

Thus, within ten short years more than $335 million was generated by a number of different financial strategies. Several million dollars were also obtained in the form of research projects from other companies (such as Kodak) and grants from the N.I.H. (National Institutes of Health). Therefore, one can assume that the company had the $40 million it required annually. This amount should not be surprising, since the development and clinical trials of erythropoietin (EPO) alone may have consumed the major portion of it. (See the chapter **Erythropoietin – the Developmental History of a New Product**, also in this section.)

Unlike Genentech, the company has not been much in the public eye since its founding. Having been in the red for five years (Table 1), it finally reported a modest profit of $548,000 from an income totaling $23 million in its fiscal year 1985 (ending 31 March 1986). (In September 1984 the CPA firm of Paine Webber predicted an income of only $5,225,000). The bulk of this income, 96%, was not product-generated, but came from cooperative research projects with large pharmaceutical manufacturers and from interest income.

The past few years, however, have brought indications that the company is being rewarded for its research efforts. In the first six months of fiscal 1986 the production output alone increased by 300%. The profit figures are clearly on the rise, with the exception of fiscal year 1988, when the company was heavily burdened with the costs of product approval, introduction, and legal disputes.

Amgen has avoided the mortgaging of as yet undeveloped products, a danger to which biotechnology companies are often subject. The company has, therefore, always cautioned against exaggerated expectations tied to the market potential of EPO. In spite of this, and based upon newer developments, even reliable economic analysts expected that by the end of 1989 Amgen's "Epogene" could realize annual sales of $500 million in the near future.

* Dr. Norbert RAU,
RauCon Bioinformatik & Consulting GmbH,
D-6912 Dielheim, Fed. Rep. Germany

Table 1 Financial profile of Amgen, Inc.

Fiscal year ending	Income, thousands of US $	Profit (loss), thousands of US $
31 March 1983	2,005	(7,079)
31 March 1984	6,116	(4,941)
31 March 1985	10,130	(7,760)
31 March 1986	23,400	548
31 March 1987	34,700	1,100
31 March 1988	53,300	1,730
31 March 1989	78,100	(8,200)
1st Qtr., ending 30 June 1989	30,200	835

Source: RauCon GmbH

Besides erythropoietin the company has numerous other products in various stages of clinical testing – because one product is not enough to build a company. These include a vaccine for hepatitis B and three immunomodulators, interleukin 2 (IL-2) and two kinds of interferon.

As early as 1985, Amgen focused on another new product, granulocyte stimulating factor (G-CSF). In 1986 the cDNA of GCSF was cloned and expressed in *E. coli*. The product is one of several substances that stimulate bone marrow cells to produce leukocytes to ward off infections. Scientists hope that this agent will be helpful in the battle against bacterial infections or certain types of cancers such as leukemia. Furthermore, it might help cancer patients tolerate the devastating side effects of radiation and chemotherapy.

G-CSF is already being tested at the Sloan-Kettering Cancer Center in New York City in cancer patients who require chemotherapy. FDA approval could possibly be obtained by 1990. Economic analysts believe that Amgen could control an annual U.S. market of about $200 million.

Amgen is working on various other products, not for the difficult medical therapy market, but rather for the research market. Towards this end, a subsidiary company was founded in 1984: Amgen Biologicals sells interferon, interleukin-2, and other substances for research.

Company Profiles II

Monsanto Company

Monsanto is a multinational company engaged in the research, manufacture, and marketing of a widely diversified line of high-quality products including chemical and agricultural products, pharmaceuticals, low-calorie sweeteners, industrial process controls, man-made fibers, plastics, and electronic materials.

The 1985 acquisition of G. D. Searle & Co. gave Monsanto entry into the international pharmaceutical industry. The NutraSweet Company was made a free-standing subsidiary in January 1986.

Monsanto, headquartered in St. Louis, MO, conducts business in more than 100 countries and has 52,000 employees worldwide. In 1989, the company's sales were $ 8.6 billion. The company operates 23 research centers and laboratories around the world.

Biotechnology is integral to Monsanto's effort to become a leader in the development and marketing of life sciences products for health care and agriculture.

Monsanto's extensive commitment to and broad-based support of biotechnology research can be traced to the early 1970s. It was the oil embargo of 1973 that threw into question the company's traditional dependence on products based on petrochemicals. The first of its many investments in smaller biotechnology companies and start-up firms was acquired in the mid-1970s.

Sponsorship of university research has given Monsanto access to some of the premier research centers in the world and quickly placed the company at the frontiers of medical and agricultural science.

In 1979, Monsanto started a major expansion of its in-house research and development staff when Dr. Howard SCHNEIDERMAN of the University of California at Irvine joined the company as Senior Vice President of Research and Development and Chief Scientist.

The company's annual research and development budget for biotechnology in 1989 was approximately $ 110 million. This represents nearly 20 percent of the total Monsanto R&D budget of $ 600 million.

Monsanto's research activities in genetic engineering fall into three general categories: plant agriculture, human health, and animal science.

Plant Agriculture

In the area of plant agriculture, Monsanto is seeking methods of crop production that stand up to disease, drought and other types of stress.

Monsanto scientists, using *Agrobacterium tumefaciens* to help transfer useful genetic information into plant cells, were the first to show that a transplanted gene can function in its host plant and later offspring. This research enabled Monsanto to develop certain plants that can tolerate Roundup herbicide (glyphosate), others that can resist certain lepidopteran insect pests, and still others that can tolerate plant viruses.

In June 1987 Monsanto gained USDA approval for the field testing of genetically engineered tomato plants. In three small-scale field tests, Monsanto planted a test plot in Illinois, 40 miles northeast of St. Louis, with about 300 tomato plants engineered to tolerate the tobacco mosaic virus, about 500 tomato plants resistent to certain insects, and about 3,700 plants engineered to tolerate Roundup herbicide. The testing represents the first field studies of genetically engineered food crops.

Recognizing the need for a fast, accurate method for monitoring genetically engineered organisms in the environment, Monsanto researchers have developed an innovative technology to mark bacteria with specific genetic materials. This process allows engineered microorganisms released into the environment to be easily distinguished from naturally occurring microbes. Field testing of the microbe marker system is to be undertaken in a joint $607,000 research program announced in June 1987 between Monsanto and Clemson University.

Human Health Care

Through collaborative agreements with Washington University in St. Louis and with Oxford University in the United Kingdom, Monsanto is focusing on the search for proteins that will provide innovative ways to treat major diseases. Two proteins currently under investigation at G. D. Searle, a subsidiary of Monsanto, are tissue plasminogen activator (tPA) and atrial peptide.

Animal Sciences

Monsanto has been a technological leader in animal nutrition since the 1950s. Today, with the aid of biotechnology, it is concentrating its research efforts on methods of enhancing overall animal production and feed efficiency. Current research includes development of bovine somatotropin (BST). BST is a protein, produced naturally in the cow's pituitary gland, that regulates milk production in cows. Independent studies demonstrate that BST can improve the milk-to-feed ratio by as much as 5 to 15 percent.

Other animal research at Monsanto includes porcine somatotropin (PST), the swine equivalent of BST. In research studies PST has increased the rate of weight gain and the leanness of market hogs, thereby offering improved economics of hog production.

The centerpiece of Monsanto's commitment to biotechnology is the $150 million Life Sciences Research Center in suburban St. Louis. This facility, which was dedicated in late 1984, employs 1,200 researchers and support personnel. The one-million-square-foot facility contains nearly 250 laboratories and 26 greenhouses, as well as 123 plant growth chambers. Computer controls can simulate any climatic condition facing agriculture, ranging from northern European wheat fields to Asian rice paddies.

Company Profiles III

Chiron Corporation

Chiron Corporation, founded in 1981 and headquartered in Emeryville, CA, near San Francisco, is building a health care business by developing products that address needs in four markets:

- diagnostics,
- adult vaccines,
- biopharmaceuticals including growth factors, and
- ophthalmics (through its fully integrated subsidiary, Chiron Ophthalmics, Inc.)

Chiron is a leader in applying biotechnology to solve health care problems not currently being met, in particular in the area of infectious diseases. For example, its discovery of the hepatitis C virus, formerly known as non-A, non-B hepatitis, marked the first time in recent history that scientists from a commercial group discovered a major infectious disease agent.

Chiron's pioneering work in hepatitis B resulted in Recombivax HB (manufactured and marketed by Merck & Co.), the first recombinant vaccine approved by the Food and Drug Administration. Chiron scientists also played important roles in sequencing the human immunodeficiency virus (HIV, the cause of AIDS) and in the discovery of hepatitis delta, a virulent form of hepatitis B.

Chiron has 580 employees in four locations. Approximately 240 of its employees work in research and development. Overall, Chiron employs about 85 PhDs.

In addition to its headquarters building in Emeryville, which houses laboratories and some manufacturing facilities, the company has a second Emeryville building where its diagnostic development and manufacturing, quality assurance, administration, and other laboratories are housed, and a fermentation plant in Manteca, 60 miles east of Emeryville. Chiron Ophthalmics is headquartered in Irvine, CA.

Chiron is commercializing some of its products by forming strategic alliances with larger health care companies to capitalize on their marketing and distribution capabilities where the partner's position creates a significant competitive advantage. Chiron also sells directly into more focused markets where it can establish a major presence, such as ophthalmics.

Diagnostics

Chiron is building a strong position in the blood screening market together with its 50-50 partner, Ortho Diagnostic Systems, a unit of Johnson & Johnson.

Following their acquisition of the blood screening business of Du Pont, the partners rank second in the industry. From this position they are launching their major product, the first test to screen blood for hepatitis C. Additionally, Chiron and Ortho have licensed Abbott Laboratories, the world's largest immunodiagnostic company, to co-market hepatitis C tests.

Chiron has a program in specialty clinical diagnostics where it is developing a series of tests based on DNA probes that amplify genetic material to provide new information about diseases.

Vaccines

Chiron is building a business in vaccines through The Biocine Company, its 50-50 joint venture with CIBA-GEIGY Limited. The Biocine Company is developing a new generation of safer vaccines for adult infectious diseases, including three that have begun human clinical trials – AIDS, genital herpes, and malaria. Its vaccines are being studied for both the prevention and the treatment of diseases.

The Biocine Company combines technologies from Chiron and CIBA-GEIGY to create a potentially powerful approach to the problems of infectious disease. Chiron contributes expertise in research and development of antigens, the recombinant proteins that mimic the appearance of virus without the genetic information that causes disease, and in vaccine formulation. CIBA-GEIGY contributes proprietary adjuvants that enhance immunogenic response, and overall research and development capabilities.

Specialty Pharmaceuticals

Chiron has been a leader in researching and developing growth factors. Chiron and its partner, Ethicon, a Johnson & Johnson company, are developing three growth factors – epidermal growth factor, platelet-derived growth factor and fibroblast growth factor – to treat topical skin problems such as burns and skin ulcers. Insulin-like growth factor-I developed at Chiron is being studied by CIBA-GEIGY for various conditions, including wasting syndromes associated with such diseases as cancers, AIDS, and type II diabetes.

Other specialty pharmaceuticals in development by Chiron and its partners include superoxide dismutase (SOD), and antibodies to tumor necrosis factor (anti-TNF), being studied by Bayer AG for efficacy in treating septic shock. Protos Corporation is an affiliate of Chiron researching and designing a new generation of therapeutic products that use modern techniques of biotechnology and molecular biochemistry.

Ophthalmics

Chiron Ophthalmics is building a market presence among its target customer base of ophthalmic surgeons through a growing line of products. Chiron Ophthalmics markets intraocular lenses, including new foldable silicone lenses, corneal collagen shields that protect the eye as they dissolve, and media for storing transplanted corneas.

Chiron Ophthalmics is developing products to extend the capabilities of ophthalmic surgeons to treat wounds with wound-healing products. The company is studying epidermal growth factor, EGF, for its potential efficacy both inside the eye, during intraocular surgeries, and on the surface of the eye. Chiron Ophthalmics also is studying the use of fibronectin, an adhesive protein, in treating topical problems of the eye.

In July 1991 Chiron Corp. announced that the company intends to merge with Cetus Corp. in the next future.

Selected Bibliography

ABELSON, P. H., Ed.: *Biotechnology and Biological Frontiers*, American Association for the Advancement of Science, Washington, DC (1984).

ALBERTS, B.; BRAY, D.; LEWIS, J.; RAFF, M.; ROBERTS, K.; WATSON, J. D.: *Molecular Biology of the Cell*, Garland Publishers, Inc., New York (1983).

ANTHONY, C.: *Bacterial Energy Transduction*, Academic Press, Inc., London (1988).

ATKINSON, B.: *Biochemical Reactors*, Pion Ltd., London (1974).

ATKINSON, B.; MAVITUNA, F.: *Biochemical Engineering and Biotechnology Handbook*, Macmillan Publishers, New York (1983).

ATTIA, A. A., Ed.: *Process Technology Proceedings*, Vol. 4: Flocculation in Biotechnology and Separation Systems, Elsevier Science Publishers, Amsterdam (1987).

BACCINI, P., Ed.: *The Landfill: Reactor and Final Storage*, (Swiss Workshop on Land Disposal of Solid Wastes, 1988: Gerzensee) Springer-Verlag, Berlin (1989).

BAILEY, J. E.; OLLIS, D. F.: *Biochemical Engineering Fundamentals*, 2nd ed., McGraw-Hill, New York (1986).

BERGMEYER, H. U.; GRASSL, M., Eds.: *Methods of Enzymatic Analysis*, Vol. I–XII, VCH, Weinheim, Germany (1983–86).

BLÖCKER, H.; FRANK, R.; FRITZ, H.-J., Eds.: *Chemical Synthesis in Molecular Biology*, VCH, Weinheim, Germany (1987).

BU'LOCK, J. D.; KRISTIANSEN, B., Eds.: *Basic Biotechnology*, Academic Press, Inc., London (1987).

BUNGAY, H. R.; BELFORT, G., Eds.: *Advanced Biochemical Engineering*, John Wiley & Sons, New York (1987).

CHMIEL, H., Ed.: *Biochemical Engineering*, Gustav Fischer Verlag, Stuttgart, Germany (1987).

COOMBS, J.: *The International Biotechnology Directory*, VCH, Weinheim, Germany (1986).

CRAFTS-LIGHTY, A.: *Information Sources in Biotechnology*, VCH, Weinheim, Germany (1986).

CRUEGER, W., Ed.: *Physical Aspects of Bioreactor Performance*, DECHEMA, Frankfurt, Germany (1987).

CRUEGER, W.; CRUEGER, A.: *Biotechnology: A Textbook of Industrial Microbiology*, Science Tech., Inc., Madison, WI (1984).

DARNELL, L.; LODISH, H.; BALTIMORE, D.: *Molecular Cell Biology*, W. H. Freeman & Co., New York (1986).

DAVIS *et al.*: *Microbiology*, 4th ed., Lippincott, Philadelphia, PA (1990). or DAVIS, B. D.; DULBECCO, R.; EISEN, H. N.; GINSBERG, H. S.; WOOD, W. B.: *Microbiology*, 3rd ed., Harper & Row, New York (1980).

DECHEMA Monographs, Vol. 105, VCH Verlagsgesellschaft, Weinheim, Germany (1987).

DEMAINE, A. L.; SOLOMON, N. A., Eds.: *Manual of Industrial Microbiology and Biotechnology*, American Society for Microbiology, Washington, DC (1986).

DEMAINE, A. L.; SOLOMON, N. A.: *Biology of Industrial Microorganisms*, Benjamin Cummings, Menlo Park, CA (1985).

DILLON, J. R.; NASIM, A.; NESTMAN, E. R.: *Recombinant DNA Methodology*, John Wiley & Sons, New York (1985).

DIXON, R. A., Ed.: *Plant Cell Culture – A Practical Approach*, IRL Press, Washington, DC (1985).

DURANG, G., Ed.: *8th International Biotechnology Symposium*, Proceedings Vol. I and Vol. II (1988).

ELVERS, B., Ed.: *Ullmanns Encyclopedia of Industrial Chemistry*, VCH Verlagsgesellschaft, Weinheim, Germany (1989).

ESSER, K.; KÜCK, U.; LANG-HINRICHS, C.; LEMKE, P.; OSIEWACZ, H. D.; STAHL, U.; TUDZYNSKI, P.: *Plasmids of Eukaryotes*, Springer Publishers, Berlin (1986).

FIECHTER, A., Ed.: *Advances in Biochemical Engineering/Biotechnology*, Springer Verlag, Berlin (1989).

FINN, R. K.; PRÄVE, P.; SCHLINGMANN, M.; CRUEGER, W.; ESSER, K.; THAUER, R.; WAGNER, F., Eds.: *Biotechnology Focus 1* and *Biotechnology Focus 2*, Hanser Publishers, Munich, New York (1988 and 1990).

FOGARTY, W. M.: *Microbial Enzymes and Biotechnology*, Applied Science Publishers, London, New York (1983).

FREEDMAN, R. B.; HAWKINS, H. C., Eds.: *The Enzymology of Posttranslational Modification of Proteins*, Vol. 2, Academic Press, Inc., London (1985).

FRESHNEY, R. I., Ed.: *Animal Cell Culture*, IRL Press, Washington, DC (1986).

GASSEN, H. G.; LANG, A., Eds.: *Chemical and Enzymatic Synthesis of Gene Fragments* (1982).

GLOVER, D. M.: *DNA Cloning – A Practical Approach*, Vol. I–III, IRL Press, Washington, DC (1987).

GODING, J. W.: *Monoclonal Antibodies: Principles and Practice*, Academic Press, Inc., London (1983).

GOLDBLITH, S. A.; REY, L.; ROTHMAYR, W. W., Eds.: *Freeze Drying and Advanced Food Technology*, Academic Press, Inc., London (1975).

HAMES, B. D.; GLOVER, D. M., Eds.: *Molecular Immunology*, IRL Press, Washington, DC (1988).

HAROLD, F. M.: *The Vital Force: A Study of Bioenergetics*, W. H. Freeman & Co., New York (1986).

HERSHBERGER, C. L.; QUEENER, S. W.; HEGEMAN, G.: *Genetics and Molecular Biology of Industrial Microorganisms*, American Society for Microbiology, Washington, DC (1989).

HIGGINS, I. J.; BEST, D. J.; JONES, J., Eds.: *Biotechnology, Principles and Applications*, Blackwell Scientific Publications, Oxford (1985).

HORAN, N. J.: *Biological Wastewater Treatment Systems, Theory and Operation*, John Wiley & Sons, Chichester, New York (1989).

HOUWINK, E. H.: *Innovations in Biotechnology*, Elsevier Science Publishers, Amsterdam (1984).

KENNETT, R. H.; BECHTOL, K. B.; MCKEARN, T. J., Eds.: *Monoclonal Antibodies and Functional Cell Lines*, Plenum Press, New York (1984).

KIRSOP, B. E.; SNELL, J. J. S., Eds.: *Maintenance of Microorganisms, A Manual of Laboratory Methods*, Academic Press, Inc., London (1984).

KLEINKAUF, H.; VON DÖHREN, H.; DORNAUER, H.; NESEMANN, G., Eds.: *Regulation of Secondary Metabolite Formation*, VCH., Weinheim, Germany (1986).

KOIVISTOINEN, P.; HYVÖNEN, L., Eds.: *Carbohydrage Sweeteners in Foods and Nutrition*, Academic Press, Inc., London (1980).

LASKIN, A. I., Ed.: *Enzymes and Immobilized Cells in Biotechnology*, Benjamin Cummings, Inc., London (1985).

LEFKOVITS, I.; PERNIS, B., Eds.: *Immunological Methods Vol. II*, Academic Press, Inc., London (1981).

LEHNINGER, A. L.: *Principles of Biochemistry*, Worth Publishers, Inc., New York (1982).

LETTINGER, G., Ed.: *Granular Anaerobic Sludge, Microbiology and Technology*, Puduc Wageningen (1988).

LYDERSEN, B. K., Ed.: *Large Scale Cell Culture Technology*, Hanser Publishers, Munich, New York (1987).

MARAMOROSCH, K., Ed.: *Advances in Cell Culture*, Vol. 4, Academic Press, Inc., London (1985).

Members of the Working Party "Bioreactor Performance" of the European Federation of Biotechnology: *Process Variables in Biotechnology*, DECHEMA, Frankfurt, Germany (1984).

MOO-YOUNG, M., Ed.: *Comprehensive Biotechnology and Bioengineering* (Vol. 1–3), Pergamon Press, Oxford (1985).

NEIDLEMAN, S. L.; Laskin, A. I., Eds.: *Advances in Applied Microbiology*, Academic Press, Inc., San Diego (Annual, Volumes **1-35** through 1990.)

NELSON, W.: *Industrial Methods of Rapid Microbiological Analysis*, VCH, Weinheim, Germany (1985).

PRÄVE, P. *et al.*: *Fundamentals of Biotechnology*, VCH, Weinheim, Germany (1987).

REHM, H.-J.; REED, G., Eds.: *Biotechnology*, Vol. 1–8, VCH, Weinheim, Germany (1981 ff).

ROELS, J. A.,: *Energetics and Kinetics in Biotechnology*, Elsevier, Amsterdam (1983).

ROSE, A. H., Ed.: *Economic Microbiology*, Vol. 7: Fermented Foods, Academic Press, Inc., London (1982).

ROSEN, B. P.: *Ion Transport in Prokaryotes*, Academic Press, Inc., London (1987).

RUSSELL, G. E., Ed.: *Biotechnology & Genetic Engineering Reviews*, Vol. 4., Intercept Ltd., Newcastle upon Tyne, UK (1986).

SATO, G. H.; PARDEE, A. B.; SIRBASKU, D. A., Eds.: *Growth of Cells in Hormonally Defined Media*, Book A, Cold Spring Harbor Laboratory, Cold Spring Harbor, NY (1982).

SCHLEIF, R.; WENSINK, P. C.: *Practical Methods in Molecular Biology*, Springer Publishers, Berlin, Heidelberg, New York (1981).

SCHOOK, L. B., Ed.: *Monoclonal Antibody Production Techniques and Applications*, Marcel Dekker, Inc., New York (1987).

SCHÜGERL, K.: *Bioreaction Engineering*, Vol. 1: Fundamentals, John Wiley & Sons, New York (1987).

SCRAGG, A. H., Ed.: *Biotechnology for Engineers: Biological Systems in Technological Processes*, Horwood, Chichester & Halsted, New York (1988).

SETLOW, J. K.; HOLLAENDER, A., Eds.: *Genetic Engineering, Principles and Methods*, Vol. 8, Plenum Press, New York (1986).

SHAW, C. H., Ed.: *Plant Molecular Biology*, IRL Press, Washington, DC (1988).

SILVER, S., Ed.: *Biotechnology: Potentials and Limitations*, Springer Publishers, Berlin (1986).

SKULACHEV, V. P.: *Membrane Bioenergetics*, Springer-Verlag, Berlin (1988).

SPRAGG, S. P.: *The Physical Behaviour of Macromolecules with Biological Functions*, John Wiley & Sons, Ltd., Chichester (1980).

STANIER, R. Y.; INGRAHAM, J. L.; WHEELIS, M. L.; PAINTER, P. R.: *The Microbial World*, 5th ed., Prentice-Hall, Englewood Cliffs, NJ (1986).

STANIER, R. Y.; ADELBERG, E. A.; INGRAHAM, J. L.: *General Microbiology*, 4th ed., MacMillan, New York (1977).

STOWELL, J. D.; BAILEY, P. J.; WINSTANLEY, D. J., Eds.: *Bioactive Microbial Products 3, Downstream Processing*, Academic Press, Inc., London (1986).

STRICKBERGER, M. W.: *Genetics*, 3rd ed., Macmillan, New York (1985).

STRYER, L.: *Biochemistry*, 2nd ed., W. H. Freeman & Co. (1981).

SUSSMAN, M.; COLLINS, C. H.; SKINNER, F. A.; STEWART-TULL, D. E.: *The Release of Genetically-Engineered Microorganisms*, Academic Press, Inc., London (1988).

THILLY, W. G., Ed.: *Mammalian Cell Technology*, Butterworth Publishers, Stoneham (1986).

TREVAN, M. D.: *Immobilized Enzymes*, John Wiley & Sons, New York (1980).

VANĚK, Z.; HOŠŤÁLEK, Z., Eds.: *Overproduction of Microbial Metabolites, Strain Improvement and Process Control Strategies*, Butterworth Publishers, Boston, MA (1986).

VIETH, W. R.: *Membrane Systems: Analysis and Design*, Hanser Publishers, Munich, New York (1988).

WALKER, J. M., Ed.: *Methods in Molecular Biology, Vol. 2: Nucleic Acids*, Humana Press, Clifton, NJ (1984).

WALTON, A. G.; HAMMER, S. K., Eds.: *Genetic Engineering and Biotechnology Yearbook 1985*, Elsevier, Amsterdam (1985).

WANG, D. I. C.: *Fermentation and Enzyme Technology*, John Wiley & Sons, New York (1979).

WARTENBERG, A.: *Einführung in die Biotechnologie*, Gustav Fischer Verlag, Stuttgart, Germany (1989).

WATSON, J. D. *et al.*: *Molecular Biology of the Gene*, 4th ed., Vol. I, Vol. II, Addison-Wesley, Amsterdam (1987).

WESTERHOFF, H. V.; VON DAM, K.: *Thermodynamics and Control of Biological Free-Energy Transduction*, Elsevier, Amsterdam (1987).

WHELAN, W. J.; BLACK, S.: *From Genetic Experimentation to Biotechnology*, John Wiley & Sons, New York (1982).

WINNACKER, E.-L.: *From Genes to Clones*, VCH, Weinheim, Germany (1987).

WISEMAN, A.: *Handbook of Enzyme Biotechnology*, 2nd ed., John Wiley & Sons, New York (1985).

WISEMAN, A., Ed.: *Topics in Enzyme and Fermentation Biotechnology 8*, Ellis Horwood Ltd., Chichester (1984).

WOODWARD, J., Ed.: *Immobilised Cells and Enzymes*, IRL Press, Washington, DC (1985).

WORK, T. S.; BURDON, R. H.: *Laboratory Techniques in Biochemistry and Molecular Biology*, Vol. I–XV, Elsevier, Amsterdam (1983).

Journals in Biotechnology and Related Topics

Title	ISSN	Publisher/Editor	Publication frequency	Subscription price/yr.
Advances in Microbial Physiology	0065-2911	Academic Press, London, UK & New York, USA	annual	
Agricultural and Biological Chemistry	0002-1369	Agricultural Chemical Society, Japan	monthly	US$ 180
Annual Review of Biochemistry	0066-4154	Annual Reviews, Inc., Palo Alto, CA, USA	annual	
Applied and Environmental Microbiology	0099-2240	American Society for Microbiology, Washington, DC, USA	monthly	US$ 220
Applied Biochemistry and Biotechnology	0273-2289	Humana Press, Inc., Clifton, NJ, USA; Ed. Howard Weetall	9/yr.	US$ 175
Applied Microbiology and Biotechnology	0175-7598	Springer Publishers, New York, USA	monthly	US$ 843
Archives of Microbiology	0302-8933	Springer Verlag, Berlin, FRG & New York, USA	monthly	
The Biochemical Journal	0264-6021	The Biochemical Society, London, UK	bimonthly	US$ 1195
Biochemical Society Transactions	0300-5127	The Biochemical Society, London, UK	bimonthly	US$ 185
Biofactors	0951-6433	IRL Press, Oxford, UK	4/yr.	US$ 125
Biological Chemistry Hoppe-Seyler	0177-3593	Walter de Gruyter, Berlin	monthly	
Biopharm	1040-8304	Aster Publishing Corp.	10/yr.	US$ 54
Bioprocess Engineering	0178-515X	Springer International, Berlin, New York	4/yr.	
Biosciences	0341-0382	Verlag der Zeitschrift für Naturforschung, Tübingen		
Biosensors	0265-928X	Elsevier, Barking, Essex, UK; Ed. W. R. Heineman	bimonthly	US$ 173
Biotech Knowledge		BioCommerce Data	monthly	£ 97
Bio/Technology	0733-222X	Nature Publishing Company, New York, USA	monthly	US$ 98
Biotechnology		Biotechnology, Martinsville, USA	monthly	US$ 112
Biotechnology and Applied Biochemistry	0885-4513	Academic Press, San Diego, CA, USA; Ed. P. N. Campbell	bimonthly	US$ 110
Biotechnology and Bioengineering	0006-3592	John Wiley & Sons, New York, USA	monthly	US$ 403
Biotechnology & Bioengineering Symposia	0572-6565	John Wiley & Sons, New York, USA	annual, symposium included in subscription; see Biotechnology & Bioengineering	

Title	ISSN	Publisher/Editor	Publication frequency	Subscription price/yr.
Biotechnology Advances	0734-9750	Pergamon Press, Oxford, UK	4 Vol./yr.	DM 395
Biotechnology Chemonomics	0734-5151	Economics of Technology, New York, USA; Ed. Robert S. Aries	monthly	US$ 290
Biotechnology International		Imsworld Publications Ltd., London, UK; Ed. Dr. A. H. Sheppard	1/yr. (three regional issues: Europe, USA & Canada, Japan & Australia)	price for 1984: US$ 500
Biotechnology Letters	0141-5492	Science and Technology Letters, Kew, UK	monthly	US$ 175
Biotechnology News	0273-3226	CTB International Publ. Co., Maplewood, NJ, USA; Ed. O. C. Brosna	30/yr.	US$ 323
Biotechnology Progress	8756-7938	American Institute of Chemical Engineers, New York, USA	4/yr.	US$ 50
Biotechnology Research Abstracts		Cambridge Scientific Abstracts, Bethesda, MD, USA	monthly	US$ 394
Biotechnology Techniques	0951-208X	Science and Technology Letters, Kew, UK	bimonthly	US$ 120
Cell	0092-8674	MIT Press, Cambridge, MA, USA	2 × /month	US$ 325
Chemical & Engineering News	0009-2347	American Chemical Society, Washington, DC, USA	weekly	US$ 55
Chinese Journal of Biotechnology	1042-749X	Allerton Press, Inc., New York, USA	quarterly	US$ 275
Computer Applications in the Biosciences	0266-7061	IRL Press, Oxford, UK	4/yr.	US$ 120
Critical Reviews in Biotechnology	0738-8551	CRC Press, Inc., Boca Raton, FL, USA; Ed. G. G. Stewart, I. Russel	4/yr.	US$ 130
Current Biotechnology Abstracts	0264-3391	Royal Society of Chemistry, UK	monthly	
Derwent Biotechnology Abstracts	0262-5318	Derwent Publications, Ltd., London, UK		
EMBO Journal	0261-4189	IRL Press, Oxford, UK	13/yr.	US$ 435
Enzyme and Microbial Technology	0141-0229	Butterworth Publishers, Stoneham, MA, USA	monthly	US$ 440
European Journal of Biochemistry	0014-2956	Springer International, Berlin, New York	semi-monthly	US$ 1643
Folia Microbiologica	0015-5632	Academia Praha, Czechoslovakia	bimonthly	US$ 153
Food Technology	0015-6639	Institute of Food Technologists, USA	monthly	US$ 60

Title	ISSN	Publisher/Editor	Publication frequency	Subscription price/yr.
Gene	0378-1119	Elsevier Publications, Amsterdam	monthly	US$ 1432
Genetic Engineering News: the information source of the biotechnology industry	0270-6377	Mary Ann Liebert, Inc., New York, USA	10/yr.	US$ 150
Genetical Research	0016-6723	Cambridge Univ. Press, Cambridge, UK	bimonthly	US$ 200
Genetics	0016-6731	Genetics Society of America, Bethesda, MD, USA	monthly	US$ 220
Journal of Antibiotics	0021-8820	Japan Antibiotics Research Assoc., Japan	monthly	US$ 230
The Journal of Applied Bacteriology	0021-8847	Blackwell Scientific Publ., Ltd., Oxford, UK	monthly	£ 130
Journal of Bacteriology	0021-9193	American Society for Microbiology, Washington, DC, USA	monthly	US$ 340
Journal of Biochemical and Biophysical Methods	0165-022X	Elsevier Publications, Amsterdam	~9/yr.	~US$ 315
The Journal of Biological Chemistry	0021-9558	American Society of Biological Chemists, Inc., Baltimore, MD, USA	3/month	US$ 490
Journal of Biotechnology	0168-1656	Elsevier Publications, Amsterdam	monthly	US$ 443
Journal of Chemical Technology and Biotechnology, Part A: Chemical Technology	0264-3413	Blackwell Scientific Publ., Ltd., Oxford, UK; Sponsor: Society of Chemical Industry	8/yr.	Part B included: £ 84
Journal of Chemical Technology and Biotechnology, Part B: Biotechnology	0142-0356	Blackwell Scientific Publ., Ltd., Oxford, UK; Sponsor: Society of Chemical Industry	4/yr.	see Part A
Journal of Chromatography – Biomedical Applications	0021-9673	Elsevier Publications, Amsterdam	~weekly	US$ 3933
Journal of Fermentation and Bioengineering	0922-338X	The Society of Fermentation Technology, Japan	bimonthly	US$ 90
The Journal of General Microbiology	0022-1287	The Society for General Microbiology, London, UK	monthly	£ 350
Journal of Hazardous Materials	0304-3894	Elsevier Publishers, Amsterdam	9/yr.	US$ 459
Journal of the Science of Food and Agriculture	0022-5142	Blackwell Scientific Publishers, London, UK	monthly	
Letters in Applied Microbiology	0266-8254	Blackwell Scientific Publications, Oxford, UK	monthly	£ 148
Microbiological Reviews	0146-0749	American Society for Microbiology, Washington, DC, USA	quarterly	US$ 115
Molecular and General Genetics	0026-3925	Springer International, Berlin, New York	monthly	US$ 1584

Title	ISSN	Publisher/Editor	Publication frequency	Subscription price/yr.
Nature	0028-0836	Macmillan Magazines, Ltd., USA	weekly	US$ 250
New Scientist	0262-4079	IPC Magazines, Ltd., England	weekly	~US$ 150
Nucleic Acids Research	0305-1048	IRL Press, Oxford, UK	2 × /month	US$ 690
PASCAL Thema. (Bulletin Signaletique) Part 216: Biotechnology (English Edition)	0761-1641	Centre National de la Recherche Scientifique; Centre de Documentation Scientifique et Technique, Paris, FRANCE	10/yr.	FF 795
Plant Biotechnology	0260-5902	University of Sheffield, Sheffield, UK; Ed. Dr. W. Middleton	monthly	US$ 95
Practical Biotechnology	0262-7884	Bailliere Tindall, London, UK	monthly	~£ 125
Preparative Biochemistry	0032-7484	Marcel Dekker, New York, USA	4/yr.	US$ 185
Proceedings of the National Academy of Sciences (PNAS)	0027-8424	National Academy of Sciences, Washington, DC, USA	2 × /month	US$ 290
Process Biochemistry	0032-9592	Turret Wheatland, Ltd., Watford, UK	monthly	£ 77
Process Engineering	0370-1859	Morgan-Grampian Publ., Ltd., London, UK	monthly	£ 48
Protein Engineering	0269-2139	IRL Press, Oxford, UK	8/yr.	US$ 190
Science	0036-8075	American Association for the Advancement of Science, Washington, DC, USA	weekly	US$ 120
Swiss Biotech	0253-9675	Dr. Felix Wüst AG, Küsnacht	6/yr.	sFr 110
Trends in Biochemical Sciences	0376-5067	Elsevier Publications, Cambridge, UK, International Union of Biochemistry	monthly	US$ 57
Trends in Biotechnology	0167-9430	Elsevier Publications, Cambridge, UK	monthly	US$ 338

Subject index